Springer Series in
Surface Sciences

14

Editor: Gerhard Ertl

Springer Series in **Surface Sciences**

Editors: G. Ertl and R. Gomer   Managing Editor: H. K. V. Lotsch

Volume 1: **Physisorption Kinetics**
By H. J. Kreuzer, Z. W. Gortel

Volume 2: **The Structure of Surfaces**
Editors: M. A. Van Hove, S. Y. Tong

Volume 3: **Dynamical Phenomena at Surfaces, Interfaces and Superlattices**
Editors: F. Nizzoli, K.-H. Rieder, R. F. Willis

Volume 4: **Desorption Induced by Electronic Transitions, DIET II**
Editors: W. Brenig, D. Menzel

Volume 5: **Chemistry and Physics of Solid Surfaces VI**
Editors: R. Vanselow, R. Howe

Volume 6: **Low-Energy Electron Diffraction**
Experiment, Theory and Surface Structure Determination
By M. A. Van Hove, W. H. Weinberg, C.-M. Chan

Volume 7: **Electronic Phenomena in Adsorption and Catalysis**
By V. F. Kiselev, O. V. Krylov

Volume 8: **Kinetics of Interface Reactions**
Editors: M. Grunze, H. J. Kreuzer

Volume 9: **Adsorption and Catalysis on Transition Metals and Their Oxides**
By V. F. Kiselev, O. V. Krylov

Volume 10: **Chemistry and Physics of Solid Surfaces VII**
Editors: R. Vanselow, R. F. Howe

Volume 11: **The Structure of Surfaces II**
Editors: J. F. van der Veen, M. A. Van Hove

Volume 12: **Diffusion at Interfaces: Microscopic Concepts**
Editors: M. Grunze, H. J. Kreuzer, J. J. Weimer

Volume 13: **Desorption Induced by Electronic Transitions, DIET III**
Editors: R. H. Stulen, M. L. Knotek

Volume 14: **Solvay Conference on Surface Science**
Editor: F. W. de Wette

F.W. de Wette (Ed.)

Solvay Conference on Surface Science

Invited Lectures and Discussions

University of Texas,
Austin, Texas, December 14–18, 1987

With 214 Figures

Springer-Verlag Berlin Heidelberg New York
London Paris Tokyo

Professor Dr. Frederik W. de Wette

Department of Physics, University of Texas at Austin,
Austin, TX 78712, USA

Series Editors

Professor Dr. Gerhard Ertl

Fritz-Haber-Institut der Max-Planck-Gesellschaft, Faradayweg 4–6
D-1000 Berlin 33

Professor Robert Gomer, Ph. D.

The James Franck Institute, The University of Chicago, 5640 Ellis Avenue,
Chicago, IL 60637, USA

Managing Editor: Dr. Helmut K. V. Lotsch

Springer-Verlag, Tiergartenstrasse 17
D-6900 Heidelberg, Fed. Rep. of Germany

ISBN 3-540-50450-8 Springer-Verlag Berlin Heidelberg New York
ISBN 0-387-50450-8 Springer-Verlag New York Berlin Heidelberg

Printing: Druckhaus Beltz, 6944 Hemsbach/Bergstr.
Binding: J. Schäffer GmbH & Co. KG., 6718 Grünstadt
2154/3150-543210 – Printed on acid-free paper

Preface

The nineteenth Solvay Conference was held at the University of Texas at Austin, from December 14 through 18, 1987. Since this year marked the 75th anniversary of the first Solvay Conference, the Scientific Committees for Physics and Chemistry of the Solvay Foundation decided to commemorate this fact with a joint Solvay Conference on the interdisciplinary subject of Surface Science.

During the past twenty years there has been enormous progress and development in a broad area of surface science. The field is scientifically challenging and interesting because of the numerous new phenomena which are associated with the lower dimensionality and the breaking of three-dimensional symmetry. At the same time, the field has enormous technological importance, with impact on such diverse subjects as microelectronics and catalysis.

The organizers decided to bring together a group of leading experts from a wide range of disciplines in surface science, to provide an overview and to discuss current problems in areas of development in this field. In keeping with the tradition of the Solvay Conferences, the total number of participants was kept low, both to maintain a small conference atmosphere and to facilitate the extensive discussions which have always been a distinctive characteristic of Solvay Conferences. These two circumstances – a broad field and a limited participation – implied that the choice of subjects had to be restricted. Thus, the conference focussed on selected experimental methods and theoretical approaches, aiming to emphasize the common problems and underlying principles that unify this very diverse field.

It is a great pleasure to thank Professor Ilya Prigogine, Scientific Director of the Solvay Institutes, as well as the Scientific Committees for Physics and Chemistry of the Solvay Foundation, for giving us the opportunity to host this conference at the University of Texas at Austin. For encouragement, and generous financial support from the University of Texas, special thanks are due to President William H. Cunningham and Vice-President Gerhard J. Fonken.

Finally, I wish to thank Drs. Anil Kulkarni and Johannes Prade for their indispensable help in monitoring the discussions, and to Amy Pearson and Pam Loflin for editorial support and their expert job in transcribing the discussions from tapes and hand-written notes.

Austin, TX, USA *F.W. de Wette*
July 1988

Group photograph of the participants in the 75th Solvay conference

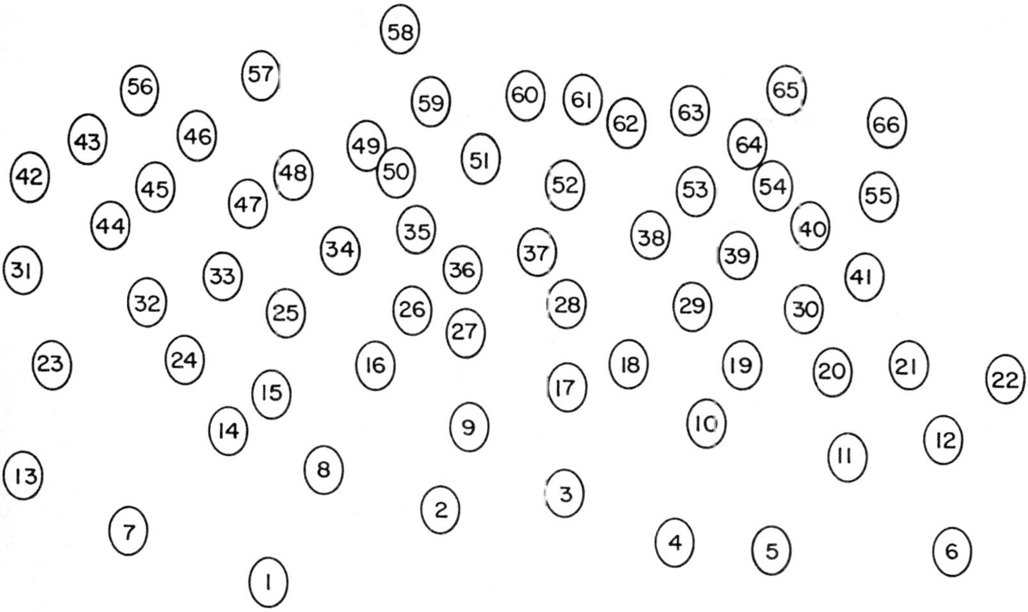

1. M.J. Cardillo
2. L. Kleinman
3. F.W. deWette
4. Xie Xide
5. Zhang Kaiming
6. I. Prigogine
7. H. Ibach
8. J.W. Gadzuk
9. M. Lagally
10. P. Marcus
11. R. Herman
12. S.G. Louie
13. G. Ertl
14. L. Van Hove
15. P. Borckmans
16. D.R. Hamann
17. N.D. Lang
18. M. Cardona
19. R.F. Wallis
20. J. Bardeen
21. J. Demuth
22. R.F. Willis
23. A.M. Bradshaw
24. G. Dewel
25. J.B. Pendry
26. R. Kubo
27. R. Smoluchowski
28. G. Somorjai
29. H. Siegmann
30. J.P. Toennies
31. J. Erskine
32. E.W. Plummer
33. R. Gomer
34. S.J. Teichner
35. Y. Ne'eman
36. K. Tamaru
37. K.-S. Sohn
38. A.C. Gossard
39. J. Golovchenko
40. K. Takayanagi
41. R.E. Allen
42. E. Tosatti
43. Zhangda Lin
44. Hua Zhongyi
45. D.A. King
46. A. Bellemans
47. W. McCormick
48. L.E. Reichl
49. I. Elskens
50. A. Nolle
51. P. Antoniewicz
52. M. Hinoul
53. C.B. Duke
54. J. Swift
55. C.Z. Wang
56. M. Bylander
57. J.G. Dash
58. W. Kohn
59. S. Akhten
60. M. Downer
61. M. Boudart
62. S. Rice
63. S.T. Ceyer
64. Y. Petroff
65. J.T. Yates
66. J. Prade

Contents

Part III Two-Dimensional Physics/Phase Transitions

Part IV Scanning Tunneling Microscopy

Part V Surface Scattering and Surface Dynamics

Part VI Chemical Reactions at Surfaces

Part VII Solid-Solid Interfaces and Superlattices

Part VIII **Surface Studies with Synchrotron Radiation**

Opening Remarks

J. Solvay

President, Administrative Commission,
Instituts Internationaux de Physique et de Chimie

The first Solvay Conference on Physics to be held outside Belgium, "Aspects of Chemical Evolution," took place in Washington in 1980 on the occasion of the 150th anniversary of the founding of Belgium as an independent state. This XVIIth Solvay Conference on Chemistry was organized jointly with the U.S. National Academy of Science.

It was followed in 1982 by the XVIIIth Solvay Conference on Physics, "Higher Energy Physics," which was held here in Austin in 1982, on the occasion of the 100th anniversary of The University of Texas.

This is therefore the third Solvay Conference to be held outside Belgium. It is, for the first time, a Joint Conference on Physics *and* Chemistry. It occurs on the occasion of the 75th anniversary of the founding of the *Instituts Internationaux de Physique et de Chimie* in 1912 by Ernest Solvay, which have since sponsored the Solvay Conferences.

As the great-grandson of Ernest Solvay, I would like to recall some of the circumstances which have led to the foundation of these Institutes.

Ernest Solvay, an entrepreneur in chemistry and a social reformer, was born in 1839, eight years after the Belgian independence. He took his first patent in 1861 on "the industrial manufacture of carbonate of soda by means of sea salt, ammonia and carbonic acid". Solvay and Company was formed as a family enterprise in 1863; but Ernest Solvay managed to give his company an international structure. He established factories in France, Germany, England, the United States, Austria and Russia—more than twenty sites at the end of the 1870's.

Ernest Solvay regarded science, in its various forms, as the key that would open the door to a richer life for man. He was also deeply interested in the problems of scientific research proper, mainly the problem of the constitution of matter in the context of the theory of gravitation. By 1870 he had developed his ideas on the subject.

Springer Series in Surface Sciences Vol. 14
Solvay Conference on Surface Science
Editor: F.W. de Wette

In 1910, he met the German chemist Walther Nernst, and had an opportunity to discuss the possibility of developing international cooperation for the study of basic problems in physics. Nernst and Planck had already discussed the possibility of holding a conference dealing with the reform of the classical physical theory; Planck felt that the time was ripe for such a conference and that one should not wait for another few years. Nernst's meeting with Ernest Solvay gave fresh encouragement for the project.

This first "Conseil Solvay" took place in Brussels from October 30 to November 3, 1911, on the theme "The Theory of Radiation and the Quanta". This was the first truly international conference as we know them today, with the usual combination of administrative council, international scientific committee deciding on a theme, rapporteurs and discussions.

Encouraged by this success, Ernest Solvay, with the counsel of Hendrik A. Lorentz, established a foundation on May 1, 1912, to be called *Institut International de Physique*. The goal of the foundation was "to encourage research which would extend and deepen the knowledge of natural phenomena". We are celebrating today the 75th anniversary of this decision. In 1913, after a series of exchanges with Wilhelm Ostwald and William Ramsay, Ernest Solvay established another foundation, the *Institut International de Chimie*, which was eventually united with the former into the *Instituts Internationaux de Physique et de Chimie*, each one having its own Scientific Committee.

During the last Solvay Conference on Physics in 1982, the Scientific Committee suggested the organization of a meeting on "Surface Science"; however, it appeared that it would be difficult to organize at Brussels, as there is not much research done on this topic at the Universities in Brussels.

This is why Ilya Prigogine established contact with Professor F.W. deWette of the Physics Department at The University of Texas. Thanks to the support of President William Cunningham and Vice-President Gerhard Fonken, it appeared possible to follow the suggestion of the Scientific Committee and to organize a meeting on "Surface Science" here at The University of Texas.

It is indeed a happy coincidence that Ilya Prigogine is both Director of the Solvay Institutes in Brussels and Director of the *Ilya Prigogine Center for Studies in Statistical Mechanics* at The University of Texas at Austin. This provides a natural and—I may even add—very strong link between these two institutions and, as President of the Solvay Institutes, I want to say how satisfied I am that this project is now realized.

It is also a testimony to the activity and international reputation of the Department of Physics at The University of Texas that this conference has attracted such a remarkable constellation of people, many of whom have been so instrumental in the immense progress achieved in surface science over the last decades.

I would like to thank the local committee, and especially its chairman, Professor F.W. deWette, for the excellent arrangements which will make this conference both a fruitful and a pleasant one.

The Table below, presenting the subjects of preceding Solvay Conferences on Physics, is taken from a paper by Professor Amaldi which was presented at the 1980 meeting.

Surface science covers today an enormous field. It includes subjects which belong traditionally to theoretical physics, others which are part of chemistry. It is therefore quite appropriate that this conference is the first Joint Solvay Conference on Physics and Chemistry.

Surfaces are the seat of a host of interesting phenomena, of both a practical and fundamental nature. Indeed, most of the action in such industrially important processes as heterogeneous catalysis, corrosion and microelectronic circuit fabrication take place at surfaces. Almost invariably, understanding these requires a detailed knowledge of the atomic structure of these surfaces, and this knowledge has been frustratingly hard to come by.

However, over the past decade, we have witnessed the development of a large number of new tools for analyzing the atomic structure of surfaces. We possess now a compendium of complementary analytical techniques for measuring the characteristics of surfaces directly, using probes such as synchrotron radiation, electrons, ions, atoms and molecules. In addition, there are accurate spectroscopic techniques for measuring the electronic and vibrational properties of surfaces. As the president of a chemical company, I am especially aware of the impact these methods are starting to have on the vast domain of heterogeneous catalysis.

Increasing emphasis on dynamic behavior has revealed deficiencies in classical kinetic treatments, through important contributions to the study of surface species dynamics and their eventual coupling to lattice events. As experimental data accumulate on more reaction systems, the equilibrium or quasi-steady state approximations routinely invoked in catalytic kinetics are often found to be unacceptable for a description of the transient characteristics of catalytic reactions. These are thus true examples of dynamical systems functioning far from equilibrium.

The physics and chemistry of far-from-equilibrium systems are today in the forefront of science. It is the main subject of research for centers such as the *Instituts Internationaux de Physique et de Chimie* at Brussels, the *Ilya Prigogine Center for Studies in Statistical Mechanics* and the *Center for Nonlinear Dynamics* directed by Professor Harry Swinney here at The University of Texas.

We may expect a lot of fruitful collaboration between academic and industrial research. With this in mind, I am very happy to open this XIXth Solvay Conference on Physics and Chemistry.

SOLVAY CONFERENCES ON PHYSICS		
Laws in Complex Systems	Constituents of Matter	Environment at Large
	1. Radiation Theory and the Quanta (1911)	
	2. The Structure of Matter (1913)	
WORLD WAR I		
	3. Atoms and Electrons (1921)	
4. The Electrical Conductivity of Metals (1924)		
	5. Electrons and Photons (1927)	
6. Magnetism (1930)		
	7. The Structure of Atmoic Nucleus (1933)	
WORLD WAR II		
	8. Spectrum of Elementary Particles (1948)	
9. Solid State (1951)		
10. Electrons in Metals (1954)		
		11. The Structure and Evolution of the Universe (1958)
	12. Quantum Field Theory (1961)	
		13. The Structure and Evolution of the Galaxies (1964)
	14. Fundamental Problems in Elementary Particles (1967)	
15. Symmetry Properties of Nuclei (1970)		
		16. Astrophysics and Gravitation (1973)
17. Order and Fluctuations in Equilibrium and Nonequilibrium Statistical Mechanics (1976)		
	18. Higher Energy Physics (1982)	
19. Surface Science (1987)		

Part I

The Structure of Surfaces

Calculation of Surface Structural Energies

D.R. Hamann

AT&T Bell Laboratories, Surface Physics Research,
600 Mountain Avenue, Murray Hill, NJ 07974, USA

Calculated energies may be used to deduce both static and dynamic structural properties of surfaces and interfaces. The classes of models used for such calculations, ranging from empirical to *ab-initio*, will be reviewed. The local density functional method will be discussed in greater detail, including the linear augmented plane wave formalism for implementing this method with great accuracy and generality. Examples from two areas will be used to illustrate the capabilities of this approach: the geometries and vibrational modes of hydrogen chemisorbed on transition metal surfaces, and the structure and formation of silicon interfaces with expitaxial silicides. Comparisons will be made with other models. In conclusion, the limitations of existing methods will be discussed.

I. Introduction

It is a great pleasure to open this Nineteenth Symposium in Physics and Chemistry, commemorating the seventy-fifth anniversary of the Solvay Institutes and devoted to surface science. In keeping with the occasion, I would like to approach my topic from a broad perspective, and begin by considering the setting of surface science. Surface science does not exist in a vacuum, so to speak, as an independent scientific discipline. There are no academic departments of surface science, to my knowledge. In terms of the backgrounds and identification of most of its participants, it is most closely tied to the disciplines of physics and chemistry, solid state physics and physical chemistry in particular. Its culture and intellectual standards are derived from those disciplines, and its problems must compete for attention with other problems from those fields.

The other component of the setting of surface science is its connection with applications. There are two major areas where we expect the new knowledge we develop to have the most impact, heterogeneous catalysis and electronics. Most commercial chemical reactions proceed on the surfaces of catalysts, and understanding surface reaction mechanisms in

Springer Series in Surface Sciences Vol. 14
Solvay Conference on Surface Science
Editor: F.W. de Wette

enough detail to permit deductive design of catalysts for specific syntheses is an oft stated, if unrealistic, goal of our field. The gap which much be bridged from our ultra-high vacuums and high-purity materials to high-pressure reactors and dirty industrial feedstocks is wide indeed. Electronics involves surface science in two ways. First of all, the active portions of solid-state devices are interfaces. Since surfaces are specialized interfaces, many of the concepts and some of the experimental techniques we have developed in surface science are transferable to such solid-solid interfaces. Secondly, many of the processes used to manufacture electronic devices, from materials growth by molecular beam epitaxy to circuit patterning by plasma etching, involve surface reactions. These often take place under conditions of vacuum and purity similar to surface science research conditions. I believe that it is possible to transfer relevant advances from science to application much more rapidly in electronics than in catalysis. The setting formed by these applications, of course, reacts back on surface science by influencing our choice of problems. If anyone doubts this, he merely has to thumb through a surface science journal and note the preponderance of studies on group VIII metals and silicon or III-V semiconductors. Of the two examples from my own work I'll discuss later, one falls clearly into the "catalysis" category, and one the "electronics."

Moving towards my topic, I'd next like to discuss the overall role of theory in surface science. Perhaps lowest on the ladder, although probably dominant in terms of total effort, is the use of theory to interpret experimental data. For some experiments the connection is obvious and direct. Low energy electron diffraction (LEED) intensities, for example, give virtually no information about atomic coordinates without detailed theoretical simulation. Ultraviolet photoemission spectroscopy (UPS) does permit you to identify surface states and bulk bands directly, but relating these features to specific bonds, reconstructions, or adsorbate levels depends on electronic structure calculations. Electron energy loss spectroscopy (EELS) of vibrational excitations may seem free from the need for theory, but as we will see in my examples, proper identification of modes is often not so straightforward. Newer uses of EELS like measuring phonon dispersion depend heavily on lattice dynamics to relate results to fundamental properties. Even scanning tunneling microscopy (STM), which produces beautiful and convincing pictures almost instantly without any analysis, may require some theory to distinguish when we are imaging geometry and when we are imaging electronic properties.

A broader task of theory is to explain and predict. Once experiment and attendant theoretical analysis have established, for example, a particular reconstruction geometry for a surface, the theorist needs to address the question of why that geometry is favored. What sort of model will lead uniquely to this structure as the most stable? Having answered this, we then can explore predictions of that model. What does it indicate for other surfaces of the material, or other materials in the same class? What will happen to the observed surface structure if the temperature is changed or in the presence of adsorbates? There are many properties other than atomic geometry we need to explain and predict, of course, ranging from Fermi level pinning energies at semiconductor surfaces to magnetism of epitaxial monolayers of metals. Probably the most difficult surface processes to deal with are surface reactions. Theory for these requires not only a detailed and accurate modeling of energies within a multidimensioned configuration space, but approximation and simulation of complex multi-atom dynamics. Reactions produced by electronic excitation such as electron or photon stimulated desorption entail yet further difficulties in that coupled electronic and atomic dynamics must be incorporated. These tasks are at or beyond the edge of our present capabilities, and very few surface reactions have been modeled in much detail.

Beyond the level of explaining and predicting surface properties on a case-by-case basis, theory must have the role of identifying unifying principles. It is these principles, rather than details of approximation and calculation that are of interest to those outside the small group of specialists who use a particular set of techniques. We must be able to generalize to use results creatively.

There is one further role for theory, especially the more general sort, that is not often recognized. Applied research in the areas impacted by surface science has traditionally progressed by empirical methods. Developing a new catalyst or etching process may be regarded, in an abstract sense, as a search in a multidimensional space whose coordinates are materials, temperatures, times, etc. A "dumb" search grows exponentially, in time required and resources consumed, with the dimension of the space. Accumulated empirical knowledge serves to reduce the effective dimensionality of the space by indicating promising subspaces to be searched. If theoretical results can further reduce the dimensionality, or perhaps even indicate an overlooked subspace excluded by existing empirical knowledge, the potential impact is great. This sort of input to application is, in my opinion, a much more realistic goal for theoretical surface science than *ab initio* design of complex processes or materials.

II. Structural Energy Methods

The total energy of a system at zero temperature as a function of all its atomic coordinates is what we mean here by its structural energy. For a surface, most static position coordinates are fixed by the known crystal structure of the bulk, and the periodicity and symmetry of the surface structure under consideration. The principal theoretical approach to surface structure questions consists of varying the independent coordinates of surface region atoms, as well as reconstruction periodicities and symmetries, to find the structure with the lowest energy. For dynamical studies, like surface phonons, this energy is needed for small excursions of all the atoms from their minimum positions.

A. Empirical Approaches

Empirical models for the structural energy are certainly the oldest and most widely applied, and there is considerable ongoing research into improving their accuracy and scope. The simplest of these are based on interatomic pair potentials. As global models, that is, models capable of describing the energy of a wide range of configurations, these are probably only adequate for collections of rare gas atoms. For dynamical calculations, harmonic pairwise interactions or "spring models" are widely applied, and probably adequate for many systems.

Classical models can be generalized beyond pair potentials to include many-atom interactions. One example is the Keating model, which includes 3-body bond-bending forces for covalently bonded solids, but only to describe small deviations from the equilibrium structure.[1] This approach has been extended in recent attempts to model more profound structural differences with 2- and 3-body potentials.[2,3] Another type of model in current use incorporates high-order multi-atom interactions, although in a very simplified way. This approach, known as "effective medium theory"[4] or the "embedded atom method"[5] is based upon the idea that an atom bonds primarily to the charge density of its neighbors. More details of this method will be discussed later in connection with hydrogen chemisorption.

Next in the hierarchy of approaches is what I call quantum models. These constitute simplified treatments of the electronic states of the system of atoms, usually based on an empirical tight-binding Hamiltonian. The matrix elements and their dependence on interatomic distance may be fitted to more fundamental calculations of the electronic states, or to simple prescriptions based on atomic energy levels, as in Hückel theory. Interactions are usually confined to nearest neighbors. The structural energy is expressed as a sum of the eigenvalues of the occupied states plus an "overcounting correction". Insofar as the spectrum is accurately reproduced by the model, this is a rigorous theory. Further approximation comes in representing the correction for the overcounting of the electron-electron interactions plus the ion-ion interactions as a sum of short-ranged pair potentials, which are chosen empirically. This approach has been applied to small deviations from equilibrium structures in lattice-dynamical calculations [6] and to a variety of surface reconstruction problems involving major bonding changes.[7]

While quantum models incorporate many-atom forces which it would be very difficult to model classically, they have their problems. These generally involve limitations of transferability between the reference structures for which the parameters are fitted and the structures to which the model is applied. The basis functions implicitly assumed in tight-binding models are orthogonal Wannier functions, and these are in fact non-local and environment-dependent. The Wannier functions at a surface, for example, cannot be the same as those in the bulk. Another problem is cutting off the matrix elements. Their fitted decay with distance is much too slow to produce an effective cutoff of higher neighbor shells, and is invariably supplemented, implicitly or explicitly, with some sort of step-function cutoff. This produces uncertainties or discontinuities in the structural energy when the near-neighbor coordination number changes as a function of some structural coordinate. Yet another recognized problem with these models is their failure to properly account for the energy of charge transfers they often predict for distorted structures.

B. First-Principles Approaches

The local density functional (LDF) method [8] is an approach to structural energies that has proven highly accurate. It is, arguably, a first-principles method, since it needs no input other than the atomic numbers of the atoms and their coordinates to calculate the energy. It contains a fundamental uncontrolled approximation in the treatment of electron correlations, however, and thus might be considered merely a sophisticated model. However, unlike the empirical models discussed previously, it avoids any dependencies on reference structures and fitting. The price paid for the accuracy and generality of LDF calculations is considerably increased computational complexity. Since the examples I'll discuss are based on LDF theory, I'd like to review the formalism briefly. It is summarized by a few equations:

$$E[\rho] = T_s[\rho] + U[\rho] + E_{xc}[\rho] , \tag{1}$$

$$\rho(\mathbf{r}) = \sum_i \psi_i^*(\mathbf{r})\psi_i(\mathbf{r}) , \tag{2}$$

$$[-\tfrac{1}{2}\nabla^2 + V_{eff}(\mathbf{r})]\psi_i(\mathbf{r}) = \epsilon_i\psi_i(\mathbf{r}) , \tag{3}$$

$$V_{eff}(\mathbf{r}) = V_c(\mathbf{r}) + \frac{\delta E_{xc}[\rho]}{\delta\rho(\mathbf{r})} , \tag{4}$$

$$V_c(\mathbf{r}) = \int \frac{\rho(\mathbf{r}')}{|\mathbf{r}-\mathbf{r}'|} d\mathbf{r}' - \sum_\alpha \frac{Z_\alpha}{|\mathbf{r}-\mathbf{R}_\alpha|} . \tag{5}$$

In Eq. (1), the total energy of the system is a three-part functional of the electron charge density ρ. T_s is a kinetic energy, U is the Coulomb electrostatic energy of ρ and the nuclear charges, and E_{xc} is the exchange-correlation energy. ρ is expressed as a sum of squared-magnitudes of single-particle wavefunctions ψ, Eq. (2), and T_s is the sum of the single particle kinetic energies of these wavefunctions. By varying E with respect to ψ, we can show ψ satisfies a Schroedinger equation, Eq. (3), with an effective potential, Eq. (4). The exchange-correlation contribution to V_{eff} is given by a functional derivative, and the Coulomb contribution by Eq. (5) which includes the nuclear charges Z_α.

It's worth recalling that density functional theory would be an exact many-body theory if the exact non-local functional E_{xc} were known.[8] We use the local approximation to E_{xc} because we don't have a better general form, and because it has been very successful.

To solve the LDF equations for anything more complex than an isolated atom, it is necessary to introduce a set of variational basis functions for ψ in Eq.(3). Two approaches dominate current practice. The first is the use of norm-conserving pseudopotentials [9] with plane wave basis functions. The pseudopotential for each atom is constructed from the full LDF potential for the atom in such a way that the valence eigenvalues and the shape and amplitude of the valence wavefunctions outside the core region are reproduced exactly. In contrast to the original valence wavefunctions, the pseudowavefunctions are nodeless, so their expansion in plane waves converges rapidly. This approach has been very successful in calculating lattice constants, structural energy differences, and phonon frequencies for semiconductors.[10] Norm-conserving pseudopotentials also offer some advantages when used with other types of basis functions, such as linear combinations of atomic orbitals (LCAO's).

The second approach is the use of augmented basis functions. This is also a means of concentrating on the valence electrons. While the full ion core potential is employed, the correct numbers of nodes are built into the basis functions to orthogonalize them to core wavefunctions. Augmented bases are much more efficient than the plane-wave pseudopotential approach for transition metals, since the highly localized d electrons are accurately represented with a relatively small basis set. The linear augmented plane wave (LAPW) method [11] has been developed into a flexible and highly accurate means of solving the LDF equations,[12] and this is the method used in the examples I'll discuss.

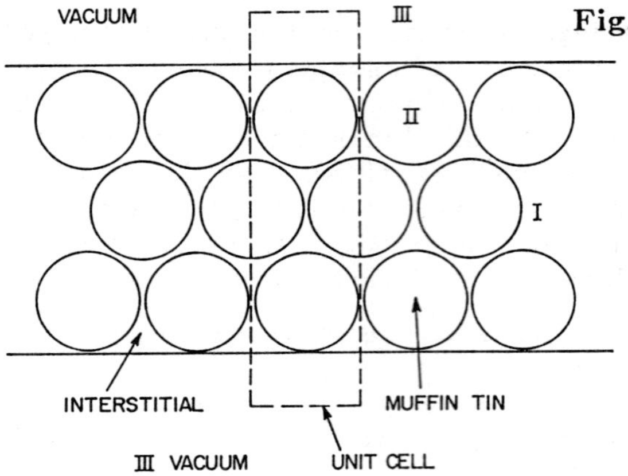

Fig. 1. Partition of space for the LAPW method.

Surface structural energy calculations are typically performed for finite-thickness slabs that extend to infinity parallel to their surfaces. To introduce LAPW basis functions, we divide space into three kinds of disjoint regions, as illustrated in Fig. 1. "Muffin tin" spheres (*II*) are centered on each atom, and a pair of "matching planes" divide vacuum regions (*III*) that run to infinity from the slab interior. The region inside these planes and outside the spheres is the interstitial region (*I*). A unit cell is introduced which is also periodic in the slab normal direction, but this is just a trick to quantize plane waves, and does not mean the slab is periodically repeated.

For each Bloch wavevector $\mathbf{k}_{\|}$, we define a set of basis functions

$$\chi_{\mathbf{K}}(\mathbf{x}) = \begin{cases} e^{i\mathbf{K}\,\mathbf{x}}, & \mathbf{x} \in I \\ \sum\limits_{lm\tau} a_{\mathbf{K}nlm\tau}\, Y_{lm}(\hat{\mathbf{x}}_n)\, \phi_{nl\tau}(x_n), & \mathbf{x} \in II_n \\ e^{i\mathbf{K}_{\|}\cdot\mathbf{x}_{\|}} \sum\limits_j b_{\mathbf{K}nj}\, u_{nj}(\mathbf{x}_{\perp}), & \mathbf{x} \in III_n \end{cases} \tag{6}$$

$$\mathbf{K} = \mathbf{k}_{\|} + \mathbf{G} \ , \ \ \mathbf{x}_n = \mathbf{x} - \mathbf{R}_n .$$

In these equations, n indexes various muffin tin and vacuum regions, $\mathbf{G}$ is a 3-dimensional reciprocal lattice vector, and Y_{lm} are spherical harmonics. There is an independent function for each $\mathbf{G}$ such that $|\mathbf{K}|$ is less than a specified cutoff. The radial functions $\phi_{nl\tau}$ are solutions of the Schroedinger equation for the spherical average of the self consistent potential in the muffin tin at an energy centered in the valence bands (τ=1), and energy derivatives of these functions (τ=2). The functions u_{nj}

are solutions of the Schroedinger equation for the planar-averaged vacuum potentials, for a sequence of normal-direction energies appropriate to the range of parallel kinetic energies of the basis functions. With these choices, each basis function is already a pretty good approximation to the eigenfunctions we want for the complete potential. The muffin-tin and vacuum pieces are joined by plane waves in the interstitial region, and the coefficients $a_{\mathbf{K}nlm\tau}$ and $b_{\mathbf{K}nj}$ are chosen to make each χ and its first derivative continuous at each boundary.

The LDF wavefunctions ψ are now simply taken to be linear combinations of the basis functions,

$$\psi_{\mathbf{k}_{\|}\nu}(\mathbf{x}) = \sum_{\mathbf{K}} c_{\mathbf{K}\nu}\, \chi_{\mathbf{K}}(\mathbf{x}). \tag{7}$$

Using this expansion, we solve Eq. (3) variationally and get a matrix eigenvalue problem for the coefficients $c_{\mathbf{K}\nu}$ and energies ϵ_i. With this basis function choice, it is possible to solve Eqs. (1-5) with no further mathematical approximations. We make only one additional physical approximation, rigid spherical atom core charge densities, and all the rest is details.[12] Admittedly, the division of space makes some of the details in evaluating ρ, the self-consistent V_{eff}, and E rather complicated. The augmentation approach can also be used with LCAO-like basis functions.[13] Such bases become quite computationally efficient when combined with further mathematical approximations such as spherical potentials, but at the sacrifice of the ability to systematically converge the LDF energy.

The last class of methods which have been widely applied to surface structural energy problems are *ab initio* quantum chemical calculations. In principle, these have a significant advantage over LDF calculations, because the accuracy with which the electron correlations are treated can be improved systematically. It is not so easy to realize this advantage in practice, since the computational complexity grows rapidly as the order of the correlations included is increased. Another issue to be considered is that these methods, developed for molecules, do not treat extended surfaces. A surface must be modeled as a finite cluster of atoms, and the convergence of results with cluster size must be established. Compounding the difficulty, computational requirements grow as a rather large power of the number of atoms. The treatment of correlation in these calculations cannot be extended to periodic systems to take advantage of translational symmetry as is done in LDF calculations. Despite my reservations, I believe that quantum chemical studies of surface problems provide an important supplement to LDF work.

C. Representative Applications

Before getting to the examples from my own research, I'd like to give you a quick (and incomplete) survey of the kinds of surface problems to which the structural energy methods discussed above have been applied.

There have been many applications to semiconductors. The Pandey chain model for the 2x1 reconstruction of the cleaved Si(111) surface [14] encountered skepticism because of its drastic modifications of bond topology and angles. The demonstration, using LDF calculations, that it had lower energy than a relaxed ideal structure [15] overcame this objection. A second important step towards establishing this model was the LDF demonstration that the barriers for its formation were small.[16] The geometry and energetics of the recently introduced dimer-adatom-stacking fault model for the equilibrium 7x7 reconstruction of Si(111) [17,18,19] were studied both using a tight-binding model for the full structure,[20] and LDF for sub-units.[21] The stability and atomic coordinates of the As-terminated Si(111) surface, which has an ideal 1x1 structure, was determined by LDF theory.[22] For the Si(001) surface, the dimers forming the basic 2x1 reconstruction were predicted to buckle, first using empirical tight-binding [23] and later LDF.[24] The vibrational frequencies of chemisorbed H on this surface were studied using quantum-chemical methods on cluster models.[25]

For III-V compound semiconductors, such as GaAs, the buckling relaxation of the (110) cleavage surface was initially studied using tight binding.[7] The same concepts that explain this buckling were later shown to predict a surface-vacancy model for the 2x2 reconstruction of III-V (111) surfaces, using both tight-binding [26] and LDF results.[27] All the LDF results on semiconductor surfaces used norm-conserving pseudopotentials.[9]

Fewer structural energy studies have been done for simple metal surfaces, perhaps reflecting the dearth of experiments on these systems. One recent example is an LDF pseudopotential study of surface phonon dispersion relations on the Al(110) surface. In this work, a harmonic pairwise-interaction model was used to extend the limited first-principles results to the extent required for full lattice dynamics.[28]

Transition metal surfaces have been the subjects of many more studies. Many quantum-chemical cluster calculations have been carried out, mainly concentrating on chemisorption. As one example, I'll mention the interesting controversy surrounding the vibrational excitations of the Ni(001) surface with chemisorbed O. Shifts with O coverage have been ascribed to qualitative changes in the Ni-O bond [29] and to lateral interactions with essentially unchanged bonds.[30] Cluster LDF calculations (relative rarities) were also performed for this system, and identified O-induced surface stress as an important component in shifting vibrational modes.[31]

While many extended-surface calculations of electronic spectra have been carried out for transition metals, energy studies are newer and fewer. The c2x2 reconstruction of W(001) has been studied using LDF-LAPW methods.[32,33] While both groups found the ideal structure unstable relative to lateral displacements of the surface atoms, results for the energy lowering differed substantially. Model calculations too numerous to mention here have also been carried out for this system. As a final example, I mention the 2x1 missing row reconstruction of Au(110). The stability of this structure and details of the atomic displacements were established by LDF-pseudopotential calculations.[34] Furthermore, the transition between stable unreconstructed (110) surfaces for some transition and noble metals (such as Ag) and this reconstruction was analyzed in terms of a critical balance between d-electron bonding energy loss and s,p-electron delocalization energy gain.[34]

III. H Vibrations on Transition Metals

Hydrogen is a prototypical adsorbate. You might say it is the "hydrogen atom" of adsorbate studies. In addition to its relative simplicity, chemisorbed H is believed to play an important role as a reaction intermediate in a wide range of catalytic reactions. Chemisorbed H drives reconstructive displacements of the substrate surface atoms of many metals, such as W, Mo, and Ni, often yielding complex phase diagrams with transitions among many different structures. Because of its light mass, H is the adsorbate most likely to display quantum effects. I have carried out a series of calculations of the vibrational frequencies of H on several transition metal surfaces, with my collaborators Peter Feibelman and Rana Biswas. Vibrational excitations are probably the most experimentally accessible quantities for chemisorbed H. They can be measured using either EELS or infrared spectroscopy. Since the frequencies are high, H modes are easily separated from bulk phonons. As a result, there is probably a greater experimental literature of surface vibrational studies for H than for any other atomic adsorbate. This was one motivating factor for our choice of this set of problems. Another was the fact that lattice dynamics is trivial because substrate atoms may be assumed infinitely massive compared to H, and their motions neglected. Finally, H often forms 1x1 structures at saturation coverage. The resulting small surface unit cell allows us to achieve a high degree of convergence in our energies, which is especially important in establishing the accuracy of our methods for this class of problems.

As an introduction, I'd like to review the terminology for sites and modes on simple surfaces. Fig. 2 shows the high-symmetry sites. H is usually found in the highest coordination site available on transition metal surfaces. For a theorist doing energy calculations, finding the lowest energy site is part of the job. Site promotion energies, which relate to surface diffusion barriers, are also of interest. Note that the so-called fcc and hcp 3-fold sites differ by the absence or presence of a second-layer substrate atom below the adsorption site, by analogy with the stacking of planes in these bulk structures.

Because of symmetry, 3-fold and 4-fold sites have only 2 inequivalent vibrational modes, sketched in Fig. 3. The vertical mode is termed the symmetric stretch (*ss*) and the two degenerate horizontal modes the asymmetric stretch (*as*). Any direction of horizontal motion has the same harmonic force constant for these sites. In carrying out our energy calculations to get force constants, we usually move the H in all inequivalent directions that preserve some point group symmetry, such as towards the nearest atop site and the nearest bridge site. For bridge-bonded H the *as* mode is inequivalent to the *wag* mode in Fig. 3, and there are 3 fundamental frequencies.

All our results for vibrational frequencies of H on pure transition metal substrates are given in Table 1, along with experimental numbers for the systems we studied. There are several different aspects of interest in each case, so I'll discuss them one at a time, in order of increasing complexity.

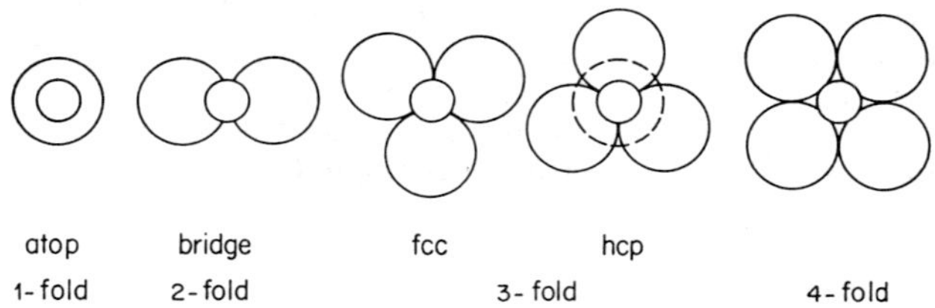

Fig. 2. High symmetry sites of low index surfaces.

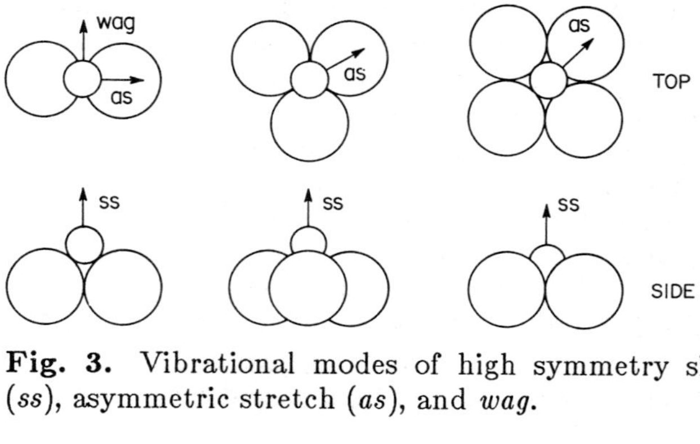

Fig. 3. Vibrational modes of high symmetry sites: symmetric stretch (*ss*), asymmetric stretch (*as*), and *wag*.

Table 1. H sites and vibration energies for the surfaces studied. Coverage is 1 monolayer (ml) except 2 for W as indicated.

Surface	Site	Modes	Theory (meV)	Experiment (meV)
Ru(0001)	3-fold fcc	as	100	138,[a] 102[b]
		ss	140	105,[a] 141[b]
W(001)	bridge 2ml	wag	77	80,[d,e]
		as	169	160,[c,d] 118[e]
		ss	141, 133[g]	130,[c,d,e] 133[f]
		opt	150	118[e]
Pt(111)	3-fold fcc	as	114	152,[h] 67[i]
		ss	166	68,[h] 112,[i] 155[j]
Rh(001)	4-fold	as	67	138[k]
		ss	92	82[k]
		mix	142	
		mix	148	152[k]

a Barteau, Broughton, & Menzel; Ref. 35
b Conrad, Scala, Stenzel, & Unwin; Ref. 36
c Ho, Willis & Plummer; Ref. 37
d Barnes & Willis; Ref. 38
e Woods & Erskine; Ref. 39
f Chabal; Ref. 40
g Weinert, Freeman & Ohnishi; Ref. 41
h Baro, Ibach, & Bruchmann; Ref. 42
i Richter & Ho; Ref. 43
j Reutt & Chabal; Ref. 44
k Richter & Ho; Ref. 45

A. H on Ru(0001)

H on Ru (0001) was probably the most straightforward case we studied.[46] This is the basal plane of an hcp metal, and we assumed the fcc 3-fold site for H based on general folklore and some other results I'll mention later. As in all our calculations, we start by picking a series of heights for H in the expected site, centered around a guessed geometry based on an H radius of 0.45Å. The binding energies for the 5 heights we used in this case are shown in Fig. 4, the curve labeled "symmetric". We then do a least-mean-squares fit of several polynomials to these results to determine the equilibrium height and the curvature at that height. Usually, we fit 3rd and 4th order functions to all the points, and then 2nd and 3rd order to subsets formed by dropping one end point or the other. We judge the consistency of the results, and add more calculated

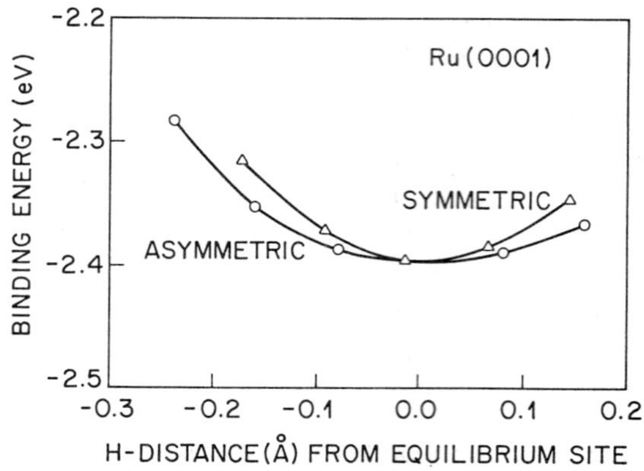

Fig. 4 Calculated binding energy for H on Ru(0001), from Ref. 46.

points if it seems necessary. We believe our quoted frequencies represent converged LDF results to $\sim 5\%$. Note that we are rigidly moving the entire monolayer of H, which corresponds to the $\mathbf{Q}_{\parallel}=0$ vibrational normal mode. This is the mode probed by infrared and most EELS experiments.

We found a H height of 1.78Å, corresponding to a H radius of 0.45Å. Keeping this height, we then moved H laterally to a series of positions along the line indicated by the arrow in Fig. 3, both toward the atop site and toward the bridge site. While there is a unique curvature in any direction from the 3-fold site, the higher anharmonic terms are inequivalent in these directions. The 3rd-order term is important in the fit, as can be seen from the "asymmetric" curve in Fig. 4. We extract the *as* frequency from this fit.

The EELS data for this system show clean, reasonably sharp, well-resolved lines, both for the fundamental modes [35] and for overtone and combination modes,[36] with consistent $1/\sqrt{2}$ deuterium shifts. Our numbers were in good agreement with the data of Barteau, et. al.,[35] which was all that was available to us at the time we performed the calculations, but they had assigned the higher frequency to the *ss* mode. They in fact could not identify either mode as dipole-active, and made their assignment primarily on the basis of a previous study of H on Pt(111).[42] We were confident that we could simply reverse the assignment and claim agreement. As we were completing our paper on this, we found Conrad, et. al.'s work, which both confirmed the numbers, and proposed the same reversed assignment on the basis of an analysis of combination and overtone excitations.[36] Some of their results are shown in Fig. 5.

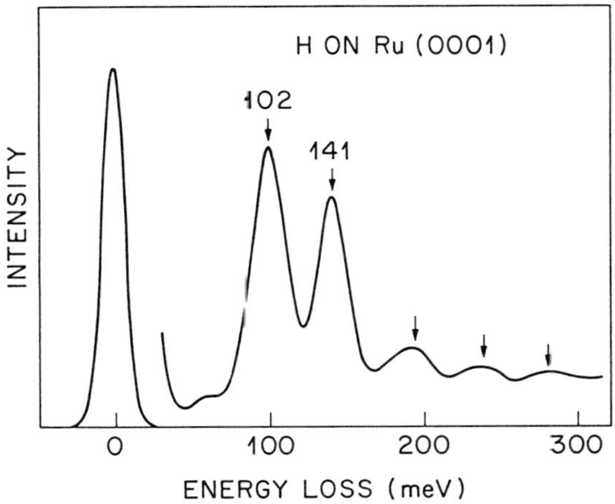

Fig. 5. Off-specular EELS spectrum for H on Ru(0001) at saturation coverage, from Ref. 36.

B. H on W(001)

H on W(001) is a very complicated system due to the reconstruction of the clean surface and the series of structures that it goes through as a function of H coverage. The saturated surface is simple, however, with 2 H monolayers residing in bridge sites in a 1x1 structure, and W in bulk-like positions. This was the earliest system in which non-dipole excitations were seen with EELS, and all 3 bridge-site modes were well established by early work, as shown in Fig. 6.[37,38] We got interested in this system when new EELS experiments challenged the *as* frequency of the old work, 160 meV, and assigned a newly-discovered 118 meV loss to the *as*.[39]

We proceeded the same way as for Ru, except after determining the equilibrium height, we fitted two independent curves for the *as* and *wag* motions shown in Fig. 3.[47] We found a height of 1.07Å, corresponding to a H radius of 0.54Å. Our frequencies for the *wag*, *as*, and *ss* are in very good agreement with the older experiments,[37,38] as can be seen from Table 1. We also tabulated an independent LDF-LAPW theoretical result for the *ss* frequency, which is consistent within our expected accuracy.

This left an unresolved issue - what was the 118 meV excitation? We realized that with 2 atoms per surface unit cell, the H on W(001) system could have an additional, previously undiscussed $\mathbf{Q}_{\|}=0$ mode, an "optic" version of the *ss*, with the 2 H's moving out-of-phase.[47] It was not clear *a priori* whether *opt* should be higher or lower than *ss*. We did the

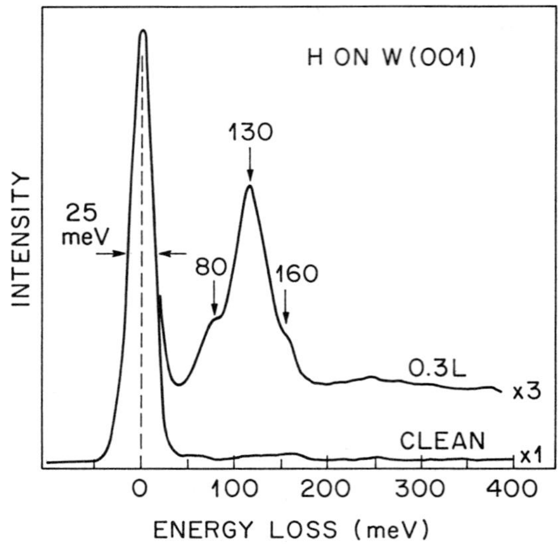

Fig. 6. Off-specular EELS spectrum for H on W(001) at saturation coverage, from Ref. 38.

calculation, and found it to be 9 meV higher, as indicated in Table 1. Woods and Erskine presently believe that their 118 meV loss is, in fact, *opt.* Reconciling this with our results remains an unresolved issue.

C. H on Pt(111)

The work of Baro, et. al., an EELS study of H on Pt(111), was the classic reference showing *as* higher than *ss* in a 3-fold site.[42] This seemed a conclusion worth challenging, and we proceeded as for Ru, placing H in the fcc 3-fold site and calculating the energies and the frequencies given in Table 1.[48] The height we found was 0.95Å, corresponding to a H radius of 0.48Å. The agreement of our calculated frequencies with Baro et. al. is poor, and while reversing modes gives reasonable agreement for *ss*, they claim to have established dipole activity for their lower mode, and none for their higher.[42]

To make life more complex, we discovered a He diffraction study and interpretation showing H to be on the hcp 3-fold site on Pt(111).[49] We did not think that vibrational frequencies would change enough from fcc to hcp 3-fold sites to resolve the disagreement. Nonetheless we had to check, so we began energy calculations for the hcp site, starting with the expected equilibrium height. We found the chemisorption energy was 0.20 eV less favorable at this site, adding the He diffraction to the list of experiments with which we apparently disagreed. To pursue this issue, we plotted the vacuum region charge density for the fcc-site system. I'll remind non-specialists that the corrugated repulsive potential in He diffraction is believed to be proportional to this charge density.[50] We found, contrary to the existing atomic-charge-superposition analysis,[49] that the most pronounced structure of this contour was a dip over the vacant hcp site rather than a peak over the H.[48] Since dips and peaks are indistinguishable in the eikonal analysis applied to the diffraction data,[50] we concluded that these data were, in fact, consistent with our fcc site.

Lacking alternatives, we took another hard look at the EELS data.[42] The spectra are much more complex than for Ru. Peaks are very broad compared to the resolution, and CO contamination obscures parts of the energy range. With a little imagination, we can find both our predicted *as* and *ss* losses under the broad upper peak in the off-specular data, Fig. 2 of Ref. 42. Our results have stimulated some new work, and an infrared study by Reutt and Chabal finds a dipole absorption at 155 meV, close to our *ss*.[44] However, a new EELS study by Richter and Ho finds an impact-scattering mode (*as* ?) at 67 meV, and a dipole-active (*ss* ?) mode at 112 meV.[43] While this assignment puts modes in what we would consider to be the right order, the numbers are way outside our uncertainty range. We could speculate about possible explanations for these diverse results, but only continued research can resolve the apparent contradictions.

D. *H on Rh(001)*

We were interested in studying a 4-fold site, and preliminary reports of some new experiments on H on Rh(001) by Richter and Ho [45] stimulated us to choose this system. We followed our usual procedures, and found 67 and 100 meV for ν_{as} and ν_{ss},[51] far from the experimental numbers of 138 and 82 meV. Having been sensitized to the site-assumption problem, we checked bridge and atop site energies, and found them 0.17 and 0.69 eV higher than the 4-fold site. While these calculations did not give us an alternative site, they did give us an idea. The 4-fold H is deeply buried in the surface, at a height of only 0.59Å. The bridge saddle point is at 1.2Å, and the atop at 1.6Å, so the H must climb considerably to follow the minimum-energy path toward these lateral positions. This suggested that the H-surface interaction potential should be something like

$$V_H = a[z - \delta z_o(x,y)]^2 + b(x^2 + y^2) \quad . \tag{8}$$

rather than the simple harmonic form ($\delta z_o = 0$).[51] A consistent expansion containing all terms permitted by symmetry to 4th order in atomic displacements is

$$
\begin{aligned}
V_H = \ & a_1 + a_2 z + a_3 z^2 + a_4(x^2+y^2) \\
& + a_5 z^3 + a_6\, z(x^2+y^2) \\
& + a_7 z^4 + a_8 z^2\, (x^2+y^2) \\
& + a_9 (x^2+y^2)^2 + a_{10} x^2 y^2 \quad ,
\end{aligned} \tag{9}
$$

where x and y are towards the bridge sites, and z is normal to the surface. We fitted this form to the energies of 20 geometries, using a global least-squares fit. The largest terms were a_6, the leading vertical-lateral coupling term suggested by Eq. (8) above, and a_{10}, the bridge-atop asymmetry term. The fitted potential is shown in Fig. 7. There are substantial departures from the harmonic form, shown for reference, over the ~1Å range spanned by the H wavefunctions.

To find the excitations of this highly anharmonic potential, we adopted a brute-force variational approach.[51] We used H wavefunctions

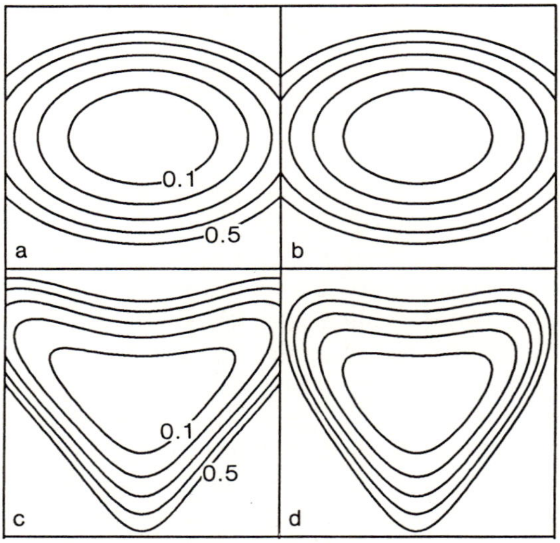

Fig. 7. Harmonic (a,b) and anharmonic (c,d) potentials for H on Rh(001), from Ref. 51. Contour levels are in eV, plotting areas are ±1.5 au from the equilibrium position in the vertical and bridge (a,c) and atop (b,d) horizontal directions.

$$\Psi_\ell(\mathbf{r}) = \sum c^\ell_{ijk}\ u^\alpha_i(x)u^\alpha_j(y)u^\beta_k(z) \quad , \tag{10}$$

where c^ℓ_{ijk} are linear variational parameters, and u are harmonic oscillator wavefunctions with nonlinear range parameters α and β. The ground state and lowest excited states are shown in Fig. 8. The *ss*, state 2, and a partner combination mode, state 5, show very substantial mixing of vertical and lateral motion. States 3 and 4, which are *as* first overtones, show an 18 meV splitting from the bridge-atop azimuthal asymmetry. The anharmonic excitation energies which we believe can best be identified with experimental losses are given in Table 1. The *ss* has been shifted down enough to be within our error range of the 82 meV experimental loss.

Our *as* at 67 meV remained essentially unshifted, and a substantial point of disagreement with the data. Richter and Ho originally identified a weak, impact-scattering loss seen at 138 meV with the *as*. We believe that this could be our state 4, at 142 meV. It cannot have a dipole by

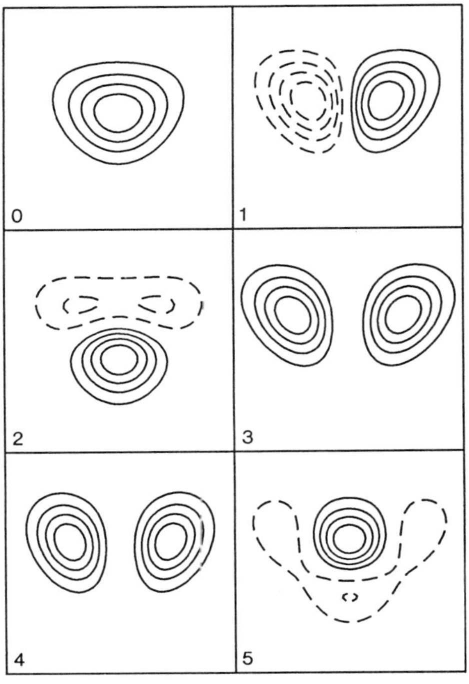

Fig. 8. Anharmonic H wavefunctions, plotted as in Fig. 7, from Ref. 51. The horizontal axis in 4 is towards the atop site; the remainder are toward the bridge site. Dashed contours are negative. The energies are 0, 67, 92, 124, 142, and 147 meV.

symmetry, and can be considered an *as* double excitation. The original experiments saw no losses in the vicinity of our prediction for the *as* fundamental excitation. Stimulated by this disagreement, Richter and Ho very recently re-did their experiment with another spectrometer with greater sensitivity in the lower energy range. A very preliminary report indicates they have found a new impact-scattering mode very close to where we predicted the *as*.

E. Spring Constant Models

Nearest-neighbor harmonic potential models have been widely used to interpret vibrational spectra for chemisorbed H.[35,39,42] They appear to provide a simple means to relate vibrational spectra to geometry. To see this relationship, consider the bridge configuration in Fig. 9. The harmonic model is

$$E = \tfrac{1}{2}k[(\delta R_1)^2 + (\delta R_2)^2], \tag{11}$$

where k is the spring constant and

$$\delta R_1 = \delta x \sin\alpha + \delta z \cos\alpha,$$

$$\delta R_2 = -\delta x \sin\alpha + \delta z \cos\alpha.$$

Thus

$$E = k \sin^2\alpha(\delta x)^2 + k \cos^2\alpha(\delta z)^2, \tag{12}$$

and

$$\frac{\nu_{as}}{\nu_{ss}} = \left[\frac{k \sin^2\alpha}{k \cos^2\alpha}\right]^{1/2} = \tan\alpha = \frac{a}{2h}. \tag{13}$$

For 3- and 4-fold sites, similar analysis gives

$$\nu_{as}/\nu_{ss} = \tan\alpha/\sqrt{2}. \tag{14}$$

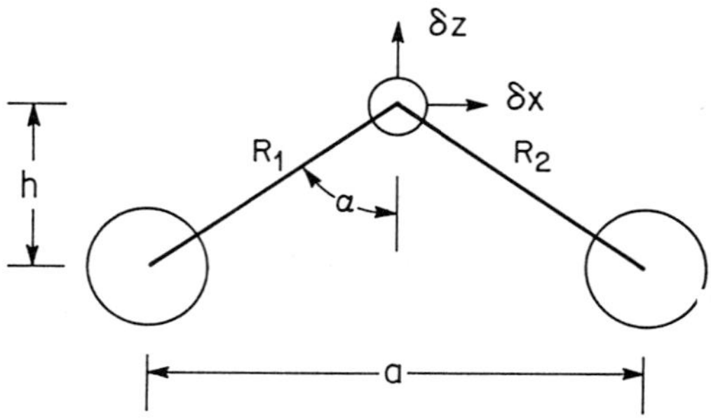

Fig. 9. Bridge site H for spring model analysis.

The spring model has been used to support mode identifications from reasonable guesses at bond lengths,[35,42] and to propose unreasonable bond lengths from modes identified by other means.[39]

To see how well the nearest-neighbor spring model correlates with our results, we have collected our calculated ratios ν_{as}/ν_{ss} and compared them with the model using Eqs. (13) and (14) above, with α taken from our calculated heights. Using a standard H radius would give essentially the same α's. The comparisons are given in Table 2. It is clear that the spring model goes from moderately inaccurate for the W bridge to terrible for everything else. (Of course this model gives a frequency of zero for the bridge-site *wag*.) The *as* is always estimated to be much too stiff compared to the *ss*. We believe the reason is that the H is bonded to delocalized electrons at a metal surface, which have been smoothed by the Smoluchowski effect.[52] This is related to the effective medium model discussed in the next section.

Table 2. ν_{as}/ν_{ss} ratios.

Surface	Spring Model	LDF-LAPW
W(001)	1.40	1.20
Ru(0001)	1.04	0.71
Pt(111)	1.19	0.69
Rh(001)	2.31	0.67

F. Effective Medium Theory

The effective medium theory (EMT) method, which I mentioned in Sec. II, has been applied to H on transition metals. As commonly applied, this approach starts with an clean surface charge density approximated as a sum of atomic charges,[4]

$$\rho_0(\mathbf{r}) = \sum_{\mathbf{R}} \rho_{at}(|\mathbf{r}-\mathbf{R}|). \tag{15}$$

The binding energy of the adsorbed atom at point $\mathbf{r}$ is then given by two terms,

$$\Delta E(\mathbf{r}) = \Delta E_{eff}^{homo}[\bar{\rho}_0(\mathbf{r})] - 2(1-f)\frac{|V_{da}(\mathbf{r})|^2}{C_d - V_0(\mathbf{r})}, \tag{16}$$

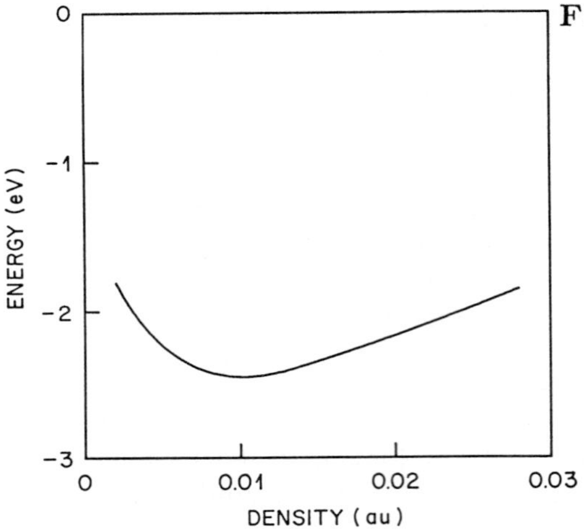

Fig. 10. Effective medium theory H embedding energy.

where ΔE_{eff}^{homo} is the energy of embedding the adsorbate in a uniform electron gas of density $\overline{\rho}_0$, and $\overline{\rho}_0$ is an average of ρ_0 which is essentially equal to $\rho_0(\mathbf{r})$ for a small adsorbate. ΔE_{eff}^{homo} for H is shown in Fig. 10. The second term is an approximate "covalent correction," in which f is the fractional filling of the substrate d band, C_d is the d band center of gravity, V_0 is the (position dependent) energy of the H 1s resonance level, and V_{da} is a coupling matrix element given by

$$|V_{da}(\mathbf{r})|^2 = 20\Delta_d\Delta_a\sum_{\mathbf{R}}\frac{1}{|\mathbf{r}-\mathbf{R}|^6}. \tag{17}$$

The constants Δ_d and Δ_a are tabulated.[4]

Without the covalent correction term in Eq. (16), H would bind equally at any point on the surface with $\rho_0 = 0.01$, the minimum in the curve in Fig. 10. It can be shown simply in this limit that the *ss* frequency is proportional to the normal gradient of ρ_0.[4] The *as* has no stiffness at a high-symmetry point. So the *as* stiffness (as well as site specificity) all comes from the covalent term. Since this is a small correction, it is qualitatively clear that EMT gives ν_{as}/ν_{ss} smaller than the spring model.

We don't have a direct comparison with EMT for all the modes of a system which I discussed, but it is instructive to compare our Rh(001) potential with EMT Ni(001) results. Fig. 11 compares the EMT potential for a single H on Ni [53] to a periodic fit to our monolayer H on Rh energies.[51] It is clear there is a high degree of similarity. To be more quantitative, we can compare the lowest "band" excitations calculated for

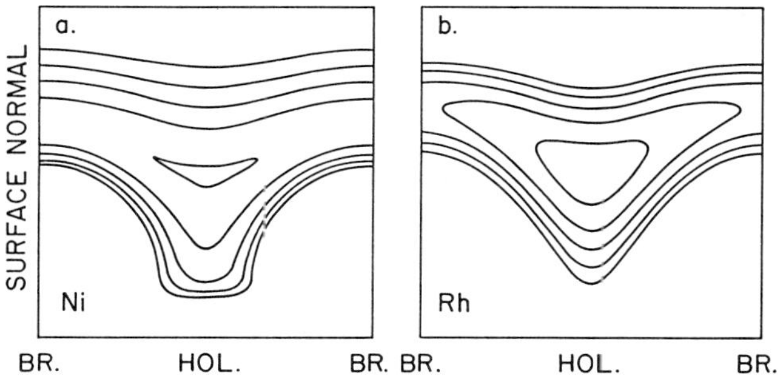

Fig. 11. Periodic H binding potentials over 1 unit cell, a) Ni(001) using effective medium theory (Ref. 53), b) Rh(001) using LDF-LAPW (Ref. 51). The abscissa runs from a bridge site through the hollow site to the next to bridge site. Contour spacings are 0.15 eV. The lowest contour is 2.7 eV for Ni and 2.74 eV for Rh.

H in the EMT potential [53] and our lowest H anharmonic excitations. We find ν_{as}/ν_{ss} is 0.75 for EMT Ni and 0.72 for our Rh.[51] For W(001), EMT gives the 4-fold site as the stable site rather than the bridge. Nonetheless, ν_{ss} for the bridge was calculated as 127 meV,[4] compared to our 141, and 130 for experiment.[37,38] It is clear that as simple models go, EMT is much better than harmonic pair potentials for describing the dynamics of chemisorbed H.

IV. Silicide Interface Structure

For a second example from a rather different area, I'd like to discuss some recent work on interfaces. Epitaxial films of the metallic silicides $NiSi_2$ and $CoSi_2$ on silicon form atomically abrupt and structurally perfect Schottky-barrier interfaces.[54,55] Metal-semiconductor contacts form an extremely important class of systems for electronic device applications, but their properties are poorly understood from a fundamental point of view. One such property, for example, is the energy difference between the semiconductor conduction band minimum and the metal Fermi level, known as the Schottky barrier height. Most metals form incommensurate polycrystalline layers on semiconductors, and the resulting poorly defined interfaces offer little possibility for microscopic modeling. The epitaxial silicides thus offer unique opportunities to develop a detailed understanding of prototypical interfaces on a microscopic basis.

To exploit the advantages offered by interfaces with well-defined atomic geometry, we clearly must determine precisely what that geometry is. $NiSi_2$ and $CoSi_2$ have fluorite structures with lattice parameters which are nearly equal to that of Si, permitting pseudomorphic growth. The fluorite structure has tetrahedrally coordinated Si, and 8-fold, cubically-coordinated metal atoms M. Diamond and fluorite lattices have common <111> planes, and a short while spent with ball-and-stick models suggests several possibilities for unreconstructed, undistorted (111) interface geometries.. There are two observed orientations for the epitaxial films on Si(111), known as type A and type B. Type A silicide has the same crystallographic orientation as the substrate; type B shares the Si substrate [111] axis, but is rotated 180° about this axis. $NiSi_2$ can be grown with either orientation by controlling growth kinetics.[56]

Experimentally determining the atomic structure of an interface is even more difficult than determining that of a surface, because the buried interface is not accessible to most structural probes. Three methods have been applied to obtain partial information. Cross-section transmission-electron-microscopy lattice imaging gave the first indications of the perfection of these interfaces. Combined with image simulation, it gave quantitative information on the displacement of the silicide <111> lattice planes relative to the extension of the substrate Si lattice beyond the interface.[54,55] These displacements were confirmed using x-ray standing wave analysis for both $NiSi_2$ [57] and $CoSi_2$.[58,59] Cherns et. al. proposed models, based on the "ball-and-stick" approach, in which the interface metal atom was either 5-fold or 7-fold coordinated.[54] They found that only the 7-fold model was consistent with their measured displacements for both A and B oriented $NiSi_2$.[54] In contrast, Gibson et. al. found that the $CoSi_2$ interface was only consistent with the 5-fold model.[55] These interfaces were also probed by high-energy ion channeling, with energy and angular analysis of the Rutherford backscattered ions.[60,61] The results appear to confirm the structural difference between the interfaces formed by the two silicides.

The 5-fold Co interface structure has seemed an outstanding puzzle to me, since it didn't make sense in terms of simple local-bonding concepts. Why should this system choose to have 3 broken metal atom bonds, when the alternative for having only one exists, at no apparent cost in strain or other energy penalty? The bulk electronic structures of $CoSi_2$ and $NiSi_2$ are very similar,[63] so there was no clue from such considerations as to why the interfaces should be different. Of course, it is a well-established experimental fact that they are. A clue to the puzzle came from a recent discovery by Hellman and Tung, who found that two distinctly different surface structures can be formed on the $CoSi_2$ (111) surface.[62] One corresponds to a 7-fold coordinated Co, but a second, Si-rich structure can be formed which is apparently energetically most stable. It is believed to consist of an ideal $CoSi_2$ surface with an added bulk-Si-like double layer. This gives the Co bulk-like 8-fold coordination, and leaves all Si 4-fold coordinated, although one surface Si has three rather acute bond angles of 70°.

I have reexamined the experimental evidence for 5-fold interface Co, and none of the experiments, in fact, can rule out two additional models in which the interface Co is 8-fold coordinated, as on the most stable surface.[62] These models were not considered in the original analysis because they entail either one broken Si-Si bond, or one overcoordinated Si. It was considered that the Si-Si bond was sufficiently strong compared to the M-Si bond that no violation of perfect tetrahedral Si bonding could occur at the interface. This is obviously an ideal sort of question to investigate with structural energy calulations, the LDF-LAPW method being particularly well suited.[12] To calculate the interface energy, I considered superlattice structures, which maintain the true periodicity of the interface in the parallel direction, but introduce an artificial periodicity in the perpendicular direction, alternating slabs of Si and MSi_2.[64] Since each unit cell of the superlattice contains two interfaces, the interface energy is half the supercell energy minus half the sum of the energies of a number of unit cells of each bulk material corresponding to the thickness of the respective slabs.

It was necessary to devise supercell structures satisfying several criteria: (1) They must reproduce the atomic geometry of the candidate interface models to a distance of several neighbor shells. (2) They must allow the most consistent possible subtraction of systematic errors by all containing the same numbers of like atoms. (3) They must be small enough to allow adequate convergence of the electronic wavefunctions within the computational capacity available. I found that all these criteria could be satisfied stacking two MSi_2 triple layers and two Si_2 double layers, for a supercell formula M_2Si_8.

I wished to investigate four interface structures: 5-fold type A, 7-fold type A, 8-fold type A, and 8-fold type B (abbreviated 5A, 7A, 8A, and 8B). We previously established that the A-B energy difference for the 7-fold Ni interface was only ~0.04 eV, with A lower.[65] Since this and the 5-fold interface involve only changes in third neighbor arrangement in changing from A to B, I economized by omitting 5B and 7B calculations. Sketches of fragments from the superlattices are shown in Fig. 12. Note that 8A and 8B involve a more important change in bonding than that discussed above; 8A has an overcoordinated Si, and 8B an undercoordinated Si. If you study the pictures carefully, you will see that only 8A is periodic in the [111] direction, and that the others seem to contradict the description I gave for supercells. 8A has a hexagonal unit cell, space group D^3_{3d}. 5A, 7A, and 8B are, however, periodic with

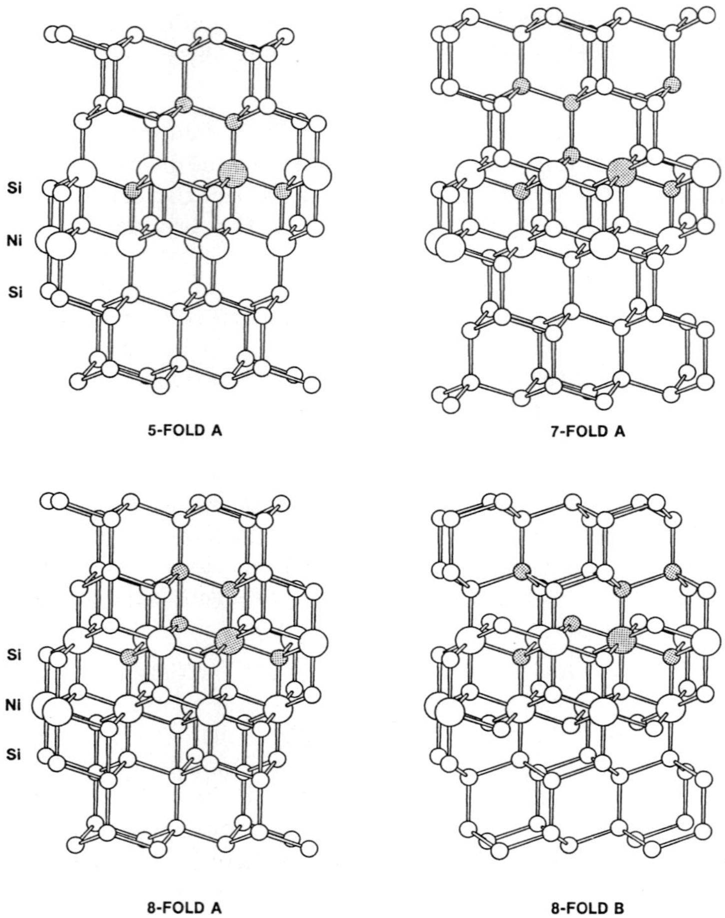

Fig. 12. Ball and stick models of superlattice fragments illustrating the interface geometries discussed. Structures are the same for Ni and Co.

rhombohedral cells, space group D^5_{3d}, formed by adding symmetry-preserving lateral translations to [111] translations.[64] This slight generalization of the supercell construct seems trivial in retrospect, but it was really a key breakthrough for me. With the narrow definition, different numbers of layers would have been necessary to achieve the different interface geometries, and the accuracy with which interface energies could be compared would have been too limited to draw firm conclusions.

The results are summarized in Table 3. It is clear that 5A (or, presumably B) structures aren't even remotely competitive. This is a

Table 3. Interface energies for the structures shown in Fig. 12.

Structure	Energies (eV) Ni	Co
5-fold A	1.52	2.20
7-fold A	0.35	0.86
8-fold A	0.78	0.68
8-fold B	0.55	0.53

confirmation of my intuitive feeling that it should be bad to break so many bonds. Coming to the real contestants, we see that 7A has the lowest energy for Ni, and 8B for Co. The Ni result is a necessary condition to establish credibility for the calculations, since the other structures are all ruled out by experiment for Ni. Only the B orientation of $CoSi_2$ is observed to grow on Si, which is consistent with the rather large calculated 8A-8B energy difference, 0.15 eV, compared to the 7A-7B difference of ~0.04 eV [65] for switchable $NiSi_2$ growth. The surface structure found for the Si-rich $CoSi_2$ surface does corresponds to the 8A interface structure,[62] but at the interface, 8A has an overcoordinated Si atom, while at the surface, it has a 4-coordinated Si with distorted bond angles. I believe the interface energy results are consistent with all structural and kinetic experiments. I have not attempted to calculate the small energy lowerings due to interface relaxation,[65] suggested to be an overall expansion of 0.07-0.11 Å for Ni,[57,60] and a contraction of ~0.05Å for Co.[58,59]

The energies alone do not give a qualitative answer to the question of why $NiSi_2$ and $CoSi_2$ have different stable interface structures. We can get a more detailed picture of bonding in the various interface structures by examining contour plots of the valence charge density, shown for Co in Fig. 13. Charge densities for the bulk silicide and for bulk Si are shown for reference. For the 7A interfaces, the charge densities within the $CoSi_2$ layers and within in the Si layers are essentially identical to the corresponding bulk densities. The interface Si-Si bond appears to be slightly weaker than in Si bulk, with the bond maximum 0.08 contour barely starting to develop. For 8A and 8B, the Co-Si bonds within the silicide layer appear to be strengthened compared to the bulk, developing an additional contour, 0.07. The bonds within the Si layers are weakened, although more so for 8A than for 8B. The charge density in the "dangling bond" above the 3-coordinated Si in 8B barely reaches 0.05.

The corresponding Ni plots, shown in Fig. 14, look very similar in general, and differ principally in those features singled out for mention

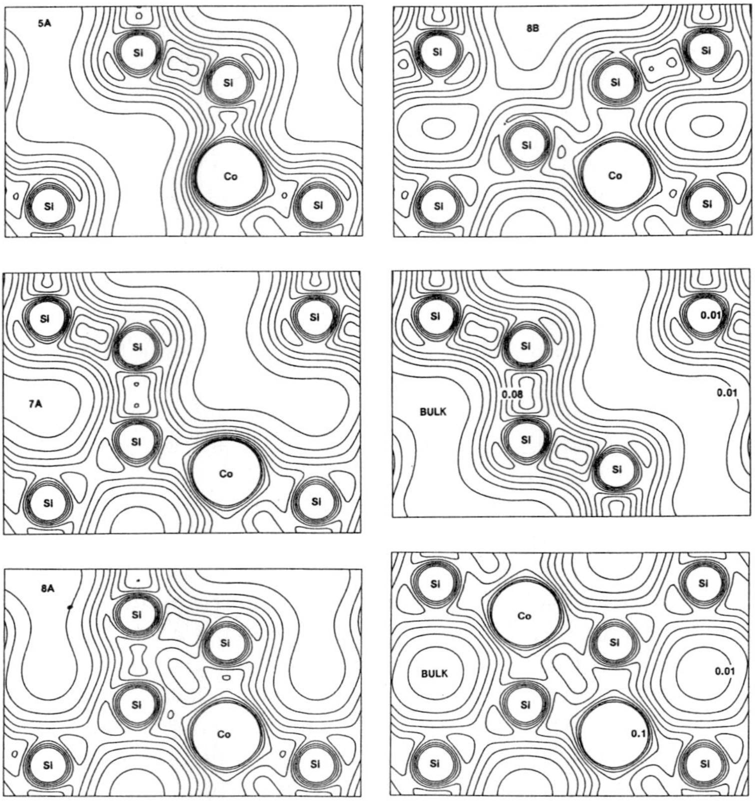

Fig. 13. Contours of constant valence charge density for the interface structures, bulk Si, and bulk $CoSi_2$. The plotting areas include the shaded atoms in Fig. 12. Contour levels are 0.01, 0.02, ..., 0.1 au (a_{Bohr}^{-3}).

above. The interface densities are closer to a "scissor and paste" joining of the bulk densities than for Co. The interface Si-Si bond isn't as weakened for 7A, and the Ni-Si bonds have not developed an additional contour for 8B. The 8B "dangling bond" region has a well developed 0.05 contour, extending as far as the 0.04 contour for Co.

The above description suggests that $CoSi_2$ is more reactive at its interface than $NiSi_2$. This can be understood in terms of an analysis of the bonding in these silicides presented earlier with the aim of explaining their unique structures among transition metal silicides.[63] We pointed out the existence of a "quasigap" separating bonding and antibonding states. For $NiSi_2$, the Fermi level falls somewhat above the gap, into the antibonding region, while for $CoSi_2$ it falls slightly below the gap. This

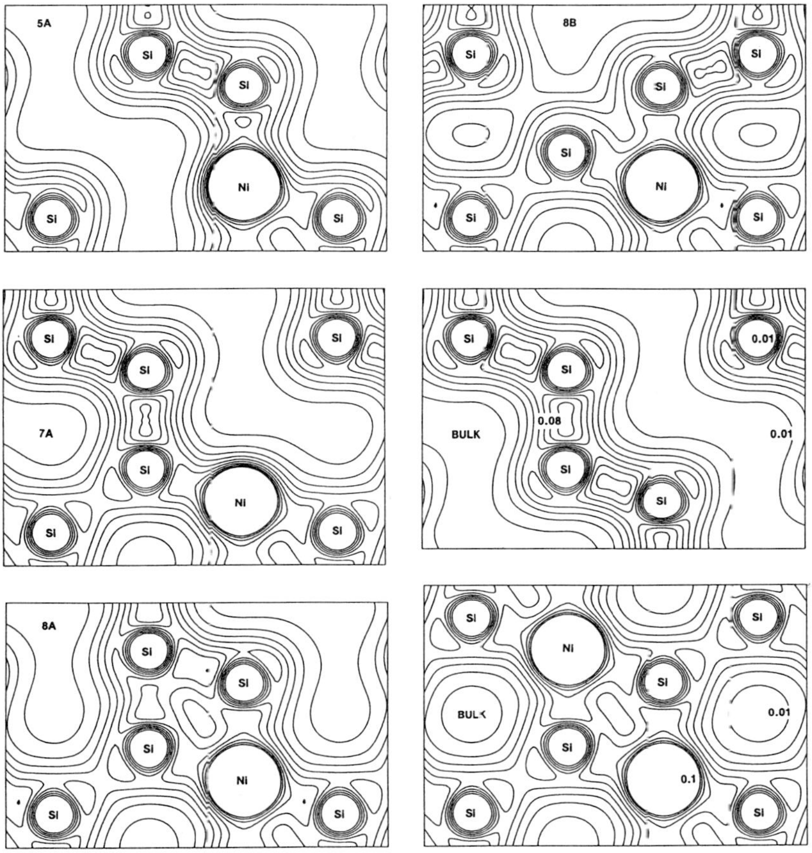

Fig. 14. Contours of constant valence charge density for the interface structures, bulk Si, and bulk $NiSi_2$. The plotting areas and coutours are as in Fig. 13.

suggests that the interface $CoSi_2$ layer can take better advantage of the extra electron made available by the 3-coordinated Si than in the case of $NiSi_2$. Atom-projected densities of states, however, did not show identifiable peaks associated with the interface bonds. It is clear that there is a delicate balance between the cost of breaking a M-Si bond and an Si-Si bond. The Si-Si distance in MSi_2 is 2.72 Å, compared to the 2.35 Å M-Si distance, and these interactions undoubtedly enter into the energetic balance as well. None of these qualitative considerations, of course, allow us to predict differences of few tenths of an eV in structural energy compared to the thousands of eV of total valence energy [12] we calculate for each supercell.

The overall thrust of this study was to identify an ambiguity in a set of experimental results and use theory to make a prediction resolving that ambiguity. It is now a challenge for the experimentalists to see what additional measurements can be devised to uniquely distinguish between 5-fold and 8-fold Co at the interface.

V. Challenges for Surface Theory

A. Extensions of Structural Energy Calculations

The organizers asked me to raise some questions rather than just describing past accomplishments of surface theory. Observations upon what needs to be done can be organized in terms of the types of structural energy methods I discussed in Sec. II.

For the empirical methods, both classical and quantum-model type, the clear challenge is to develop a systematic approach to modeling. We need some objective criteria for defining a "best" model at any given level of complexity, and mapping out its realm of applicability. Fitting to the properties of a material or set of materials will require a large data base, either from experiment or from believable first-principles calculations. Rana Biswas and I undertook such a task in attempting to develop a better model for Si, and found it to be a difficult and frustrating undertaking.[3] There must be better approaches than ours, but they need to be developed. Modeling is usually seen as an adjunct activity to solving some specific problem that requires the use of the model. Perhaps it needs to be treated as more of a research topic in its own right.

The next area that needs attention is extending the reach of LDF theory. One kind of problem of great fundamental interest which is beyond all currently popular LDF calculational methods is that of an isolated adsorbate on an extended surface. Aside from single-atom chemisorption on a structureless jellium adsorbate,[66] the only serious attempt to develop such capabilities has been that of Peter Feibelman. He has implemented a Green's function approach using norm-conserving pseudopotentials, localized orbital basis functions, and a finite-thickness slab as a substrate.[67] Calculating energies and forces for such a system is extremely demanding, and the calculations are extremely complex. His choice of approach may be the best, or something else may be more efficient and accurate. We'll only find out if some competition can be developed.

Another direction in which the reach of LDF needs to be extended is that of dynamics. Conventional energy and force calculations are much too cumbersome to allow a dynamically evolving configuration of atoms to be simulated for a useful amount of time. A bold step in this direction was taken recently be Car and Parrinello, who introduced a streamlined procedure for evolving LDF wavefunctions, self-consistent LDF potentials, and atomic coordinates simultaneously.[68] Applications of this approach have been limited so far, and the only surface structural application of which I am aware has been to a Ge grain boundary.[69] Further development of this or alternative methods is clearly called for. Perhaps we will not reach the point of being able to find the 7x7 structure by simulated annealing of an initially disordered Si(111) surface, but many less ambitious tasks can benefit from an approach not limited to a few high-symmetry guesses.

The third challenge for LDF theory is to quantify its limitations. There are really two issues here. LDF energies for any structure beyond simple bulk materials are never converged to errors small compared to the structural energy differences of interest. Well-designed calculations maintain the maximum possible consistency in calulating the large energies which are to be subtracted. Nevertheless, there is more wishful thinking than hard science in the supposed cancelation of systematic convergence errors. We need, as a practical matter for assessing the calcuations being done today, to develop more quantitative measures of these subtractions. The second issue is the absolute errors of LDF theory. We have a good idea of the errors for some isolated atoms, but the folklore is that the theory works much better for extended systems. Does it? Are there modifications of presently used functionals which will uniformly improve calculated energy differences without sacrificing the computational advanages of a local, state-independent potential?

At the end of my list of structural energy methods came those which could systematically improve the treatment of electron correlations, with quantum-chemical theories as an example. I think it goes without saying that attempts at direct translation of these approaches to extended systems will be fruitless. It is known that important correlations in extended systems involve selective summations of perturbation theory to infinite order, like the plasmon contributions to the correlation energy of the electron gas.[70] Considerable progress has recently been made in applying such selectively-summed perturbation approaches to the problem of excitation energies in semiconductors and insulators.[71,72] (The eigenvalues of Eq. (3) are often interpreted as band structure energies, but this is without any real justification, since density functional theory, even without the local approximation, is not a theory of excitation energies. The LDF "band gap problem," the observation that valence-conduction band gaps come out systematically too small, reflects such an interpretation.) The accuracy of the recent calculations is close to the level of experimental uncertainty for the few systems treated so far. It may be that a parallel level of many-body approximation for the correlation energy would be more accurate than LDF theory. It would be nice to find out.

There are other routes which might be followed to systematically improve correlation energies. Numerical approaches to the evaluation of the energy of correlated many-particle wavefunctions such as quantum Monte-Carlo methods have been successful for simple systems such as liquid He [73] and the free electron gas.[74] The extension of these techniques to solids and surfaces awaits some innovations and a lot of hard work. Finally, we might consider the proposal that high-energy quantum field theory methods combined with the renormalization group will offer the ultimate solution to the electronic structure problem.[75]

B. Real Challenges

All the above is in the direction of doing what we do now bigger and better. Stepping back to the broader view of Sec. I, however, I think we should look at things a little differently and more critically. One problem with the results I've been describing, my own work included, is understanding the results of these massive calculations. Making the calculations more complex may make them more accurate, but isn't likely to help our understanding. Part of the problem is that much of this work involves calculating large numbers and subtracting them to find the small difference we're interested in. It would be much better to calculate the difference of interest directly. Formulating the problem to isolate the energies which change would, in and of itself, foster greater understanding. To emphasize this point, suppose the only approach attempted to create a theory of superconductivity was to calculate the total free energy of a metal in the normal state and in the superconducting state with sufficient accuracy to subtract them. We'd still be waiting for the BCS theory!

Another impediment to understanding is the case-by-case mentality that pervades much of surface science. The reconstruction of Si(001) is one problem, and that of Si(111) is a separate problem. Each structure of each adsorbate on each low-index transition metal surface is a separate problem. We must remember the setting of surface science I discussed earlier. I don't know how other parts of physical chemistry operate, but I know that solid state physicists outside of surface science most frequently focus on phenomena displayed by classes of materials. The kind of generalization this implies is crucial for another part of our setting, our input to the applied areas.

The last challenge I'll cite is, once again, in contrast to other areas of solid state physics. That is the challenge of generating some intellectual excitement in surface science. Within this decade, there have been three major solid state problems requiring qualitatively new understanding: the integer and fractional quantum Hall effects, heavy-fermion materials and their superconductivity, and the new high T_c cuprate superconductors. I'm confident that there are some problems posing a comparable degree of intellectual challenge out there in surface science. Perhaps our traditional approaches, both theoretical and experimental, bias us against finding them. We, the Solvay conferees, are all established and comfortable in our field. We should recognise, however, that the long term health of our science depends on attracting the brightest young people, and that we are up against some stiff competition.

References

1. P. N. Keating, Phys. Rev. *145*, 637 (1966).
2. F. Stillinger and T. Weber, Phys. Rev. B *31*, 5262 (1985).
3. R. Biswas and D. R. Hamann, Phys. Rev. Lett. *55*, 2001 (1985); Phys. Rev. B *36*, 6434 (1987).
4. J. K. Nørskov and N. D. Lang, Phys. Rev. B *21*, 2131 (1980); J. K. Nørskov, Phys. Rev. Lett. *48*, 1620 (1982); P. Nordlander, S. Holloway, and J. K. Nørskov, Surface Sci. *136*, 59 (1984).
5. M. S. Daw and M. I. Baskes, Phys. Rev. B *29*, 6443 (1984).
6. C. M. Varma and W. Weber, Phys. Rev. B *19*, 6142 (1979).
7. D. J. Chadi, Phys. Rev. B *19*, 2074 (1979); Phys. Rev. Lett. *41*, 1062 (1978).
8. P. Hohenberg and W. Kohn, Phys. Rev. *136*, B864 (1964); W. Kohn and L. J. Sham, *ibid. 140*, A1133 (1965).
9. D. R. Hamann, M. Schlüter, and C. Chiang, Phys. Rev. Lett. *43*, 1494 (1979); G. B. Bachelet, D. R. Hamann, and M. Schlüter, Phys. Rev. B *26*, 4199 (1982).
10. M. T. Yin and M. L. Cohen, Phys. Rev. Lett. *45*, 1004 (1980); Phys. Rev. B. *26*, 3259 (1982); *ibid. 26*, 5668 (1982).
11. O. K. Andersen, Phys. Rev. B *12*, 3060 (1975).
12. D. R. Hamann, Phys. Rev. Lett. *42*, 662 (1979); L. F. Mattheiss and D. R. Hamann, Phys. Rev. B *33*, 823 (1986).
13. O. K. Andersen and R. V. Kasowski, Phys. Rev. B. *4*, 1064 (1971).
14. K. C. Pandey, Phys. Rev. Lett. *47*, 1913 (1981).
15. K. C. Pandey, Phys. Rev. Lett. *49*, 223 (1982).
16. J. E. Northrup and M. L. Cohen, Phys. Rev. Lett. *49*, 1349 (1982).
17. E. G. McRae, Phys. Rev. B *28*, 2305 (1983).

18. G. Binnig, H. Rohrer, C. H. Gerber, and E. Weibel, Phys. Rev. Lett. *50*, 120 (1983).
19. K. Takayanagi, Y. Tanishiro, M. Takahashi, and S. Takahashi, J. Vac. Sci. Technol. *A3*, 1502 (1985).
20. G.-X. Qian and D. J. Chadi, Phys. Rev. B *35*, 1288 (1987).
21. J. E. Northrup, Phys. Rev. Lett. *57*, 154 (1986); D. Vanderbilt, Phys. Rev. Lett. *59*, 1456 (1987).
22. R. I. G. Uhrberg, R. D. Bringans, M. A. Olmstead, R. Z. Bachrach, and J. E. Northrup, Phys. Rev. B *35*, 3945 (1987); J. R. Patel, P. E. Freeland, M. S. Hybertsen, D. C. Jacobson, and J. A. Golovchenko, Phys. Rev. Lett. *59*, 2180 (1987).
23. D. J. Chadi, J. Vac. Sci. Technol. *16*, 1290 (1979); Phys. Rev. Lett. *43*, 43 (1979).
24. M. T. Yin and M. L. Cohen, Phys. Rev. B *24*, 2303 (1981).
25. K. Raghavachari, J. Chem. Phys. *81*,2717 (1984); Y. J. Chabal and K. Raghavachari, Phys. Rev. Lett. *53*, 282 (1984); *ibid.* *54*, 1055 (1985).
26. D. J. Chadi, Phys. Rev. Lett. *56*, 769 (1986).
27. E. Kaxiras, K. C. Pandey, Y. Bar-Yam, and J. D. Joannopoulos, Phys. Rev. Lett. *56*, 2819 (1986) and *57*, 923 (Erratum); *ibid.* *57*, 106 (1986); Phys. Rev. B *33*, 4406 (1986).
28. K.-M. Ho and K. P. Bohnen, Phys. Rev. Lett. *56*, 934 (1986).
29. T. H. Upton and W. Goddard, Phys. Rev. Lett. *46*, 1635 (1981).
30. C. W. Bauschlicher and P. S.Bagus, Phys. Rev. Lett. *52*, 200 (1984).
31. J. E. Müller, M. Wuttig, and H. Ibach, Phys. Rev. Lett. *56*, 1583 (1986).
32. C. L. Fu, A. J. Freeman, E. Wimmer, and M. Weinert, Phys. Rev. Lett. *54*, 2261 (1985).
33. D. Singh, S.-H. Wei, and H. Krakauer, Phys. Rev. Lett. *57*, 3292 (1986).
34. K.-M. Ho and K. P. Bohnen, Phys. Rev. Lett. *59*, 1833 (1987).
35. M. A. Barteau, J. Q. Broughton, and D. Menzel, Surface Sci. *133*, 443 (1983).
36. H. Conrad, R. Scala, W. Stenzel, and R. Unwin, J. Chem Phys. *81*, 6371 (1984).
37. W. Ho, R. F. Willis, and E. W. Plummer, Phys. Rev. Lett. *40*, 1463 (1978).
38. M. R. Barnes and R. F. Willis, Phys. Rev. Lett. *41*, 1729 (1978).
39. J. P. Woods and J. L. Erskine, Phys. Rev. Lett. *55*, 2595 (1985).
40. Y. J. Chabal, Phys. Rev. Lett. *55*, 845 (1985).
41. M. Weinert, A. J. Freeman, and S. Ohnishi, Phys. Rev. Lett. *56*, 2295 (1986).
42. A. M. Baro, H. Ibach, and H. D. Bruchmann, Surface Sci. *88*, 384 (1979).

43. L. J. Richter and W. Ho, to be published.
44. J. E. Reutt and Y. J. Chabal, to be published.
45. L. J. Richter and W. Ho, J. Vacuum Sci. Techol. *A5*, 453 (1987).
46. P. J. Feibelman and D. R. Hamann, Surf. Sci. *179*, 153 (1987).
47. R. Biswas and D. R. Hamann, Phys. Rev. Lett. *56*, 2291 (1986).
48. P. J. Feibelman and D. R. Hamann, Surface Sci. *182*, 411 (1987).
49. J. Lee, J. P. Cowin, and L. Wharton, Surface Sci. *130*, 1 (1983); I. P. Batra, J. A. Barker, and D. J. Auerbach, J. Vacuum Sci. Tehnol. *A2*, 943 (1984); I. P. Batra, Surface Sci. *137*, L97 (1984).
50. N. Esbjerg and J. K. Nørskov, Phys. Rev. Lett. *49*, 807 (1980).
51. D. R. Hamann and P. J. Feibelman, Phys. Rev. B *37*, (1988).
52. R. Smoluchowski, Phys. Rev. *60*, 661 (1941).
53. M. J. Puska, R. M. Nieminen, M. Manninen, B. Charkraborty, S. Holloway, and J. K. Nørskov, Phys. Rev. Lett. *51*, 1801 (1983).
54. D. Cherns, G. R. Anstis, J. L. Hutchinson, and J. C. H. Spence, Philos. Mag. A *46*, 849 (1982).
55. J. M. Gibson, J. C. Bean, J. M. Poate, and R. T. Tung, Appl. Phys. Lett. *41*, 818 (1982).
56. R. T. Tung, J. M. Gibson, and J. M. Poate, Phys. Rev. Lett. *50*, 429 (1983).
57. E. Vlieg, A. E. M. J. Fischer, J. F. van der Veen, B. N. Dev, and G. Materlik, Surface Sci. *178*, 36 (1986).
58. A. E. M. J. Fischer, E. Vlieg, J. F. van der Veen, M. Clausnitzer, and G. Materlik, Phys. Rev. B *36*, 4769 (1987).
59. J. Zegenhagen, K. G. Huang, B. D. Hunt, and L. J. Schowalter, Appl. Phys. Lett. *51*, 1176 (1987).
60. E. J. van Loenen, J. W. M. Frenken, J. F. van der Veen, and S. Valeri, Phys. Rev. Lett. *54*, 827 (1985).
61. A. E. M. J. Fischer, T. Gustafsson, and J. F. van der Veen, to be published.
62. F. Hellman and R. T. Tung , to be published.
63. J. Tersoff and D. R. Hamann, Phys. Rev. B *28*, 1168 (1983).
64. D. R. Hamann, Phys. Rev. Lett. *58*, (1988).
65. D. R. Hamann and L. F. Mattheiss, Phys. Rev. Lett. *54*, 2517 (1985).
66. N. D. Lang and A. R. Williams, Phys. Rev. B *18*, 616 (1978); Phys. Rev. Lett. *37*, 212 (1976).
67 P. Feibelman, Phys. Rev. B *35*, 2626 (1987).
68. R. Car and M. Parrinello, Phys. Rev. Lett. *55*, 2471 (1985).
69. M. C. Payne, P. D. Bristowe, and J. D. Joannopoulos, Phys. Rev. Lett. *58*, 1348 (1987).
70. M. Gell-Mann and K. Brueckner, Phys. Rev. *106*, 364 (1957).
71. M. S. Hybertsen and S. G. Louie, Phys. Rev. Lett. *55*, 1418 (1985); Phys. Rev. B *34*, 5390 (1986).

72. R. W. Godby, M. Schlüter, and L. J. Sham, Phys. Rev. Lett. *56*, 2415 (1986); Phys. Rev. B *35*, 4170 (1987); *ibid. 36*, 6497 (1987).
73. W. L. McMillan, Phys. Rev. *138*, A442, (1965).
74. D. M. Ceperley and B. J. Alder, Phys. Rev. Lett. *45*, 566 (1980).
75. S. R. White, J. W. Wilkins, and K. G. Wilson, Phys. Rev. Lett. *56*, 412 (1986).

Discussion

C.B. DUKE: In compound semiconductor surfaces universal scaling laws for establishing the structures of the same surface of different materials have been established. E.g., for Zincblende (110), Wurtzite ($11\bar{2}0$), Wurtzite ($10\bar{1}0$), rutile (100) and rutile (110). Thus, item two on your last slide has been successfully addressed in this class of systems. I would also like to ask a question. The above analyses reveal that each "surface" exhibits a unique type of chemical bonding for a wide variety of specific materials. Other workers (e.g., Prof. Tiller of Stanford) argue that many-body force law models are applicable to multiple semiconductor surfaces, in spite of the different detailed surface electronic structures (i.e., bonding). Would you comment on the use of multi-body effective force models to evaluate quantitatively semiconductor surface and interface geometries?

D.R. HAMANN: Historically, other people started using slab/supercell methods because they could be done with existing 3–dimensional band structure approaches. An operational difficulty Appelbaum and I had with the semi-infinite calculations was that sharp surface resonances made adequate energy sampling for surface self-consistency very difficult. Since we established that screening lengths are very short, even in semiconductors, it is in fact practical to do self-consistency and total energies on slabs. The ideal way to study the electronic spectrum would be to do self-consistency on a slab and join that surface potential to a semi-infinite bulk.

J.W. GADZUK: First, I would like to say that I think you very accurately identified, in your last slide, the malaise of surface science, to the extent that there exists a malaise (i.e., case–by–case mentality; need for intellectual excitement, etc.).

Secondly, I would be interested in your comments on the evolution of large scale surface electronic structure calculations, particularly with regard to the shift from semi-infinite, matching schemes (bulk, selvedge, vacuum) to periodic slab models. Why has this shift occurred from the philosophical and/or practical point of view?

D.R. HAMANN: Mark Stiles and I have just completed the development of a variational method for semi–infinite calculations that can be applied to transitions metals, or any system.

L. KLEINMAN: (Addressed to Gadzuk) Another reason the Appelbaum-Hamann semi-infinite crystal method fell into disuse was that it required the use of a local pseudo-potential. Therefore, it could not be applied to transition metals or even to accurate calculations for semiconductors beyond the third row of the periodic table.

W. KOHN: 1) The good theoretical results for the vibrational frequencies of H on metal surfaces recall the similarly good results for compressibilities of metals, calculated with local density functional theory. The local quasi–free electron picture *between* atoms appears to be quite successful.

2) Surface scientists should claim high T_c superconductivity as part of their field: the 2-D CuO layers appear to be essential!

D.R. HAMANN: I'd like to claim this field for surface science, but I don't believe many surface scientists have been involved.

P. MARCUS: I should like to call attention to a universal rule of the kind Dr. Hamann wanted that explains simply a large body of complex surface phenomena. I refer to a simple electrostatic model of relaxation in clean metals which permits calculation of the large and deep relaxations that are observed on open surfaces of metals. Our model is a simple, empirical modification of the Finnes-Heine model in which rigid nets of ions float in a background of rigid electron charge.

A New Model for the Cleaved Si (111)-(2x1) Surface

*D. Haneman and M.G. Lagally**

Department of Condensed-Matter Physics,
University of New South Wales, Kensington NSW 2033, Australia
*Permanent Address: University of Wisconsin-Madison,
Madison, WI53706, USA

There is currently general agreement about the structures of Si(100) and annealed Si(111) (7x7) surfaces, but the situation for the cleaved Si(111) (2x1) surface is less satisfactory [1]. A Si crystal in the (111) orientation can be viewed as a series of tetrahedra with one bond vertical and the other three forming the base, resulting in a structure that consists of double layers of closely spaced atomic planes (d = 0.78 Å) separated from each other by the Si interatomic distance d = 2.35 Å. All of the existing models use as their basis the assumption that in cleavage, the one bond that is oriented normal to the (111) surface breaks, so that the termination is the closely-spaced-layer pair that then reconstructs to form the (2x1) surface. Of these models, there is much support for the π-bonded chain structure [2-4]. Although a wide variety of measurements are found to be consistent with or are interpreted in terms of it [1,3], low-energy electron diffraction (LEED) analysis using this structure has not provided a satisfactory fit to the data. The model also appears to have conceptual difficulties. To form the π-bonded chain structure necessitates breaking bonds between the second and third layers below the surface, which requires shear forces for which there appears to be no physical basis [1]. Based on energy calculations [4], the structure itself appears to be too stable to transform (as it does) to the completely different (7x7) structure at temperatures as low as 650°K. It is possible, therefore, that the correct structure for the (2x1) surface has not yet been identified. In this paper we provide justification and plausibility arguments for a cleavage process and structure, recently suggested by HANEMAN [1]. Although detailed analysis will be required, we believe that what we propose is at least not inconsistent with available data.

The impetus for the cleavage model came from the recognition [1] that the feature that gives a fit to optical data is principally the π-bonded chain along <110> in the surface [2]. These chains can be far more naturally obtained than in the favored model discussed above if the cleavage occurs between the closely spaced planes of the double layer of atoms rather than between the double layers themselves, as all previous models have assumed. Such a cleavage, however, requires the scission of three bonds (as every atom in this double layer makes three bonds with other atoms in the layer), rather than one bond if cleavage occurs between the double layers. Such a process may therefore on the face of it be unlikely.

It is important to recognize that such scission between the planes of the double layer does, in fact, occur in the behavior of dislocations in Si. It has been known for almost twenty years [5] and has been reaffirmed a number of times [6-8], that dislocations in Si dissociate into partial dislocations separated by an intrinsic stacking fault. In Si,

Springer Series in Surface Sciences Vol. 14
Solvay Conference on Surface Science
Editor: F.W. de Wette

dislocations reside on the (111) plane and are glissile on that plane [9]. Their preferred orientation is along <110>. This means that an edge dislocation in Si can be thought of as an extra half-plane of atoms inserted into the crystal in such a manner that the last row of atoms of this plane lies along <110> in the (111) plane. Atoms in this last row will have unsaturated bonds and the dislocation will create a strain field around it. This strain field is apparently sufficiently large so that the dislocation dissociates into partial dislocations. Partial dislocations are ones that are separated by an intrinsic stacking fault. Therefore in Si atoms rearrange around the site of the original dislocation to form an intrinsic stacking fault that is then bounded at each end by a partial dislocation in which the misfit of the atoms is less than in the original dislocation. By this process, the total energy is minimized, even with the addition of the stacking fault energy, which is very small [10] in Si. The separation of partial dislocations (the width of the stacking fault) is typically of the order of 30-80 Å, and depends on the minimization of the total energy of this defect [5-7].

It can be shown that to create an intrinsic stacking fault requires rearrangement of atoms in the closely spaced double layer, and that therefore the scission (and reforming) of three bonds per atom is required. It is easiest to see this by imagining the motion of an atom (actually a row of atoms along <110>) from one triangle formed by the atoms of the underlying hexagonal array of a (111) plane to the immediately adjacent one, and having this process repeat for subsequent atoms (rows) until the stable stacking fault width has been achieved. The driving force is the stress caused by the initial dislocation. In some materials with the diamond structure, the motion of partial dislocations can be observed at room temperature [11]. If the stress due to a dislocation is sufficient to cause three-bond scission in Si, it appears reasonable that stress can be applied during cleavage to cause the same effect. All that appears necessary in a stress concentrator, which may in fact be a dislocation or other defect.

One can attempt to make three-bond scission plausible on the basis of microscopic interactions in Si. Using force constants for bending and stretching in Si [12], one can estimate the strain in the bonds, using the appropriate equations for the bond-bending and bond-stretching interactions [13], for a force applied normal to the (111) surface (the typical cleavage experiment). If a given tetragon of atoms is stretched in a <111> direction, it can be shown that the major distortion occurs from the increase in spacing between the atoms in the double layer, because of the weak bond bending force constant, and little from the axial stretch along the <111> bond. The weak bond bending force constant can be thought of as the consequence of an ability for easy distortion of the charge cloud that represents this bond, and the ability to reform a bond with a different atom [14].

The theoretical maximum strain in Si has been estimated to be between 5% and 17% [15], with experimental measurements on whiskers yielding 4% [16]. The latter are believed to be limited by defects, in particular steps at the whisker surface [16]. Based on simple interatomic potentials, it seems that a typical bond can be stretched to greater than 10%. The process of cleavage then is envisioned as follows. The application of stress, probably with the addition of a stress concentrator such as a defect, raises the strain in the vicinity of a Si tetragon to high levels. The bonds normal to the surface stretch. The three bonds between atoms in the double layer shear, which they can easily do. As the charge clouds of the half bonds making up such a bond separate [14], they begin to see preferentially the charge clouds of atoms in the same plane (the upper or lower half of the double layer) and rehybridize with those atoms, forming the surface reconstruction that is observed. In a purely mechanistic way, one could view the surface with the bonds broken but before reconstruction as shown in Fig. 1a. There are rows of atoms along <110> with three broken bonds each. Of course, the surface never finds itself in this situation, as bonds reform and rehybridize as they break. Figure 1b shows a possible structure [17]. Adjacent rows of atoms move toward each other parallel to <112> to form the (2x1) reconstruction. Two broken bonds are immediately eliminated by reforming with neighbors in the surface plane, creating the chain structure as shown in Fig. 1b. The remaining broken bond can form π bonds as suggested by Seiwatz [17] and later by PANDEY [2]. Although the positions of the surface atoms must be determined by fitting to data from structural determinations such as LEED and ion scattering, a starting point is to let the atoms move toward each other in the <112> direction until they have the Si-Si bond length (2.35 Å). They then form nearly tetragonal bond angles in the chain, and also with the underlying atoms. All surface atoms are therefore in five-membered rings. It should be noted that the underlying symmetry of the atoms in the two rows that move toward each other is different, allowing for asymmetries such as tilting of the chain.

The model appears to be consistent with the results of optical data and at least not inconsistent with STM results. It is consistent with surface energy measurements [18], in that the final reconstructed surface has about one broken bond per atom. Conclusions about other experiments must await a direct comparison. The model has additional advantages, e.g., 1) ready conversion to a (1x1) structure on adsorption; 2) feasible

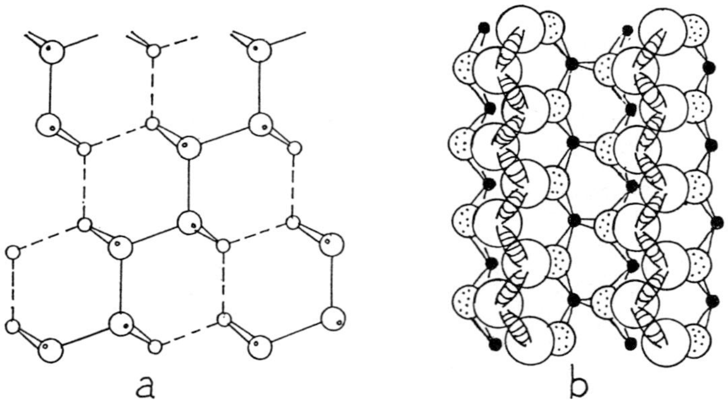

Figure 1 Model of the cleaved Si(111) (2x1) surface. a) Bond termination at the surface if cleavage occurs through the three bonds of the double layer, no reconstruction, view along <110>; b) possible model for the reconstructed surface. Two of the three broken bonds in a) are used to form chains, the third can form π bonds along the chains. The chains are parallel to <110>, with <112> perpendicular to the chains. In a) two planes of atoms are shown and indicated by different-size circles; in b), atoms are shown in decreasing size from the surface, and π bonds between surface atoms are shown schematically. In the bulk termination, the atoms in the outer two layers lie on top of each other.

conversion to the complex (7x7) structure; 3) explanation of the preferred direction of the surface corrugation relative to cleavage direction, and 4) the observation [19] of less damage in the surface for cleavage along <112> relative to cleavage along <110> [20].

The work of MGL was partly supported by ONR, Electronics Program. We acknowledge useful discussions with D. J. H. Cockayne.

References

1. For a recent review of Si surfaces, see D. Haneman, Repts. Prog. Phys. 50, 1045 (1987).

2. K. C. Pandey, Phys. Rev. Letters 47, 1913 (1981).

3. J. A. Stroscio, R. M. Feenstra, and A. P. Fein, J. Vac. Sci. Technol. A5, 838 (1987).

4. J. E. Northrup and M. L. Cohen, Phys. Rev. Letters 49, 1349 (1982).

5. I. L. F. Ray and D. J. H. Cockayne, Philos. Mag. 22, 853 (1970); D. J. H. Cockayne and A. Hons, J. Physique, Coll. C6, 40, 11 (1979) and references therein.

6. P. B. Hirsch, J. Physique, Coll. C6, 40, 27 (1979); A. Ourmadz, G. R. Anstis, and P. B. Hirsch, Philos. Mag. A48, 139 (1983).

7. P. B. Hirsch, J. Electr. Micros. 118, 3 (1980).

8. H. Alexander, J. C. H. Spence, D. Shindo, H. Gottschalk, and N. Long, Philos. Mag. A53, 627 (1986).

9. J. Hornstra, J. Phys. Chem. Solids 5, 129 (1958); J. P. Hirth and J. Lothe, Theory of Dislocations, (McGraw-Hill, New York 1968).

10. There are a number of measurements and calculations. For a first-principles calculation and listing of earlier work, see M. Y. Chou, M. L. Cohen, and S. G. Louie, Phys. Rev. B32, 7979 (1985).

11. D. J. H. Cockayne, private communication; D. J. H. Cockayne, A. Hons, and J. C. H. Spence, Micron 11, Suppl. 1, 32 (1980).

12. P. N. Keating, Phys. Rev 145, 637 (1966).

13. J. A. Appelbaum and D. R. Hamann, Surface Sci. 74, 21 (1978).

14. E. O. Kane, Phys. Rev. B31, 7865 (1985).

15. For a review, see S. S. Brenner, In: Fiber Composite Materials, ASM, Metals Park (1985).

16. D. M. Marsh, In: Fracture of Solids, ed. by D. C. Drucker and J. J. Gilman, (Interscience, New York 1963). See also R. L. Mehan and J. A. Herzog, In: Whisker Technology, ed. by A. P. Levitt, (Wiley-Interscience, New York 1970).

17. This structure is similar to that obtained by arbitrarily adding adatom chains onto the ideal (111) surface, as suggested by R. Seiwatz, Surface Sci. 2, 473 (1964). To create adatoms requires thermal treatment sufficient to give atomic transport at the surface, and thus a structure involving adatoms is not directly applicable to a cleaved surface. In the present model for the cleavage process, the chains are an integral part of the structure, being the normal outermost layer.

18. J. J. Gilman, J. Appl. Phys. 31, 2208 (1968).

19. L. Smit, R. M. Tromp, and J. F. van der Veen, Surface Sci. 163, 315 (1985).

20. A more extensive exposition of these concepts will be presented elsewhere. D. Haneman and M. G. Lagally, J. Vac. Sci. Technol. A6 (1988) (submitted).

Discussion

D.R. HAMANN: Silicon is really more like a free electron metal than a material with highly local bonds. This is shown by the low ratio of the average bonding–antibonding energy gap of $\sim 4\ eV$ to the electron gas Fermi energy of $\sim 12\ eV$. It is also shown by the small energy difference between the 4-fold diamond and 6-fold structures, $\sim 0.25\ eV$ from local density calculations. The liquid also has 6-fold coordination. I think this supports your model, since the Si is probably in something like a 6-fold geometry while the bonds are breaking and reforming. Question: What is the difference in topography between your model and Pandey's?

M. LAGALLY: As in Pandey's model, there is a chain structure along $< 110 >$ that can form π-bonding. The new model has 5-fold rings, while Pandey's has 5- and 7-fold rings. Because of asymmetry provided by the second and third layer, tilting of the π-bonded chains is feasible. These differences come about because this model uses an essentially different starting point for creation of the surface structure.

J.E. DEMUTH: In looking for evidence of this model, have you considered or examined the defect structures found on these cleaved surfaces by STM? R. Feenstra at IBM has found unusual defects of the chains where they

J.B. PENDRY: The CO_2 occupies two distinct configurations: both sit with carbon directly above a nickel atom, one with the axis of the molecule parallel to the rows, the other perpendicular. The bending is of the order of 10° and is in the upward direction.

A.M. BRADSHAW: Our NEXAFS data taken together with Freund's group suggest that the bent anionic CO_2 species is oriented perpendicular to the Ni(110) surface with the molecular plane parallel to both the [001]

twist over and may originate from the dynamics of formation. There are also different structures at cleavage steps.

M. LAGALLY: We have not looked into these defect features in the terrace. The observed features at cleavage steps appear to be explainable within this model.

P. MARCUS: In a recent publication we have carried out a LEED analysis of the Si(111) 2×1 structure. The structure differs from the Pandey chain model only in that the chains are tilted by 12°. It gives only a moderately good fit to the data. Possibly your structure with tilted chains would fit better.

M. LAGALLY: A structure analysis is certainly necessary. Our proposal does allow for tilting the chains, because of the asymmetry of the underlying double layer.

"Diffuse LEED and the Structure of Random Adsorbates"

J.B. Pendry

Discussion

J. T. YATES, JR.: Could you comment in more detail on the structure of chemisorbed CO_2 on Ni(110)? What is the degree of bending of the molecule from the linear structure?

J.B. PENDRY: The CO_2 occupies two distinct configurations: both sit with carbon directly above a nickel atom, one with the axis of the molecule parallel to the rows, the other perpendicular. The bending is of the order of 10° and is in the upward direction.

A.M. BRADSHAW: Our NEXAFS data taken together with Freund's group suggest that the bent anionic CO_2 species is oriented perpendicular to the Ni(110) surface with the molecular plane parallel to both the [001] and [1$\bar{1}$0] azimuths. The experimental observations are, however, made complicated by the occurrence of a reorientation effect, whereby a preferential alignment along the [1$\bar{1}$0] azimuth seems to take place as a function of time.

M.G. LAGALLY: Can you comment on the likely resolution of diffuse-intensity LEED in terms of determining the positions and coordinates of atoms in a disordered array on a surface relative to the determination of coordinates of ordered layers by conventional LEED?

J.B. PENDRY: The sensitivity of DLEED appears to be at least as great as conventional LEED. The R-factors we get are low by conventional LEED standards and there are prospects for, perhaps, even higher resolution with the new technique.

G.A. SOMORJAI: You have well demonstrated the importance of restructuring of the substrate to explain the adsorbate surface structure. This is very likely to happen whenever the adsorbate substrate bond is of similar strength as the substrate-substrate atom bonds. What fraction of the solved surface structures show such a substrate restructuring effect upon adsorption?

J.B. PENDRY: It is too early to say how pervasive substrate reconstruction is. I suspect that a large fraction of the poor-R-factor adsorbate structure determinations may be explained by inclusion of substrate reconstruction in the model.

M. BOUDART: The reconstruction of metal clusters ($\sim$ 40 atoms) by adsorbates has been documented by wide angle x-ray scattering and extended x-ray adsorption fine structure, for instance in the work of Pierre Gallezot in Villeurbanne. This phenomenon is of great importance in catalysis because of the fact that catalysis takes place normally under conditions of high surface coverage.

Diffuse LEED, Tensor LEED, and the Structure of Random Adsorbates

J.B. Pendry

The Blackett Laboratory, Imperial College, London SW7 2BZ, UK

Extensive progress has been made in collaboration with the Erlangen group of Müller and Heinz in treating diffraction from the simplest class of disorder: a lattice gas of adsorbed atoms on an otherwise ordered substrate [1,2]. By using an interplay of Bloch's theorem for the substrate and local treatments for the adsorbate, accurate calculations of the diffusely scattered intensities can be made. The central element of the theory is that in order to appear in the diffuse part of the pattern an electron must have struck an adsorbate atom and therefore its experience of the disorder is a strictly local one, just as in a SEXAFS experiment.

SEXAFS has been very effective in those systems to which it can be applied. Now we have a complementary technique which is laboratory based and offers virtually unlimited data for structural analysis. Diffuse LEED data are actually more easily acquired than conventional diffracted data now that whole screen digitisation techniques are becoming standard.[3]

Further increase in the power of the method has been achieved by the introduction of tensor LEED methods [4] which allow the analysis of extremely complex structures to be made as perturbations of a simpler reference structure. So fast are these methods that the analysis can often be performed on a PC, once the tensor has been calculated for the reference structure. By making one full dynamical calculation for a reference structure we express diffraction from modifications of this structure in first order perturbation theory. This gives a theory analogous to the Xray case in that it can be separated into a structure factor and a form factor which is a tensor in the dynamical regime. The linear dependence on the structure factor gives a computational savings between 100 and 1000 over conventional methods.

Our inspiration lies in Xray scattering theory [5] where an atom scatters Xrays from state $\underline{K}$ to state $\underline{K}'$ with form factor $f_j(|\underline{K}'-\underline{K}|)$. The beauty of this weak scattering theory is that the total scattered amplitude in state $\underline{K}'$, $A(\underline{K}')$, is calculated by first order perturbation theory and is therefore a linear function of the individual atomic scattering amplitudes:

$$A(\underline{K}') = \sum_j f_j(|\underline{K}'-\underline{K}|)\exp(i(\underline{K}'-\underline{K}).\underline{R}_j) \tag{1}$$

This linear property leads to a complete factorisation of the problem into form factors and structure factors. The latter contain all the geometric information we desire in a simple form, in fact as a sum of exponentials. Thus the scattering from a trial structure can be calculated very simply and rapidly and compared with experiment.

Springer Series in Surface Sciences Vol. 14
Solvay Conference on Surface Science
Editor: F.W. de Wette

In contrast LEED theory [6] involves complex dynamical processes that weave the form factors and the structure factors together in an inseparable way. We must look at the problem in a different way if we are to approach the simplicity of the Xray picture. Our objective is to find as close a parallel with the Xray theory as we can. In particular we wish to separate the form factor and the structure factor so that we do not have to repeat the calculation of the form factor for each trial structure.

The first step is to define a reference structure the scattering from which we shall treat from a full dynamical standpoint. In the Xray case our starting point was empty space in which Xrays propagate as plane waves, and the atoms introduce a weak perturbation, scattering the Xrays from one plane wave state to another. In the LEED case we start from the reference structure. The crucial point is that having made this investment in a time consuming calculation, if we then move the atoms only a small amount the perturbation will be a weak one and can be calculated by first order perturbation theory as for the Xray case. In practice these displacements can be as much as 0.4Å and for many systems only one reference structure need be calculated, preferably the unreconstructed surface. Systems where a larger range of displacements needs to be considered may require more than one reference structure.

The theory comes in three levels of sophistication. At the simplest level, which treats the smallest displacements, a displacement of the j-th atom by $\underline{\delta R}_j$ changes the potential in the surface by

$$\delta V_j = \underline{\delta R}_j \cdot \underline{\nabla} V_j(\underline{r}-\underline{R}_j) \tag{2}$$

Let $|\underline{k}_p\rangle$ be the exact LEED state of the undistorted system produced by an incident beam of energy E and momentum $\underline{k}_p$ and parallel to the surface. Then in the presence of a distortion the amplitude of the diffracted wave, energy E, momentum $\underline{k}'_p$ parallel to the surface, is increased by

$$A(\underline{k}'_p) \approx \sum_j \langle \underline{k}'_p | \delta V_j | \underline{k}_p \rangle = \sum_{i,j} T_{ij} \underline{\delta R}_{ij} \tag{3}$$

where the tensor, T, analogous to the form factor in Xray scattering, is defined by

$$T_{xj} = \langle \underline{k}'_p | \underline{\nabla}_x V_j(\underline{r}-\underline{R}_j) | \underline{k}_p \rangle \tag{4}$$

and similarly for for T_{yj} and T_{zj}.

The $\underline{\delta R}_j$ play the role of the "structure factors". Once T is known, I/V spectra from a large number of distortions of the reference structure can be generated by resumming (3). leading to vast increases in computational efficiency over conventional methods.

The second level of sophistication recognises that if $\underline{\delta R}_j$ becomes comparable with the electron's wavelength the approximation will break down. However there is a better way to calculate the change in the scattering power of a displaced atom. In an angular momentum basis we can use the translation theorem [7,8] to expand the atomic t matrix as:

$$t_{lm,l'm'}(\underline{R}+\underline{\delta R}) = t_{lm,l'm'}(\underline{R}) + \delta t_{lm,l'm'}(\underline{R}) \quad (5)$$

where

$$\delta t_{lm,l'm'}(\underline{R}) = - t_{lm,l'm'}(\underline{R}) + \sum_{LM,L'M'} G_{lm,LM}(\underline{\delta R})\, t_{LM,L'M'}(\underline{R})\, \tilde{G}_{l'm',L'M'}(\underline{\delta R}) \quad (6)$$

The G's are translation operators containing Clebsch Gordan coefficients. We can now develop a renormalised perturbation theory in δt_j which takes account of the change in atomic scattering factor exactly. Equation (3) is replaced by

$$A(\underline{k}_p') \approx \sum_j \langle \underline{k}_p' | \, \delta \tilde{t}_j \, | \underline{k}_p \rangle = \sum_{j;lm,l'm'} \tilde{T}_{j;lm,l'm'}\, \delta t_{j;lm,l'm'} \quad (7)$$

where

$$\tilde{T}_{j;lm,l'm'} = \langle \underline{k}_p' | \underline{R}_j ; lm \rangle \langle \underline{R}_j ; l'm' | \underline{k}_p \rangle , \quad (8)$$

$|\underline{R}_j;lm\rangle$ being the state of angular momentum lm centred on $\underline{R}_j$. Equation (7) for the change in amplitude is still factorised into a form factor and a structure factor the latter, given by (6), is a purely geometric quantity. This is our preferred formula and is almost as convenient as (3) to evaluate. The third level of approximation includes corrections to δt from scattering by atoms surrounding the one displaced which we have found represents an unnecessary level of sophistication.

The utility of this new approach depends on how far atoms can be displaced whilst maintaining the validity of first order perturbation theory upon which tensor LEED (TL) is based. For practical structure determinations TL must be able to treat atomic displacements of several tenths of an Ångström. The complexity of dynamical processes contributing to diffraction of low energy electrons means that the potential value of TL can be judged only by the direct comparison of I/V spectra generated using FD methods and the new perturbation scheme. To do this we have implemented the TL technique as a modification of the standard Van Hove and Tong [9] package of computer codes for three model surfaces. Agreement between the tensor LEED theory and exact calculations is highly satisfactory for atomic displacements of around 0.2Å although 0.4Å some peaks and minima are displaced.

Diffuse LEED has been used with striking success to determine the local adsorption geometry of a disordered oxygen overlayer on W(100) [10]. In this previous study it was assumed that the tungsten substrate was unreconstructed so that only the oxygen adsorption height was varied during the structure search. However tungsten shows a variety of complex reconstructions and therefore we have used TL to extend the previous analysis to include substrate reconstruction. Displacements of the substrate produce an additional contribution to the diffuse intensity which can be evaluated by TL. We considered displacements of the four top layer tungsten atoms surrounding each oxygen atom in the hollow sites and simultaneously varied the adsorption height. We find a best fit structure for an adsorption height of 0.58+/-0.1Å with the four tungsten atoms adjacent to the oxygen atoms puckered inwards in the plane of the top layer by 0.25Å. The R-factor minimum at R = 0.05 represents a 50% reduction of the R-factor from its value for the unreconstructed substrate - a very significant improvement by any measure.

Tensor LEED offers substantial savings of computer time over conventional methods, in many circumstances approaching three orders of magnitude, and therefore we believe that TL will make some of the most complex reconstructed surfaces accessible to the LEED technique. The technique is directly applicable to other electron spectroscopies operating in the LEED energy range and has proved capable of extending the potential of diffuse LEED.

References

1. JB Pendry and DK Saldin: Surf. Sci. 145 33 (1984)
2. K Heinz, DK Saldin and JB Pendry: Phys. Rev. Lett. 55 2312 (1985)
3. K Heinz, K Müller: in Structural Studies of Surfaces ed. G Höller, Springer Tracts in Mod. Phys. 91 (springer, New York, 1982)
4. PJ Rous, JB Pendry, DK Saldin, K Heinz, K Müller and N Bickel: Phys. Rev. Lett. 57 2951 (1986)
5. RW James: "The Optical Principles of the Diffraction of Xrays" (London-Bell, 1962)
6. JB Pendry: "Low Energy Electon Diffraction" (Academic, London & New York, 1974)
7. PM Marcus in "Determination of Surface Structure by LEED", pp. 93-128, Eds. PM Marcus and F Jona (Plenum, New York & London, 1984)
8. PJ Durham, JB Pendry & CH Hodges: Comput. Phys. Commun. 25 193 (1982)
9. MA Van Hove and SY Tong: "Surface Crystallography by LEED" (Springer, Berlin, 1979)
10. K Heinz, DK Saldin and JB Pendry: Phys. Rev. Letts. 55 2312 (1986)

Surface Structure Study by High-Resolution Electron Microscopy and Diffraction

K. Takayanagi

Tokyo Institute of Technology, Physics Department, Oh-okayama, Meguro-ku, Tokyo 152, Japan

Structures of reconstructed surfaces and adsorbed layers are studied in-situ by UHV electron microscopy and diffraction. Transmission electron diffraction is useful for determining long range ordered structures such as Si(111)7x7 and also short range ordered structures using kinematical approximation. High-resolution electron microscopy at UHV can reveal local structures at atomic level resolution in transmission mode and at sub-nm resolution in reflection mode. Observations of the phase transition of Si(111)-7x7 to 1x1 and adsorption processes give finer information on structural details.

1. SURFACE STRUCTURE ANALYSIS

Structure is a fundamental subject in surface science. In this paper recent advance in structure studies of reconstructed surfaces and adsorbates done by UHV electron microscopy and diffraction is briefly summarized[1]. Details are reported separately.

1.1 Electron Diffraction

Surface structures are studied by electron diffraction and microscopy using incident electron beam energy of 100 kV and/or 200 kV. In electron diffraction method, transmission electron diffraction (TED) from a surface layer of an ordered structure gives superlattice reflections in addition to the reflections from the bulk layers. The intensity of the reflections are analysed by the kinematical diffraction theory (Fourier transform) to derive the structure[2,3], although the dynamical diffraction of the electrons passing through the bulk layers can not be ignored at all. This is a great advantage in solving a structure based on the diffraction data. Another advantage is that TED intensities obtained by the incident electron beam normal to the surface (actually the beam has to be slightly inclined to the surface normal to avoid strong Bragg reflections from the bulk layers) give "projected structure" in the plane of the surface. To obtain the three dimensional structure TED for inclined electron beam incidences is used.

A new application of TED is for analysis of disordered structures giving diffuse intensity. When the diffuse intensities can be analysed kinematically, the short range order of the structure, and then the interaction potential, can be known by simple Fourier transform.

Springer Series in Surface Sciences Vol. 14
Solvay Conference on Surface Science
Editor: F.W. de Wette

1.2 Electron Microscopy

In microscopy, transmission and reflection electron microscopy[4] are known as useful methods. In the profile imaging[6-7], atomic arrangements in vertical sections of the surface are known. In the high-resolution plan view imaging, it is aimed to detect individual atoms arranged in the plane of the surface[8].

Particularly important usage of the microscopic technique is to see local structural variation. Therefore, time dependent phenomena such as phase transition processes and adsorption processes are well studied by "in-situ" microscopy. In our study the variation of the structures are recorded on VTR with time resolution of 1/30 sec.

2. SURFACE STRUCTURES

2.1 Structure analysis by TED

Structure analysis by TED is made in a following manner: "Partial" Patterson map is calculated by Fourier transform of TED intensities of the superlattice reflections. Models are deduced from atomic correlations derived from the Patterson map. Models are selected by comparison of TED intensity (R-factor). Fourier map (Difference Fourier map) and/or R-factor optimization are done for decision or for refinement of the model structure.

Two subjects are considered. The first is to understand the reconstructed structures of nxn unit cells (n=odd) and 2xm unit cells (m=even) and the second is to understand adsorbed structures of $\sqrt{3}$x$\sqrt{3}$ unit cells and those of resembling structures.

Si(111)7x7:

1) Dimer adatom stacking-fault(DAS) model has been derived[9,10].
2) Dynamical calculation of the TED intensities has shown validity of kinematical approximation[3].

Si(111)-5x5-Ge:

1) Phase transition temperature of 7x7 structures formed by dissolving Ge into Si surface increases with the coverage of Ge-deposit and the temperature reaches to 920°C at coverage of the formation of 5x5 structure[cf.11].
2) Two monolayer(ML) of Ge deposit is sufficient for the formation of 5x5 structure[12].
3) Three dimensional islands are observed on the 5x5 structure by Ge deposit of over 2 ML.
4) The 5x5 structure transforms to "1x1" by excess Ge atoms or by heating.
5) The 5x5 structure has DAS arrangement[13-15], adatoms and stacking-fault layer atoms (all or part of them) are occupied by Ge.

Si(111)-5x2, $\sqrt{3}$x$\sqrt{3}$, 6x6-Au:

1) 5x2 structure is disordered in bi-fold direction for every 5-fold period[16].
2) $\sqrt{3}$x$\sqrt{3}$ structure (high temperature phase) has short range order which varies depending on the coverage of Au.

3) The long range ordered arrangement of the $\sqrt{3}\times\sqrt{3}$ structure consists with Au trimers surrounded by honeycomb lattice of Si. Atomic distance of Au in the trimer is around 0.6a (a=0.384 nm) and close to ionic diameter.
4) The $\sqrt{3}\times\sqrt{3}$ structure transforms to other $\sqrt{3}\times\sqrt{3}$ structure (low temperature phase) of different short range order.

Si(111)-"5x5"-Cu:
1) "5x5" is incommensurate to the Si 1x1 lattice[16,17].
2) A closely packed (111) layer of Cu is epitaxed on Si(111)layer with rotation angles of about 3.5°.

2.2 Observation of Structure Change by Electron Microscopy.

A high-resolution UHV electron microscope of 200 kV accerelation[18] is used. Clean surfaces are prepared "in-situ" in the microscope and phase transition and adsorption processes are observed directly or recorded on videotapes. 16 mm cine films are reproduced from the video and are presented at the conference.

A. High-Resolution Reflection Electron Microscopy

Si(111)7x7:
1) The reconstructed 7x7 structure is seen by the interference fringes of spacing 2.3 nm (for the [110]incidence)[18,cf.19].
2) No "1x1" disordered area at base of atomic step[cf.20].
3) Positions of dimers of DAS structure coincide with the step[cf.21].

Si(111)7x7 to "1x1":
1) "1x1" phase nucleates at the lower side of the step[22].
2) 7x7 structure is formed even at the edge of the upper side of the step.
3) Phase boundary between the 7x7 and the "1x1" looks sharp[23,cf.24].

B. High-Resolution Transmission Electron Microscopy

Au(110)2x1:
1) Missing-row model is confirmed[18,cf.25].
2) Bi-fold periodicity changes locally.
3) During deposition of Au on a faceted Au surface, growth of double layer high steps occurs on the (110) surface, while a monolayer high step growth occurs on the (111) surface.

Au cluster growing on graphitized carbon:
1) Single atom and clusters consisting of several tens of Au atoms are seen[18].
2) Clusters of sizes larger than 2 nm have icosahedral structure, but no structural change to the f.c.c. (reported by Iijima and Ichihashi[26]) has been observed at room temperature substrate and for reduced imaging electron beam.

3. CONCLUDING REMARKS

Structure analysis by UHV electron microscopy is useful. Not only long range ordered structures but also short range ordered

ones, which is physically interesting to know the interaction, are studied increasingly in future.
Studies on dynamic processes are also pursued further for understanding "surfaces".

References

1. K.Takayanagi: In **Proc. Asia Pasific Symposium on Surface Physics"**, ed. by Xie Xide (World Scientific, Singapole 1987) p.23
2. Y.Tanishiro, K.Takayanagi, S.Takahashi, M.Takahashi and K.Yagi: In **Proc. XI th Int. Cong. on Electron Microscopy**, Kyoto, 1986 (Jpn. Soc. of Electron Microscopy, 1986)p.1340
3. K.Takayanagi: Jpn. J. Appl. Phys. **22**, L4(1983)
4. K.Yagi, K.Takayanagi and G.Honjo: In **Crystals**, ed. by H.C.Freyhardt, vol.7 (Springer-Verlag, Berlin 1982)p.47
5. L.D.Marks and D.J.Smith: Nature, **303**, 316(1983)
6. J.O.Bovin, Nature **317**, 417(1985)
7. T.Hasegawa, K.Kobayashi, N.Ikarashi, K.Takayanagi and K.Yagi:Jpn. J. Appl. Phys., **25**, L366(1986)
8. K.Takayanagi, K.Kobayashi, K.Kodaira, Y.Yokoyama and K.Yagi: In **Proc. 8th Int. Conf. in HVEM**, (Berkeley,1983)p.47
9. K.Takayanagi, Y.Tanishiro, M.Takahashi and S.Takahashi: J. Vac. Sci. and Technol., **A3**, 1502(1985)
10. K.Takayanagi, Y.Tanishiro, S.Takahashi and M.Takahashi: Surface Sci., **164**, 367(1985)
11. T.Ichikawa and S.Ino: Surface Sci., **136**, 267(1984)
12. H.J.Gossmann, J.C.Bean, L.C.Feldmann and W.H.Gibson: Surface Sci., **138**, L175(1984)
13. R.S.Becker, J.A.Golobchenko and B.S.Swartzentruber: Phys. Rev., **B32**, 8455(1985)
14. K.Kajiyama, K.Takayanagi, Y.Tanishiro and K.Yagi: In **Proc. XIth Cong. on Electron Microscopy**, Kyoto, 1986 (Jpn. Soc. of Electron Microscopy,1986)p.1341
15. K.Takayanagi, Y.Tanishiro and K.Kajiyama: J. Vac Sci and Technol., **B4**, 1074(1986)
16. K.Takayanagi: In **Mat. Res. Soc. Sympo.**, ed. by J.M.Gibson, G.C.Osbourn and R.M.Tromp, vol.56(Materials Research Society, Pittsburgh,1986)p.129
17. T.Ishitsuka, K.Takayanagi, Y.Tanishiro and K.Yagi: In **Proc. XIth Int. Cong. Electron Microscopy**, Kyoto, 1986(Jpn. Soc. of Electron Microscopy,1986)p.1347
18. K.Takayanagi, Y.Tanishiro, K.Kobayashi, k.Akiyama and K.Yagi: Jpn. J. appl. Phys., **26**, L957(1987)
19. Y.Tanishiro, K.Takayanagi and K.Yagi: J.Microscopy, **142**, 211(1986)
20. D.Haneman: Phys. Rev. Letters, **25**, 1370(1982)
21. R.S.Becker, L.A.Golovchenko, E.G.McRae, B.S.Swartzentruber: Phys. Rev. Letter, **55**, 2028(1985)
22. see references in [5]
23. Y.Tanishiro et al. : in preparation
24. Y.Tanishiro, K.Takayanagi and K.Yagi: Ultramicroscopy, **11**, 95(1983)
25. H.Jagodzinski, W.Morits and D.Wolf: Surface Sci., **77**, 233(1978)
26. S.Iijima and T.Ichihashi: Phys. Rev. Letters, **56**, 616(1986)

Discussion

D.R. HAMANN: The DAS model has many independent parameters describing atomic relaxation. Relaxation has been predicted by theoretical calculations. What results have you found for relaxation in the course of refining your diffraction?

K. TAKAYANAGI: As briefly discussed in our paper (*Surface Sci.* **164** (1985) 367), the reliability factor was reduced by 5% when atoms in the stacking fault layer and dimers are displaced to positions given by peaks in the Fourier map. The amounts of the displacements are less than $0.03a$ ($a = 3.84\AA$), the order of which agrees with the results of LEED and x-ray diffraction, given in the literature.

Y. PETROFF: Have you done electron microscopic experiments on Ge(111)2×1 above $750°C$? The reason for my question is that McRae from Bell Labs has reported LEED experiments showing that above $T = 760°C$ a few atom double layers lose completely the lateral crystalline order.

K. TAKAYANAGI: We have not done experiments on Ge(111)2×1. We made the following observations on the stabilized Si(111) surface: the 5×5 structure formed by Ge deposition of under 2ML transformed to the "1×1" structure at $\simeq 830°C$. The "1×1" structure turns in to 7×7 (not clean surface) due to dissociation of Ge from the surface region.

The 5×5 structure formed above 2ML depositions, which coexist with three dimensional islands, transformed to "1×1" at lower temperatures, depending on the deposited amount of Ge. The "1×1" structure is considered to be a disordered phase similar to the "1×1" for the clean Si(111) surface above $830°C$.

C.B. DUKE: In response to Dr. Hamann's question: at the 1987 American Vacuum Society meeting (Anaheim, Nov. 2-5), two refinements of the DAS structure were reported. A dynamical LEED analysis was used to refine displacement normal to the surface of the adatom and stacking–fault layers (S.Y. Tong, *et al.*). An analysis of glancing–angle x–ray scattering (I. Robinson,*et al.*) led to refinements parallel to the surface in these layers. These relavations are of the order of a few tenths of angstroms. Accounts of these analyses appear in the conference proceedings: *J. Vac. Sci. Technol.* **A6** (1988).

K. TAKAYANAGI: In TED analysis of nearly normal incidence, only the displacements parallel to the surface can be defined. Refinement of the DAS model resulted in small displacements of dimer atoms (and other atoms also) within a tenth of an angstrom. Where R–factor reduced 5% from that of the original DAS model (*Surface Sci.* **164** (1985), K. Takayanagi *et al.*). In this respect, TED is useful to determine the small displacements of atoms, although the accuracy might be not so high as x–ray methods. For the determination of the normal displacements we did not pursue detailed analysis.

Complete Surface Structures

P.M. Marcus [1] *and F. Jona* [2]

[1] IBM Research Division, T.J. Watson Research Center, P.O. Box 218, Yorktown Heights, NY 10598, USA
[2] SUNY at Stony Brook, NY 11794, USA

1. INTRODUCTION

Our experience with surface structure determination by LEED intensity analysis has led us to the concept of complete surface structures. This concept is forced on us by the fact that a good fit to LEED data requires fixing all atom positions that deviate significantly from bulk positions. Hence many cases of deep structure, i.e., deviations extending several atomic layers below the top layer have been found. Especially studies of multilayer relaxation of open metallic surfaces[1] have emphasized the need to specify many structural parameters, even on clean metallic crystals. In addition other experimental techniques have shown capability as deep probes, notably ion scattering[2], and several theoretical approaches to relaxation in metals [3-6] have given insight into and consistent descriptions of multilayer relaxations which reinforce the experimental conclusions. Deep structure is also found in semiconductor surfaces, where the driving force is the formation of strained surface bonds, particularly angular strains, which in turn strain deeper layers to reduce the total strain energy[7]. Complete structure determinations on semiconductor surfaces have not been as successful as on metal surfaces, so we do not discuss semiconductors explicitly.

In this paper we discuss some developments in our LEED studies of relaxation on open metallic surfaces and also theoretical modelling of relaxations which have helped to understand the driving force and observed complex patterns of metal relaxation. We give some examples of deep structure found experimentally and theoretically. Then we consider the desirable features of a probe suitable for determining complete surface structures and the extent to which LEED possesses these features. LEED does well on all features except the computation of intensities for complex structures with many nonequivalent atoms, i.e., requiring independent specification of its position and not fixed by symmetry in the complete surface structure. However we offer hope of overcoming this deficiency by noting that a computational procedure exists that scales linearly with the number of nonequivalent atoms rather than cubically as present methods do.

2. ON RELAXATION OF METAL SURFACES

Substantial relaxations on metal surfaces were not found in the early LEED work because that work focused on the simpler low-index more closely packed surfaces where the relaxations are small. However refinement of the LEED technique, development of other surface probes and of various theoretical approaches have provided a solid foun-

Springer Series in Surface Sciences Vol. 14
Solvay Conference on Surface Science
Editor: F.W. de Wette

dation for quantitative evaluation of multilayer relaxation on a large number of metal surfaces (ref. 1 lists measured relaxations for 39 surfaces on 17 metals by 5 techniques).

Two developments in our own work have been particularly helpful in revealing and understanding deep relaxations. The first development is the LEED analysis of a sequence of six surfaces of bcc Fe, all of which were successful in finding a high-quality fit to the data[8]. These structures revealed a strong dependence on surface openness or roughness (the inverse of the packing density), with relaxation increasing at first rapidly with openness, but showing a tendency to level off above openness 4. The second development is the application of a simple electrostatic model to the calculation of relaxations on many surfaces which gives good agreement in depth with measured relaxations. We have called this model the modified point-ion model[9] to indicate its relation to the original point-ion model of Finnis and Heine[10], but modified by introduction of an empirical parameter. A discussion of the physical basis for this model can be found in refs. 1 and 9, and we will only note here that it minimizes the electrostatic energy of rigid nets of ions floating in a rigid background of electronic charge. The model makes two useful contributions to a discussion of relaxations, i.e., 1) the driving force of metallic relaxation is shown to be essentially the surface Madelung forces due to the asymmetric charge distribution produced when a bulk crystal is truncated (before relaxation), 2) the model is sufficiently simple to make computation practicable for very open surfaces; in ref. 1 results of calculation with the model are shown up to openness 7 and indicate clearly saturation of relaxation; in a more recent paper[11] some results have been obtained for the limit of infinite openness, i.e., for the isolated step.

We now illustrate the phenomenon of deep relaxation in Figs. 1 and 2 with examples from both experiment and theory on a common scale and featuring the absolute displacement of each atom from its bulk position. For each structure the bulk positions are shown by solid circles (projected on the plane of the paper) on one scale (the length of 1Å is drawn separately) and the displacements due to relaxation are shown as arrows with lengths enlarged four times over the bulk scale (actual displacement magnitudes in Å are marked).

We show in Fig. 1(a) relaxations in Al(110) from LEED data[12], because it has become a benchmark for the study of relaxations both experimentally and theoretically; it was the first surface for which single-layer relaxation was found and later multilayer relaxation. Its modest relaxation at openness value 1.8 (interlayer spacing changes[12] of $-8.6\%,+5.0\%,-1.6\%,+0.1\%$) provides a basis for judging the relaxations of the other examples. In Fig. 1(b) the remarkably large relaxation of Fe(111) (for a low-index surface) is shown from a recent LEED analysis[13], but is understandable from the large openness value of 2.9; note that the displacements do not show simple alternation. An example of parallel relaxation[14] is shown in Fig. 1(c) (top view) for Fe(211), which is similar to fcc Al(110), but has lower symmetry and greater openness value 2.1; the perpendicular relaxations are shown in the side view.

In Fig. 2 the relaxation in Al(110 (panel (a)) is compared on the same scale (reduced by 1/3 from Fig. 1) with relaxation in Au(110)2×1 (panel (b)) from LEED analysis[15]; much of the detail of this missing-row model of the reconstruction has been confirmed by ion scattering[16]. We see the striking increase in atom displacement when the surface

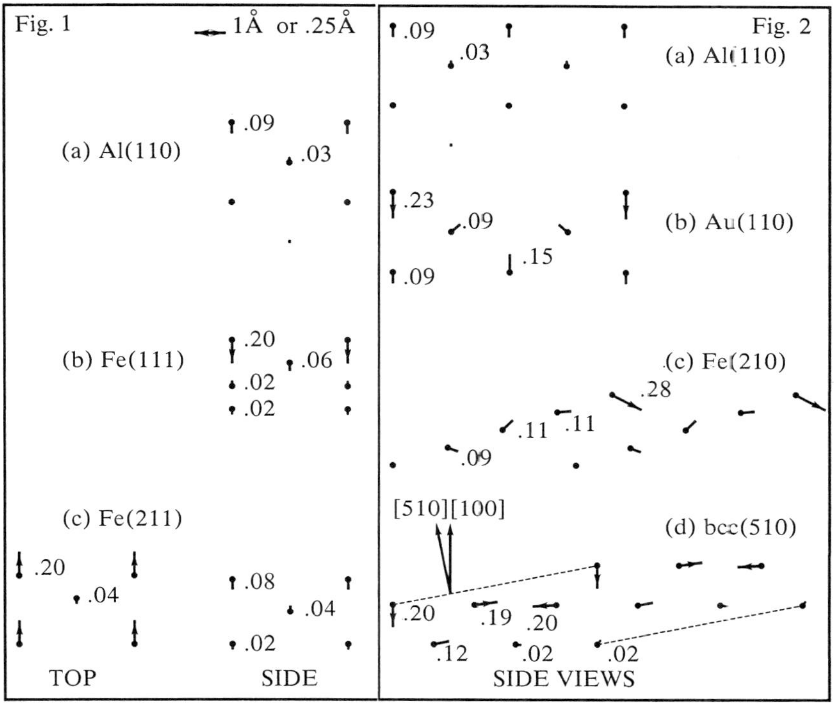

Figs. 1 and 2 show bulk atom positions (solid dots) on a common scale (1Å length shown at the top of Fig. 1) and displacements to relaxed structure on a common scale four times larger. The actual displacement magnitudes are given in Å. Smaller displacements do not have arrow heads; very small displacements are indicated by a dot.

is opened up. Figure 2(c) shows Fe(210), the most open Fe surface that has been studied experimentally[17] at openness value 3.8, which shows large displacements extending to the fourth layer. To illustrate relaxations for still more open surfaces, we take an example from calculations with the modified point-ion model[18] applied to bcc(510) with openness 4.3. The figure is drawn with the lattice spacing of Fe (a=2.866Å) and is oriented to show the surface as a stepped (100) surface with a tread of 2½ lattice spacings; the light lines show the actual (510) surface with relaxations to the sixth layer. Note the large vertical displacement 0.20Å of the step edge atom, which forces the atoms below it aside (≃0.4Å in the isolated step).

3. DESIRABLE FEATURES OF A DEEP-STRUCTURE PROBE AND LEED

We discuss five desirable features in a probe that is particularly suited to study surface structure in depth and comment on how well LEED shows them:

1. Appropriate depth sensitivity. Probe sensitivity should not be restricted to the top layer, but should not respond to more than 8 or 10 layers, so that the signal is not

swamped by bulk response. This scale of depth sensitivity is found in LEED in the energy range up to a few hundred eV because of the strong attenuation of coherent electrons in this energy range; a typical amplitude free path is about 10Å. The strong attenuation also loses structural information, since the intensity vs. energy curves (spectra) are smeared over ranges corresponding to the size of the imaginary part V_{oi} of the potential in the crystal, typically 4 to 6 eV. However we show in Fig. 3 that considerable distinctive structure remains even with realistic attenuation.

Thus Fig. 3(a) shows the fine structure in a LEED spectrum for Cu(001) at a hypothetical small attenuation of V_{oi}=0.01eV (amplitude free path for 100eV electrons $\ell=4\times10^3$Å); Fig. 3(b) shows the smoothing for V_{oi}=1eV, ℓ=40Å (somewhat greater than the effect of lattice thermal motion) and Fig. 3(c) shows the effect of a realistic attenuation of V_{oi}=4eV, ℓ=10Å, which fills in the valleys between peaks. An order of magnitude decrease in intensity is produced by each change. Figure 3(c) retains much geometrical information in the shapes as well as the positions of the peaks.

2. Sensitivity to small displacements of the lattice. Displacements of magnitude 0.01 or 0.02Å are chemically significant and reflect differences in the environment of a

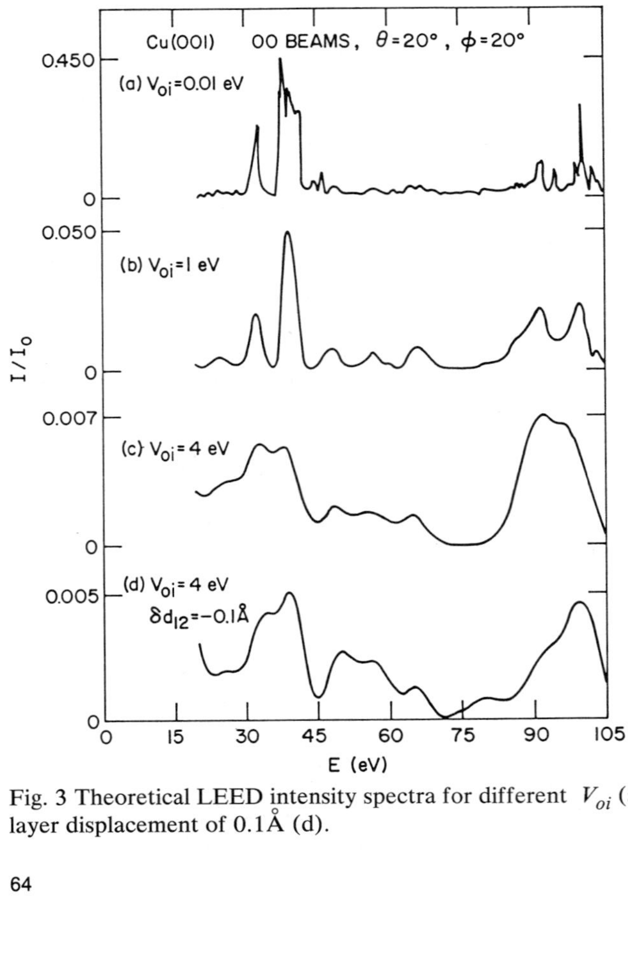

Fig. 3 Theoretical LEED intensity spectra for different V_{oi} (a), (b) and (c), and for first layer displacement of 0.1Å (d).

given bond, hence could be used to give meaning to the requirement of "significant deviations from bulk" used in Sect. 1 to define complete surface structures. As an example of the magnitude of environmental effects on bond lengths consider the C-Cl distance in C_2Cl_6 (1.74±0.01Å), CCl_4 (1.766±0.003Å), CH_2Cl_2 (1.7724±0.0005Å), CH_3Cl (1.784±0.003Å)[19]. We should like to have sensitivity of this magnitude, but LEED in practice has not quite achieved it. However that this sensitivity is potentially available is demonstrated by Fig. 3(d), which shows the effect of an inward displacement of the first Cu layer by 0.1Å. Comparison with 3(c) shows that gross changes occur, so that clearly sensitivity to 1/5 to 1/10 of this change is possible.

3. A known scattering law for atoms, independent of the atomic environment. Such *a priori* information, equivalent to a form factor for the scattering, means that the scattering potential is not one of the quantities to be fitted to experiment in the structure determination. LEED achieves this above about 50eV, since scattering is then from the deeper part of the potential in the ion core and does not depend on the valence electron distribution. This insensitivity of scattering phase shifts to valence electrons has been demonstrated[20] by calculating the phase shifts as functions of radial distance in the atom to show that above 50eV they are either flat or small outside the ion core.

4. Possibility of a large data base. A large data base is required to eliminate accidental agreement with wrong structures and to improve statistical precision; it will grow with the complexity of the structure. A large data base is easily produced in LEED (with automatic intensity measurement) because the diffracted intensity is a function of six independent variables: three continuous (the incident energy and polar and azimuthal angles) and three discrete (two beam indices and the electron spin - with appropriate spin-sensitive source and detector).

5. Readily computable intensities. For effective implementation of the iterative fitting process to data an efficient computation of intensities is needed. Present methods for LEED are adequately efficient and accurate only for small numbers of nonequivalent atoms N_{neq}, up to about 10, but computation time scales as the cube of N_{neq} and prevents application of LEED analysis to many interesting complex systems. A satisfactory solution exists in principle, since it can readily be shown that a procedure exists that scales as N_{neq} in consequence of the rapid decrease of scattering amplitude at each successive scattering due to attenuation and spherical wave spreading[21-22]. Hence scattering at an initial atom will proceed through a finite cascade of subsequent scatterings to a converged result. This cluster of scatterings is roughly the same in number for all initial atoms in a given lattice, but will have to be repeated using each nonequivalent atom as an initial atom, i.e., scales as N_{neq}. Implementation of such a procedure represents the real hope of extending the power of LEED to find complete structures to complex systems.

REFERENCES

1. F. Jona and P. M. Marcus in *Proceedings of the 2nd International Conference on Surface Structure*, Amsterdam, ed. by J. van der Veen, (Springer 1988); contains a tabulation of all relaxation measurements as of early 1987)
2. J. F. van der Veen, Surface Sci. Rept. **5**, 199 (1985); a review of ion scattering.
3. R. N. Barnett, U. Landman and C. L. Cleveland, Phys. Rev. Lett. **57**, 1339 (1983), Phys. Rev. B**27**, 6534 (1983), Phys. Rev. B**28**, 1685 (1983).

4. K. M. Ho and K. P. Bohnen, Phys. Rev. B**32**, 3446 (1985); A. Eguiluz, Phys. Rev. B**35**, 5473 (1987) (first-principles band calculations).
5. P. Jiang, P. M. Marcus and F. Jona, Solid State Commun. **59**, 275 (1986); see also refs. 1, 9, 18 (electrostatic model).
6. M. S. Daw and M. I Baskes, Phys. Rev. B**29**, 6443 (1984); S. P. Chen, A. F. Voter and D. J. Srolovitz, Phys. Rev. Lett. **57**, 1308 (1986) (embedded atom method).
7. J. A. Appelbaum and D. R. Hamann, Surf. Sci. **74**, 21 (1978).
8. J. Sokolov, F. Jona and P. M. Marcus, Solid State Commun. **49**, 307 (1984); for more recent discussion of trends with openness see refs. 13 and 1.
9. P. M. Marcus, P. Jiang and F. Jona in *Proceedings of the 2nd International Conference on Surface Structure*, Amsterdam, ed. by J. van der Veen, (Springer 1988); discusses electrostatic models.
10. M. W. Finnis and V. Heine, J. Phys. F**4**, L37 (1974).
11. P. Jiang, F. Jona and P. M. Marcus, (submitted to Physical Review).
12. J. N. Andersen, H. B. Nielsen, L. Petersen and D. L. Adams, J. Phys C**17**, 173 (1984).
13. J. Sokolov, F. Jona and P. M. Marcus, Phys. Rev. B**33**, 1397 (1986).
14. J. Sokolov, H. D. Shih, U. Bondi, F. Jona and P. M. Marcus, J. Phys. C**17**, 371 (1984).
15. W. Moritz and D. Wolf, Surface Sci. **163**, L655 (1985).
16. M. Copel and T. Gustafsson, Phys. Rev. Lett. **57**, 723 (1986).
17. J. Sokolov, F. Jona and P. M. Marcus, Phys. Rev. B**31**, 1929 (1980).
18. P. Jiang, F. Jona and P. M. Marcus, Phys. Rev. B**35**, 7952 (1987).
19. *CRC Handbook of Chemistry and Physics*, 67th ed., ed. Robert C. Weast, (CRC Press, Boca Raton, Florida, 1986-1987), pp.F-161, 162.
20. P. M. Marcus and F. Jona, Applications of Surface Sci. **11/12**, 20 (1982).
21. J. B. Pendry, in *Determination of Surface Structure by LEED*, ed. P. M. Marcus and F. Jona, (Plenum Press, New York, 1984), pp. 1-15.
22. P. M. Marcus, F. Jona and M. Torrini, *Proceedings of the International Seminar on Surface Structure Determination by LEED and Other Methods*, Erlangen, W. Germany, 1986, pp. 60-63 (unpublished but available).

Discussion

W. KOHN: You have described relaxations of surface planes of different degrees of "openness". For a given metal, is the depth d below the surface (rather than the number of lattice planes), at which significant atom displacements (say, 0.02 Å) occur, reasonably independent of the openness? And if so, what are the characteristic values of d?

P. MARCUS: The relaxation does fall off in roughly the same physical distance from the surface on different surfaces, although for an open surface, many mathematical layers are counted in that distance. The decrease to insignificant values (less than 0.01 Å) takes place in about 3 or 4 Å from the physical surface for metals.

S.A. RICE: Can you indicate, in a few words, the current status of the calculation of the multiple scattering correction to LEED intensities? Is the observed intensity pattern dominated by single scattering or multiple scattering? How robust are the calculated intensities to changes in structural parameters? What is the accuracy of geometric parameters so determined?

P. MARCUS: Multiple scattering calculations are almost always required for accurate LEED intensity spectra; coherent 100 volt electrons have 5 or 6 collisions before their amplitudes are attenuated so much that they do not contribute to diffracted intensities. The calculated intensities are highly sensitive to small structural changes, a characteristic feature of a diffraction phenomenon. In good fits to adequate data bases the precision is in the range of $\pm 0.02 Å$.

J.B. PENDRY: The limiting factor in surface structure determinations by LEED is not the theory, but the ability to prepare a surface of the requisite quality.

P. MARCUS: I would modify the comment to say that the limit is probably set by experimental uncertainties not just in preparing the surface, but also in fixing the angles, correcting for instrumental broadening and diffuse background.

J.T. YATES, JR.: Could you comment on the technological implications of the reconstruction of metal surfaces in areas such as catalysis, corrosion chemistry, adhesion or lubrication?

P. MARCUS: I am not aware of any direct technological implications, although knowledge of structure is certainly needed to give basic understanding and quantitative grasp on all physical and chemical behavior at surfaces.

D.A. KING: My question relates to the current status of refinement of the models discussed, particularly Au{110}(1×2). We have analyzed the Pd{110}(1×1) surface and obtain a very good R-factor with a 3-layer multilayer relaxation. However, when we analyze the Pd{110}(1×2) formed with a small amount of adsorbed Cs, we do find best agreement with a 3-layer reconstruction model similar to that published recently for Au{110}(1×2). But the R factor is *not* as good as for the (1×1) analysis. Is this because (in general) our samples are insufficiently ordered; or the structure still needs further refining?

P. MARCUS: In general we find that failure to get a good R factor fit is most likely to be due to a missing feature in the structural model. However, bad data can also lead to a poor fit, such as data contaminated by a superposition of several ordered structures. I would expect disorder to have less serious effects on achieving a good fit.

G.A. SOMORJAI: The contractions observed at surfaces are well correlated with the openness at the surface, as you point out. One may then think of extrapolating the contraction at surfaces to the short interaction distance for diatomic molecules. Would such extrapolation be possible to predict interatomic bonds in small clusters or may one expect different interatomic distances for small clusters with 4 to 12 number atoms?

P. MARCUS: Our results apply to the spacings of layers at the clean metal surface, and will not approach diatomic values for open surfaces. An adsorbate bond on the surface usually takes some value between the diatomic molecule and a bulk value for the same two atoms. A cluster on the surface will have many bonds that are only slightly affected by the presence of the surface.

C.B. DUKE: There are many diverse models of relaxations at clean metal surfaces from many–body force laws through embedded atom methods to local density calculations. What is the status of these various models relative to the one mentioned in your talk for the systems which you have considered? Specifically, can you compare your model's results with those on $A\ell$, Ni, and $Ni_3A\ell$ from the Los Alamos group?

P. MARCUS: Our model of relaxation is an electrostatic model, i.e. the ions are driven from bulk positions purely by Coulomb forces, and it is a modification of the idea of Finnes and Heine. We fit experiment well in depth and on many surfaces of the same material, by choice of a single empirical parameter. We do as well as any and better than most of the theoretical models in fitting experiment and have the advantage of great simplicity so that very open surfaces can be handled—in fact, approaching the isolated step.

D.R. HAMANN: When the LEED R-factor is poor, existing methods give no indications of how to improve the trial structure. Do you foresee the possibility of the direct calculations of a LEED "force" to point towards a better geometry?

P. MARCUS: The use of a gradient of the R-factor with respect to structural parameters has long been suggested as a systematic pointer toward better geometry. The basic calculation of intensities for a complex structure is still too lengthy to make calculation of a gradient efficient, and for simple structures it isn't needed since a minimum of R can be found by taking the parameters in pairs.

J.B. PENDRY: The search of a many–parameter space in structural determination can in part be done by direct methods based on inversion of the "tensor" in tensor–LEED methods, to give a density of scattering power distribution in real space.

General Discussion

G.A. SOMORJAI: [To D.R. Hamann] I have four questions.

(1) Have you considered the location of H atoms under the surface? How is the energy of that location compared with H adsorption sites on the surface?

(2) The H atom has a very low activation energy for surface diffusion on most surfaces: about 2 kcal/mole. In this circumstance, wouldn't there be a fluctionality of bonding within the time scale of the experiment? This may not permit the use of localized bonding models to be successful.

(3) Have you effected the H–H interaction as you consider the H–coverage dependence of H bonding to the substrate? As you well know from studies of CO adsorption, the molecule may seek new sites of higher coverages because of the adsorbate–adsorbate interaction.

(4) Have you considered H–induced reconstruction of the substrate in your calculation?

D.R. HAMANN:

(1) Feibelman and I compared sub–surface H on Ti(0001) and on the bimetallic surface Ni on W(100). For the geometries we considered, the sub-surface sites were unfavorable compared to the surface sites. For some metals, gas–phase H is absorbed into the bulk, so sub–surface sites must be favorable.

(2) For low–coverage H, tunneling is important, as Gomer's experiments showed. For the saturated surfaces I considered, H–H repulsions rule this out. The whole wave layer would have to tunnel together, which is highly improbable.

(3)&(4) I haven't carried out any calculations for surfaces where H–H interactions, or H–induced substrate motion influence the structure. The same methods could be applied, and it would be interesting (but expensive) to study these systems.

R. GOMER: [On Somorjai's comment to D.R. Hamann.] Activation energies of H diffusion on transition metals are not $\sim$ 2 Kcal/mole but 4–6 Kcal/mole.

P. MARCUS: [Referring to Somorjai's question.] In a recent publication we have given a structure obtained by LEED analysis of H on Ge(111). A good 1×1 structure is formed by exposure of the cleaved Ge(111) 2×1 surface to atomic H. A contraction of the first Ge spacing of about 12% is found, which is less contracted than clean Ge(111); the contraction of the first spacing of clean Ge(111) has been calculated theoretically to be about 40% by Maxiras and Joannapolis. It would be interesting to see if surface phonon dispersions are affected by this residual relaxation after H is added.

M. CARDONA: For a complete description of IR spectra of H on metals one needs not only the vibrational frequencies but also the dynamical charges. What is the experimental and theoretical state of this problem? Can one say whether the dynamical charge on the hydrogen is positive or negative? Can one get experimental information on the sign through the Fano–interaction with electronic continua?

D.R. HAMANN: We have calculated the dynamic effective charge for H on Pt, W, and Rh. It is typically in the range 0.02 – 0.05e, and has a sign which seems to be correlated with the electronegativity difference between H and the substrate. For Rh, the dynamic charge was zero, probably because the site is so deep. Experimentally, charges are in the same range, but theory and experiment are off by about a factor of 2. The only observed Fano lineshapes I know of, for the wag overtone of H on W(100), has zero dynamic charge.

H. IBACH: Since there has been so much discussion about the particular feature of hydrogen on surfaces, I would like to draw the attention to another feature which is particular to hydrogen. For some unknown reason, hydrogen vibrational modes are always significantly broadened with a FWHM of the order of 50–100 cm^{-1}. This even holds for monolayer coverages where hydrogen sits in well–defined sites.

W. KOHN: [Addressed to Ibach.] I wonder if it is simply the low mass of the hydrogen. What you are seeing if you have complete coverage is really a surface phonon mode and you would expect that all associated frequencies would be relatively large. This change of dispersions, broadening of frequencies, all that, it seems to me, would be enhanced by a small mass.

H. IBACH: Well, my comment applies to spectra which are Q–resolved so that we actually look at a particular Q vector; and of course we see dispersion also, but this is a very weak effect in most cases for hydrogen modes. But still, all along the dispersion line, the line is broad.

A.M. BRADSHAW: You showed a hierarchy of theoretical descriptions of surfaces with the quantum chemical *ab initio* methods occupying the pinnacle. On the other hand, your list of selected applications did not appear to show an appropriate example of the latter. Does this represent an evaluation on your part of the cluster approach? Those of us involved in studies of the adsorption of molecules and molecular fragments are particularly interested in the results from calculations of this kind. Slab calculations appear difficult to perform for such systems.

D.R. HAMANN: I did include 2 examples studied by quantum chemical cluster methods: H on Si(100) and O on Ni(100). In general, I believe these methods can give reliable results when they are adequately converged both with respect to basis functions and cluster size. Many questions of dynamics and reconstruction can only be addressed by extended surface calculations, however.

L. KLEINMAN: Quantum chemists can't handle a chromium dimer with twelve electrons, so there's no hope of them doing anything on the surface.

D.R. HAMANN: But there was no such bias and omission. Two of the systems I had on there were addressed with quantum chemical methods: the vibrations of hydrogen on Si(100) surface were studied by Krishna Raghavachari with quantum chemical methods and the oxygen on Ni was studied by Bill Goddard with quantum chemical methods, as well as by others with other methods.

W. KOHN: Just a quick follow-up question for Dr. Ibach: Have studies been made of this linewidth that you spoke of with deuterium also, and is there anything interesting coming out of this? Is the width independent of the mass?

H. IBACH: The width does depend on the mass. The line width is reduced for deuterium, which indicates that the line broadening for hydrogen is related to its small mass.

W. PLUMMER: The multilayer relaxation of surfaces is a physical property at surfaces in search of a simple physical explanation. As Dr. Duke has pointed out, many of the theories are model dependent. My comment relates to the LEED measurements being made at room temperature and the relationship of these measurements to those for $T = 0$. A student of mine, Art Baddorf, has shown that for Cu(110), the measured anharmonicity of the potential well for the surface atoms is very temperature dependent. Therefore, it is important that all structural studies be done as a function at temperature and extrapolated back to $T = 0$. Also, LEED theory for finite temperature should take into account anharmonic potentials.

P. MARCUS: LEED measures the average position of the atoms at the temperature of the measurement. It is not necessary to include anharmonic potentials in the theory. However, anisotropy of lattice motion and correlated atom motion are possible significant corrections. Extrapolation to $T = 0$ is important only if you wish to check a structure from another source at $T = 0$.

J.W. GADZUK: One must be very careful in using the results of quantum chemical cluster-calculations to infer properties of extended chemisorption systems, particularly when interested in properties requiring realistic treatment of screening and/or low energy electronic excited states.

C.B. DUKE: Following up on Prof. Plummer's question, for semiconductors (even nearly–equal–mass cases like GaAs and InSb), temperature is an important variable in determining second and deeper layer relaxations. Also, if the $< u^2 >$ on the surface are different from those in the bulk, the calculated LEED lineshapes become sensitive to $(< u^2 >_s - < u^2 >_b)$. I should think that these phenomena would lead to important refinements of the deep relaxations in clean metals. In addition, the energy gain due to the deep static relaxation is $E_g < kT$ at room temperature. Would one not expect anharmonicity rather than static relaxation in such cases?

P. MARCUS: In our experience change of the Debye temperature does not have much effect on the structure analysis of clean metal surfaces. Anharmonicity would be a still smaller effect.

C.B. DUKE: When you put an anharmonic potental in, is there a difference between the surface and the bulk? Is it a big difference between the surface and the bulk?

P. MARCUS: LEED is not specially sensitive to temperature motion and...

C.B. DUKE: Only if all the scatterers have the same phases. The minute you have different phases, then that changes, Paul.

P. MARCUS: You mean different kinds of atoms in the same structure...

C.B. DUKE: Right. Different kinds of atoms and different vibrations on the surface and in the bulk. In all those beautiful single–scatterer scaling models, the same structure in the intensity profiles just goes up and down and then disappears at high enough temperature and that's really the problem; since surface atoms have different atomic vibrations these simple scaling models are not pertinent.

P. MARCUS: Well, of course in our multi–layer relaxation we had only one kind of atom and so...

C.B. DUKE: Different atoms vibrate differently and you've got those little teeny energies for the deeper layer relaxations. These energies are less than the kT at the temperature at which you're making the experiment so how do you really believe that these things have a static displacement?

P. MARCUS: But the temperature is producing an averaging over a whole distribution of positions which comes out very close to the static position.

W. PLUMMER: The averaging doesn't take into account the anharmonicity.

P. MARCUS: Well, all right...

Part II

Surface Science and Catalysis

The Activation of Chemical Bonds at Surfaces

J.T. Yates, Jr.

Surface Science Center, Department of Chemistry,
University of Pittsburgh, Pittsburgh, PA 15260, USA

I. Introduction

The activation of chemical bonds in molecules adsorbed on surfaces is of fundamental interest in many types of surface chemistry. For example, all heterogeneous catalytic processes proceed by way of bond activation in adsorbed species. Chemical vapor deposition onto semiconductor surfaces to produce thin film semiconductor materials depends upon bond activation at the surface. Surface corrosion processes generally involve activated molecular surface reactions. Thus, it is important to understand the activation of chemical bonds in surface species so that better understanding and control of surface chemical processes may be achieved.

In this lecture, a variety of activated surface processes will be discussed with two primary objectives. The first is to illustrate the activation of chemical bonds in surface species starting from very low activation energy molecular events such as hindered molecular rotation on a chemisorption site and proceeding to higher energy activated processes such as surface diffusion and desorption, and finally to chemical bond scission in polyatomic adsorbates, caused either by thermal or by electronic excitation processes. The second objective of the lecture is to use the examples chosen to illustrate certain methods of research which can give detailed molecular level insight into the details of bond activation in surface chemistry.

II. Molecular Rotation in Adsorbates on Their Adsorption Sites on Single Crystal Surfaces

We begin by illustrating the use of electron stimulated desorption (ESD) to observe the hindered rotation of an adsorbed species on a single crystal substrate. When an electron having an energy of about 100 eV strikes a molecule adsorbed on a surface, an electronic excitation process occurs [1]. Positive ions originating from the adsorbate molecule are often released, along with negative ions, electronically excited

Springer Series in Surface Sciences Vol. 14
Solvay Conference on Surface Science
Editor: F.W. de Wette

(metastable) neutral species, and ground state neutrals. The processes causing this effect are initiated by rapid Franck-Condon transitions from the ground electronic state of the system to repulsive upper states. Ejection of positive molecular ions [2] and excited metastable neutral species [3] generally occurs in specific angular directions which correlate with the direction of the chemical bond being broken in the process. Thus the observation of the ion ejection directions can give information on the chemical bond directions exhibited by the adsorbate prior to electronic excitation.

The phenomenon of ejection along the chemical bond direction is called electron stimulated desorption ion angular distribution (ESDIAD) and the apparatus used for these measurements is termed an ESDIAD apparatus [4]. For the first years of ESDIAD research [2], photographic images of the ESDIAD patterns on phosphor screens were analyzed to obtain symmetry information, and to make crude estimates of ion ejection angles. Recently, an improved version of the ESDIAD method has been developed in which the ion trajectories are analyzed by digital counting procedures [5-7]. This method yields superior images over the method in which data is collected on a phosphor screen and permits accurate analysis of the ion angular distributions as well as the cross section for ion and metastable species production, using pulse counting methods.

A schematic diagram of the digital ESDIAD apparatus employed at the University of Pittsburgh is shown in Fig. 1. A single crystal containing the adsorbed species to be studied is irradiated with electrons from an electron gun. The positive ions ejected from the adsorbate on the crystal are intercepted by a pair of hemispherical grids and then by a pair of flat grids, which transfer the ions to the front surface of a pair of microchannel plates. For each ion entering the front surface of the microchannel plates, about 10^5 electrons are ejected from the backside of the second channelplate in spatial registry with the ion initiating the process. The electron pulses are collected on a resistive anode device, and the x and y coordinates of each pulse are calculated by a position computer and displayed on a dual parameter multichannel analyzer. Thus, as the ion counts are collected, a statistical distribution of ion ejection angles is acquired with increasing signal-to-noise quality. These ion trajectories coincide with the angular distribution of the chemical bonds being ruptured by ESD. For a molecule undergoing zero point or thermally activated motions, the ESDIAD process will catch the molecule in a portion of its trajectory, and will deliver a "snapshot" of its spatial orientation at the time of the electronic excitation. There is, however, some increase in polar angle due to the interaction of the positive ions with their image charges in the metal substrate [8,9].

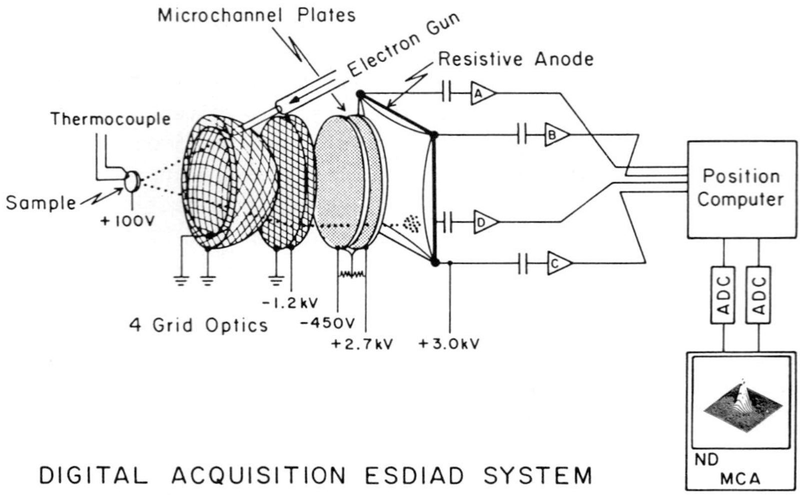

Fig. 1. Digital acquisition ESDIAD apparatus. A schematic drawing showing the crystal, grids, microchannelplate pair, and resistive anode drawn approximately to scale [5,6]

We have employed digital ESDIAD to make the first observations by a direct method of the thermal onset of hindered rotation of a chemisorbed molecule on its adsorption site. Spectroscopic methods to observe surface rotor species have been employed, but the direct study of the thermally-induced rotation of adsorbed species has been an elusive issue in surface chemistry research. The candidate molecule for study is the PF_3 molecule chemisorbed on Ni(111). All evidence involving temperature programmed desorption [10], ultraviolet photoelectron spectroscopy [11], LEED [11], and theoretical calculations [12,13] show that PF_3 is not dissociated on Ni(111) below 400 K, and that it is bonded to atop Ni sites through the P atom via donor-acceptor bonds with the surface [14].

Figure 2 shows the ESDIAD pattern for F^+ emission from chemisorbed PF_3 molecules on Ni(111) as a function of temperature for two PF_3 coverages. In the first column, experiments for low PF_3 coverages and at two temperatures are shown: at 85 K, six PF_3-derived F^+ beams are observed due to the presence of a mixture of two classes of three-fold symmetric PF_3 molecules which possess azimuthal orientations corresponding to the principal crystallographic directions in the Ni(111) surface. This is shown in Fig. 3, where a schematic diagram of the surface structure of Ni(111) and the ESDIAD ion beam directions is illustrated.

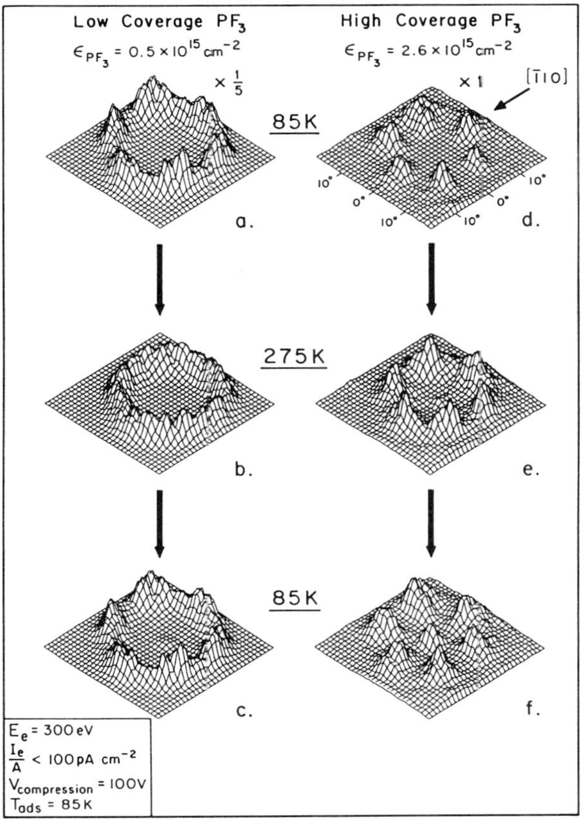

Fig. 2. Temperature and coverage dependence of PF_3 hindered rotation on Ni(111): F^+ ESDIAD [16]

Upon heating the surface containing the PF_3 to 275 K (Fig. 2), the six F^+ beams become diffuse and a ring pattern of F^+ emission is produced as shown in Fig. 2b. Cooling back to 85 K reproduces the six beam pattern as shown in Fig. 2c. This experiment therefore shows the activation of the hindered rotational motion of chemisorbed PF_3 molecules at low coverages, where intermolecular forces are small. The coverage for this series of experiments is ~ 0.04 PF_3/Ni.

At the saturation coverage of PF_3 (0.25 PF_3/Ni), a p(2 x 2) LEED pattern is observed, and as shown in Fig. 4, close packing of (possibly interlocked) PF_3 species exists. In this high-coverage case, as shown in Fig. 2d, e, and f, the heating experiment does not cause molecular rotation of PF_3, and the six beam F^+ ESDIAD pattern exists over the temperature range 85-275 K as intermolecular forces cause an enhancement of the barrier to rotation. The general problem of interlocking gear

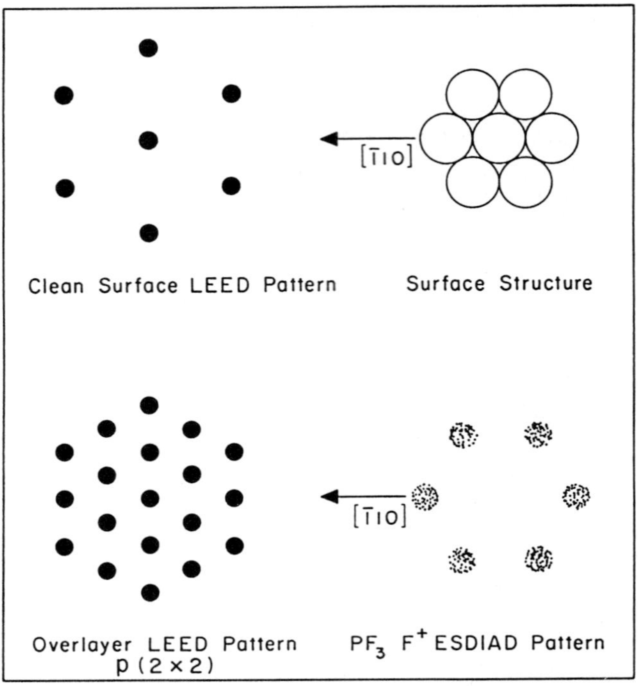

Fig. 3. Surface structure definitions [10]

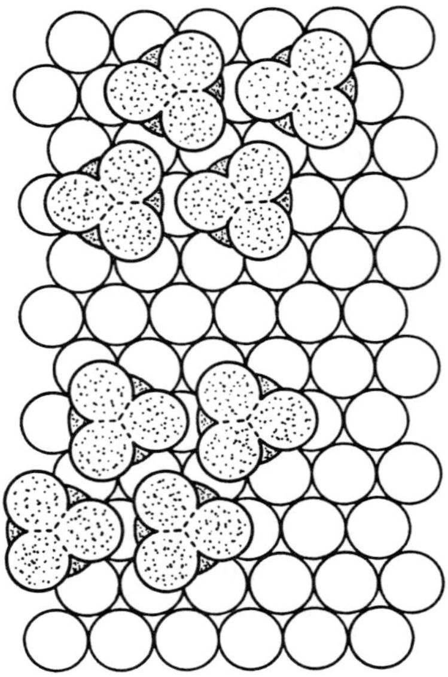

Fig. 4. The orientation and spacing of PF_3 molecules at 0.25 PF_3/Ni on Ni(111) for a p(2 x 2) surface structure [16]

rotors on a surface has been theoretically treated by ZWANZIG [15], and may apply to this experiment.

The hindered rotor problem is a problem that has been addressed theoretically for over a century, and a review of the important papers in this area is given in reference [16]. For the purposes of analysis of the data for PF_3 hindered rotation reported here, the method of KOEHLER and DENNISON was used [16]. In our study, the quantum mechanical problem was solved for 99 hindered rotor states for a number of barriers in the range 80 - 150 cm^{-1}. An ensemble-averaged set of azimuthal probability densities was derived, summing over the 99 states. This analysis is shown for an 80 cm^{-1} barrier for three temperatures in Fig. 5, and it may be seen clearly that the contrast between the valley and peak regions decreases with increasing temperature, as the theoretical ESDIAD pattern broadens azimuthally. In Fig. 5, the valley integration region and the peak integration regions correspond exactly to comparable regions studied in the experimental ESDIAD patterns, where averages of intensities in the 6 valley regions and 6 peak regions were measured over the temperature range 85-359 K. A comparison between the theoretical relative valley-to-peak intensities and the experimentally measured results is shown in Figure 6. On the basis of the comparison, we estimate that the barrier to PF_3 rotation is 80 ± 20 cm^{-1} [16].

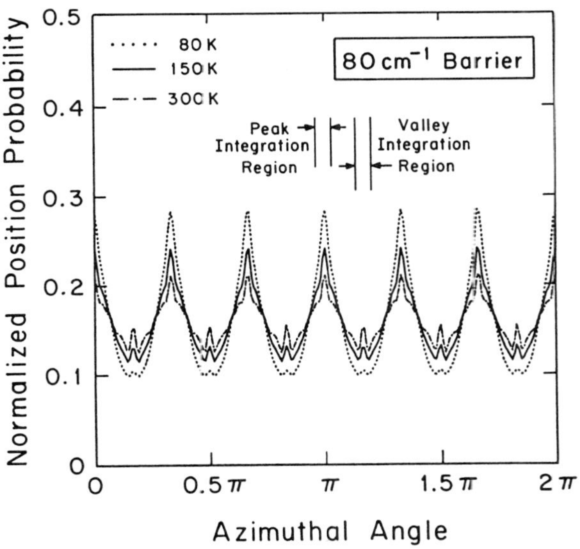

Fig. 5. Statistical thermal population of hindered rotor states: PF_3/Ni(111) [16]. Here 99 hindered rotor wave functions are summed together with appropriate weighting of each

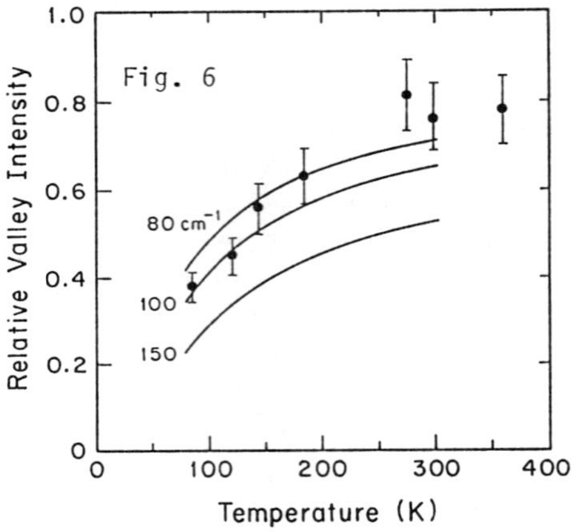

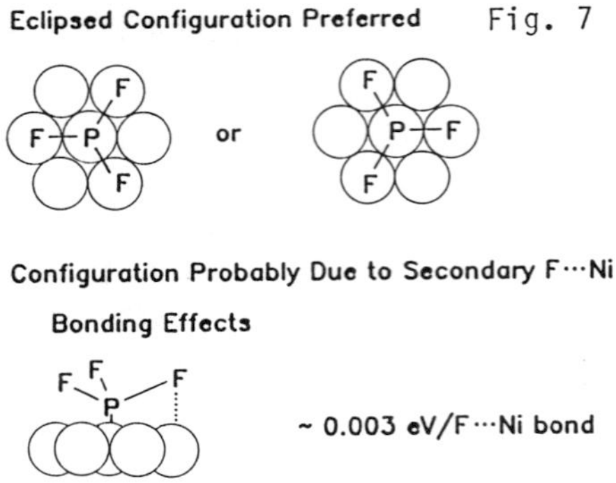

Fig. 6. Comparison of calculated thermal effects to observed thermal effects in ESDIAD: PF_3/Ni(111) [16]

Fig. 7. Summary of findings regarding the chemisorption and hindered rotation of PF_3 on Ni(111) [16]

The strength of the Ni...PF_3 bond for the chemisorbed PF_3 species has been determined by temperature programmed desorption to be about 1.1 eV [10]. This ESDIAD experiment has shown that the bonding potentials above the Ni(111) surface will orient a PF_3 molecule azimuthally in the eclipsed configuration, where F atoms prefer to orient over top of Ni atoms rather than over the two fold bridge sites, as shown in Fig. 7. The effect is small (energy difference/F...Ni bond ~ 0.003 eV per F atom). At the present time, we believe that this must be due to secondary F...Ni bonding effects since the orientation energy is a very tiny fraction of the total bond energy in the chemisorption system.

These experiments represent the lowest energy bond activation processes which we have studied. It has been shown that the hindered rotation of the chemisorbed PF_3 molecule is the first bond excitation event to be observed; presumably, at higher temperatures, other low frequency adsorbate motions will begin to be excited, leading ultimately to thermal dissociation of the chemisorbed molecule near 400 K. To the best of our knowledge, these are the first measurements of the thermal activation of the hindered rotation of a surface species using a DIRECT (non-spectroscopic) experimental method to observe the rotational excitation.

III. Adsorbate Migration Effects

The surface chemical processes which are responsible for adsorption, surface reaction, and evolution of surface species into the gas phase all may involve the migration of adsorbates from place to place on a surface. Thus, it is well known that adsorption often proceeds by way of a mobile precursor species leading to characteristic adsorption kinetics. Surface reactions involving Langmuir-Hinshelwood type reaction steps also involve the migration of adsorbate species to favorable configurations in which new chemical bonds may be formed. Finally, the desorption of surface species can involve migration steps prior to the desorption event. These topics have recently been reviewed in some detail [17].

One method to study site interchange between adsorbed species involves the study of single crystal surfaces containing known defect densities intentionally introduced by cutting the single crystal in directions to expose atomic steps and terraces. If, as is generally found, the chemical bonding of an adsorbate atom or molecule to the step differs from the bonding on the terrace, one has designed a system which can in principle be used to study adsorbate exchange between the different classes of sites. The detailed character of the metal surface sites on stepped single crystal surfaces may be categorized most simply by reference to the metal-metal coordination number of individual metal atoms in the surface. Some theoretical understanding of the electronic properties of surface transition metal atoms at step defect sites is available [18].

We begin by the investigation of the adsorption of oxygen onto two Pt surfaces [19,20]. It is well known from spectroscopic studies that at low temperatures, oxygen adsorbs non-dissociatively on Pt surfaces as peroxo species [21,22]. Upon heating the surface to about 120-200 K, a portion of the molecular oxygen desorbs and a portion dissociates to atomic oxygen. In Fig. 8, one may see a comparison between the oxygen desorption behavior of a planar Pt(111) crystal and a stepped Pt(112) crystal. On Pt(111) a single α-O_2 desorption process is observed near 120 K; the desorption process is abruptly discontinued as the peroxo O_2 species begin to dissociate to produce atomic oxygen as indicated by the sharp almost vertical trailing edge in the O_2 desorption trace [19,20,22]. The atomic oxygen trapped on the surface recombines starting at about 600 K, yielding the β-O_2 desorption process. In contrast to the relatively simple behavior of Pt(111), the stepped Pt(112) crystal liberates three α-O_2 states, and two distinct β-O_2 states, as shown in the top portion of Fig. 8. The three molecular states

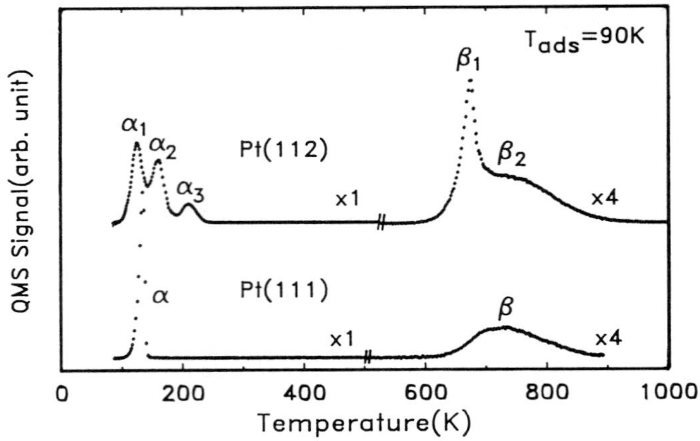

Fig. 8. Temperature programmed desorption of oxygen from Pt(111) and Pt(112) [19]

of oxygen are associated with the three kinds of Pt atoms present in the Pt(112) plane. Pt atoms at the top edge of the steps exhibit Pt-Pt coordination numbers of 7; those in the middle of the terrace exhibit coordination numbers of 9 (identical to a Pt(111) surface atom); the Pt atoms at the base of a step possess Pt-Pt coordination numbers of 10. The α_1-O_2 species desorbs with an activation energy of 7.7 kcal/mole; α_2: 9.8 kcal/mole; α_3: 13.1 kcal/mole [19]. It is believed that α_3-O_2 is adsorbed at the base of the Pt steps on the Pt atoms possessing coordination number = 10 [19]. It is also believed that the majority of the atomic oxygen production occurs from α_3-O_2, and this is responsible for its lower yield of molecular oxygen in the α_3-state compared to α_2-O_2 and α_1-O_2 species [19,20].

In order to investigate the exchange via surface migration between the various α-O_2 states present on Pt(111), an experiment was performed in which the surface was populated to full coverage with $^{18}O_2$ at 90 K [19]. An interrupted desorption experiment was carried out in which both the α_1- and the α_2-O_2 species were desorbed; then upon cooling the crystal to 90 K, the α_1 and α_2 states were repopulated with a second oxygen isotope, $^{16}O_2$. Thermal desorption from the isotopically-dosed layer was carried out while monitoring the two isotopic types of O_2, as shown in Fig. 9. It is seen that no isotopic exchange between the molecular states occurs; α_1-O_2 and α_2-O_2 retain only $^{16}O_2$; α_3-O_2 is only $^{18}O_2$, with no mixed $^{16}O_2$ species. Furthermore, all of the atomic O has been produced from the first-adsorbed oxygen, $^{18}O_2$.

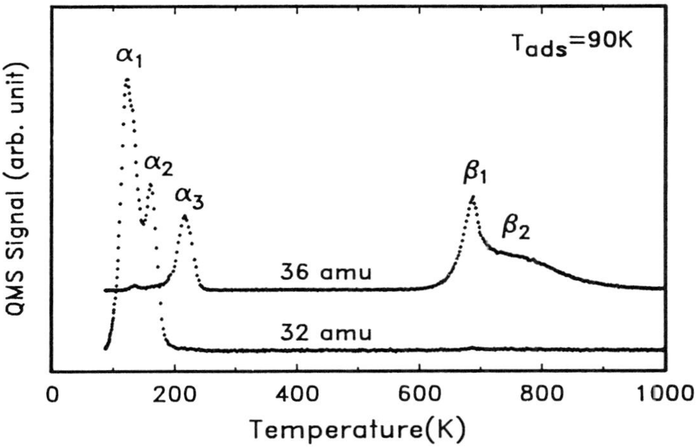

Fig. 9. Isotopic separation of molecular states of oxygen on Pt(112) [19]

The dramatic lack of adsorbate site exchange for the peroxo species on Pt(112) which is observed in the isotopic experiments above suggests that for energetic or entropic reasons, the elementary step involving direct molecular desorption is preferred over steps in which migration of the peroxo species from one site to another occurs. This is a somewhat surprising finding, since in general it is found that the activation energy for surface migration is small compared to the activation energy for desorption. The isotopic labeling experiment is also telling us that the three separate α-O_2 states observed here for Pt(112) probably correspond to the three types of Pt atoms found in the Pt(112) surface- that is, in this case the characteristic desorption behavior of peroxo oxygen can be used to assay the distribution of Pt atomic surface sites having different Pt-Pt coordination numbers. If future studies on other Pt single crystals confirm the results obtained here, it may be possible to employ molecular oxygen desorption for the study of defect site distributions on Pt surfaces of technological importance, i.e. in heterogeneous catalysts and on Pt electrodes.

We have also studied the thermal desorption of CO from Pt(112) compared to Pt(111) [23,24]. In Fig. 10, for Pt(112), it is seen that initially a high temperature CO desorption state is evolved near 500 K. Once this state has been fully populated, a second lower temperature desorption state begins to become evident. The high temperature state corresponds to adsorption on the steps, while the low temperature state corresponds to adsorption on the terraces of the Pt(112) crystal.

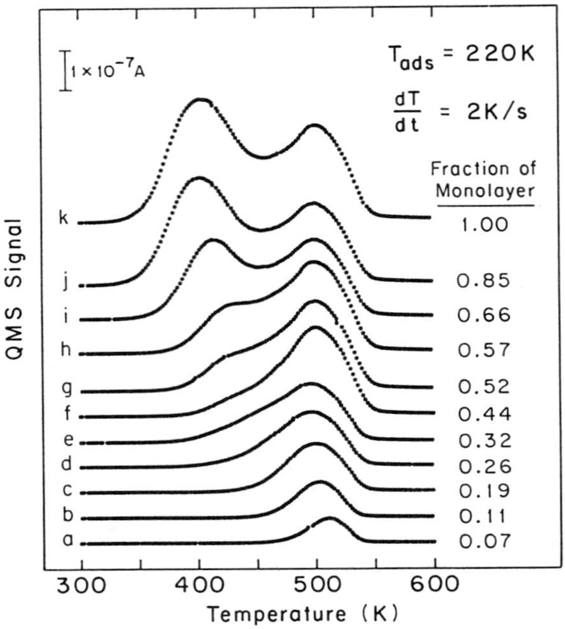

Fig. 10. Temperature programmed desorption for CO from Pt(112) as a function of CO coverage. $^{12}C^{18}O$ was used for these experiments [23]

Careful studies of the kinetics of CO thermal desorption as a function of surface coverage have been made [24], using the leading edge of the temperature programmed desorption traces for kinetic measurements in a method called the TTPD (threshold temperature programmed desorption) method [24,25]. Figure 11 shows a plot of the activation energy for desorption of CO from Pt(112) compared to Pt(111) as a function of CO coverage [24]. It can be seen that the step sites involve an activation energy of about 40 kcal/mole in the limit of zero coverage, whereas the terraces involve CO desorption activation energies which are quite similar to those measured for Pt(111). The difference in activation energy for desorption between CO at the step and CO on a Pt(111) surface in the limit of zero coverage is 7 kcal/mole. These results are similar to earlier measurements made by others [26-28].

We asked the same question about CO on the Pt(112) surface as was asked for oxygen on the same surface. Does migration between adsorption sites occur prior to desorption? Again, an isotopic dosing experiment was carried out in which 1/3 of a monolayer of $^{13}C^{16}O$ was adsorbed onto the step sites, followed by 2/3 of a monolayer of $^{12}C^{18}O$, which

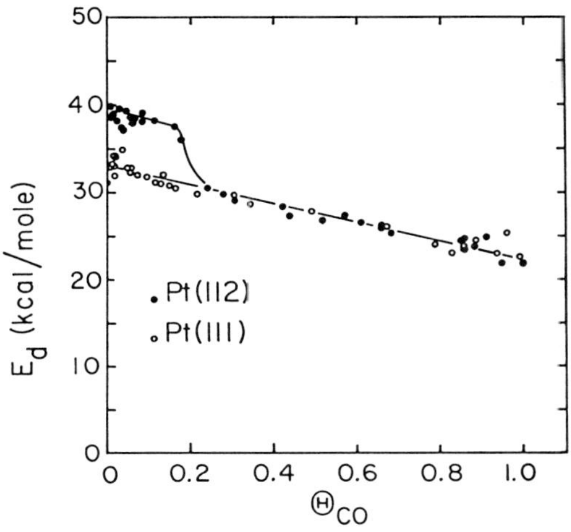

Fig. 11. Activation energy for CO desorption from Pt(111) and Pt(112) [24]. The TTPD method [25] was used in this analysis

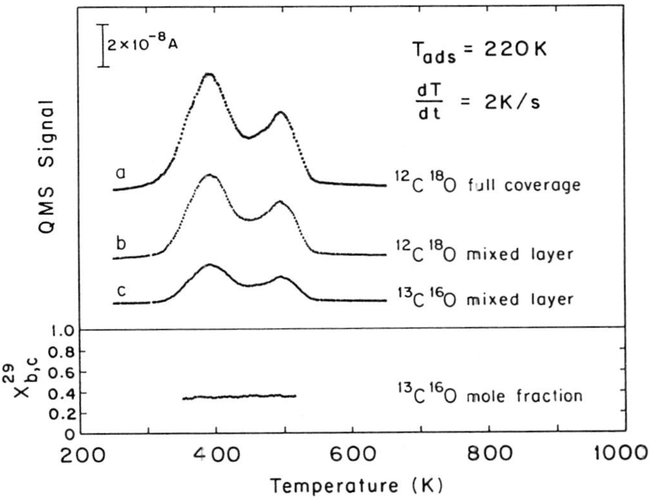

Fig. 12. Isotopic CO desorption from Pt(112) [23]

saturated the terrace sites. Thermal desorption from this isotopic composite layer was carried out using multiplex mass spectrometry with the results shown below in Fig. 12. It can be seen that no memory for the initial adsorption of CO on the step sites is evident; both isotopic CO traces are similar (Fig. 12 b and c). In addition, the mole fraction of $^{13}C^{16}O$ remains constant at about 0.33 over the entire desorption range as shown in the lower portion of Fig. 12. For comparison, a full coverage desorption experiment for pure $^{12}C^{18}O$ is shown in Fig. 12a. The conclusion from this experiment is that complete site exchange occurs between CO species on the stepped Pt surface prior to desorption.

This result is consistent with the measured migration activation energy (7 kcal/mole) for CO on Pt by POELSEMA, VERHEIJ and COMSA [29], in which He scattering methods were employed.

IV. Desorption from the Bulk

Although the majority of the surface phenomena which have been studied to date have involved species whose chemistry occurs in the monolayer regime on top of the surface, there are now a number of reports concerned with the involvement of foreign atoms within the bulk of the metal or semiconductor surface. The adsorption of hydrogen is particularly related to bulk behavior, and a number of studies have been concerned with the properties of hydrogen originating from the bulk. Among the well-known systems are H/Pd [30-39], H/Ru [40-43], H/Cu and H/Ni [44-50]. Two methods are used for these studies. The first involves the use of thermal methods and possibly high pressures of H_2 to effect entry of significant quantities of H into the bulk, and many permeation-type studies of this type have been carried out. The second method involves ion implantation in which significant quantities of hydrogen can be artificially introduced into the solid substrate by non-thermal routes [50].

In the study to be described here, H_2^+ has been introduced into a Ni(111) crystal by implantation, and thermal desorption methods have been used to observe the liberation of the implanted hydrogen. The work builds on the well-studied adsorption of hydrogen on Ni(111) [51,53].

Figure 13 shows the desorption of hydrogen from the bulk mixed with desorption from the surface. Following the saturation chemisorption of hydrogen (see lower trace), two desorption states of H_2 are observed, β_1 and β_2. These desorption states have been well studied [52,53]. Implantation of hydrogen ions with 3.0 keV energy/H_2^+ (= 1.5 keV/H) leads to an increase of the quantity of hydrogen retained by the crystal, and an additional desorption feature below the β_1 desorption temperature is observed. As the total H_2^+ dose is increased, the implanted H_2 desorption feature dominates the desorption spectrum. This method for observing and directly measuring the quantity of implanted hydrogen is in fact an absolute method, since the area of the desorption trace for any level of implantation may be related to the area of the desorption trace for a chemisorbed hydrogen monolayer whose absolute coverage is known [51]. Thus the absolute efficiency of hydrogen ion implantation was measured in this work [50].

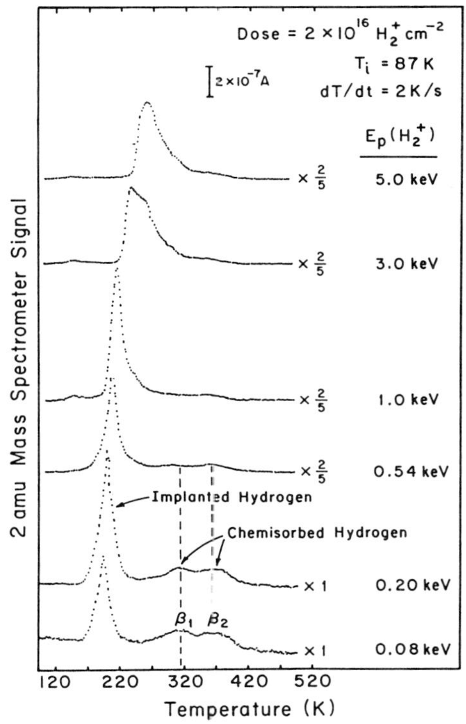

Fig. 13. Temperature programmed desorption of hydrogen implanted in Ni(111) as a function of dose [50]. The desorption of one monolayer of chemisorbed hydrogen is shown in the lower trace and this is a background feature for all implantation experiments as well as a means of absolute calibration of the quantity of implanted hydrogen in the Ni(111) crystal

The implantation of hydrogen into Ni(111) as a function of bombardment energy was studied at constant H_2^+ fluence. At higher energies, Ni vacancy defect sites are produced more efficiently within the bulk of the Ni, and migration of H atoms through the bulk is impeded by the presence of these defects. This effect can be seen in Fig. 14, where integral curves of the total amount of hydrogen in the bulk (and on the surface as adsorbed hydrogen) are shown versus the temperature during temperature programmed desorption. It is easily seen that both the quantity of implanted hydrogen, and the temperature of desorption increase with increasing implantation energy. The shift to higher temperature has been shown to be essentially unrelated to the greater average depth of hydrogen implantation for the higher energies [50].

A very interesting dynamical effect was discovered in the above studies of the desorption of implanted hydrogen from the bulk. In our work, a differentially-pumped quadrupole mass spectrometer is employed. The mass spectrometer is situated behind a small aperture, and can be used as a crude measurement tool for the study of the angular distribution of desorbing species. As shown in the inset to Fig. 15, increasing

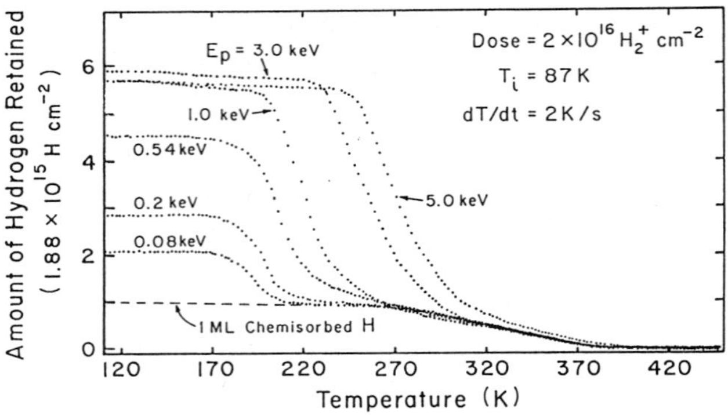

Fig. 14. Amount of hydrogen retained by a Ni(111) crystal after ion implantation as measured by temperature programmed desorption [50]. The production of vacancy defect trap sites is thought to be responsible for the increasing temperature of desorption as the implantation energy is increased

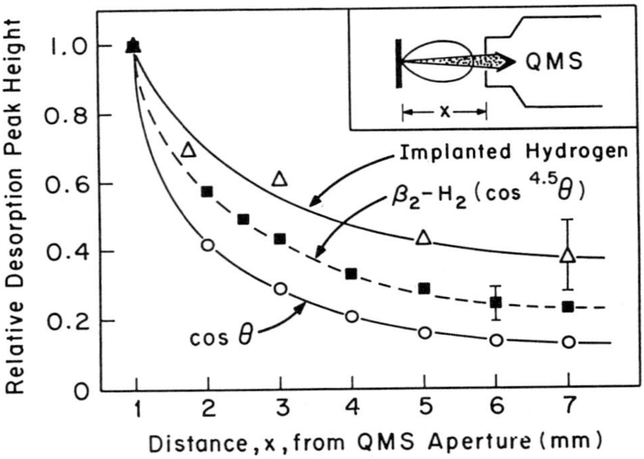

Fig. 15. Desorption intensity versus crystal distance from the quadrupole mass spectrometer aperture [50]

the distance x between the crystal and the aperture will cause a fall off in the intensity of the desorption signal. A desorption process which yields a cosine like angular distribution will be observed to fall off rather rapidly with increasing distance; a more sharply focused angular distribution of desorbing species will yield a lower rate of fall off. We have calibrated the apparatus using desorption processes from Ni(111) which have been accurately measured by STEINRÜCK, et al [54], and calibration curves for a cosine θ and a cosine$^{4.5}$ θ angular distribution are shown in the figure. It is interesting to note that for the implanted hydrogen, an angular distribution which is much sharper than cosine$^{4.5}$ θ is observed [50].

The forward-directed desorption of H_2 from the implanted Ni(111) crystal is consistent with the known energetics of hydrogen dissolution in Ni. As shown in Fig. 16A, the dissolution of hydrogen into Ni is endothermic or in other words the potential energy of a dissolved H atom exceeds that of 1/2 H_2(g) by about 0.17eV. Thus hydrogen migration through the solid will occur at elevated potential energy as shown in the figure. The presence of vacancy traps will increase the activation energy of hydrogen migration as shown in Fig. 16A, and this is observed experimentally as shown above in Fig. 14. When a migrating H atom arrives at the surface of the Ni(111) crystal it will likely also be in a region of elevated potential energy relative to 1/2H_2(g). Recombination with a surface hydrogen atom, which is always present in these experi-

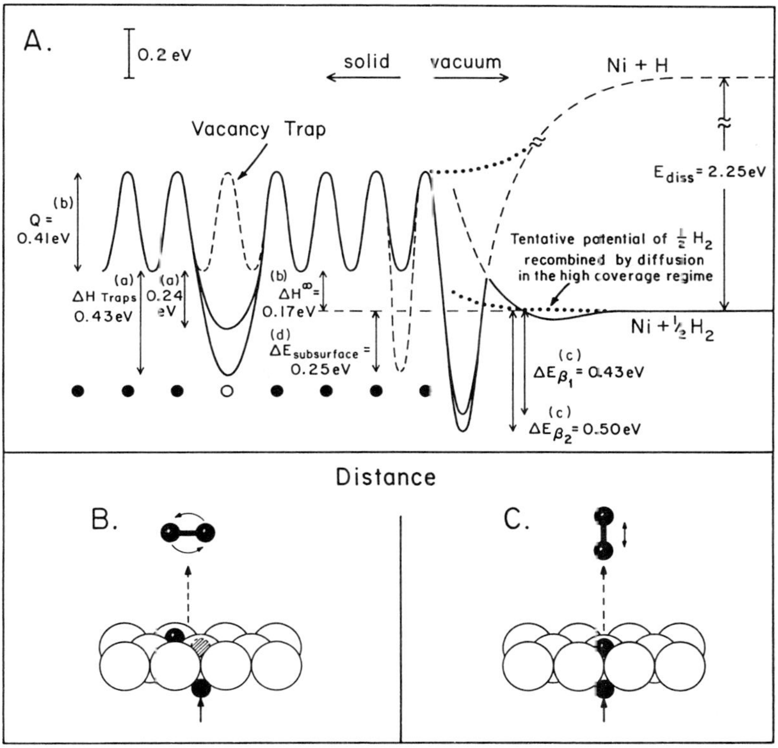

Fig. 16. Schematic potential energy diagram for H in and on a Ni(111) crystal [50]

ments, will produce H_2 which will gain kinetic energy as it moves away from the region of elevated potential energy into the gas phase. Thus the H_2 produced from H atoms in the bulk will be ejected with excess normal kinetic energy, and according to the theory of VAN WILLIGEN [55], an angular distribution which is sharply focused along the normal direction will result. Thus, the experimental results obtained in Fig. 15 are an expected consequence of the energetics of H dissolution in Ni. The experimental results we have obtained are consistent with earlier permeation experiments of COMSA and DAVID [46,48]. In Fig. 16B and 16C, schematic pictures of the recombination of bulk H with surface H are shown, and the possibility of either rotational excitation of desorbing H_2 or vibrational excitation of desorbing H_2 are indicated. We have here then, the indication that the activated breaking of Ni-H bonds in the near surface region may lead to highly non-thermalized desorption processes.

V. Chemical Bond Scission in Adsorbate Species

The investigation of the bond scission processes which occur in adsorbed species forms a major portion of the activity in surface science research which is directed to providing a better understanding of the mechanism of heterogeneous catalysis. A primary goal of this kind of work is to obtain information about the reaction channels which are present in the chemisorbed layer. The question is, which bonds break and in what sequence? Once this type of information is available, more detailed questions about the barrier heights and the detailed dynamics of the elementary surface steps may be probed using a variety of modern methods of surface chemical physics research.

We have devised an experimental procedure for rapidly performing a survey of the reaction channels experienced by a molecule incident on a single crystal surface. Essentially, a single crystal is exposed to a continuous, collimated molecular beam of reactant molecules while the crystal temperature is being raised. A multiplexed mass spectrometer located in a differentially pumped shield is directed to the crystal by means of a small aperture located near the crystal surface. The mass spectrometer samples the reflected reactant molecules as well as all product molecules during the period in which the crystal temperature is being programmed upwards. Thus as new reaction channels become kinetically favorable, the onset of consumption channels of the reactant is observed which is often coordinated with the onset of production channels of product species. The method is called Scanning Kinetic Spectroscopy (SKS) [56], and bears some similarity to temperature programmed reaction spectroscopy (TPRS) pioneered by the MADIX group

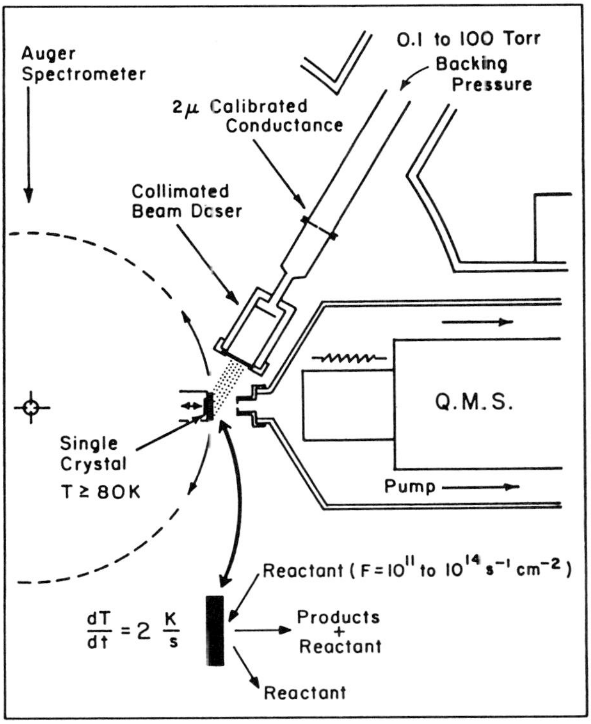

Fig. 17. UHV apparatus for scanning kinetic spectroscopy (SKS) [56]

[57]. A somewhat similar method was reported by GOSCHNICK and GRUNZE [58] in 1981. The apparatus for SKS measurements is shown in Fig. 17.

We begin with a discussion of the interaction of CH_3OH with Ni(111) [56,59]. For convenience in discussion, a scheme of elementary surface processes in the decomposition of CH_3OH is shown in Figure 18. Here it may be seen that CH_3OH adsorbs molecularly at low temperatures, then decomposes via O-H bond scission to produce a surface methoxy species. The methoxy species undergoes C-H bond scission, and the hydrogen atoms captured by the surface drive a recombination reaction in which methanol is reformed and liberated by a process designated β. The selective decomposition of $CH_3O(a)$ to $CH_2O(g)$ is not observed on Ni(111) but does occur on Cu(111) [60]. The final steps of the chain of reactions is the liberation of H_2 and CO. CH_4 production is not observed for Ni(111) [56,59] nor is C-O bond scission observed by isotopic exchange studies involving $^{13}CH_3{}^{16}OH$ plus $^{12}CH_3{}^{18}OH$ molecules [61].

Adsorption → ← Desorption	$CH_3OH(g) \underset{\alpha}{\rightleftarrows} CH_3OH(a)$
CH_3O Formation	$CH_3OH(a) \xrightarrow{\gamma_1} CH_3O(a) + H(a)$
Recombination	$CH_3O(a) + H(a) \xrightarrow{\beta} CH_3OH(g)$
CH_3O Decomposition	$CH_3O(a) \longrightarrow CO(a) + 3H(a)$
Selective Decomposition	$CH_3O(a) \xrightarrow{A} H(a) + CH_2O(g)$
H_2 Desorption	$2H(a) \xrightarrow{B} H_2(g)$
CO Desorption	$CO(a) \xrightarrow{C} CO(g)$

Fig. 18. Elementary steps observed for CH_3OH decomposition on Ni(111) and Cu(111) [56,60]

In Fig. 19 is shown the actual result of an SKS experiment involving CD_3OH interacting with Ni(111). The experiment begins with the establishment of a monolayer of undissociated CD_3OH on Ni(111) at 187 K. In the upper trace, the beam of CD_3OH is turned onto the saturated crystal at 187 K, and the reflected signal quickly reaches a constant level (P_4). Following the establishment of the steady state condition, the crystal temperature program is initiated, and an α-CD_3OH desorption state is observed as the monolayer begins to depopulate. In the α-desorption region, CD_3OH begins to break its O-H bond, but this event cannot be detected in this SKS experiment. At about 280 K the β process is observed. This process is the recombination process for the reaction of methoxy with hydrogen, additional hydrogen being liberated onto the surface by means of C-D bond scission in the CD_3O species on the surface. It is interesting to note that the onset of the β-recombination process coincides almost exactly with the onset of D_2 desorption, i.e., the recombination of CD_3O with D and D with D are controlled by the same process, namely the availability of D from the C-D bond scission of some methoxy species. It should be noted that the unique isotopic methanol product, CD_3OD, occurs in the β-recombination channel, and this species can only originate from C-D bond scission in neighboring methoxy species.

As open sites begin to be produced on the surface at higher temperatures by both hydrogen and CO evolution, the consumption of CD_3OH

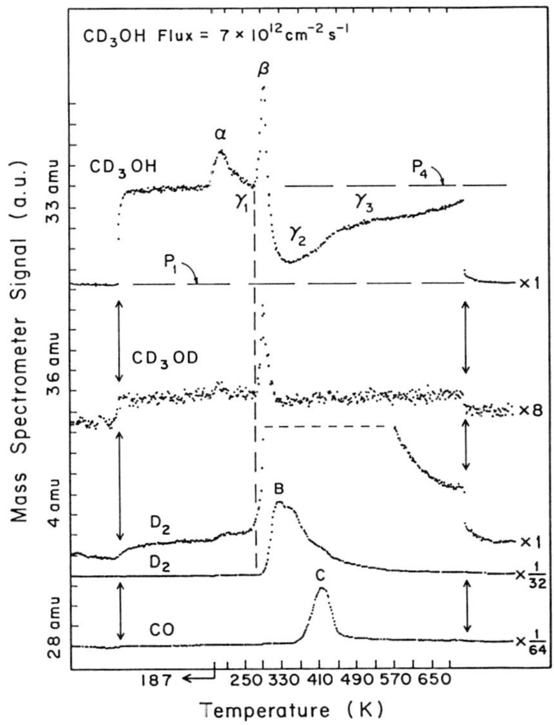

Fig. 19. Scanning kinetic spectrogram: CD_3OH + Ni(111) [56]

from the beam is enhanced, and region γ_2 is reached in the upper SKS trace of Fig. 19. The SKS method has supplied a clear view of several of the elementary processes at work under reactive conditions and has been used in a number of applications for obtaining this class of information in survey experiments. The method provides the surface chemist with a single crystal temperature programmable chemical reactor for survey investigations of interrelated surface reactions.

Another interesting aspect of the SKS method is its ability to understand elementary surface processes by using the deuterium kinetic isotope effect (DKIE) [62]. In Fig. 20 is shown a generalized potential energy diagram for a chemical reaction. The two levels for the reactant schematically represent the zero point energies for a molecule labeled with hydrogen or deuterium. For the transition state, one vibrational coordinate in the reactant becomes a translational coordinate in the transition complex, and the splitting of the zero point energy due to isotopic composition will differ in general from that in the reactant. Thus the enthalpy of activation, $\Delta H^{\ddagger}$, will differ for the two isotopic

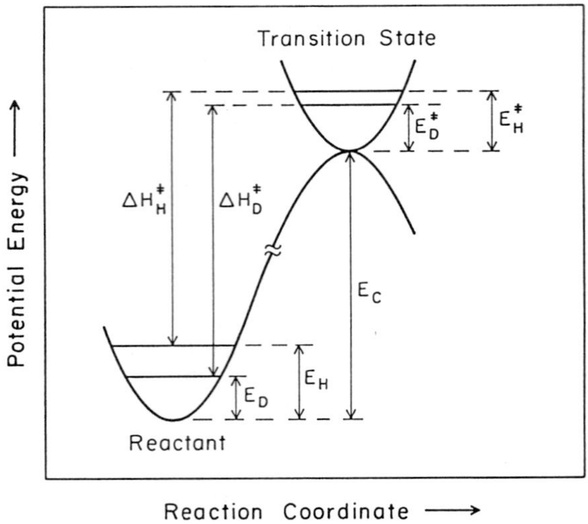

Fig. 20. Deuterium kinetic isotope effect- schematic potential energy diagram [62]

compounds, with the deuterium-labeled compound exhibiting a slightly higher $\Delta H^{\ddagger}$. This means that the deuterium labeled compound will possess a lower rate constant for reaction. More sophisticated treatments of the DKIE can be found in appropriate textbooks [63,64].

The involvement of the DKIE in the elementary steps associated with methanol decomposition on Ni(111) is graphically seen in Fig. 21, where the initial region of the SKS experiment with CD_3OH and CH_3OH are compared. It can be seen that the first evolution of methanol at about 210 K occurs at essentially the same temperature for the two isotopic versions of the methanol molecule. This is the desorption of intact methanol, and its rate does not depend upon the zero point energies of the methyl groups since bond breaking in these groups is not involved in the desorption of the intact molecule. However, the second feature in the SKS experiments, which is the β-feature of Fig. 19, is strongly affected by deuterium labeling in the methyl group. A difference of about 18 K is observed in comparing the two β-processes, with the CD_3OH exhibiting the slower (higher temperature) β-process. This observation indicates that C-D bond breaking must be involved in the rate-determining step for this β-process. The evolution of methanol in the β recombination process is controlled by the availability of surface H(D), coming from the scission of C-H(D) bonds [56]. Thus, the DKIE observed here is entirely expected on the basis of the assignment of the β-process to a recombination process.

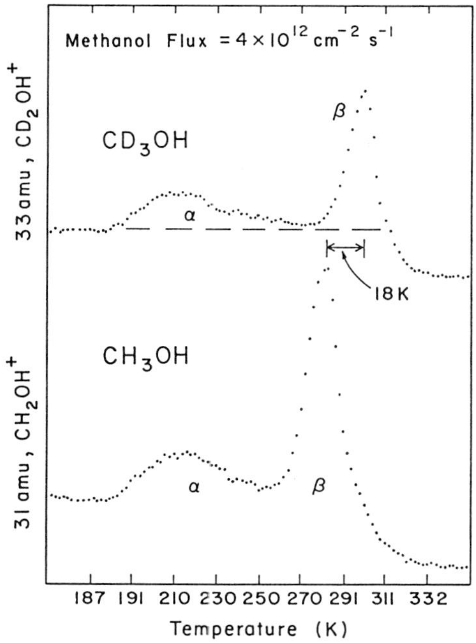

Fig. 21. C-H(D) bond breaking isotope effect in the SKS study of methanol interacting with Ni(111) [56]. The 18 K retardation of the second methanol evolution step (β) is caused by a rate determining step involving C-H(D) bond breaking, which promotes the recombination process

The utility of the DKIE for study of the rate determining step in surface ractions is further illustrated by the example given below. The adsorption and decomposition of ethanol on Ni(111) has been studied recently, and a summary of the sequence of bond breaking steps is given in the schematic diagram of Fig. 22. As shown in the upper part of the

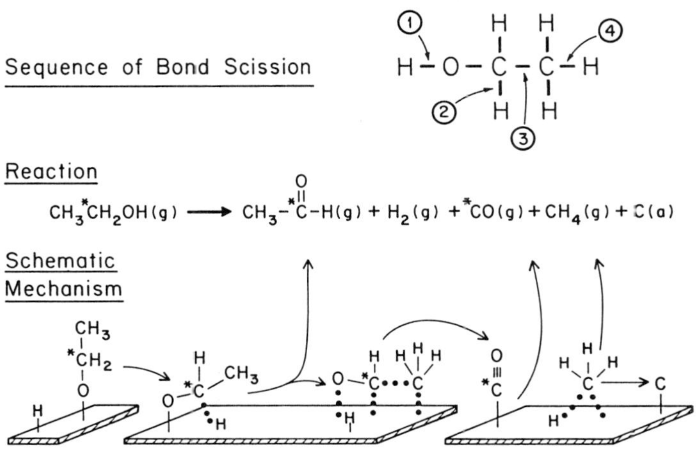

Fig. 22. Ethanol decomposition on Ni(111) [65]

figure, a sequence of 4 bond breaking steps is observed, with the first being the breaking of the O-H bond to form adsorbed ethoxy and adsorbed hydrogen. The sequence of the four steps is shown in the schematic mechanism given in the lower portion of Fig. 22. All of these steps have been observed with the help of deuterium and other isotopic labeling in various places within the molecule [65], but for our purposes here, we illustrate only one application of the DKIE for understanding the kinetic mechanism of reaction.

The initial step in the reaction has been observed at 240 K, through the DKIE, by measuring the rate of ethanol adsorption at 240 K using various deuterium isotopic labeling positions. For these experiments, the growth of the carbon Auger intensity has been measured in precise adsorption experiments from a molecular beam of ethanol of known flux. The solid curve represents the adsorption rate for three different isotopic ethanol molecules containing D in the methyl group, or D in the methylene group, or no D at all. To within the experimental accuracy of the experiments, all of these isotopically labeled molecules behave identically, and their adsorption rates are indistinguishable. However, when D is placed in the hydroxyl group, a significant retardation of the kinetics of adsorption is observed, as shown by the dashed curve. These results show that the rate controlling step in the adsorption of ethanol at 240 K involves the scission of the O-H bonds to produce surface ethoxy species. At lower temperatures, it would be expected that the adsorption of the ethanol would not involve O-H bond scission (ethanol remains intact), and the effect of D labeling in this group would disappear, as has been seen in other work involving a similar situation [66].

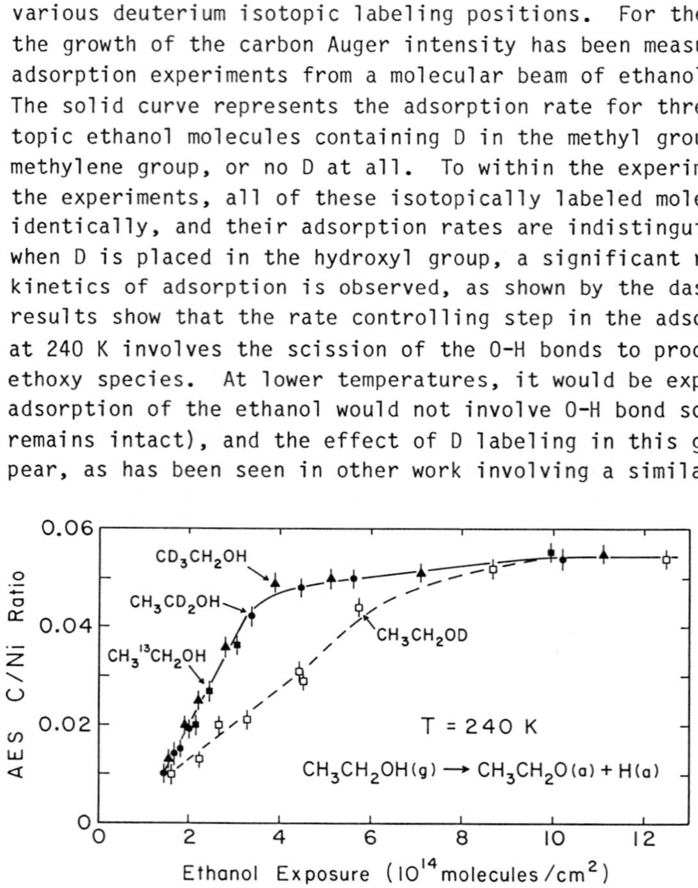

Fig. 23. Deuterium kinetic isotope effect on the rate of ethanol decomposition on Ni(111) [65]. The results show that at 240 K, D labeling in the O-H group is effective in decreasing the rate of adsorption, indicating that O-H bond scission occurs in the adsorption step

Thus, we have illustrated the sequential bond scission steps which occur during chemisorption and dissociation of polyatomic molecules. These processes generally occur in sequences which develop as the temperature of the surface is increased. The selective use of deuterium isotopic labeling, combined with studies of the SKS or thermal desorption spectrum of the reactive molecule can often be used to judge both the sequence of the reaction steps and the temperature at which each elementary step can be activated.

VI. Electronic Excitation Effects in Adsorbed Molecules- The Artificial Production of Cracked Surface Species

It has been shown, through numerous examples above, that the thermally activated breaking of chemical bonds in adsorbed species controls surface chemistry in adsorption, desorption, surface migration, and internal bond scission. From the point of view of control of surface chemistry, we are therefore at the mercy of the natural activation energy barriers for these thermally induced chemical processes.

In this last section, I wish to show how one may circumvent the natural sequence of reactive events produced by thermal means, and how large quantities of "cracked" surface species may be prepared selectively on surfaces. This involves the use of electronic excitation of the adsorbed species on the surface by means of electron bombardment. We have already shown above that electron bombardment will cause the ESD of positive ions which may be used to determine the bond direction in adsorbates through the ESDIAD phenomenon. This ion production channel is generally a minor channel compared to the desorption of neutral and excited neutral species. In addition to the production of ions and the desorption of the intact molecule by ESD processes, we are also beginning to see evidence for the cracking of adsorbates with the retention of the cracked products by the surface.

As an illustration, let us return to the ESDIAD behavior of PF_3 chemisorbed on Ni(111) [10,16]. Here it was found, as shown above in Fig. 2 - 4, that at 85 K, PF_3 molecules exist as azimuthally oriented species with PF bond directions parallel to the principal crystallographic axes of the Ni(111) surface. This produces a particular six-fold F^+ ESDIAD pattern shown, for example, in Fig. 2a and 2c. The same F^+ ESDIAD pattern is shown in Fig. 24a, for a saturated PF_3 overlayer.

It was found that rather heavy electron bombardment of the saturated PF_3 overlayer caused significant changes in the ESDIAD pattern as shown

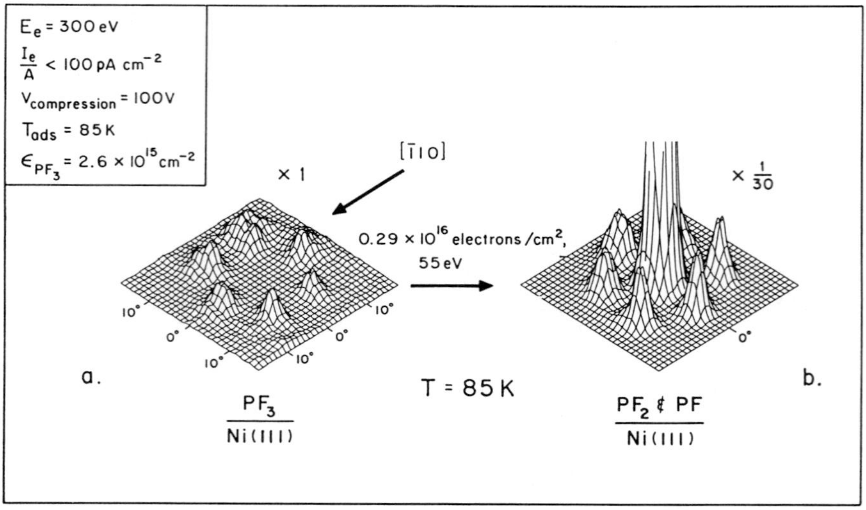

Fig. 24. Electron induced conversion of PF_3 to PF_x species on Ni(111): F^+ ESDIAD [10]

in Fig. 24. The original six-fold F^+ ESDIAD pattern is destroyed and is replaced BY A NEW SIX-FOLD PATTERN, ROTATED 30°, as shown in Fig. 24b. The new six-fold pattern contains an intense central normal beam of F^+ ions. The observations made here suggest that new species are being produced by the act of heavy electron bombardment, and that these new surface species exhibit different P-F bond directions from the original PF_3 species. As will be shown in Fig. 25, the changes observed are consistent with the production of adsorbed PF_2 and PF species from PF_3 by electron impact.

The changes in the F^+ ESDIAD pattern can be readily understood from the diagrams in Fig. 25, where logical bonding sites for PF_3, PF_2 and PF are shown. In order to maintain approximate tetrahedral coordination to phosphorus atoms, PF_2 species will seek two-fold bridging sites, of which three types coexist on Ni(111). Occupancy of the three types of two-fold Ni bridge sites by PF_2 will define six P-F bond directions as shown in the middle section of the figure. THESE SIX DIRECTIONS ARE ROTATED 30° WITH RESPECT TO THE PF_3 DIRECTIONS. In addition, continued disruption of the surface species by electron bombardment will produce PF species, which on the basis of maintenance of approximate tetrahedral bonding about the P atom, will therefore seek three-fold bridge sites, as shown in the lower section of Fig. 25. The PF species will direct F^+ ions normal to the surface in this bonding site, and the central F^+ ESDIAD beam is therefore assigned to the PF species [10]. One should

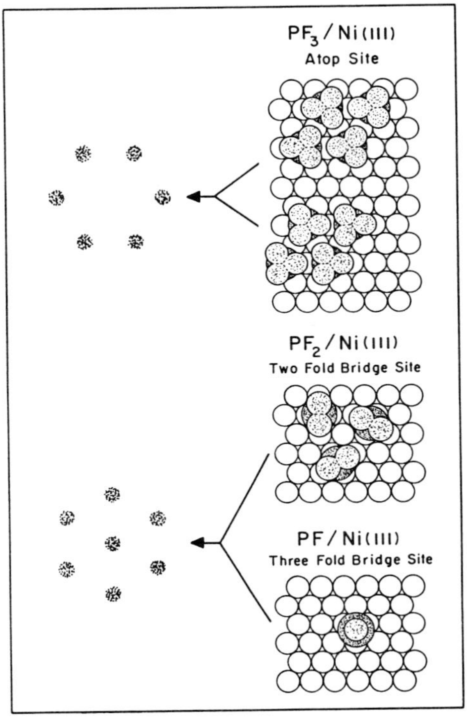

Fig. 25. Summary of PF_3 and PF_x surface structures and resultant F^+ ESDIAD patterns [10]

note that as one proceeds through the series PF_3, PF_2 and PF, the intensity of the F^+ beams from these species increases dramatically. This is probably due to the decrease in the rate of neutralization of the F^+ ions as the polar angle of ion emission decreases from species to species [10]. Control experiments have shown that surface F atoms have a very low F^+ ion yield, and do not interfere with these assignments [10]. The experiments described here have been done using temperature programmed desorption methods to observe the loss of PF_3 species by electron bombardment of the entire front face of the Ni(111) crystal, and observations of the destruction of the PF_3 layer can be made in this manner also. It has been found on the electron bombarded surfaces that recombination processes involving fragment species to produce new high temperature desorption states of PF_3 are present [10]. Thus, with this example, it is seen that the artificial production of cracked molecular products from ESD of a parent molecule can occur with rather high efficiency, and that these products can be captured by the surface in specific bonding configurations.

This phenomenon of the artificial production of cracked surface species which are captured by the surface has been seen previously in a number of cases, including the dissociation of chemisorbed CO [67], to produce C(a) and O(a) species. In our own work, the conversion of NH_3 surface species to NH_2 species has been observed by ESDIAD [68]. A halo H^+ ESDIAD pattern from NH_3 adsorbed on Ni(110) is originally observed [68,69], and is postulated to be due to the rotation of NH_3 about its vertical C_{3v} axis. Conversion to NH_2 occurs upon electron bombardment, producing a two-fold H^+ ESDIAD pattern which is oriented in a plane perpendicular to the rows and channels present on the Ni(110) surface as shown in Fig. 26. Ultimately, extensive electron bombardment destroys all N-H bonds, leaving behind adsorbed atomic nitrogen [68]. ERTL and coworkers [70] have recently studied this ESD process using HREELS, and have also observed the production of an N-H adsorbed surface species.

It may be concluded from these investigations that it is possible to artificially produce rather large concentrations of reactive surface

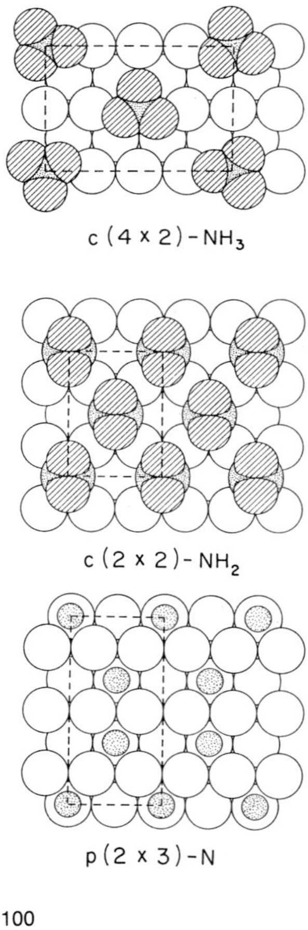

Fig. 26. Sequence of structures observed by the electron induced conversion of chemisorbed NH_3 to NH_2 and N [68]. Both ESDIAD and temperature programmed desorption methods were used

species by the action of ESD. This new method of preparative surface chemistry can have two major consequences:

1. The ability to chemically modify surfaces by the ESD production of cracked molecular species which bond in different ways from the parent molecule (and perhaps in most cases, with stronger bonding). Such methods therefore offer the possibility of making new types of protective overlayers for technological applications.

2. The ability to artifically make large surface concentrations of intermediate species which are normally present at low surface concentrations during catalytic reactions. This would permit the more-or-less leisurely observation of the properties of these species using spectroscopic and kinetic tools for detailed surface science investigations.

VII. Summary

In this lecture I have attempted to give a broad overview of the various types of chemical bond activation processes which have been observed on single crystal surfaces. We have seen a number of interesting phenomena which have general significance to the development of our ability to understand and control chemistry at surfaces. These concepts are listed below:

1. The adsorbate species and their fragments (produced thermally and by electron impact) are similar in structure to ligands known to exist in molecules [71].

2. The "cluster-surface" analogy [72] applies to surface species stability and structure.

3. Both molecular orientations and molecular motions in adsorbates may be effectively studied by the digital ESDIAD method.

4. The metal-metal coordination number is of fundamental significance in determining the desorption and dissociation kinetics of adsorbates on single crystal surfaces.

5. The endothermic absorption of hydrogen in Ni(111) results in a sharply-focused desorption profile for hydrogen exiting from the surface. This is consistent with a surface H atom recombination process in a region of elevated potential energy. These ideas are consistent with the earlier postulates of VAN WILLIGEN and COMSA.

6. The bond scission processes in adsorbates can be unraveled using the deuterium kinetic isotope effect in conjunction with thermal spectroscopies such as scanning kinetic spectroscopy, temperature programmed desorption, and temperature programmed reaction spectroscopy.

7. The artificial production of catalytic intermediates by electron stimulated activation of molecular adsorbates may be a possible route to deeper understanding of these intermediate species through the artificial increase of their surface concentration above that normally obtained via thermal routes.

VIII. Acknowledgments

I would like to thank all of my students and senior guestworkers for participation in the work reported here. I also acknowledge support from AFOSR, NSF Division of Materials Research, and ARO, as well as the Solvay Foundation for support to attend this meeting.

IX. References

1. (a) Desorption Induced by Electronic Transitions, DIET I, eds. N.H. Tolk, M.M. Traum, T.E. Madey, J.C. Tully: Springer Series in Chemical Physics, Vol. 24, (Springer-Verlag, Heidelberg 1983)
(b) Desorption Induced by Electronic Transitions, DIET II, eds. W. Brenig, D. Menzel: (Springer Verlag, Heidelberg 1985).
2. T.E. Madey, J.T. Yates, Jr.: Surface Sci. 63, 203 (1977); T.E. Madey: J. Vac. Sci. and Technol. A4, 257 (1986).
3. M.D. Alvey, M.J. Dresser, J.T. Yates, Jr.: Phys. Rev. Lett. 56, 367 (1986).
4. J.J. Czyzewski, T.E. Madey, J.T. Yates, Jr.: Phys. Rev. Lett. 32, 777 (1974)
5. M.J. Dresser, M.D. Alvey, J.T. Yates, Jr.: Surface Sci. 169, 91 (1986).
6. M.J. Dresser, M.D. Alvey, J.T. Yates, Jr.: J. Vac. Sci. and Technol. A4, 1446 (1986).
7. See R.H. Stulen in reference 1b, pg. 130.
8. T.E. Madey: In Inelastic Particle-Surface Collisions, eds. E. Taglauer, W. Heiland (Springer, Berlin 1981, pg. 80)
9. W.L. Clinton: Surface Sci. 112, L791 (1981).
10. M.D. Alvey, J.T. Yates, Jr.: J. Amer. Chem. Soc., in press.
11. F. Nitschke, G. Ertl, J. Küppers: J. Chem. Phys. 74, 5911 (1981).
12. G. Doyen: Surface Sci. 122, 505 (1982).
13. H. Itoh, G. Ertl: Z. Naturforsch, 37, 346 (1982).

14. This type of bonding is also consistent with the binding mode of PF_3 in molecular complexes. See J.F. Nixon: In Advances in Inorganic and Radiochemistry, eds. H.J. Emeleus, A.G. Sharpe (Academic Press, New York, Vol. 29, pg. 62 1985), and references therein.
15. R.W. Zwanzig: J. Chem. Phys. 87, 4870 (1987).
16. M.D. Alvey; J.T. Yates, Jr.: J. Chem. Phys. 87, 7221 (1988).
17. M. Grunze, H.J. Kreuzer: Kinetics of Interface Reactions, Springer Series in Surface Science, 8 (Springer-Verlag, Berlin 1987).
18. L. Falikov, G.A. Somorjai: Proc. National Academy of Science, 82, 2207 (1985).
19. H.R. Siddiqui, A. Winkler, X. Guo, P. Hagans, J.T. Yates, Jr.: Surface Sci. (in press, 1988).
20. A. Winkler, X. Guo, H.R. Siddiqui, P.L. Hagans, J.T. Yates, Jr.: Surface Sci. (in press, 1988).
21. H. Steininger, S. Lehwald, H. Ibach: Surface Sci. 123, 1 (1982).
22. J.L. Gland, B.A. Sexton, G.B. Fisher: Surface Sci. 95, 587 (1980).
23. H.R. Siddiqui, X. Guo, I. Chorkendorff, J.T. Yates, Jr.: Surface Sci. 191, L813 (1987).
24. H.R. Siddiqui, X. Guo, J.T. Yates, Jr.: Surface Sci., to be published.
25. J.B. Miller, H.R. Siddiqui, S.M. Gates, J.T. Yates, Jr., J.C. Tully, M. Cardillo: J. Chem. Phys. 87, 6725 (1987).
26. R.W. McCabe, L.D. Schmidt: Surface Sci. 66, 101 (1977).
27. D.M. Collins, W.E. Spicer: Surface Sci. 69, 85 (1977).
28. H. Hopster, H. Ibach: Surface Sci. 77, 109 (1978).
29. B. Poelsema, L.K. Verheij, G. Comsa: Phys. Rev. Lett. 49, 1731 (1982).
30. R.J. Behm, V. Penka, M.G. Cattania, K. Christmann, G. Ertl: J. Chem. Phys. 78, 7486 (1983).
31. K.H. Rieder, M. Baumberger, W. Stocker: Phys. Rev. Lett. 51, 1799 (1984).
32. T. Engel, H. Kuipers: Surface Sci. 90, 162 (1979).
33. M.G. Cattania, K. Christmann, G. Ertl: Surface Sci. 126, 382 (1983).
34. W. Eberhardt, F. Greuter, E.W. Plummer: Phys. Rev. Lett. 46, 1085 (1981).
35. W. Eberhardt, S.G. Louie, E.W. Plummer: Phys. Rev. B28, 465 (1983).
36. T.E. Felter, R.H. Stulen: J. Vacuum Sci. and Technol. A3, 1566 (1985).
37. T.E. Felter, S.M. Foiles, M. Daw, R.H. Stulen: Surface Sci. 171, L379 (1986).
38. G.D. Kubiak, R.H. Stulen: J. Vacuum Sci. and Technol. A4, 1427 (1986).
39. G.E. Gdowski, T.E. Felter, R.H. Stulen: Surface Sci. 181, L147 (1987).

40. J.T. Yates, Jr., C.H.F. Peden, J.E. Houston, D.W. Goodman: Surface Sci. 160, 37 (1985).
41. P. Feulner, H. Pfnür, P. Hoffmann, D. Menzel: Surface Sci. 173, L576 (1986).
42. C.H.F. Peden, D.W. Goodman, J.E. Houston, J.T. Yates, Jr.: Surface Sci. 184, L405 (1987).
43. M.Y. Chou, J.R. Chelikowsky: Phys. Rev. Lett. 59, 1737 (1987).
44. G. Comsa, R. David, K. Rendulic: Phys. Rev. Lett. 38, 775 (1977).
45. G. Comsa, R. David: Chem Phys. Lett. 49, 512 (1977).
46. G. Comsa, R. David, B.J. Schumacher: Surface Sci. 85, 45 (1979).
47. G. Comsa, R. David, B.J. Schumacher: Surface Sci. 95, L210 (1980).
48. G. Comsa, R. David: Surface Sci. 117, 77 (1982).
49. G. Comsa, R. David: Surface Sci. Rept. 5, 145 (1985).
50. I. Chorkendorff, J.N. Russell, Jr., J.T. Yates, Jr.: Surface Sci. 182, 375 (1987).
51. K. Christmann, R.J. Behm, G. Ertl, M.A. Van Hove, W.H. Weinberg: J. Chem. Phys. 70, 4168 (1979).
52. J.N. Russell, Jr., S.M. Gates, J. T. Yates, Jr.: J. Chem. Phys. 85, 6792 (1986) (also see references therein).
53. J.N. Russell, Jr., I. Chorkendorff, A-M. Lanzillotto, M.D. Alvey, J. T. Yates, Jr.: J. Chem. Phys. 85, 6186 (1986) (also see references therein).
54. H.P. Steinrück, A. Winkler, K.D. Rendulic: J. Phys. C. 17, L311 (1984).
55. W. Van Willigen: Phys. Lett. A 28, 80 (1968).
56. S.M. Gates, J.N. Russell, Jr., J.T. Yates, Jr.: Surface Sci. 159, 233 (1985); J. Vacuum Sci. Technol. A3, 1472 (1985).
57. R.J. Madix, J. Benziger: Annual Rev. of Physical Chemistry 29, 285 (1978); R.J. Madix: In Chemistry and Physics of Solid Surfaces, ed. by R. Vanselow, Vol. 2 (Chemical Rubber Co., Cleveland, 1979).
58. J. Goschnick, M. Grunze: J. Vacuum Sci. Tech. 18, 561 (1981).
59. S.M. Gates, J.N. Russell, Jr., J.T. Yates, Jr.: J. Catalysis 92, 25 (1985).
60. J.N. Russell, Jr., S.M. Gates, J.T. Yates, Jr.: Surface Sci. 163, 516 (1985).
61. J.N. Russell, Jr., I. Chorkendorff, J.T. Yates, Jr.: Surface Sci. 183, 316 (1987).
62. S.M. Gates, J.N. Russell, Jr., J.T. Yates, Jr.: Surface Sci. 146, 199 (1984).
63. L. Melander, W.H. Saunders: Reaction Rates of Isotopic Molecules (Wiley, New York, 1980).
64. L. Melander: Isotope Effects on Reaction Rates (Ronald, New York, 1960).
65. S.M. Gates, J.N. Russell, Jr., J.T. Yates, Jr.: Surface Sci. 171, 111 (1986).

66. I. Chorkendorff, J.N. Russell, Jr., J.T. Yates, Jr.: J. Chem. Phys. 86, 4692 (1987).
67. W. Erley, H. Wagner: Surface Sci. 74, 333 (1978).
68. C. Klauber, M.D. Alvey, J.T. Yates, Jr.: Surface Sci. 154, 139 (1985); M.J. Dresser, M.D. Alvey, J.T. Yates, Jr.: Surface Sci. 169, 91 (1986).
69. C. Klauber, M.D. Alvey, J.T. Yates, Jr.: Chem. Phys. Lett. 106, 477 (1984).
70. I.C. Bassignana, K. Wagaman, J. Küppers, G. Ertl: Surface Sci. 175, 22 (1986).
71. M.R. Albert, J.T. Yates, Jr.: The Surface Scientist's Guide to Organometallic Chemistry (ACS Books, Washington, DC, 1987).
72. This general concept was first devised by Professor Earl Muetterties. For an historical review, see reference 71.

Discussion

J.G. DASH: In your interpretation of the freezing out of the PF_3 rotation at 85K, you attribute the freezing out to the interaction between single molecules and the solid surface. But since 85K is a very low temperature for lateral interactions between the molecules, it would seem most likely that the PF_3, even though at low coverage, is condensed into a 2D–solid phase. That is, the inhibition of rotation can well be due to PF_3-PF_3 interactions.

J.T. YATES, JR.: We have made measurements in two PF_3 coverage regimes using both LEED and digital ESDIAD. At a coverage of $\sim 0.04 PF_3/Ni$, the onset of hindered rotation is observed as the temperature is raised. Under these conditions evidence for a laterally ordered PF_3 overlayer is absent (no overlayer LEED pattern). Thus the postulate of local ordered islands which involve intermolecular forces causing the hinderance of rotation is unlikely for the strongly adsorbed PF_3 molecules (binding energy to Ni site = $1.2eV$). The effect which you describe does in fact occur at the high coverage limit ($0.25 PF_3/Ni$; $p(2 \times 2)$ overlayer) where the barrier to rotation is increased, preventing rotation in the temperature range studied. We believe this effect may be attributed to $PF_3 \ldots PF_3$ interactions in the ordered overlayer.

D.A. KING: My point relates to your observation that no diffusion of $\alpha_3 - O_2$ on Pt(112) occurs before desorption. Firstly, the effect has been previously observed with adsorbed molecules: $\gamma - N_2$ on W (Ehrlich) and Ar-CO on W (Gomer). Secondly, for the case of O_2 adsorption in the α states on Pt, there is a large bond order change in O_2 on adsorption (double to single bond), involving a large bond energy change. Since the molecule is only weakly adsorbed, most of this energy is converted to the two $O-Pt$ bonds formed per molecule. These must be highly directional, hence diffusion will involve breaking these strong directional bonds, returning the $O_2(a)$ to a doubly bonded species; this energy may be (must be) close to that required for direct desorption. This is easily understood in the framework of a one-dimensional potential energy curve.

J.T. YATES, JR.: Essentially you are saying that the transition complexes for desorption and for diffusion are similar. This idea is not inconsistent with the kinetics of adsorption of O_2 on Pt(112), where, at 90K, $S-O_2 \simeq 1.0$ [work in press] over a wide coverage range, indicative of a mobile molecular precursor.

H. IBACH: Is there a natural line width in ESDIAD or is the width of the ion distribution entirely due to the vibrational motion of the atoms?

J.T.YATES, JR.: Of course zero point motion in the adsorbates represent the natural limiting line width of an ESDIAD ion beam. In addition, the influence of the ion trajectory and the ion velocity on the efficiency of its neutralization will also give a contribution to the linewidth and line shape in ESDIAD.

G. ERTL: (a) The activation barrier for rotation of PF_3 adsorbed on Ni(111) might (at least in part) be caused by the "backbonding" of metallic d-orbitals with the empty levels localized along the P^+F bonds and the change of hybridization upon rotation. Steric hindrance appears too improbable since with chemisorbed systems, *repulsive* (e.g., dipole-dipole) interactions will frequently overcompensate the van der Waals attraction and thus prevent condensation at low coverage.

(b) With hydrogen diffusing through Pd and subsequently recombining at the surface (which is analogous to the presented example of hydrogen implantation in Ni) it was indeed found by Zare as well as by Zacharias that the desorbing H_2 molecules were vibrationally excited.

(c) This question refers to the three states of adsorbed O_2 on Pt(211): Were these states equally propulated already at low coverages (indicating non-thermal equilibrium) or were they sequentially filled? Could any effect of the heating rate on their relative populations be observed?

J.T. YATES, JR.: (a) It is of course not possible for us to determine the origin of the forces orienting PF_3 azimuthally on Ni(111) at low PF_3 coverages. Either specific directional effects in back bonding into the P–centered 7e orbital or secondary F...Ni bonds are possible explanations for the azimuthal ordering of PF_3. At high PF_3 coverages ($\theta_{PF_3} = 0.25 PF_3/Ni$; $p(2 \times 2)$ overlayer) the interaction between neighboring PF_3 molecules, whether attractive or repulsive, could be responsible for an increase in the rotational barrier to values in excess of the $80 cm^{-1}$ barrier measured at low PF_3 coverages.

(b) While we have no direct evidence for either vibrationally or rotationally excited H_2 from the desorption of H_2 originating from hydrogen implanted in the bulk Ni(111)–oriented surface, we believe that these kinds of internal excitations are to be expected on general dynamical grounds.

(c) We believe, on the basis of comparison of the sticking coefficient of O_2 on Pt(111) and Pt(112), that adsorption on Pt(112) occurs initially from the mobile–precursor O_2 molecule in its collision with the step, due possibly to effective momentum loss there from the translating precursor species. Indeed, the $\alpha_3 - O_2$ state is observed to fill first in temperature–programmed desorption experiments, followed by $\alpha_2 - O_2$ and the $\alpha_1 - O_2$. We have not yet studied the relative yields of the three $\alpha - O_2$ states using different heating rates.

J.W. GADZUK: ESDIAD, as usually practiced, measures the angular distributions of desorbing *ions*. The ions are a very small fraction of the total desorbates (neutrals are much more copious). In your observations of *neutral* desorbate angular distributions, have you seen any effects that are in conflict with the picture obtained from straightforward ESDIAD measurements?

J.T. YATES, JR.: Our detection of neutral desorbing molecules is limited to electronically excited neutrals which can excite our channel plate detector. To date, CO on Ni(110) has been shown to produce a metastable excited neutral, probably $a^3\pi$ – CO(6eV). This species exhibits a very sharp angular distribution not unlike that observed for th positive ions in ESDIAD. It was found that the $a^3\pi$ – CO from Ni(110) was ejected in a normal direction, consistent with the normal Ni–CO bond being ruptured. There was some evidence that the $CO^+ \rightarrow a^3\pi - CO$ transitions were involved, and if so, then the trajectory of the $a^3\pi - CO$ should resemble that of CO^- except for anisotropic effects in neutralization due to the corrugation of the Ni(110) surface. These effects were observed giving a slight anisotropy in $a^3\pi$ emission. [Reference: M.D. Alvey, M.J. Dresser, and J.T. Yates, Jr., *Phys. Rev. Lett.* **56**, 367 (1986).]

D.R. HAMANN: For the PF_3 hindered rotor, how important are the quantum corrections? Would the classical Boltzmann distribution have given equally good agreement with the data?

J.T. YATES, JR.: Comparison of the quantum calculation to a 2 level Boltzmann calculation having an $80cm^{-1}$ energy separation indicates that the classical calculation underestimates the F^+ valley (rotor) to peak (oriented molecule) ratio by 34% at 100K and 5% at 300K. This would suggest that for the quantum system, penetration of the bound oscillator wavefunctions from the peak regions to the valley regions is significant. Looked at another way, the error in the estimate of barrier height at 100K is about $30cm^{-1}$ if a two–level Boltzmann calculation is used.

R. F. WILLIS: Concerning the 6–peak pattern for chemisorbed PF_3: my point is that quantum mechanical tunneling of the H atoms in a double-well molecular structure would produce similar 6 fold symmetry and/or a ring image.

J.T. YATES, JR.: The NH_3 molecule, chemisorbed on several transition metal single crystal surfaces, produces an H^+ ring pattern in ESDIAD. At 29K on Ni(110) the pattern does not freeze out into a sectored pattern. Thus we are not able to determine in this case whether there is a low barrier to rotation for NH_3 or whether tunneling is responsible for the ESDIAD–H^+ ring. Studies with ND_3 to check for the possibility of tunneling have not been done. [Reference: K. Klauber, M.D. Alvey and J.T. Yates, Jr., *Chemical Physics Letters* **106**, 477 (1984); M.J. Dresser, M.D. Alvey, and J.T. Yates, Jr., *Surface Science* **169**, 91 (1986).]

J.E. DEMUTH: (1) Have you considered the possibility of different molecular O_2 species on your stepped Pt surface as found for example for O_2 on Pd(111) by Imbihl *et al.* in this same temperature regime?

(2) Although we interpret many of our results in terms of models of bonding or reaction paths on idealized crystal surfaces, can we rule out the possibility of active sites (e.g., random defects, steps, etc.) on these properties? For both low coverage experiments or at high pressure– high temperature conditions, such sites may dominate what we see. Could you comment on this issue?

J.T. YATES, JR.: (1) Yes, the observation of desorption kinetic rates with $E_d = 7 - 13$kcal/mole suggests that the O_2 species differ chemically; probably they could be discriminated on the basis of their vibrational spectrum, which we have not measured.

(2) Our experiments on macroscopic stepped crystals having different controlled step densities are an attempt to answer this question systematically. However, all such experiments ordinarily involve a background of random defects which must also be considered in detailed interpretation.

Collision Induced Dissociation and Desorption: CH_4 on Ni(111)

J.D. Beckerle, A.D. Johnson, Q.Y. Yang, and S.T. Ceyer

Department of Chemistry, Massachusetts Institute of Technology, Cambridge, MA 02139, USA

We have recently shown that there is a barrier to the dissociative chemisorption of CH_4 on Ni(111) and that the translational energy associated with motion normal to the surface (from here on referred to as the normal component) and the vibrational energy of the incident CH_4 molecule are effective in surmounting it [1]. The role of translational and vibrational energy in promoting dissociation is interpreted in terms of a deformation model where the barrier arises largely from the energy required to deform the methane molecule sufficiently to allow a strong attractive interaction between the carbon and the Ni surface atoms.

As corroboration of this model, we have shown that the dissociative chemisorption of CH_4, molecularly adsorbed on a Ni(111) surface at 46 K, is induced by the impact of incident Ar atoms. The impact of the Ar atom pounds the molecularly adsorbed methane molecule into the configuration for the transition state that leads to dissociation. This observation represents the discovery of a new mechanism for dissociative chemisorption: collision induced dissociation of adsorbates [2].

These experiments are performed in an apparatus previously described in detail [3,4]. The Ni(111) crystal is maintained at 46 K in an ambient atmosphere of 4×10^{-7} or 2×10^{-6} torr of methane. Under these conditions, about 0.05 or 0.33 of a monolayer of molecularly adsorbed methane is maintained, respectively. A monoenergetic Ar atom beam, produced by high pressure expansion of a 3% mixture of Ar in He from a nozzle at a temperature variable between 700-1100 K, impinges on the methane covered surface at normal incidence. The ambient CH_4 is then pumped away and the crystal is heated to 74 K to remove any molecularly adsorbed CH_4. Raising the crystal temperature to 74 K results only in desorption of molecular methane; no thermally induced dissociation of methane occurs. The derivative carbon Auger signal, which is proportional to the amount of dissociation, is then recorded. The procedure is repeated on a clean surface for other incident angles up to 60° from the normal angle. When the crystal temperature is maintained at 74 K, which is above the desorption temperature of molecular methane, no carbon deposition is observed. This eliminates the possible contributions to the carbon signal from the dissociation of energetic gas phase methane molecules [1] produced by collisions with high energy Ar atoms.

A dissociation rate per unit area is calculated from the absolute carbon coverage derived from the carbon Auger signal. From the dissociation rate, a cross section for dissociation can be extracted since the methane coverage (0.33 of a monolayer of Ni atoms) and the Ar atom flux are known. For a total Ar atom kinetic energy of 53 kcal/mole, the cross section for collision induced dissociation is observed to increase exponentially as the normal component of the energy of the incident Ar is increased to 35 kcal/mole. For higher values of the normal kinetic energy, the dissociation cross section

Springer Series in Surface Sciences Vol. 14
Solvay Conference on Surface Science
Editor: F.W. de Wette

increases less rapidly. Experiments carried out as a function of the incident angle at other total kinetic energies reveal that the cross section for dissociation does not strictly depend on the normal component of the Ar atom kinetic energy. In other words, normal kinetic energy scaling, of the kind observed for translational activation, breaks down. It has been shown, however, that the breakdown of normal energy scaling in the case of collision induced activation is expected and is the result of the range of impact parameters of the Ar-CH_4 collisions that contribute to the dissociation cross section [5].

This was accomplished in the following manner. The amount of energy transferred from the argon atom to the methane molecule in the normal direction was calculated from hard sphere, classical mechanics as a function of the impact parameter and incident angle. Since the methane molecule is then accelerated into the surface with its newly acquired energy, its probability for dissociation should be the same as in the case of translational activation. Therefore, the dissociation probability measured after translational activation is written as a function of the amount of energy transferred to the methane and integrated over all impact parameters to yield a calculated value for the collision-induced cross section for dissociation. These calculated cross sections agree very well with the experimentally determined cross sections. Therefore, translational activation of methane and collision induced activation of methane are completely consistent with each other. They are simply different methods for providing deformation energy to the methane molecule but the mechanism for dissociation is the same.

The product of the collision-induced dissociative chemisorption event is identified by high resolution electron energy loss spectroscopy. The vibrational spectrum of a monolayer of methane at 46 K before Ar atom bombardment exhibits frequencies which are unperturbed from the gas phase values within the resolution of this technique (±20 cm^{-1}). The loss observed at 1305 cm^{-1} is assigned to the ν_4 mode, the loss at 1550 cm^{-1} to the ν_2 mode and the losses at 2895 cm^{-1} and 3015 cm^{-1} to the ν_1 and ν_3 modes, respectively. The vibrational spectrum after exposure of the methane monolayer at 46 K to a beam of Ar atoms is typical of that assigned previously to an adsorbed methyl radical [1]. Therefore, the same product is observed for methane dissociation activated by translational and vibrational excitation or by impact with a gas phase species.

If the energy of the inert gas atom incident on the adsorbed methyl radicals is increased above 50 kcal/mol, collision induced dissociation of the adsorbed methyl radical occurs. The vibrational spectrum of the adsorbed methyl radical gradually disappears as the bombardment proceeds and a new spectrum emerges. The impact of the inert gas atom forces the hydrogens closer to the Ni atoms of the surface so that strong Ni-H bonds are formed, thereby cleaving the C-H bond. The product of the dissociation of CH_3 is tentatively assigned to a mixture of CH and CH_2 species.

In competition with collision induced dissociation, we have observed another process, the collision induced desorption of molecular methane [6]. It is the simultaneous occurrence of this process, which is roughly an order of magnitude more probable than dissociation, that necessitates the presence of a constant background pressure of CH_4 during the dissociation measurements. The continuous flux of methane incident on the surface allows the dissociation probability measurements to be made at constant coverage.

The rate of collision induced desorption was measured on a monolayer of methane adsorbed on the surface of 40 K. A surface temperature below that used in the dissociation experiment is necessary to increase the residence time of the adsorbed methane. The methane monolayer is then exposed to the Ar beam while the methane partial pressure is monitored with the mass spectrometer. The partial pressure decreases as the methane monolayer is desorbed. Since the desorption rate is at least an order of magnitude higher than that for dissociation, depletion of the methane overlayer due to dissociation can be neglected. At low coverage, where an incident Ar atom interacts with a single methane molecule, the desorption rate decays with a first order dependence on coverage. The experiment is then repeated with different kinetic energies of the Ar atom and different incident angles.

From the absolute desorption rate, a collision induced desorption cross section is calculated. Cross sections for collision induced desorption at low coverage are about an order of magnitude smaller than hard sphere, gas kinetic cross sections. The small magnitude of the desorption cross section suggests that direct collisions of the incident Ar atom with the adsorbed methane molecule are predominantly responsible for the desorption event. Large impact parameter surface-mediated interactions, such as desorption due to hot spots or a drumhead effect [7], are likely not important.

The desorption cross section at low coverage is observed to approximately double as the incident angle is increased from 0° to 70°. This effect is partially due to the increase in the collision cross section of an adsorbed molecule as the incident angle increases. However, the magnitude of the increase in the desorption rate is dependent on and is largest for high total energies of the incident Ar atom. This observation indicates that the dynamics of the collision process also plays a role and is the result of two competing effects. As the incident angle is increased, the normal component of the kinetic energy decreases but this is the component that is responsible for desorption. So as the incident angle increases, the desorption cross section should decrease because there is less energy in the normal direction to push the molecule away from the surface. However, as the incident angle increases the extent of trapping decreases. Trapping is the result of a secondary collision of the methane with the Ar atom because the Ar atom moves slowly compared to the motion of the methane into and back from the surface. At large incident angles, the Ar atom has sufficient kinetic energy in the parallel direction to move aside, thereby clearing a space for the methane molecule to desorb. The complicated dependences of the desorption cross section on energy and incident angle are the result of these competing factors.

Perhaps more important than the physics behind these processes is the observation of these processes and what the knowledge of their existence means for understanding the complex environment of high pressure, heterogeneous catalysis. Collision induced chemistry and desorption are likely playing important roles in the high pressure catalytic reaction. This is because in a high pressure reaction, a catalyst is covered with adsorbate and the adsorbate covered catalyst is continually bombarded by gas phase molecules. With the observation of these processes, no mechanism for a high pressure reaction can now be considered complete without an assessment of the role of collision induced chemistry and desorption as potential major steps. These observations are now cause for reexamination of the mechanisms of catalytic reactions in which inert gas effects on the rates of these reactions have been noted [8]. Collision induced chemistry and desorption are plausible explanations why the surface chemistry at high pressures is often different from the surface chemistry at low pressures.

This work is supported by the National Science Foundation (CHE-8508734) and the Synthetic Fuels Center of the Energy Lab at MIT.

References

1. M.B. Lee, Q.Y. Yang and S.T. Ceyer, J. Chem. Phys. 87, 2724 (1987); M.B. Lee, Q.Y. Yang, S.L. Tang and S.T. Ceyer, J. Chem. Phys. 85, 1693 (1986); S.T. Ceyer, J.D. Beckerle, M.B. Lee, S.L. Tang, Q.Y. Yang and M.A. Hines, J. Vac. Sci. Tech. A5, 501 (1987)
2. J.D. Beckerle, Q.Y. Yang, A.D. Johnson and S.T. Ceyer, J. Chem. Phys. 86, 7236 (1987)
3. S.L. Tang, M.B. Lee, Q.Y. Yang, J.D. Beckerle and S.T. Ceyer, J. Chem. Phys. 84, 1876 (1986); S.L. Tang, J.D. Beckerle, M.B. Lee and S.T. Ceyer, J. Chem. Phys. 84, 6488 (1986)
4. S.L. Tang, Ph.D. Thesis, Massachusetts Institute of Technology, 1985. M.B. Lee, Ph.D. Thesis, Massachusetts Institute of Technology, 1987.
5. J.D. Beckerle, A.D. Johnson and S.T. Ceyer, in preparation
6. Y. Zeiri, J.J. Low and W.A. Goddard, III, J. Chem. Phys. 84, 2408 (1986)
7. E. Vilallonga and H. Rabitz, J. Chem. Phys. 85, 2300 (1986)
8. A.E. Silva, R.R. Hudgins, and P.L. Silveston, J. Catal. 49, 376 (1977) and references therein.

Discussion

D.A. KING: What evidence do you have to eliminate the notion that the collision-induced processes are influenced by surface phonon excitation (local hot-spots)?

S.T. CEYER: Several pieces of evidence eliminate a hot spot or surface mediated process.

(1) The experimental energy and incident angle dependence of the desorption cross section agrees very well with that calculated from a classical trajectory simulation of the direct methane–argon collision treating methane and argon as hard spheres.

(2) If hot spots were important, then the energy transferred from the Ar to the nickel lattice would be important. This would be largest at normal incidence where the desorption cross section should be largest. However, the experimental observation is that the cross section is largest at glancing incidence.

(3) If surface mediated processes were important, you expect the magnitude of the cross section to be larger than a gas phase collision cross section. However, the experimental observation is that the cross sections are at least an order of magnitude below the gas phase cross section. Since we know that the desorption probability is close to one, this implies that only small impact parameters or in other words, direct collisions of Ar with methane are important in collision induced desorption.

R. F. WILLIS: You attribute the change in linearity of your "kinetic curve" specifically to a deformation mechanism. Would you see the same effect simply by 'exciting' the adsorbed molecule with any radiation source? My point is: Can you distinguish between *momentum* and *energetics* here, and is it momentum which is important or the cross-section?

S.T. CEYER: The barrier to decomposition is interpreted as the energy required to deform the molecule. You would see the same effect if you excite the fourth overtone of the ν_4 mode of physisorbed methane.

The change in the linearity of the dissociation cross section vs $E_\perp$ of the Ar atom and the breakdown of normal energy scaling in the Ar kinetic energy is due to the range of impact parameters that contribute to the dissociation cross section. By the excellent comparison of the experimental dissociation cross section with that calculated from integration of the dissociation probability from the translational activation measurements over impact parameters, we have shown that it is the energy transferred to the methane that is important in activating methane. This was verified by experiments where we used Ne and Kr as the collision partner. The picture is the following: the Ar collides with the methane and transfers energy according to the kinematics of the collision. The Ar atom then moves away. The methane-nickel collision is then identical to the case of translational activation. The methane with its newly acquired energy in the normal direction is then accelerated into the surface, deforms and dissociates.

Associated Effect in Adsorbed Layer

K. Tamaru

Science University of Tokyo, Kagurazaka, Shinjuku-ku, Tokyo, Japan

It is generally accepted that the rate of adsorption of gases on metals depends upon the pressure of the gas and its coverage, whereas that of desorption depends upon the coverage only. This is the reason why the desorption rate is, in many cases, measured by means of the temperature programmed desorption technique under vacuum.

The rate of desorption and that of adsorption of CO on various transition metals, such as Pd, Rh, and Ni, were measured while adsorption was taking place by means of a technique of switching to a labelled species without changing the ambient pressure. The rate of desorption (v_-) and that of adsorption (v_+) were found to follow the following equations:

$$v_- = k_d(1 + kP^\alpha)\theta \tag{1}$$

$$v_+ = k_a P(1 + K\theta/(1-\theta)) \ , \tag{2}$$

where θ and P are coverage and pressure of CO, respectively, and k_d, k_a, K and α are constants [1]. In Fig. 1, v_-/θ is plotted against pressure.

It has been known in some cases that the exchange reaction between ambient gas molecules and adsorbed species takes place more rapidly than the desorption rate [2]. In the case of hydrogen (or deuterium) irreversibly adsorbed on chromia at liquid nitrogen temperature, for instance, the adsorbed species may be molecularly desorbed only when its isotope gas molecules come into contact as Burwell and Stec reported [3]. In the case of CO chemisorbed on transition metal surfaces such as ruthenium, it may be replaced readily with the ambient labeled CO, although the mere evacuation can hardly remove all the CO chemisorbed at reaction temperature such as 400-500K [4].

The kP term in (1) is seemingly explained in terms of "associated effect" or "adsorption-assisted desorption" in which the desorption is accelerated by the interaction with the adsorbing precursor state species. This new kinetics of desorption shows clearly that the dynamic behavior of adsorbed species or reaction intermediate should always be examined in the presence of adsorbing gas, not under the conditions far from the adsorbing conditions.

Springer Series in Surface Sciences Vol. 14
Solvay Conference on Surface Science
Editor: F.W. de Wette

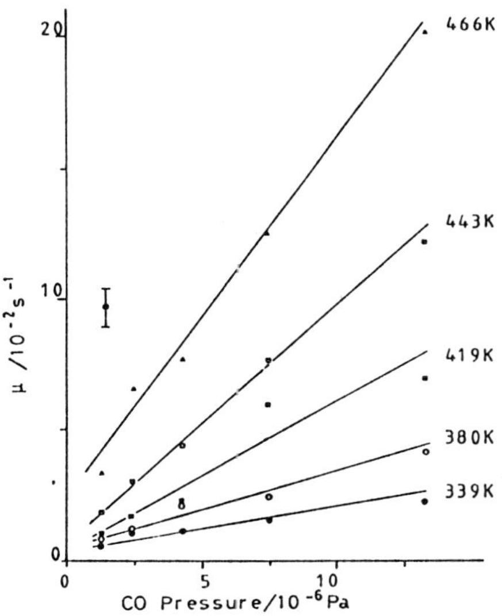

Fig.1 ABSOLUTE DESORPTION RATE CONSTANT AS A FUNCTION OF PRESSURE & TEMPERATURE: CO/Pd

More important is the effect of "adsorption-assisted decomposition" which is sometimes observed in catalytic systems. In the case of methyl alcohol decomposition, for example, on chromia it was concluded by means of transient infrared spectroscopic technique that the first step is the dissociative adsorption of alcohol to form methoxide and proton (on top of surface oxygen ion to form OH) and the methoxide then converts into a formate ion which slowly decomposes to form the reaction products. A very interesting phenomenon is the kinetic behavior of the formate ion on the catalyst surface. The formate ion, formed while the reaction is taking place, is very stable *in vacuo*. However, in the presence of methanol, the formate ion behaves as if it is a reaction intermediate at the same coverage: that is, when $^{12}CD_3OD$ is introduced over $D^{13}COO(a)$, the formate ion is replaced by ^{12}C-formate ion to form ^{13}CO in the reaction products [5]. Such behavior is shown in Fig. 2. In this connection it is to be noted here that the infrared spectrum of the formate ion exhibits a certain shift in the absence and presence of methanol. It is suggested that the $C-H$ bond in the formate ion on the catalyst surface is weakened by the presence of methanol, which makes the formate ion more reactive.

If a bond inside the adsorbed species is weakened by adsorption of methanol or reactant, it may be called "adsorption-assisted decomposition", whereas, if the bond between the adsorbed species and catalyst surface is weakened, "adsorption-assisted desorption" takes place.

In the case of formic acid decomposition on transition metal surfaces, it is sometimes reported that the decomposition takes place via formate ion

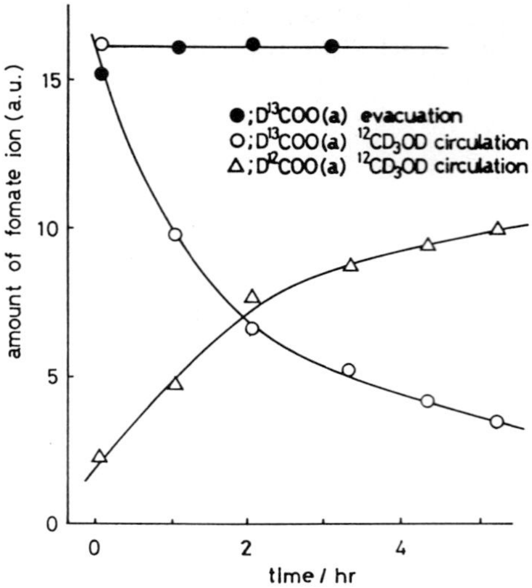

Fig.2 Decomposition of Formate Species at 533K (i) under vacuum and (ii) under metanol atmosphere

chemisorbed on the catalyst surface and the stability of the formate ions (or the heats of formation of formates) is directly associated with the catalytic activity of the metals: If the formate ion is too unstable the rate of its formation is rate determining and becomes slow, whereas if it is too stable, again its decomposition becomes slow. In this manner, a volcano shape appears when the catalytic activity of various metals is plotted against the heat of formation of formates [6]. However, when Block studied the rate of formic acid and its labeled compounds, he observed the following order of the decomposition rate over a silver catalyst [7]:

$$HCOOH > HCOOD > DCOOH > DCOOD \quad .$$

We have also measured those rates over a Ni catalyst and obtained the following relative rates: 1 : 0.74 : 0.59 : 0.20 [8]. If the decomposition of formate ion is rate determining, the rate of the decomposition should be classified into two groups: $HCOO(a)$ and $DCOO(a)$, which is not the case.

When the mechanism of the decomposition of formic acid is studied by measuring the rate of decomposition of surface formate species in the presence and absence of formic acid in the ambient gas, the rate of decomposition of surface formate was remarkably accelerated by the presence of formic acid in the ambient gas, as given in Fig. 3. In this manner it was demonstrated that the reaction takes place through the interaction between the formic acid molecule and the surface formate ion [8]. Since heterogeneous catalysis takes place only in the presence of reactants, in most cases, on a two-dimensional catalyst surface, the behavior of each of the reactants, reaction intermediates, reaction byproducts and reaction products

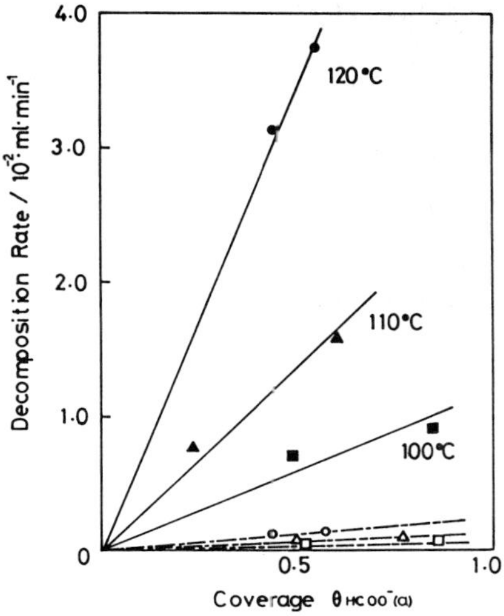

Fig.3 The decomposition rate as a function of θ and T over Ni/SiO_2 catalyst. —— under 11 torr formic acid ; - - - under vacuum; (●, ○) T=120°C, (▲, △) 110°C, (■, □) 100°C.

which are being adsorbed on the working catalyst surface should be examined under the reaction conditions, not under the conditions far from the reaction conditions.

References

1. T. Yamada, T. Onishi and K. Tamaru, *Surf. Sci.* **133**, 533 (1983). T. Yamada and K. Tamaru, *Surf. Sci.* **138**, L155 (1984); **139**, 463 (1984); **146**, 341 (1984); *Z. Physik. Chem. (N.F.)* **144**, 195 (1985).
2. See, for example, K. Killer, A.C. Zettlemoyer and H. Leidheiser, Jr., *J. Chem. Phys.* **52**, 589 (1970); A. Bossi et al., *J. Catal.* **65**, 16 (1980).
3. R.L. Burwell, Jr., and K.S. Stec, *J. Colloid Interf. Sci.* **58**, 54 (1977).
4. N.W. Cant and A.T. Bell, *J. Catal.* **73**, 257 (1982).
5. K. Yamashita, S. Naito and K. Tamaru, *J. Catal.* **94**, 353 (1985).
6. W.M.H. Sachtler and J. Fahrenhort, *Actes du 2me Congr. Internl. Catal. 1960*, Paris (1961) 182.
7. J. Block, *Z. Elektrochem.* **63**, 182 (1959).
8. K. Takahashi, E. Miyamoto, K. Shoji, and K. Tamaru, Annual Meeting of Chemical Society of Japan, Fukuoka, October 1987.

Discussion

J.T. YATES, JR.: Your experiments in which CO displaces chemisorbed hydrogen from polycrystalline Ni(100) may be related to earlier studies of this process in which CO induces the production of weakly–bound hydrogen states (Σ states). [Reference: D.W. Goodman, J.T. Yates, Jr., and T.E. Madey, *Surf. Sci.* **93**, L135 (1980).] This phenomenon is not observed on other Ni crystal planes which have been investigated by J.M. White and his group.

D.A. KING: Have you conducted experiments with rare gases such as xenon to examine the hypothesis suggested by the data of Ceyer (presented at this conference), that desorption is collision-induced?

K. TAMARU: Yes, we did, but could not observe any collision–induced effect in the case of CO desorption.

S.T. CEYER: If the CO pressure above the Ni catalyst is 10^{-7} torr, then I do not think that collision–induced desorption is a viable explanation of your observations. We have looked at the collision–induced desorption of CO on Ni(111) by Ar and have found that at least 60–70 kcal/mole in Ar translational energy are necessary for significant desorption rates. I believe that the flux of CO molecules with energies sufficient to induce collisionally the desorption of CO are too low at 10^{-7} torr to explain your effects. In your experiment, it is more likely that there is a local opening in the moving CO adsorbed layer in which a gas phase CO molecule can adsorb. This decreases the CO binding energy of the neighboring CO molecule due to adsorbate–adsorbate repulsion, thereby allowing the CO to desorb. In the formic acid decomposition carried out at 11 torr, did you look at the effects of inert gases on the decomposition rates?

K. TAMARU: In the case of formic acid decomposition the effect of inert gases on the decomposition rates has not been studied.

J.T. YATES, JR.: The displacement of one isotopic form of adsorbed CO by another has been investigated on Ni(100) [Reference: J.T. Yates, Jr. and D.W. Goodman, *J. Chem. Phys.* **73**, 5371 (1980)]; at ~300K by Klier [Reference: K. Klier, A.C. Zettlemeyer, and H. Leidheiser, Jr., *J. Chem. Phys.* **53**, 589 (1970)].

At about 200K, CO(g) displaces chemisorbed CO with an efficiency of ~1 : 10 according to our measurements. The production of transient local regions of high coverage, leading to decreased CO binding energy was postulated.

S.T.CEYER: What were the pressures that were involved for the formic acid decomposition?

K. TAMARU: In that case it is rather high. It is around 11 torr.

R. GOMER: I worry about decomposition of CO or mere surface reconstruction at the relatively high temperatures and pressures at which you observe pressure–enhanced CO desorption rates on transition metals.

K. TAMARU: We have studied not only chemisorption, but also the decomposition of CO on transition metals for which we observed the pressure enhanced decomposition rates. Actually, on some of these metals such as Pd, the CO decomposition did not take place under the experimental conditions we studied, whereas it took place on Ru at higher temperatures, especially on step sites, as we have reported in *Surface Science.*

M. BOUDART: If adsorption assisted desorption (AAD) is not due to energy transfer ("hot spot") or collision induced desorption, can it be explained by adsorbate-adsorbate interactions? If so, will AAD take place at low surface coverage θ? What is the lowest value of θ for which you have actually observed AAD?

K. TAMARU: The rate of desorption of CO chemisorbed was estimated under various pressures of CO and was more rapid under higher pressures of CO at the *same coverage* of chemisorbed CO. The AAD could be observed at any coverage of the chemisorbed CO approximately in proportion to the pressure of the ambient CO gas and its coverage. Since molecular beam experiments demonstrate the scattering towards random direction, we assumed that AAD takes place through the interaction between the CO molecules in its precursor state and the chemisorbed state to assist the desorption of the latter.

D.A. KING: If you have also eliminated adsorbate-adsorbate interactions, are you suggesting that the effect is due to pressure dependence of the population in the very short-lived precursor states?

K. TAMARU: Yes, I am.

W. KOHN: You have suggested that precursor states of the ambient CO may play an essential role. You showed, in support of this idea, that the angular distribution of the reflected CO is very broad. However, this is consistent not only with trapping in a precursor state, but also with several other mechanisms. Do you agree?

K. TAMARU: Yes, I do, in particular, in the case of inelastic collision.

The Surface Chemical Bond of Adsorbed Organic Monolayers and Adsorbate-Induced Restructuring of Surfaces

G.A. Somorjai and S.T. Oyama *

Department of Chemistry, Materials and Chemical Sciences Division,
Lawrence Berkeley Laboratory, University of California,
Berkeley, CA 94720, USA
*Present address: Department of Chemical Engineering,
Clarkson University, Potsdam, New York, 13676, USA

1. THE STRUCTURE AND BONDING OF ORGANIC MOLECULES ON SURFACES

Recent compilations of structural and spectroscopic data attest to the growing number of well-characterized adsorbates on single crystal surfaces [1,2]. LEED (low energy electron diffraction) surface crystallography studies often combined with HREELS (high resolution electron energy loss spectroscopy) have made possible the determination of the precise structure of these adsorbates. The bond lengths, bond angles, symmetry of the adsorption site and number of molecules in the surface unit cell can be obtained readily.

One structural observation that arises from these studies is that there is a very close relationship between hydrocarbon species adsorbed on surfaces and organic ligands attached to multinuclear metal clusters [3]. In many cases structural studies show almost exact correspondence between atomic coordination and arrangement for both systems (Figs. 1 & 2). The implications reach far beyond the similarity, or perhaps the near identity, of the surface chemical bond with the chemical bonding in the organometallic clusters. If the description of bonding in clusters can be applied to surfaces, then molecular reaction concepts like acidity/basicity, nucleophilicity, oxidative addition and ligand displacement can also be applied to surface reactions.

Benzene usually adsorbs with its ring parallel to the metal surface. Its C-C bond length expands and exhibits alternating distortion imposed by the bonding to the metal substrate at a given site (3-fold or bridge). Similar distortion can be found for a benzene molecule in a tri-ruthenium complex [4]. Thus, the similarity of the surface chemical and organometallic cluster bonds appear to hold for aromatic adsorbates as well.

These organic adsorbates are very stable and long-lived on most metal surfaces because of their strong chemisorption bonds. As a result they may not be important as catalytic reaction intermediates, unless the reactions take place at high temperatures. However, the adsorbates may play important secondary roles in catalytic mechanisms that involve hydrogen storage and transfer, blocking of undesirable sites, or the creation of weak bonding sites for rapid turnover or desorption [5].

Springer Series in Surface Sciences Vol. 14
Solvay Conference on Surface Science
Editor: F.W. de Wette

Structure			Vibrational Frequencies (cm^{-1})		
Surfaces	Proposed Surface Geometry	Cluster Analogue	Characteristic Dipole-Active Modes	Surface	Cluster
Pd(111) Pd(110)	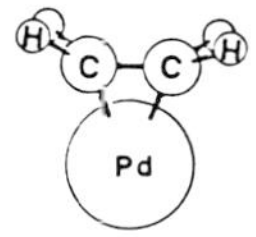	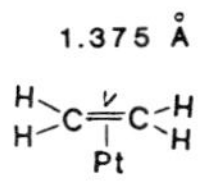	$\nu_s(CH_2)$ $\nu(CC)$ + CH_2 scissor	3000 1520	3031 1515
Rh(111) Pt(111) Pd(111) Rh(100)	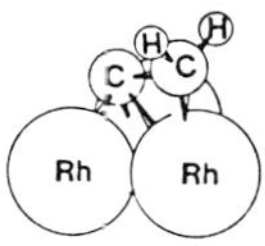	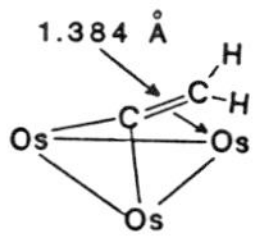	$\nu_s(CH_2)$ $\nu(CC)$ + CH_2 scis + CH_2 wag	2945 1450	2986 1470
gas phase	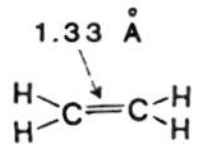		$\nu_s(CH_2)$ $\nu(CC)$	3026 1623	

XBL 8610-4091

Figure 1. The metal surface/cluster analogy for coordinated ethylene (C_2H_4) and vinylidene (CCH_2). (Ref. 3).

Structure			Vibrational Frequencies (cm^{-1})		
Surfaces	Proposed Surface Geometry	Cluster Analogue	Characteristic Dipole-Active Modes	Surface	Cluster
Rh(111) Pt(111) Pd(111) Ru(001) Rh(100)	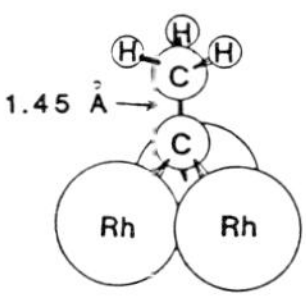	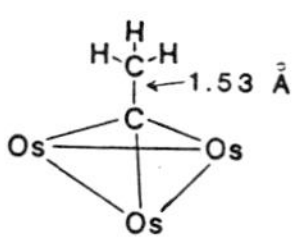	$\nu_s(MC)$ $\nu(CC)$ $\delta_s(CH_3)$ $\nu_s(CH_3)$	435 1121 1337 2880	401 1163 1356 2888
Rh(111) Ru(001) W(110)	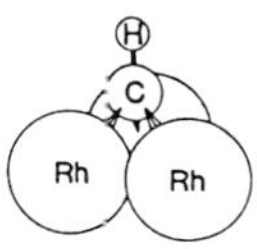	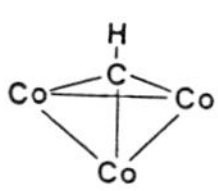	$\nu_s(MC)$ $\nu(CH)$	~740 29[illegible]0	715 3041
gas phase	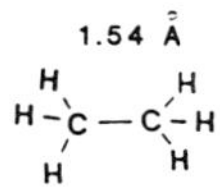		$\nu(CC)$ $\delta_s(CH_3)$ $\nu_s(CH_3)$	995 1379 2954	

Figure 2. The metal surface/cluster analogy for ethylidyne (CCH_3) and methylidyne (CH). (Ref. 3).

2. ADSORBATE-INDUCED RESTRUCTURING OF SURFACES

2.1 Surface Thermodynamic Driving Forces

The reconstruction of a clean surface, that is, the formation of a surface structure that is different from the projection of the bulk unit cell unto the surface, is a well documented phenomenon on metals, semiconductors and insulators. It is generally acknowledged that the driving force for reconstruction is the lowering of the surface free energy, γ. In addition it has been shown [6] that for reconstruction to occur, the variation or anisotropy of γ for the crystallographic orientations involved must be large enough to overcome the interfacial free energy, ΔF_{inter}, between the reconstructed surface layer and the bulk .

The chemisorption of adsorbates can cause reconstruction in two ways. First, by forming stronger chemical bonds with the surface atoms than the atoms have with each other. Second, by forming stronger chemical bonds on one crystal face than on another crystal face of different structure and, thus, increasing the anisotropy of γ for the faces. Examples of these two types of adsorbate-induced reconstruction are presented below.

An example of the first type is oxygen chemisorption on the Fe(110) face (Fig. 1). Upon adsorption of oxygen the surface unit cell of the clean (110) surface changes by rotation of iron atoms around a central atom to form a pseudo hexagonal structure similar to the hexagonal FeO(111) structure [7,8]. This happens because the iron-oxygen bond is stronger than the iron-iron bond. Oxygen adsorption on W or on any other surface where the newly formed metal adsorbate bond is stronger than the bonds between the substrate atoms leads to this type of adsorbate-induced reconstruction. Indeed, often this type of reconstruction is the first step toward bulk compound formation. Changes in surface structure such as the above probably occur within the time scale of adsorption, since the restructuring involves only local rearrangements of bonding.

Often adsorption induces the formation of chemical bonds that are strong but do not replace or locally modify bonding between substrate atoms. Chemisorption of this type can also induce restructuring if the bonding depends strongly on the structure of the surface. For example, the chemisorption of sulfur on the Ni(100) surface produces a stronger adsorbate-substrate bond than the adsorption on the Ni(111) face. On the (100) face sulfur is located in 4-fold sites, while on the (111) face sulfur is found in 3-fold sites, the highest coordination sites available. It appears that the bonding of sulfur to the surface is stronger when it is surrounded by four metal atoms instead of three. Because of this sulfur chemisorption lowers γ more for the Ni(100) face than for the Ni(111) face and this results in anisotropy. On the clean metal surfaces the denser packing of metal atoms lowers γ more for the Ni(111) face than for the Ni(100) face [9]. As a result, in the presence of chemisorbed sulfur the Ni(111) surface reconstructs to a metal surface with (100) orientation and with an ordered sulfur structure. Reconstructions of this type may involve diffusion of atoms and thus can require more time than the first type of reconstruction.

Reconstruction of surfaces driven by this type of relative change in surface free energies induced by chemisorption is very common and is probably responsible for the massive restructuring observed in chemical reaction studies. Several LEED studies [10] monitored the chemisorption-induced restructuring of a large number of Pt crystal faces upon oxygen adsorption and in the presence of carbon on the metal surfaces. The high Miller index stepped and kinked surfaces showed entirely different surface structures in the presence of these adsorbates. Steps of single atom height on the clean Pt surface transformed to steps of several atoms in height. Many surfaces faceted and transformed into other crystal faces of different atomic structures. The nature of the restructuring was also very different when carbon was the adsorbate as opposed to oxygen. It is not certain what aspect of the bonding characteristics of the two adsorbates caused the differences (electron acceptor-donor properties, sizes, valencies), however, the different results obtained exemplify the complexity and diversity of the adsorbate-induced restructuring process.

2.2 The Time Dependence of Adsorbate-Induced Reconstruction

Adsorbate-induced reconstruction is involved in many dynamic surface phenomena. Its role in adsorption-desorption has been cited extensively in the literature [11].

It is convenient to divide adsorbate-induced restructuring processes into three classes according to the time period in which they occur: 1) those that occur during the adsorption process in characteristic times of 10^{-12} - 10^{-4} s [12]; 2) those that occur within the time scale of surface reactions (10^{-3} - 10^{3} s); and 3) those that occur on a much longer time scale (10^{3} - 10^{6} s). It is not possible to provide sharp borderlines between these subgroups since the time of reconstruction depends exponentially on temperature. It also depends on the pressure through its influence on the the surface concentration of adsorbates. Moreover, the kinetics of restructuring will be affected by the size of the particles involved. These may range from metal clusters of 10 atoms to single crystals of macroscopic dimensions. It is clear that, where mass transport is involved, small clusters will be disrupted more easily than larger crystals. Some examples of these classes of restructuring were given earlier; further examples are described below.

2.2.1 Adsorbate-induced reconstruction in short time scales

Aside from being reconstructed, clean surfaces may be relaxed, that is, atoms on the surface may have interlayer spacings that are different from those in the bulk. When atoms or molecules adsorb, relaxation is immediatedly relieved. Strong adsorption causes reconstruction in the time scale of adsorption. For example, the adsorption of hydrogen induces the reconstruction of W or Pd crystal surfaces, while its desorption reverses the changes in structure [11]. Similarly, the adsorption of CO reconstructs the Pt or Ir 5x20 (100) surface to the 1x1 (100) surface (Fig. 3); its desorption again reverses the process [10].

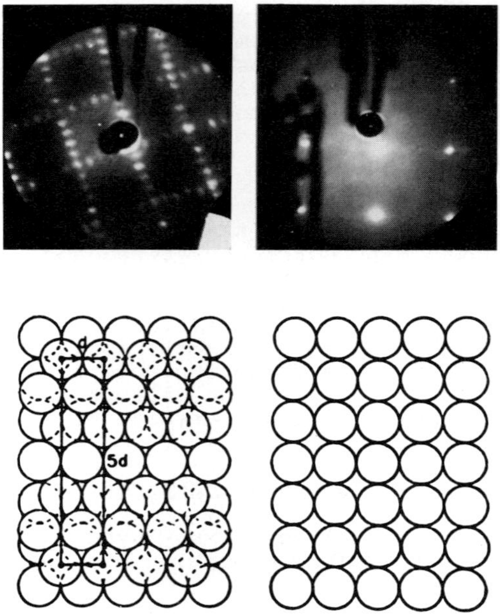

Figure 3. Structures of the 5x20 reconstructed and 1x1 unreconstructed surfaces of Pt(100). The 5x20 structure is similar to the 5x1 structure actually shown

Surface reconstruction during adsorption may cause an apparent structure insensitivity of catalytic reactions. For example, when CO and H_2 adsorb on nickel in the course of methanation at 773 K, drastic changes occur in the surface structure with the development of facets and terraces [13]. This may account for the reported structure insensitivity of the reaction on the (100) and (111) planes [14]. The classification of catalytic reactions as structure sensitive or insensitive must take into consideration the surface structural changes that occur upon adsorption and reaction.

2.2.2 Adsorbate-induced reconstruction on an intermediate time scale

There are several examples of surface structural transformations that are intimately coupled with chemical reactions. One of these is the oxidation of CO to CO_2 on the platinum (100) face at low pressure [15,16]. CO removes the 5x20 reconstruction of the bare surface and forms a 1x1 metal surface structure. Oxygen chemisorbs at defect sites on this surface and reacts rapidly with adsorbed CO. This removes CO from the surface and leaves patches of bare 1x1 metal which reconstruct to the 5x20 face by an activated process. Oxygen has a very low sticking probability on this surface and CO oxidation rates are small. The surface oscillates between the two structural forms and this gives rise to the observed temperature and reaction rate fluctuations (Fig. 4) Another reaction on Pt(100) that exhibits oscillatory behavior involving surface reconstruction is that between NO and CO [17]. However, not all oscillatory reactions are driven by the same adsorbate-induced restructuring

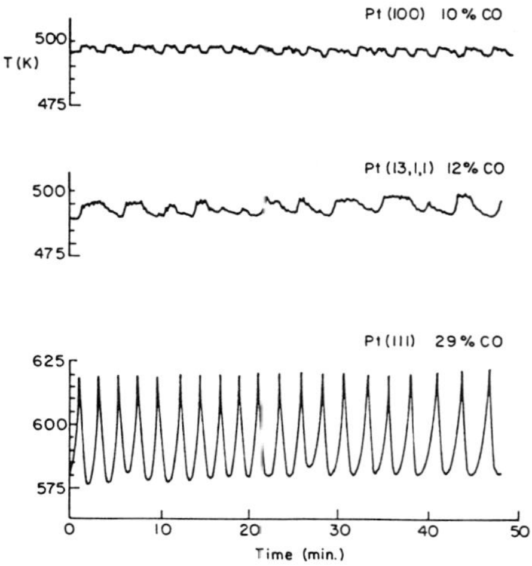

Figure 4. Typical oscillating temperature variations obtained on platinum crystals. (Ref. 16)

mechanism. Oscillations on Pt(111), which does not reconstruct, have been attributed to a cyclic oxidation and reduction of the surface [16].

2.2.3 Adsorbate-induced reconstruction that is slow compared to surface reaction times

Frequently, adsorbate-induced reconstruction involves massive atom transport processes that produce new crystal faces. Thus, when the adsorbate is removed the restructuring is controlled by surface diffusion and occurs on a much longer time scale than the desorption or the turnover time of the catalytic surface reaction. This slower restructuring time is readily observed in changes of particle shape or surface structure during chemical reactions or during catalyst poisoning or regeneration. The case of rhodium provides a suitable example [18]. When rhodium particles are heated in hydrogen at elevated temperatures electron microscopy shows that they sinter and grow (Fig. 5). Upon reaction with oxygen the chemisorption causes a drastic modification of shape as the particles flatten. Upon heating at lower temperature in hydrogen the flattened platinum particles break up into small clusters and form metal particles of higher dispersion. Finally, upon reheating in hydrogen at high temperatures they sinter again to form bigger clusters. This sequence can be repeated many times. It mimics changes in catalyst particle size and dispersion during catalytic reaction, sintering and regeneration.

Long term surface restructuring processes are usually identified by the occurrence of long induction periods before steady-state catalytic rates are reached. The induction periods depend on pressure, temperature and particle size of catalysts. This is perhaps the most difficult and complex regime in the adsorbate-induced reconstruction area. It includes such processes as the build-up of stable hydrocarbon overlayers on metal surfaces, and the formation of carbides, nitrides or oxides by

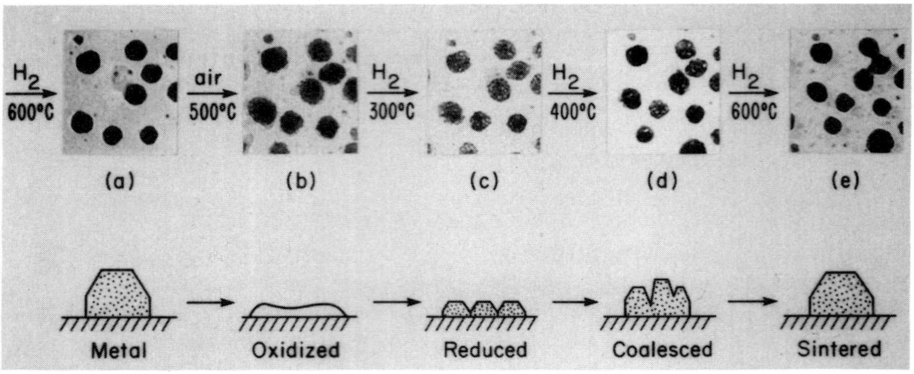

Figure 5. Changes in rhodium particle morphology with various treatments. (Ref. 18)

slow solid state reactions. Nevertheless, it is also one of the most important areas for developing an understanding of ways to fabricate stable catalysts and surface reaction systems.

SUMMARY

Surface structures do not remain unchanged during adsorption or surface chemical reactions. Surface structural changes range from small displacements in atomic positions due to relaxation to the formation of new surface unit cells with massive atom movement due to adsorbate-induced reconstruction. The extent of these changes depends on the adsorbate-surface bond strengths and the anisotropy in surface free energy of different crystal faces. Different time scales associated with adsorption, reaction and slow diffusional changes are identified.

ACKNOWLEDGEMENTS

This work was supported by the Director, Office of Energy Research, Office of Basic Energy Sciences, Materials Sciences Division, U.S. Department of Energy under Contract Number DE-AC03-76SF00098.

REFERENCES

1. P. A. Thiery, J. Elec. Spec. Rel. Phenom. 39, 273(1986)
2. J. Darville, J. Elec. Spec. Rel. Phenom. 39, 311(1986)
3. B. E. Bent, PhD Thesis, University of California, Berkeley, November, 1986.
4. M. P. Gomez-Sal, B. F. G. Johnson, T. Lewis, P. R. Raithby and A. H. Wright, J. Chem. Soc. Chem. Commun., 1682(1985)
5. S. M. Davis, F. Zaera and G. A. Somorjai, J. Catal. 77, 439(1982); S. M. Davis, F. Zaera and G. A. Somorjai, J. Catal. 85, 206(1984); F. Zaera and G. A. Somorjai, J. Am. Chem. Soc. 106, 2288(1984)

6. H. P. Bonzel and K. Dückers, in "Chemistry and Physics of Solid Surfaces", Vol. 7, (Springer, Berlin 1988); H. P. Bonzel, N. Freyer and E. Preuss, Phys. Rev. Let. 57, 1024(1986)
7. W. Erley and H. Ibach, Solid State Commun. 37, 937(1981)
8. M. Langell and G. A. Somorjai, J. Vac. Sci. Tech. 21, 858(1982); L. Papagno, L. S. Caputi, G. Chiarello and P. Delogu, Surf. Sci. 175, L767(1986)
9. J. F. Nicholas, Aust. J. Phys. 21, 21(1968)
10. D. W. Blakely and G. A. Somorjai, Surf. Sci. 65, 419(1977); M. A. Van Hove, R. J. Koestner, P. C. Stair, J. P. Bibérian, L. L. Kesmodel, I. Bartos and G. A. Somorjai, Surf. Sci. 103, 189(1981); M. A. Van Hove, R. J. Koestner, P. C. Stair, J. P. Bibérian, L. L. Kesmodel, I. Bartos and G. A. Somorjai, Surf. Sci. 103, 218(1981); G. Comsa, G. Mechtersheimer and B. Poelsema, Surf. Sci. 119, 159(1982); G. Comsa, G. Mechtersheimer and B. Poelsema, Surf. Sci. 119, 172(1982)
11. K. Christmann, Z. Phys. Chem. Neue Folge 154, 145(1987)
12. The limiting value of 10^{-12} s is a collision time obtained from the mean molecular velocity (~ 10^{4} cm s^{-1}) and the char-acteristic adsorbate length (~ 10^{-8} cm); the value of 10^{-4} s is a relaxation time of surface vibration by multiphonon excitation.
13. D. A. Wesner, F. P. Coenen and H. P. Bonzel, in Proceedings ICSOS-II, Amsterdam, June 1987
14. D. W. Goodman, Acc. Chem. Res. 17, 194(1984)
15. G. Ertl, Ber. Buns. Phys. Chem. 90, 284(1986)
16. R. C. Yeates, J. E. Turner, A. J. Gellman and G. A. Somorjai, Surf. Sci. 149, 175(1985)
17. S. B. Schwartz and L. D. Schmidt, Surf. Sci. 183, L269(1987)
18. L. D. Schmidt and C. Lee, in "Catalyst Deactivation", ed. by E. E. Peterson and A. T. Bell (Marcel Dekker, 1988)

Discussion

C. B. DUKE: The relaxation of surface reconstruction by adsorption and the formation of surface chemical compounds also are common phenomena in semiconductor surface chemistry (e.g., H, group VII species on Si and compound semiconductor polar surfaces). In many cases, however (e.g., GaAs(110)-p(1×1)-Sb) new types of chemical bonding exhibiting neither bulk nor molecular analogs occur. Moreover, these are believed to be locally thermodynamically stable, resulting from the two-dimensional (as opposed to three-dimensional) geometry of the adsorbate-substrate system. Such compounds seem different from the organometallic molecular-cluster-like structures which you described. What are some examples of such structures on single-crystal metal surfaces? (As opposed, for example, to new alloy compounds which exist in small-particle form.)

G.A. SOMORJAI: There are indeed surface compounds that exist only in two dimensions, that is, they do not have three dimensional phase diagrams.

One example perhaps is the behavior of (111) face of Cu–Al alloy single crystal, [*Surface Sci.* **165**, 345 (1986)]. The surface is ordered and exhibits a $(\sqrt{3}\times\sqrt{3})$ surface unit cell with one Al and one Cu atoms per unit cell, while the bulk is disordered and has a very different atomic composition. Unfortunately there are very few surface studies of systems where equilibrium exists between the surface and the bulk and the surface is well characterized.

J.C. TULLY: I was impressed by the close agreement you demonstrated between bonding in isolated organometallic compounds and bonding at surfaces. This appears to contradict some statements in this morning's session concerning the inadequacy of finite cluster calculations for describing bonding at surfaces.

G.A. SOMORJAI: Some surface properties are more likely to be controlled by nearest neighbor interactions and bonding than other properties. I suspect that bond distances, bond angles and vibrational frequencies of adsorbed molecules are among these properties. Perhaps ionization potentials or electron affinities would not be predictable by finite cluster calculations.

A.M. BRADSHAW: Returning to Bill Gadzuk's remarks this morning, it should perhaps be said that the cluster aproach is not expected to give a wholly accurate description of the adsorbate/substrate system. For example, the vibrational frequencies of the internal adsorbate modes may be wrong by up to a few tens of wavenumbers due to the fact that the cluster cannot simulate accurately enough the screening of the dynamic dipole which occurs on the extended surface. The same will be true of ionization potentials, although this is less important because in practice we still apply Koopmans' theorem. Nevertheless, the essential features of the chemistry are expected to emerge in a cluster calculation, thus fully justifying in my view the approach.

The Atomic Structure of the Liquid-Vapor Interface of a Metal: An Example of the Influence of Strong Density Dependence of the Interatomic Potential

S.A. Rice

Department of Chemistry and the James Franck Institute,
The University of Chicago, Chicago, IL60637, USA

Experimental studies and Monte Carlo simulations have established that the atomic structure of the liquid-vapor interface of a simple metal is significantly different from that of a Lennard-Jones system. In the latter the density along the normal through the interface (the longitudinal density distribution) monotonically decreases from the liquid to the vapor side of the interface; near the triple point of the system this variation in density occurs over a range of two to three atomic diameters [1]. In the former the longitudinal density distribution shows stratification on the liquid side of the interface and decreases rapidly, within a fraction of an atomic diameter, to the gas density on the vapor side of the interface [2,3]. The differences between the atomic structures of the liquid-vapor interfaces of a simple metal and the Lennard-Jones fluid can be traced to the different forms of the interatomic potential energy functions in the two cases. The potential energy of a Lennard-Jones system is rather accurately represented as a sum of density independent pair interactions. In contrast, the potential energy of a metal is strongly density dependent. In the pseudopotential representation the potential energy of a metal has two terms; a sum of density dependent pair interactions and a density dependent one body potential. it is the electron density dependence of the one body potential that is most important in a region where the liquid is inhomogeneous, as in the liquid-vapor interface. In a homogeneous liquid metal the one body potential is constant everywhere and, although it dominates the binding energy of the system, has no influence on the atomic distribution. In an inhomogeneous system, such as the liquid-vapor interface, the variation in density forces a variation in the one body potential which then acts as a force constraining the ions to remain in the region where the density is high. That is, the variation through the liquid-vapor interface of the one body potential in the pseudopotential for the metal is equivalent to the presence of a relatively steep continuous wall. It is the packing of the rest of the liquid against this "wall" which generates the stratification of the longitudinal density distribution in the liquid-vapor interface.

The first Monte Carlo simulations of the liquid-vapor interface of a metal, by D'EVELYN and RICE [2,3], assumed that the system pseudopotential could be calculated as if the electron and smoothed ion distributions in the interface were coincident, i.e. that electroneutrality prevails at each point in the interface. This

Springer Series in Surface Sciences Vol. 14
Solvay Conference on Surface Science
Editor: F.W. de Wette

approximation is suggested by the results of the selfconsistent jellium calculations of ALLEN and RICE [4]. More recently, HARRIS, GRYKO and RICE [5] generalized the D'Evelyn-Rice procedure so that the electronic structure, hence the system pseudopotential, is treated more accurately. In particular, the local electroneutrality approximation was replaced with a density functional theory treatment of the electron gas, including solution of the selfconsistent Kohn-Sham equations [6] to determine the electron distribution at an inhomogeneous smoothed ion distribution-vapor interface. A comparison of the results of simulations carried out with and without the local electroneutrality approximation shows that the stratification of the interface is rather insensitive to the treatment of the electronic structure provided that the one body potential in the pseudopotential is electron density dependent, as it always is in the pseudopotential representation of a metal. The Monte Carlo simulations have been tested for dependence of the predictions on the size of the system, the geometry of the system, and analytic form of the pseudopotential [7]. It is found that the prediction that the liquid-vapor interface of a metal is stratified for about three atomic diameters into the bulk liquid is robust. Fig. 1 displays the predicted longitudinal density distribution in the liquid-vapor interface of Cs at 100°C.

The only suitable experimental technique for studying the longitudinal density distribution in the liquid-vapor interface is the variation of x-ray reflectivity with angle of incidence. LU and RICE [8], and later BOSIO and OUMEZINE [9], have measured the x-ray reflectivity of Hg, and SLUIS and RICE [10] have measured the x-ray reflectivity of Cs. The results obtained (see Figs. 2 and 3) are consistent with the results of the Monte Carlo simulations.

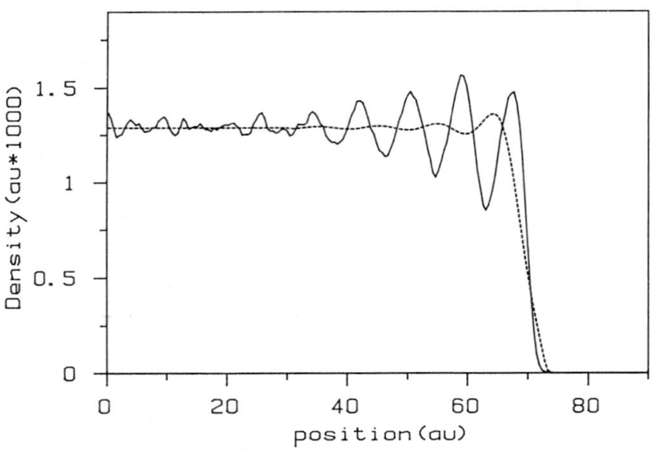

Fig. 1. The longitudinal density distribution in the liquid-vapor interface of Cs at 100°C

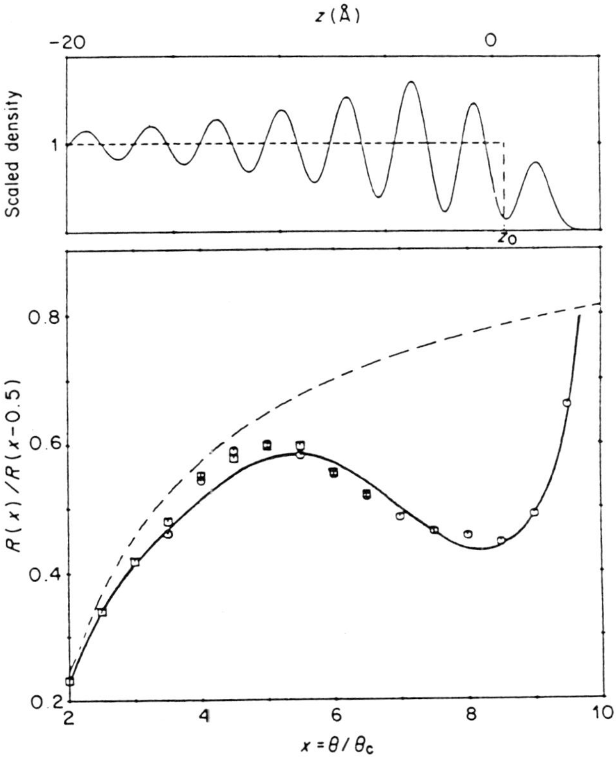

Fig. 2. The measured ratio of x-ray reflectivity for two adjacent angles of incidence as a function of angle of incidence for Hg. The upper panel shows two longitudinal density profiles, one an analytic form fit to the Monte Carlo simulations of D'Evelyn and Rice [3] and the other a zero thickness interface

Monte Carlo simulations of the transverse pair correlation function in the stratified liquid-vapor interface of a metal, by HARRIS and RICE [11], show that the conventional point local density approximation often used to calculate the properties of an inhomogeneous liquid leads to very serious errors. A much better approximation is that proposed by FISCHER [12], namely that an appropriate local density can be defined by averaging the point local density over a volume centered at the point of contact of two pseudoatoms and with diameter equal to the distance of closest approach of two pseudoatoms. This approximation preserves, in lowest order, the force associated with the inhomogeneous density distribution. When applied to the liquid-vapor interface of a metal, the appropriate local density is nearly the same as the bulk density everywhere except on the vapor side of the outermost layer in the stratified interface (the spacing of the strata is about an atomic diameter and the

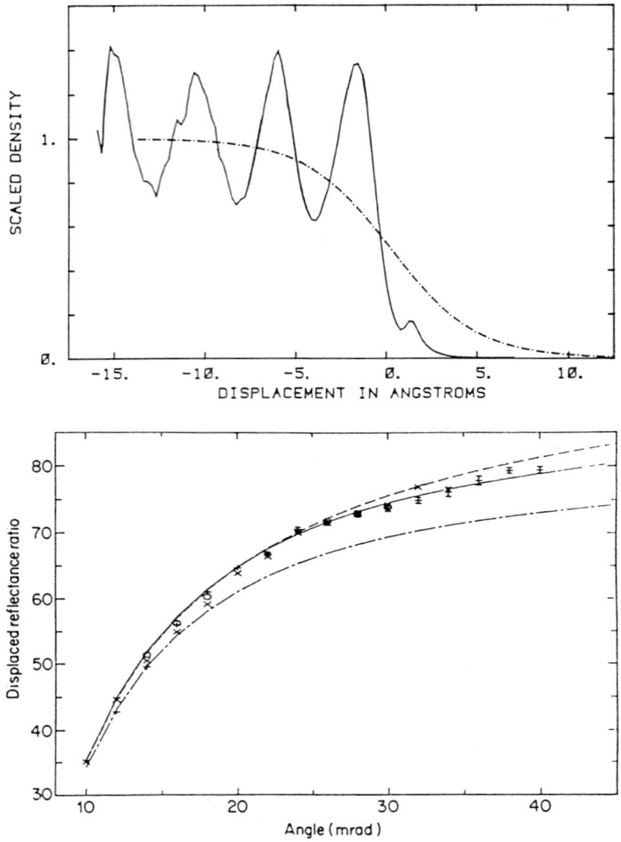

Fig. 3. The measured ratio of x-ray reflectivity for two adjacent angles of incidence as a function of angle of incidence for Cs. The upper left hand panel shows two longitudinal density profiles, one from the Monte Carlo simulations of D'Evelyn and Rice [2] and the other from a van der Waals model of the interface.

excursions above and below the bulk density are about equal in magnitude). A consequence of this interpretation is that the transverse pair correlation function in the interface should be very nearly the same as that of the bulk liquid despite the stratification of the interface.

The transverse structure factor in the liquid-vapor interface can be studied experimentally using grazing incidence in-plane x-ray diffraction [13]. The results of such a study of Hg, by THOMAS, BARTON, NOVAK and RICE [14], are shown in Fig. 4. It is found that the surface transverse structure factor is, as

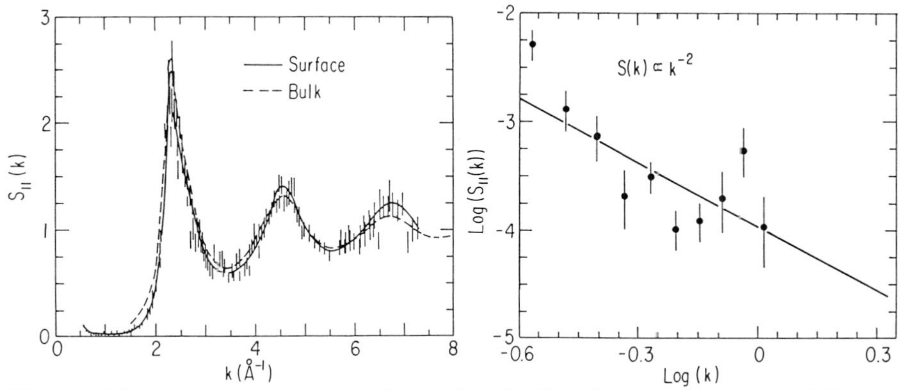

Fig. 4. The transverse structure factor for the liquid-vapor interface of Hg. The right hand panel is a blow up and replotting in different form of the small k region

predicted, very nearly the same as the bulk structure factor except at very small momentum transfer and on the small momentum transfer edge of the first peak. The very small momentum transfer behavior is consistent with the expected k^{-2} divergence of the transverse structure factor of a two dimensional liquid as k→0 [15], and the deviation in the shapes of the bulk and surface structure factors on the leading edge of the first peak is consistent with the Fischer definition of the local density for the Hg liquid-vapor interface and the results of the simulations of D'Evelyn and Rice.

GRYKO and RICE [16], and later HARRIS, GRYKO and RICE [17], have studied the longitudinal and transverse atomic distributions in several NaCs alloys. The two most interesting results they obtain are:

(1) The variation of composition with position in the interface is very different in metallic [17] and van der Waals mixtures [18]. In the NaCs alloys, the Cs segregates at the surface forming an almost pure monolayer which sits atop a layer slightly depleted in Cs, and underneath this layer the bulk composition is achieved. Put another way, the interface concentration is different from the bulk concentration for only about one atomic diameter into the bulk liquid. In the model van der Waals mixture generated by turning off the one body potential in the pseudopotential, the extent of composition segregation is less than when the one body potential is nonzero and, more important, the interface concentration is different from the bulk concentration for about five atomic diameters into the bulk liquid (Fig. 5).

(2) The transverse pair correlation function in the outermost layer of the stratified liquid-vapor interface of a NaCs alloy, which is pure Cs, does not depend on the bulk composition of the alloy and is identical to the transverse pair correlation function in the outermost layer of the stratified liquid-vapor interface of pure Cs.

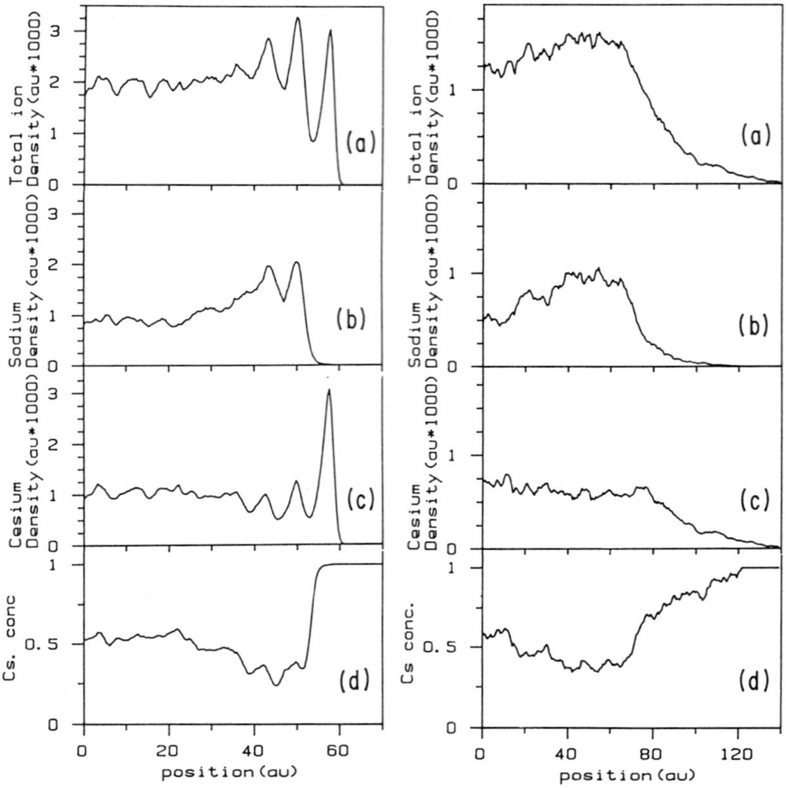

Fig. 5. The composition variation in the longitudinal density distribution in the liquid-vapor interface of an equimolar NaCs alloy

Given that the liquid-vapor interface of a metal is stratified, it is to be expected that (a) its electrical and optical properties will deviate from those predicted using a simple isotropic Drude model which assumes the existence of a homogeneous metal separated from its vapor by a geometric boundary at which the system density drops discontinuously from the bulk to the vapor and (b) the work function of the system will deviate from that predicted for a homogeneous metal. All of these effects have been inferred from a variety of measurements. BLOCK and RICE [19], and later GUIDOTTI and RICE [20], showed that a consistent interpretation of the optical constants of liquid Hg obtained from reflectivity and ellipsometric measurements could be given if it is supposed that the conductivity is anisotropic in the liquid-vapor interface, with the in plane component of the conductivity about five times larger than the normal component. MOHANTY and RICE [21] showed how the Boltzmann equation for the stratified interface could be solved and found that, for the kind of stratification predicted by the Monte Carlo simulations, qualitative arguments were sufficient to establish that the in plane component of the conductivity should be larger than the normal component. GOODISMAN [22] showed that unless stratification of the liquid-vapor interface is included in the analysis of the work function it is not possible to account for the observed value for Hg; when it is included the agreement between theory and experiment is very good. EARNSHAW [23] has shown that the unusual damping of capillary waves on the surface of Hg can be accounted for if the features of the liquid-vapor interface predicted by D'EVELYN and RICE are incorporated in the analysis; the damping of capillary waves on the surface of a dielectric liquid is well accounted for without appeal to the structure of the interface.

References

1. F.F. Abraham, D.E. Schreiber and J.A. Barker, J. Chem. Phys. **62**, 1958 (1975).
2. M.P. D'Evelyn and S.A. Rice, J. Chem. Phys. **78**, 5225 (1983).
3. M.P. D'Evelyn and S.A. Rice, J. Chem. Phys. **78**, 5081 (1983).
4. J.W. Allen and S.A. Rice, J. Chem. Phys. **68**, 5053 (1978).
5. J. Harris, J. Gryko and S.A. Rice, J. Chem. Phys. **87**, 3069 (1987).
6. A.G. Equiluz, D.A. Campbell, A.A. Maradudin and R.F. Wallis, Phys. Rev. **B30**, 5449 (1984).
7. See, for example, S.A. Rice, J. Gryko and U. Mohanty, in *Fluid Interfacial Phenomena*, Ed. C.A. Croxton, Wiley, New York (1986) p. 255.
8. B.C. Lu and S.A. Rice, J. Chem. Phys. **68**, 5558 (1978).
9. L. Bosio and M. Oumezine, J. Chem. Phys. **80**, 959 (1984).
10. D. Sluis and S.A. Rice, J. Chem. Phys. **79**, 5658 (1983).
11. J. Harris and S.A. Rice, J. Chem. Phys. **86**, 5731 (1987).
12. J. Fischer and M. Methfescel, Phys. Rev. **A22**, 2836 (1980).
13. U. Mohanty and S.A. Rice, J. Chem. Phys. **79**, 2482 (1983); G.H. Vineyard, Phys. Rev. **B26**, 4146 (1982).
14. B. Thomas, S. Barton, F. Novak and S.A. Rice, J. Chem. Phys. **86**, 1036 (1987).
15. M.S. Wertheim, J. Chem. Phys. **65**, 2377 (1976); D. Bedeaun and J.D. Weeks, J. Chem. Phys. **82**, 972 (1985); M.H. Kalos, J. Percus and M. Rao, J. Stat. Phys. **17**, 111 (1977).
16. J. Gryko and S.A. Rice, J. Chem. Phys. **80**, 6318 (1984).
17. J. Harris, J. Gryko and S.A. Rice, J. Stat. Phys. **48**, 1109 (1987).
18. G.A. Chapela, G. Saville, S.M. Thompson and J.S. Rowlinson, J.C.S. Faraday II **73**, 1133 (1977).

19. A. Bloch and S.A. Rice, Phys. Rev. **185**, 933 (1969).
20. D. Guidotti and S.A. Rice, Phys. Rev. **B15**, 3796 (1977).
21. U. Mohanty and S.A. Rice, J. Chem. Phys. **79**, 5652 (1983).
22. J. Goodisman, J. Chem. Phys. **82**, 560 (1985).
23. J. Earnshaw, Phys. Lett. **92A**, 40 (1982).
24. S. Barton, B. Thomas, E. Flom, F. Novak and S.A. Rice, Langmuir, in press.
25. Z. Cai, J. Harris and S.A. Rice, to be submitted to J. Chem. Phys.

Discussion

D.R. HAMANN: Does your simulation method treat the full three–dimensional self-consistency of the electrons for each ion configuration, or only the one–dimensional problem of electrons in the planar-averaged ionic potential?

S. A. RICE: The self consistent electron and ion distribution calculations are for a slab configuration of $\sim$ 1000 ions. We assume (and later confirm) there is isotropy in the directions (x, y) parallel to the surface, so we take planewave type solutions in x and y. The Schroedinger equation is solved in the z direction for each (smoothed) configuration of the ions. The calculation is repeated when the ions are moved in the simulation process. At the end of the calculation we verify that the ion distribution is isotropic in x and y and stratified in z (normal to the surface).

E. TOSATTI: With reference to your data, showing that the liquid Hg surface has sharper peaks in the transverse g(r) than the liquid bulk Hg, I would like to mention recent simulations of the liquid gold surface, carried out in Trieste [1], using the well–tested many–body ("glue") scheme [2]. Preliminary results indicate sharper short–range correlations than in the bulk liquid gold, as well as the k^{-2} capillary wave behavior, quite analogous to your experimental results. ([1] P. Carnevali, F. Ercolessi, S. Iarlori, and E. Tosatti, unpublished; [2] F. Ercolessi, E. Tosatti and M. Parrinello, *Phys. Rev. Letters* **57**, 719; *Surface Science* **177** (1986) 314.)

S.A. RICE: I am very interested in your results.

Y. PETROFF: Due to the long time that you need to obtain the x-ray reflectivity curve, do you have problems keeping the surface clean?

S.A. RICE: (Answer included with response to J.C. Thompson's question below.)

J.C. THOMPSON: Is there not a flux of atoms across the surface – i.e., is it not "boiling" as you take data so that the surface is constantly renewed?

S.A. RICE: Yes—in our work on the x–ray reflectivity of Hg the system was continuously pumped.

No—in our work on the transverse structure factor in the liquid–vapor interface of Hg, after careful preparation of the sample the system was sealed.

In all cases the sample chamber is first pumped to $\sim 10^{-11}$ torr of O_2 and monitored by field emission microscopy, the sample is forced to grow surfaces in the vacuum (by being forced through a capillary so that each drop grows fresh surface in the vacuum). Note that Hg is not very reactive–the reproducibility of our data suggests our samples remained clean for the duration of the experiment.

Discussion on Observation of Phase Transition and Adsorption Processes in Real Space (*16 mm movie*)

K. Takayanagi

E. TOSATTI: It would be interesting to have a better characterization of your Au clusters regarding the state of excitation. We are currently studying the energetics of clusters (W. Andreoni, F. Ercolessi and E. Tossati (unpublished) in the 100–1000 atom range, and we find fcc, icosahedra, decahedra and glassy clusters all very near in energy at T=0, with marginal prevalence of fcc. What state of excitation should we consider for a realistic comparison with your results?

K. TAKAYANAGI: The structure assigned from TEM is icosahedra for Au clusters greater than those consisting more tan 55 atoms. Smaller clusters showed a very complicated change of their image. Dark dots, image of a single atom or an atomic row, in such small clusters, however, arrange locally with 5–fold symmetry, indicating that icosahedral configuration is locally formed in the custer (*J.J. Appl. Phys Letter, 1987*). Clusters in the 100–1000 atom range have icosahedra as a stable phase at around room temprature. They fluctuate between icosahedral and f.c.c. structure above $\simeq$ 500°C of substrate temperature (clusters sit on graphitized carbon). The clusters are, therefore, not excited to f.c.c. We cannot, however, answer what excited state other than icosahedra they are in, before the analysis of our VTR film.

M.G. LAGALLY: In your images of the nucleation of the (1 × 1) phase at the bottom of steps in the (7 × 7) surface, the width of the (1 × 1) seems to be constant and follows the contour of the step edge. Is this an indication of strain at the step edge? How else could one explain this?

K. TAKAYANAGI: The width of the "1 × 1" phase is not constant. For better understanding, I describe the formation process of the 7 × 7 domain on cooling following the TEM observation. The 7 × 7 domains nucleate at the upper side of the step. These domains form "out of phase boundaries" when they touch with each other because of lack of phase correlation between the domains. As the temperature is further lowered, the 7 × 7 domains grow inwards until they touch the next step, the out–of–phase boundary remains across the terrace. On heating, the reverse process occurs: the "1 × 1" phase nucleates at the bottom of a step and extends through the out–of–phase boundary.

Springer Series in Surface Sciences Vol. 14
Solvay Conference on Surface Science
Editor: F.W. de Wette

Strain at the step edge might be a cause of the nucleation of "1×1" phase at the bottom but it is not necessarily the cause.

J.E. DEMUTH: Your images of semiconductor surfaces looks extremely perfect (aside from the steps and terraces). I would like to ask how much of this "perfection" is the result of the image reconstruction process which typically uses particular diffraction beams and essentially Fourier filters non periodic components (i.e., point or local defects or irregularities).

K. TAKAYANAGI: Fringe images we observed in REM are interference fringes of three diffraction spots (*Jpn. J. Appl. Phys.*, 1987), and the resolution for detecting irregularities is of the order of one–half of the 7×7 unit cell, $\simeq$ 1nm. Therefore, it is difficult to say whether the surface has point defects within the 7×7 unit cells (seen in STM images). The image shows that the 7×7 unit cells are regularly ordered (no phase shift) within each domain.

In TEM (*J. V.S.T.*, 1985) of higher resolution, the regularity is also seen, except in contaminated areas. It should be noticed that our microscope images are for 500 $\sim$ 600°C. Some vacancies, which may exist to a value of thermal equilibrium, could form defects in specimens of room temperature observed by STM, depending on ways of cooling.

Part III

Two-Dimensional Physics/Phase Transitions

Phase Transitions at Interfaces: Roughening, Surface Melting, and Triple Point Wetting

J.G. Dash

Department of Physics, University of Washington, Seattle, WA 98195, USA

1. INTRODUCTION

Roughening, surface melting, and wetting are three closely related topics of great current interest in surface science. They involve the fundamental properties of thin films, and they are important phenomena in modern interface technology. Each has a long, nearly independent scientific history reaching back to the 19th Century, and we are only now learning how closely they are related. We are finding that much of the early work is inaccurate and just plain wrong, but a portion is bearing up surprisingly well. The state of knowledge in all three fields has improved considerably in the past few years, but some of the original speculations are still being tested.

A significant part of the recent advances has come from studies of physisorbed films. One of the principal motivations for modern studies of adsorbed films is an abiding interest in the roles of dimensionality in condensed matter. Research during the past few decades has shown that physisorbed monolayers can be quite close approximations of two dimensional matter. These systems are testing grounds for theory and challenges to experimental techniques, and they have played an important part in developing our current understanding of lower dimensional condensed matter physics. While studies of monolayers continue, recent years have seen a shift in the center of interest, to multilayer films. Multilayers are intrinsically much more complicated than either monolayers or bulk phases. The third dimension plays an important role in most multilayer properties, yet it is not yet equivalent to the in-plane dimensions. In some sense, the multilayer film has an intermediate dimensionality, between two and three. With increasing thickness the film begins to develop a distinct interior region and two interfaces. New phenomena emerge, and they evolve, with increasing thickness, to become the surface transitions of bulk matter. Although these phenomena are not unique to adsorbed multilayers, films are extremely useful in studying them. We find them particularly helpful in studies of surface roughening and surface melting. A particular type of wetting transition, known as triple point wetting, is closely related to surface melting and provides important insight into the nature of the melted layer.

Springer Series in Surface Sciences Vol. 14
Solvay Conference on Surface Science
Editor: F.W. de Wette

2. MULTILAYER AND BULK INTERFACE PHENOMENA

The systematics of multilayer phenomena and the evolution of bulk behavior are discussed in an important paper by PANDIT, SCHICK, and WORTIS[1]. Although several discoveries since its publication have introduced substantial changes in the theory, the original work presents a valuable classification scheme for a comprehensive picture of film and bulk surface phase transitions.

The basic scheme allows one to trace the phases and phase transitions of adsorbed films, as thickness is increased from monolayers to bulk, as a graph of chemical potential $\mu(T)$ versus temperature. Figure 1a is such a phase diagram, describing a system which undergoes *complete wetting.* At low temperature the film deposits as a succession of distinct solid layers, detected by sharp steps in a *class I* isotherm (Fig. 1b). Layer-by-layer deposition continues ad infinitum as the pressure rises to the saturation P_o and μ reaches μ_o; the film grows asymptotically to bulk, producing a flat and smooth interface with the coexisting vapor. The layer transitions are terminated at critical temperatures, above which the isotherm steps become rounded, and the layers lose their individuality. These critical points progress systematically to the coexistence boundary, where they end at a roughening temperature T_R. Above this temperature the surface is irregular but the basic structure of the material is still a solid. At still higher temperature melting occurs: the bulk transition is at the triple point T_t, the termination of a melting line in the film.

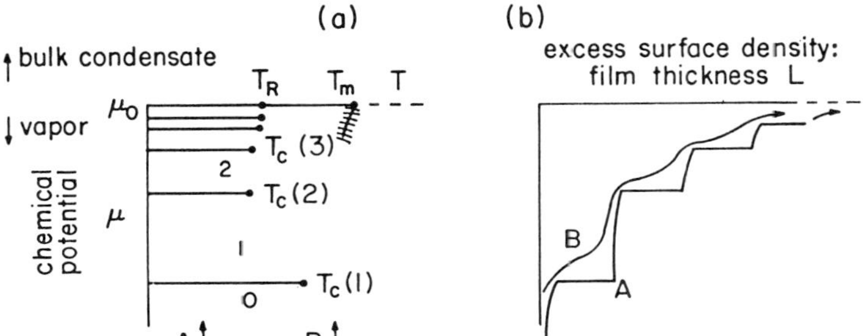

Fig. 1(a). Surface phase diagram of a model system exhibiting complete wetting[1]. Surface phases are designated 0, 1, 2, ... indicating the number of completed layers at $T = 0$. The sequence of layer transitions, terminating at critical points $T_c(L)$, approach the roughening temperature T_R as $L \to \infty$. The model is constructed from the 2D Ising model, with no provision for the propagation of long range order between layers; in more realistic models the first layer phase boundary continues up to meet the bulk boundary. (b) Isotherms corresponding to trajectories A, B indicated in (a)

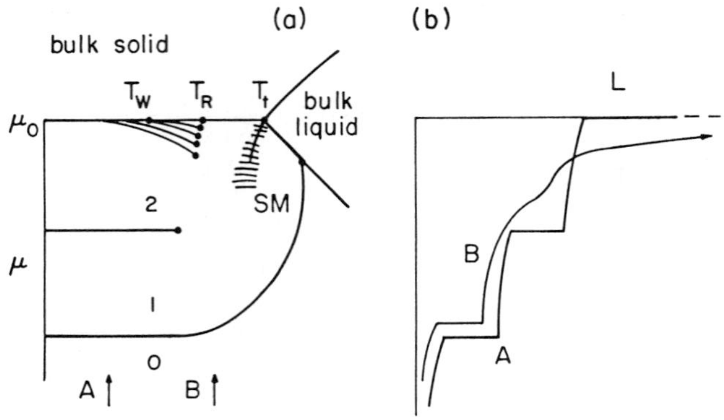

Fig. 2(a). Surface phase diagram of a film with a transition between incomplete and complete wetting at T_W. The hatched band SM extending below the triple point T_t represents the region of surface melting, which in this system is not a line of phase transitions, but a region of rapid development of liquid like character, as seen in recent studies of surface melting. (b) Isotherms corresponding to trajectories A and B indicated in (a)

A number of real systems have phase diagrams similar to Fig. 1, but others are more complicated, being incompletely wet over some range of the solid or liquid domain. In this range a film of only limited thickness can be deposited, after which bulk material is nucleated. Such a film typically undergoes a wetting transition at some temperature $T_W \leq T_t$, above which it can be thickened asymptotically to bulk. Figure 2a shows a phase diagram for a system with $T_R > T_W$. Typical isotherms in two principal temperature regions below T_t are shown schematically in Fig. 2b. Recent experiments have given evidence for both types of phase diagrams. The point of showing these diagrams is to illustrate that each of the two disordering transitions of bulk solid surfaces, roughening and surface melting, have phase boundaries extending along another thermodynamic variable. The chemical potential can be varied by changing the film thickness, and this provides additional means for studying the phenomena. In the following discussion we will present some recent results of experiments that demonstrate the usefulness of physisorbed multilayers in such studies.

3. SURFACE ROUGHENING

3.1 Theory and Computer Simulations

The smoothness of crystal facets determines their speed of growth and the morphology of the crystal. The theory of the growth of perfect crystals was begun by GIBBS[2], and extended by CURIE[3], WULFF[4], VOLMER[5], STRANSKI[6], and others. It was recognized that the growth of a crystal facet is primarily by the addition of atoms at the edges of a step or terrace, so that, when all stepped surfaces have been completed, the crystal can continue to grow only by the nucleation of new terraces, requiring an appreciable supersaturation. Two mechanisms were proposed for avoiding the nucleation barrier. FRANK[7] realized that growth could continue if a surface contained screw dislocations, which present self sustaining edges or steps. FRENKEL[8] had argued that the structure of a step should contain a large number of kinks at finite temperature, and concluded that all crystal surfaces at finite temperature should have a certain roughness produced by thermal fluctuations. BURTON and CABRERA[9] accepted the thesis for stepped faces, but they showed that the thermal excitation of steps and clusters on low index facets should be unimportant except perhaps close to the melting point. BURTON et al.[10] then presented their celebrated theory of the roughening *transition*, occuring at a definite roughening temperature T_R. They confirmed that the behavior of low index faces is different from the behavior of stepped surfaces, which will "...remain flat, and not acquire any steps, below a certain transition temperature; above this temperature the surface becomes essentially rough, a large number of steps appearing."

Burton, Cabrera, and Frank recognized that the structure of the crystal surface is a cooperative phenomenon, involving not only the excitation of individual surface atoms, but also the interactions between them. They mapped the problem on to a two-level model of the surface, and transcribed it to the recently solved two dimensional Ising model. But they recognized that roughening is actually a multi level problem, and they introduced what has come to be known as the solid–on–solid (SOS) model. The model allows for the minimum number of configurational degrees of freedom that are deemed necessary to describe roughening, but it excludes all internal defects and thermal excitations. It continues to be extremely useful in modern theory, and is the basis for many analytical and numerical studies of the transition.

The theory is now highly developed, and several reviews[11] outline the present status of the subject. The roughening transition is continuous, in the class of the planar XY model. The evolution of roughening is described in terms of a correlation function for the height difference between two locations at lateral separation r on the interface:

$$G(r) \equiv < [h(r) - h(0)]^2 > \quad . \tag{1}$$

$G(r)$ gives a measure of the diffuseness of the interface due to the fluctuations in height between different regions of the surface. Well below T_R the interface width and the lateral correlation length ς are finite. At $T > T_R, G(r)$ diverges as $\ln(r)$, and ς is infinite. Renormalization group methods have shown that the correlation length and the free energy have characteristic square root exponents as the transition is approached from below:

$$\varsigma_r = \text{const.} \exp[c/(T_R - T)^{1/2}]; F = \text{const.} \exp[-c'/(T_R - T)^{1/2}] \quad . \tag{2}$$

The thermodynamic quantities have only essential singularities at the transition, hence there is no heat capacity peak at T_R. However, a rounded anomaly is predicted to occur near the transition temperature, due to a rapid continuous change in the surface disorder. The anomaly is similar to the theoretical behavior of other systems describable by the planar XY model. The roughening transition differs from the continuous transition of other systems described by the XY model, in that the anomaly occurs at a temperature *below* the transition temperature[12]. This is because ς diverges as $T \rightarrow (T_R)^-$, whereas in other transitions such as two dimensional melting, the correlation length diverges as the transition is approached from higher temperature.

Several studies have been carried out by numerical methods[13-15]. Rosato, Cicotti and Pontikis investigated the (110) face of a fcc Lennard-Jones solid by molecular dynamics simulation[14]. They found that although the surface retained its solid like character to temperatures very close to the triple point, there was a large surface concentration of vacancy adatom pairs, accounting for high values of surface diffusivity. The activation energy of surface defects and the magnitude of the surface order parameter decrease slowly at low temperature, but more rapidly in the range above 0.8 of the bulk melting point. They identified the changeover with a roughening transition, with $T_R = 0.8T_m$. As we note below, there is increasing experimental evidence that this relation between the roughening and triple point temperatures may be rather general for the low index faces of molecular crystals.

3.2 Crystal Shape

The experiments most often quoted in the early discussions bearing on the problem of roughening were measurements of growth rates of different crystal facets. There are large differences in the rates of growth of different facets, which depend individually on the densities of steps and the local supersaturation. At relatively low temperature, growth is oriented and the surface is flat; above the roughening transition, the growth tends to be isotropic, and the surface is rounded. Since different facets have different roughening temperatures, some faces may be absent, depending on the temperature at which the crystal had been prepared. Changes of the equilibrium shapes of crystals are describable in terms of the temperature dependence of the polar plot of interfacial energy known as a ***Wulff** diagram*[4,16], as illustrated in Fig. 3.

The first experiments indicating changes of growth morphology that were consistent with the prediction of a roughening transition were done in the 1970's[17,18]. These studies, on very small molecular crystals, confirmed that the crystal habits changed within very narrow temperature intervals, where certain facets disappeared and were replaced by smoothly rounded surfaces. The dates are surprisingly recent; in spite of long standing observations of variations seen in natural crystals, well known to mineralogists for many years. For example, a book[19] on empirical aspects

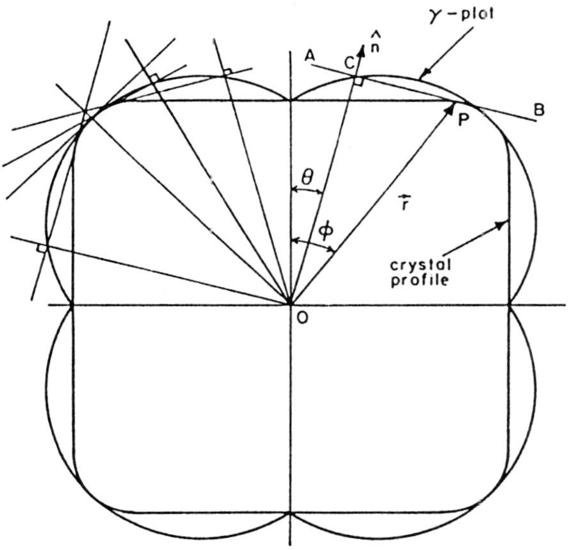

Fig. 3. A polar diagram of the surface free energy $\gamma(\hat{n})$ as a function of the surface orientation. The equilibrium crystal shape is described by the region enclosed by planes perpendicular to $\hat{n}$ at all points contacting $\gamma(\hat{n})$.

of crystal growth states that there are *"...almost innumerable examples of crystals whose different faces show differences in their surface perfection. It is well known that epidote and topaz, among various minerals have, as a rule, well-faceted terminations.....yet, at the same time, having very poor, rounded-off 'prism' zones."* It is now generally recognized that the rounding is associated with roughening and that the roughening temperature is peculiar to each face. However, in large crystals (with the exception of helium, discussed below) the actual morphology is the dynamical shape, which is not identical to the equilibrium form. Equilibrium depends upon surface diffusion and dissipation of the latent heat; with increasing size the transport processes become incapable of achieving equilibrium in even geological time spans. But equilibrium shapes are obtained in experimentally reasonable times in crystallites of very small size, and studies of microscopic Au, In, and Pb crystals have shown temperature dependent morphology similar to theoretical predictions[20].

The problem of slow equilibrium is almost totally absent in the growth of solid ^4He from the liquid phase below 1.4 [K]. This is due to three factors: the liquid is superfluid, the latent heat of fusion is near zero, and diffusion in the solid is extremely rapid and temperature independent. Extensive studies have been carried out in Moscow, Paris, and Haifa. The experiments have provided critical tests of the theory, and they are in extremely good agreement with essential features. As an example of the clarity of profile and the sensitivity of shape to temperature in the neigh-

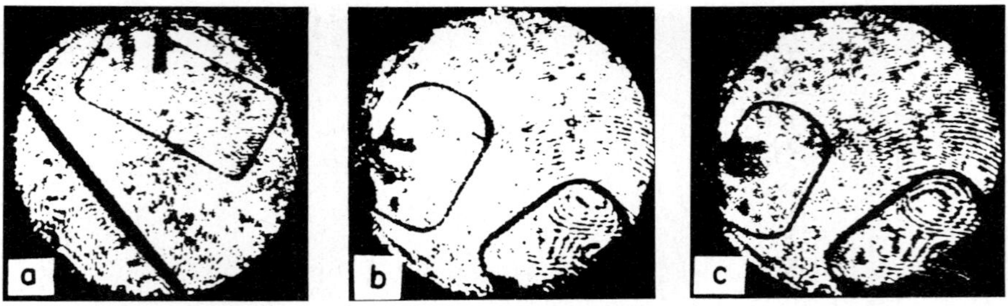

Fig. 4. Profiles of ^{4}He crystals exhibiting progressive rounding of $(10\bar{1}0)$ facets as the roughening temperature is approached[21]. The temperatures are (a) 0.75 K; (b) 0.77 K; (c) 0.84 K. The roughening temperature is approximately 0.95 K

borhood of the roughening temperature, Fig. 4 reprints a series of photographs from one of the several reviews of the field[21,22], showing the progressive rounding of a facet as the crystal is warmed slightly. The reader is referred to the reviews for further details.

3.3 Adsorbed Films

Several experiments on physisorbed films have explored different aspects of roughening transitions. One of the most direct involves vapor pressure isotherms, measurements of the equilibrium pressure of a film as a function of its thickness. Films which wet a substrate at low temperature can be deposited layer by layer, and if the substrate is quite uniform the layering can be distinguished in the stepwise character of the isotherm. Several film–substrate combinations can exhibit a substantial number of distinct layer steps at low temperature, including Kr[23], methane[24], oxygen[25], carbon tetrafluoride[26] (all on basal plane graphite), and methane on MgO[27]. A recent set of measurements of CF_4 by ellipsometry[26] is shown in Fig. 5. The nearly vertical portions (vertical in the ideal system) are due to phase condensation in the partially completed individual layers; after completion, the pressure rises strongly until a new layer can begin to form. The sharp steps must disappear or become rounded when the layers lose their individuality on passage through the roughening transition. Such a changeover is predicted on the basis of theoretical models of multilayer adsorption[1,28] and observed in vapor pressure isotherms of methane on graphite[24] and Ar and Kr on lamellar halides[29,30]. The changeover temperature in both systems is slightly greater than $0.8T_t$, close to the empirical relation between T_R and T_t for close packed facets of molecular crystals in observations of growth forms[18] and in computer simulations of rare gas crystals[14].

Calorimetry on adsorbed multilayers have revealed specific heat maxima similar to the predicted anomalies associated with roughening transitions. The films are Ar, Ne and Kr adsorbed on graphite[31,32], measured

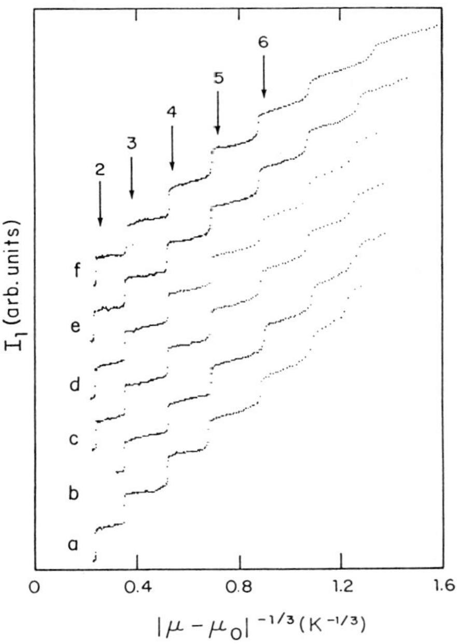

Fig. 5. Isotherms of CF_4/graphite obtained by ellipsometry[26]. Layer step numbers are given at the top; the temperatures range from 88.0 K (a) to 94.1 K (f). The bulk triple point is 89.5 K

over thicknesses ranging from 3 to 40 layers. The anomaly begins to appear as the coverage is increased above four layers, and by six layers it develops into a form which remains essentially independent of further increases in thickness. Its shape is consistent with a continuous transition, and the shape and entropy are qualitatively similar to theory and a model calculation[12]. The peak temperature is about 0.8 of T_t.

3.4 Dynamical Studies

Surface diffusion is an important transport mechanism in sintering and crystal growth. Roughness can have a strong effect on the mobility of atoms along the surface, as shown by computer simulation[14]; consequently, the surface diffusion constant can be used as a probe of roughening. Surface diffusion measurements have been made of many metals by examing dynamical changes of the shape of field emission tips[33]. Activation energy changes are seen in Cu, Mo, and W at $T/T_m \cong 0.8$. It is noteworthy that the same ratio is found in bcc W and Mo as in fcc Cu, because on the basis of earlier experiments at lower temperature it had been suggested that such changes would only occur in close packed metals.

Detailed investigations of surface disorder on surfaces vicinal to the close packed facets of Cu and Ni have been carried out by helium beam diffraction[34,35]. On stepped (11m) interfaces at low temperatures there is a regular array of unit height steps of low index terraces spaced m units apart. As temperature rises the steps begin to meander and nested islands appear. Because of the preexisting steps, disorder occurs at lower temperature than on dense facets. Since the step–step repulsive energy, which stabilizes the step array at low temperature, falls off with separation, T_R is predicted to decrease with increasing m. (The origin of the step–step repulsion is not known; possible mechanisms are believed to be elastic strains and surface dipole interactions.) The roughening temperatures of vicinal surfaces of Cu and Ni show the predicted trend: $T_R(113) \cong 0.32T_t$; $T_R(115) \cong 0.27T_t$ in Cu, and $T_R(115) \cong 0.26T_t$ in Ni. Theory predicts that $T_R(11m)$ depends upon a single energy parameter related to the binding energy in the material. The calculated ratios for Cu and Ni are brought into good agreement with the measurements by reasonable values for the energy parameters.

4. SURFACE MELTING

On September 8, 1842 Michael Faraday recorded in his diary[36] some thoughts about the surface properties of ice, and he then began a series of investigations that were to last twenty years. His first notes, still perceptive and incisive in today's light, begin the first scientific investigation into what we now know as surface melting.

"When wet snow is squeezed together, it freezes into a lump (with water between) and does not fall asunder as so much wetted sand or other kind of matter would do."

"In a warm day, if two pieces of ice be laid one on the other and wrapped up in flannel, they will freeze into one piece."

"All this seems to indicate that water at 32° will not continue as water, if it be between two surfaces of ice touching or very near to each other."

"The ice probably acts as a nucleus, but it appears that the effect of one surface of ice on water is not equal to the joint effect of two."

But although Faraday's experiments[36,37] and others by Tyndall[38] were proof to them of a liquid film at equilibrium, some contemporaries were unconvinced. James Thomson, Lord Kelvin's brother, countered with an explanation growing out of his own earlier work, which had predicted the negative slope of the melting line. He argued that regelation (sintering is the modern term) of ice results from the lowering of the melting temperature due to the increased pressure at the points of contact[39]. Thomson's theory came to be the accepted explanation for the ready sintering and low friction coefficient of ice. For example, a standard text for Honors Course students[40] states, *"In skating, the pressure of the skate liquefies the ice under it, and the liquid so formed acts as a lubricant,"* The textbook acknowledges that the mechanism is not effective if *"the ice is at a temperature far below freezing."* (My emphasis). The question is, how far? Any physics undergraduate can show that pressure melting can work only within a tenth of a degree or so of the freezing point. In 1939 Bowden and Hughes recognized the problem, and they proposed that the liquid film is due to frictional heating[41]. Later work shows that, while friction has some effect in certain circumstances, it can not play a role in all: for example, friction has nothing to do with sintering. More than a hundred year later, Faraday's views began to be reexamined[42], and now it is increasingly clear that he was basically correct. Moreover, the phenomenon is not confined to water: it is predicted to be typical of most solids. But although recent studies are beginning to explore the phenomenon in other substances, the effect in ice is still incompletely understood, and current experiments are only now beginning to answer questions raised or implicit in Faraday's original work.

Surface melting has been thought to be predicted by Lindemann's phenomenological model of melting[43], through the softening of vibrational modes near a surface. But although the Lindemann model may provide an interpolative rule, its application to surfaces suffers from the same basic deficiencies encountered in applying it to ordinary bulk solids[44]. Another long standing confusion has been a tendency to equate surface melting with roughening[45]. In the simplest formulations of surface melting and roughening, the nature of the surface disorder is qualitatively distinct. In the rough surface, the atoms are still located at the positions of the crystalline space lattice, while in surface melting the layer has typical liquid like short range order. In any case, the premelted layer is in principle different from the true liquid phase, since it is outside the stability range of the equilibrium phase, and it is often call the *quasi liquid layer (QLL)*. In this schematic view the interface between solid and liquid is abrupt, but in reality the gradation extends over an appreciable distance. Nevertheless, even in these situations, there may be fundamental differences between the two types of disorder, since roughening and surface melting can be observed as distinct successive transitions on certain well characterized surfaces. The thermodynamic motivation for surface melting as a general phenomenon was stated by STRANSKI[46] and FRENKEL[47] and outlined more explicitly by KRISTENSEN and COTTERILL[48]. Recent papers describe the basic thermodynamic principles[49] and detailed statistical mechanics of some theoretical models[50], but as discussed below, the theory is still incomplete. In the next sections we summarize the basic mean field model, then describe recent experiments that confirm its essential features; we then discuss later work going beyond the simple theory.

4.1 Mean Field Theory

If the boundary between a solid and its vapor can be wetted by the melt liquid, then the system free energy at $T < T_t$ can be lowered by converting a layer of solid to liquid (more exactly, as noted above, a quasi liquid[51]). Differences between the properties of the QLL and the equilibrium liquid are neglected in the elementary theory, but will be specifically addressed in the discussion of recent experimental results. The conversion causes the solid–vapor interfacial coefficient γ_{sv} to be replaced by the sum $(\gamma_{s\ell} + \gamma_{\ell v})$. The coefficients approach the $\gamma's$ of bulk phases in the limit of very thick liquid layer thickness L_ℓ but are functions of L_ℓ for finite thickness, and this dependence causes an effective interaction between the interfaces of the melted layer. For power law forces between interfaces varying as (distance)$^{-p}$, the mean field free energy change per unit area, due to the truncation of interactions across the interface, is

$$\delta F_{\text{surf}} = -\Delta\gamma[1 - \alpha L_\ell^{1-p}] , \tag{3}$$

where $\Delta\gamma \equiv (\gamma_{sv} - \gamma_{s\ell} - \gamma_{\ell v})$ and α is a constant on the order of unity. There is also a change δF_m proportional to L_ℓ due to solid–liquid conversion: minimizing $\delta F_m + \delta F_{\text{surf}}$ with respect to L_ℓ establishes the equilibrium thickness of liquid. The variation of thickness with temperature arises from δF_m, which is zero at T_t and increases at lower T. Expanding $(\mu_s - \mu_\ell)$ to first order about T_t,

$$L_\ell = \lambda t^{-1/p}; \lambda \equiv [\Delta\gamma\alpha(p-1)/T_t n_\ell \partial(\Delta\mu)/\partial T]^{1/p} , \tag{4}$$

where $\Delta\mu \equiv (\mu_s - \mu_\ell), t \equiv (T_t - T)/T_t$, and n_ℓ is the surface density of the liquid phase. The power law is a consequence of long range interactions between the liquid–solid and liquid–vapor interfaces. In the case of dispersion forces, where pair potentials vary as $(r)^{-6}$, the interfacial exponent $p = 3$. (We ignore retardation effects, which begin to be important at distances greater than several tens of atomic diameters.) Short range forces, i.e. those varying exponentially with separation, produce logarithmic temperature dependence[50,52]. The surface term in this case has the form

$$\delta F_{\text{surf}} = -A\Delta\gamma[1 - \alpha' exp(-L_\ell/\varsigma)] , \tag{5}$$

which leads to a logarithmic dependence,

$$L_\ell = \varsigma \mid \ell nt \mid + \lambda'; \lambda' = \ell n(\alpha'\Delta\gamma/\varsigma\partial(\Delta\mu)/\partial T) . \tag{6}$$

The theory is appropriate also to melting at an interface with a foreign wall, in which case the γ's involve the wall in place of the vapor phase.

4.2 Surface Melting of Lead

An impressive series of measurements have been made on Pb crystal surfaces by proton backscattering[53]. The method is very sensitive to surface melting, because the backscattered intensity at certain angles is extremely low in good crystals due to blocking by the ordered rows of atoms. The signal is not strongly affected by surface roughening, where the atoms are still on the lattice positions[54]. The scattering is then primarily due to

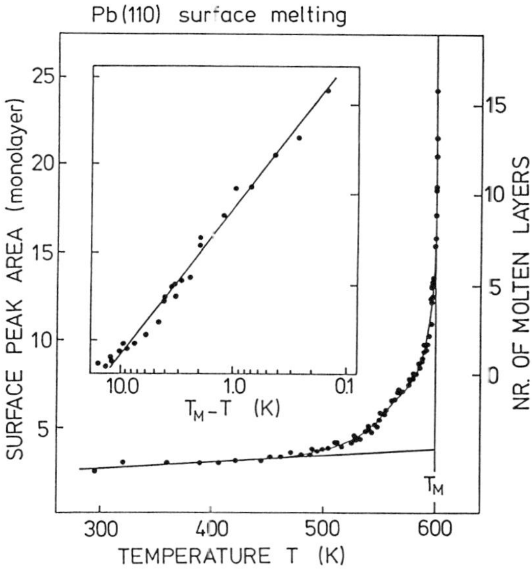

Fig. 6. Surface melting of (110)Pb crystal, obtained by proton backscattering[53]. The surface peak area is proportional to the thickness of the disordered surface layer; the thickness scale is obtained from energy loss measurement

surface structural disorder such as in a quasi liquid layer. The intensity from the (110)face, illustrated in Fig. 6 shows logarithmic dependence for a disordered layer thickness up to about 15 atomic layers, over a range of reduced temperature of nearly two decades. Thus the experimental temperature dependence indicates short range forces: the effective potential between the two surfaces of the liquid is exponential.

Short range forces may seem surprising in view of the ubiquity of dispersion forces: All substances, whether metallic or non conducting, exert long range interactions[55]. Thus, two objects of Pb at separations of a few Angstroms or more will normally attract each other by power law forces. But in the surface melting study the space between the two liquid interfaces is not vacuum or air but the liquid metal itself: The conductivity of the medium screens the charge fluctuations, producing the exponential type of interaction characteristic of the interior of all metals[56]. The empirical screening length in the Pb study corresponds to about 3 atomic distances. This range is somewhat greater than typical values quoted for good metals at low temperature, i.e. one atomic distance. However, since the experimental system is a liquid near its melting point, it is quite disordered, which increases the scattering of the electrons and increases the screening length[57].

4.3 Surface Melting of *Thick* Rare Gas Multilayers

Measurements on thick adsorbed Ar[31] and Ne[32] films have yielded detailed agreement with the theory of surface melting of molecular solids governed by dispersion forces. The experimental method in these studies is calorimetry, which shows strong precursors to the triple point anomalies beginning one or two degrees below T_t. Calorimetry was used previously to detect surface melting in thin metal plates[58] but the very small area to volume ratio of the samples prevented a determination of the temperature dependence of L_ℓ. Adsorbed multilayer films are particularly useful in calorimetric studies of surface melting, for one can produce much thinner samples with very large relative surface areas, and therefore large relative signal strength. Another advantage of multilayers for such studies is that their thicknesses can be controlled and measured to high precision, allowing additional experimental tests of the theory.

Argon was the first choice because it is one of the few adsorbates that can form thick uniform incommensurate solid films on graphite at low temperature. The data show many thickness–dependent features, particularly complex in the thinner films, where discrete layer phenomena are evident. The few–layer results will not be discussed here: we focus instead on the thicker range, where a strong peak develops near T_t. Figure 7 presents data for films thicker than five layers. Each shows a gradual rise, beginning a few K below the triple point to a peak at $T_p < T_t$ (83.80 K) and a more rapid fall at higher temperature. The peak height grows and T_p increases with L. The peak of the thickest sample ($L = 40$) is impressively high ($C/Nk > 350$) and narrow, with the same type of asymmetry seen at lower coverage. The peaks evidently tend toward a normal first order melting transition with increasing thickness.

When surface melting occurs, the heat capacity will have a contribution due to the progressive conversion from solid to liquid. In thick samples, the phases should be homogeneous, except for gradients close to the interfaces. If finite thickness, interfacial and gradient effects are expressed as a lumped correction δC, the heat capacity at constant total N is the sum

$$C = Nc_s + N_\ell(c_\ell - c_s) + T\Delta s(dN_\ell/dT) + \delta C \ , \qquad (7)$$

N_ℓ, N_s and $(c_\ell - c_s)$ are the populations and specific heats of uniform phases. The terms due to premelting should be distinguishable by their T dependence: N_ℓ and dN_ℓ/dT vary with the difference $\Delta T = (T_t - T)$, while the remaining *background* is a function of T. In the limit of infinite thickness N_ℓ diverges as $\Delta T \rightarrow 0$, hence premelting dominates near T_t, and of the two premelting terms, the second, involving (dN_ℓ/dT) becomes far stronger. The background–corrected data of the four thickest films are shown as a function of ΔT, in Fig. 8. The films exhibit power law behavior ΔT^{-r} over about one decade in ΔT. There is no obvious trend of r with thickness, and the values are quite close: the average for the four films,

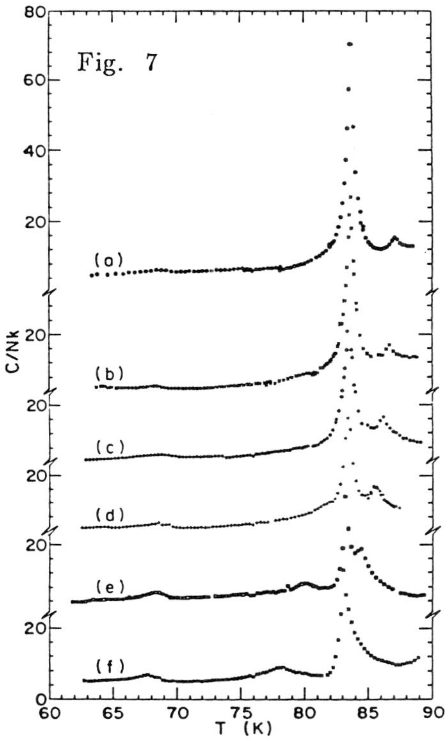

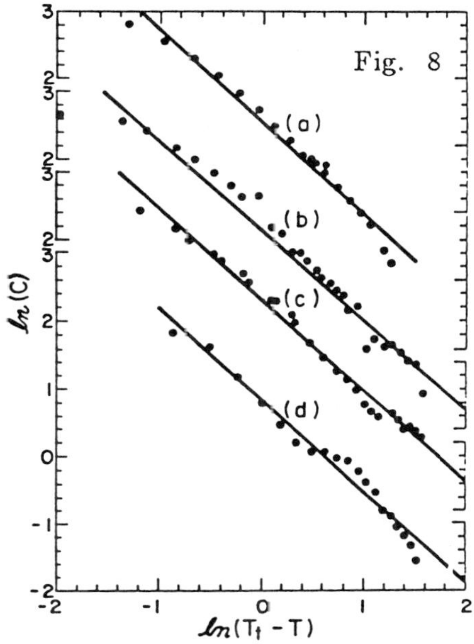

Fig. 7. Specific heat of argon/graphite films, corrected for desorption, in dimensionless units[31]. N, Avogadro number; k, Boltzmann constant. Thicknesses in layers at 83 K are as follows: curve a, 10.5; curve b, 8.9; curve c, 8.0; curve d, 7.1; curve e, 6.2; curve f, 5.3. The rises beginning at $\approx$ 80 K are due to surface melting, and the weak anomalies at 68 K, to surface roughening. The sharp cusps above 83 K are due to the melting of a compressed monolayer next to the substrate

Fig. 8. Background-corrected heat capacities of the four thickest films versus relative temperature[31]. The empirical exponents r are as follows: curve a, 1.38; curve b, 1.32; curve c, 1.33; curve d, 1.35. Letters refer to the same thicknesses as in Fig. 7

after minor corrections is 1.36 ± 0.05, experimentally indistinguishable from 4/3, the characteristic value for (dN_ℓ/dT) in a system governed by dispersion forces.

Finite film thickness in the range above $L \approx 5$ appears to have no significant influence on the exponent, but the monotonic drift of T_p toward T_t can be understood as a simple effect of finite thickness. The trend is

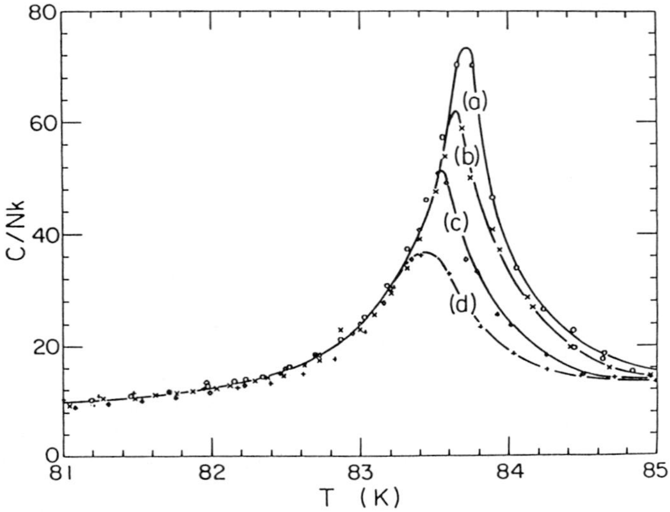

Fig. 9. Expanded view of the Ar multilayer melting peaks, showing the common envelope and the progression of peak temperatures with increasing thickness[31]. The thicknesses indicated by the letters are the same as in Figs. 7 and 8

shown in Fig. 9, which superposes the specific heats near T_t of several samples. The data are enclosed in a common envelope, but each curve drops away at some temperature T_p which increases with L. This can be understood as follows: As melting proceeds the thickness of liquid increases rapidly as T approaches T_t, until its progress is truncated by reaching the substrate interface. If there is a highly compressed thin solid layer next to the substrate, the truncation occurs before the entire film is melted, i.e. at some finite value $L_\circ$. Since C is dominated by the term involving dN_ℓ/dT, truncation causes a strong drop in C. Thus we expect that the T_p's should follow a power law with an exponent characteristic of dispersion forces;

$$(L - L_\circ) = \text{const.}(T_t - T_p)^{-1/3} \ . \tag{8}$$

Figure 10 shows that the peak temperatures do obey Eq. (8), with exponent -0.35, and $L_\circ$ corresponding to 1 layer. The empirical value for $L_\circ$ indicates that a single layer closest to the substrate has a melting temperature appreciably different from the others. This in turn suggests that it is the melting of this single layer that causes the small peak at $T > T_t$. Its shift to higher T in thicker films is attributed to compression by the substrate field.

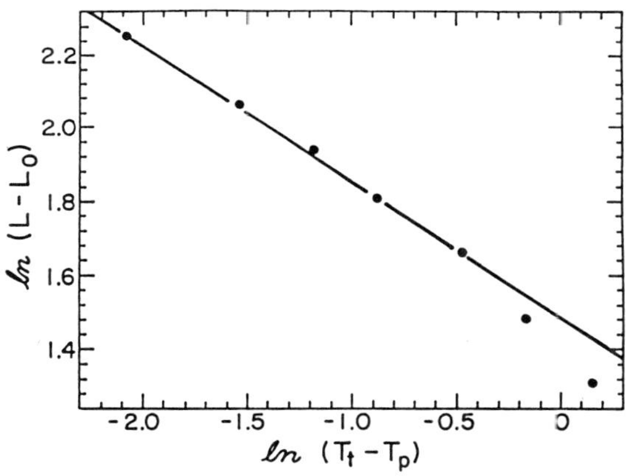

Fig. 10. Variation of peak temperature T_p with thickness, showing that $(T_t - T_p)$ varies as a power law with thickness

4.4 Why Does Mean Field Theory Work?

The mean theory assumes smooth interfaces...it ignores the effects of thermal fluctuations, which cause capillary waves in the liquid–vapor surface and roughening of the solid–vapor boundary. It is therefore surprising that the Ar and Pb results should agree so well with the classical theory. However, there is a nice proof that the mean field approximation should be valid.

The proof involves the concept of the *upper marginal dimensionality* d^*, introduced some years ago in the renormalization–group analysis of phase transitions[59-61]. When the physical dimensionality d of a system is larger than d^*, the system exhibits classical behavior, so that the critical exponents will have their mean field values. The upper marginal dimensionality is that value of d which satisfies the hyperscaling relation

$$\nu d = 2 - \alpha \ , \tag{9}$$

with classical exponents ν and α. It has been shown[62] that for surface melting governed by an effective potential between interfaces $v(L) \sim L^{-p}$, the marginal dimensionality is

$$d^*(p) = (3p + 2)/(p + 2) \ . \tag{10}$$

Since $p = 2$ for the interfaces of three dimensional van der Waals systems, $d^* = 2$; thus the classical exponent is predicted.

In contrast, $d^* = 3$ for surface melting governed by short–range forces[62], such as those apparently involved in the surface melting of

Pb. In systems where $d^* = d$ there are logarithmic corrections near the phase transition, but they are probably not noticeable except in extremely precise measurements.

4.5 Thinner Films: the Profile of the Crystal-Melt Interface

The basic theory assumes that the interface between solid and liquid is abrupt, and that the properties of the liquid are independent of its thickness. The assumption is unrealistic; indeed, very recent studies show effects due to finite interface thickness. Krim, Bouzidi, and Coulomb examined O_2 adsorbed on graphite by neutron diffraction[63] and Bienfait studied methane by inelastic neutron scattering on MgO[64], while GAY and SUZANNE have studied the latter system by neutron diffraction[65]. Both studies give clear evidence of the formation of a liquid film at temperatures below the bulk triple point. Since the systems are composed of van der Waals molecules, one might expect power law temperature dependence, as in the Ar films, but the O_2 and methane results show approximately logarithmic growth (see Figs. 11 and 12). KRIM et al.[63] suggested that logarithmic dependence is associated with the films' thinness: the thickness of the liquid was only one to four molecular layers, whereas in the Ar study the 1/3–power law was obtained from liquid thicknesses of seven to nine layers. Their publication was soon followed by a calculation of a Lennard–Jones model exhibiting a crossover from logarithmic to power law growth as the liquid thickness is increased above a few layers[66] and by an experimental study of Ne films demonstrating the effect[32].

Logarithmic growth results from exponential interactions: in the very thin van der Waals films one expects that the short range nature comes from *effective interactions* in contrast to the real exponentially screened forces in metals. The effective interactions arise from the extension of solid like structure and properties a short distance into the melt liquid. The exponential decay of the solid order parameter introduces some exponential

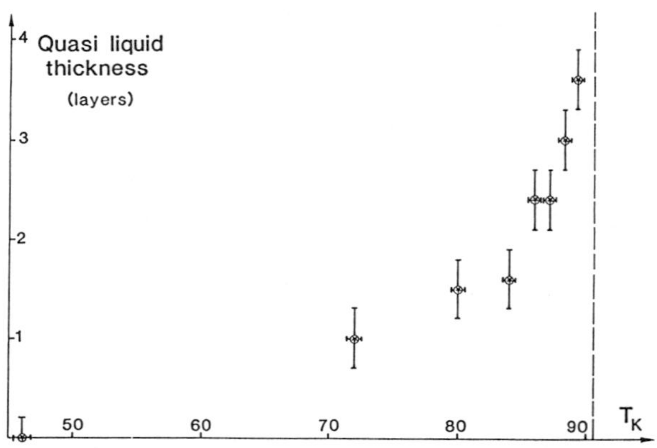

Fig. 11. Surface melting of CH_4/MgO, obtained by inelastic neutron scattering[64]

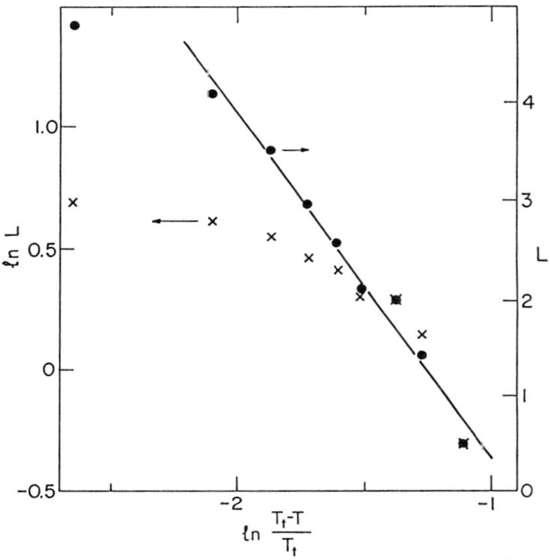

Fig. 12. The temperature dependence of the effective thickness of surface-melted liquid in methane/MgO films, obtained by neutron diffraction[65]. The logarithmic dependence is clearly favored

character into the effective interaction between the interfaces bounding the liquid. At distances larger than the correlation length of the order parameter the growth settles into the dispersion force power law.

In a study of Ne following the initial calorimetric work on Ar multilayers, average power law exponents were obtained from the thickness range $7 < L < 24$. It was found that the average exponent changes from about 1 at $L = 7$ to $\sim 4/3$ at $L = 24$, consistent with the expected crossover between short range and long range behavior.

But in spite of the agreement, the test is only a qualified success. The precision of the empirical exponents is not great in view of the small difference between the two theoretical values. The fitting range is limited because the exponents are not in fact constant throughout the transition. It would be desirable to relate the crossover to the physical properties of the system; furthermore, a basic assumption of the analysis, that of two sharply distinct phases, is incorrect. The analysis below avoids these deficiencies.

Theories and computer simulations of molecular systems have shown that the correlation length of the solid order parameter in simple molecular systems is on the order of a few molecular diameters[67-70]. For particular relevance here, there is a gradual change in the entropy density $s(z)$ with distance z from the solid boundary. An approximation describes the gradation as an exponential attenuation of solid-like properties with distance from the solid boundary. More realistic calculations show that the change begins a short distance on the solid side of the boundary, so that the profile has a sigmoid shape in the central region, while the trend at greater distances on the liquid side is probably exponential: a simple theory approximates the entire range by a hyperbolic tangent[70]. The data can be described quite well by both the exponential and hyperbolic tangent functions, as outlined below.

We first assume a simple exponential form; the entropy density $s(z)$ varies with distance z from the solid as

$$s(z) = s_s + \Delta s(z) = s_s + \Delta s_m(1 - \exp[-(z/\varsigma'])\ , \qquad (11)$$

where Δs_m is the transition entropy density of bulk phases and ς' is the coherence length of the solid order parameter. We neglect the diffuseness of the liquid–vapor interface, a reasonable approximation if ς' is much greater than the interface width, so that Eq. (11) with constant ς' is assumed to extend throughout the liquid, to a sharp cut off at the vapor boundary. The total transition entropy due to melting a finite thickness of film is an integral of $\Delta s(z)$; with $L'_\circ$ as a fitting parameter for the number of unmelted layers, the total entropy of melting $(L - L'_\circ)$ layers is

$$\Delta S = A\rho\Delta s_m(L - L'_\circ - \varsigma'[1 - exp\{(L'_\circ - L)/\varsigma'\}])\ , \qquad (12)$$

where A is the area and ρ is the areal density of a liquid layer. L is known from the total thickness, A and ρ by vapor pressure isotherms and Δs_m by bulk measurements. With the two empirical parameters $L'_\circ$ and ς' Eq. (12) can be fitted to the data for both Ne and Ar, with standard deviations of about 8%. Slightly better fits can be made with a hyperbolic tangent profile $\Delta s(z) = (1/2)\Delta s_m[1 + \tanh(z/\varsigma)]$. At $z >> \varsigma, \Delta s$ approaches the bulk value as $exp(-2z/\varsigma)$, so $\varsigma/2$ is the effective coherence length at large distances. 2ς is the interface *width*; the size of the central region where Δs undergoes its major change. Integration for ΔS also requires a parameter, defined as $L_\circ$, to position the center of the interface relative to the substrate. The total transition entropy is

$$\Delta S = \rho A\Delta s_m(L - L_\circ + (1/2)\varsigma\ell n[\frac{1 + e^{-2(L-L_\circ)/\varsigma}}{1 + e^{-2L_\circ/\varsigma}}]);. \qquad (13)$$

Figure 13 displays the experimental $\Delta S(L)$ of Ne and Ar, with fitted curves according to Eq. (13). The best fit parameters are Ne: $\varsigma = 3.0, L_\circ = 6.0$; Ar: $\varsigma = 1.9, L_\circ = 5.0$. Standard deviations between the data and the fitted curves are 6%(Ne) and 7%(Ar). We also show the bulk entropy for the same thickness range,i.e. how $\Delta S(L)$ at large L tends to parallel the bulk lines but is shifted: the shift is approximately 7 layers in Ne and 5 layers in Ar. The shifts are due to a combination of substrate and

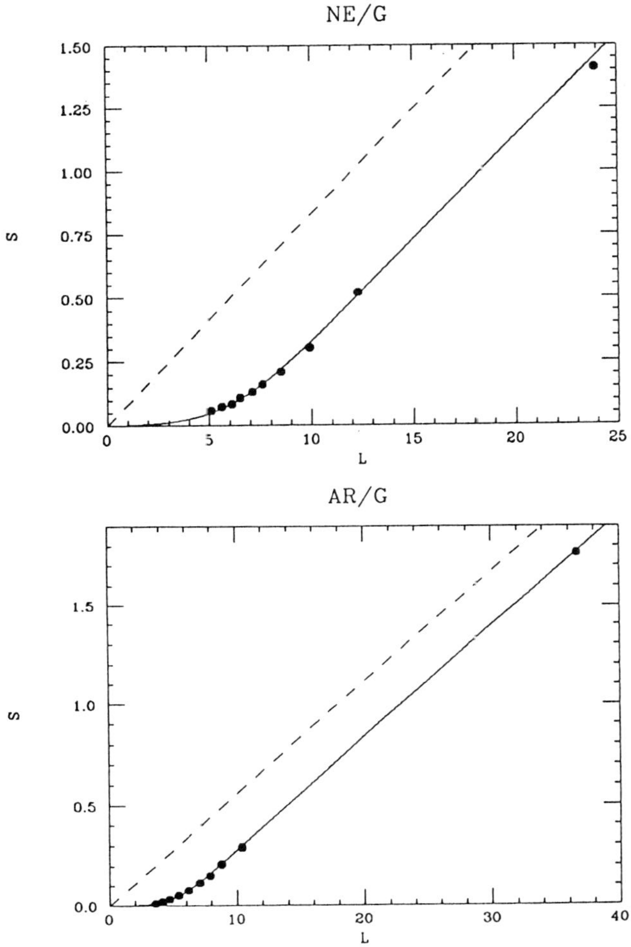

Fig. 13. Total entropy in the melting peaks of Ne (a) and Ar (b) films as functions of thickness. The broken lines correspond to the melting of equivalent quantities of bulk material, i.e. without surface corrections. The solid lines are calculated due to the finite width of the crystal-melt interface, and show the suppression of the total melting entropy due to the intrusion of solid order in the liquid adjacent to the solid boundary

crystal–melt interface effects. A cusp shaped peak in the heat capacity at temperatures above the main anomaly is attributed to the melting of one substrate–field compressed layer; added evidence for compression is that the cusp contains only 0.2-0.3 of the transition entropy of an equal amount of bulk. Thus $\Delta S(L)$ from the main anomaly is due to the melting of $L-1$ layers; the remaining suppression is associated with the interface width: 6 layers in Ne and 3.8 in Ar.

The experimental widths are comparable to theoretical estimates and computer simulations. However, the difference between the Ne and Ar values is unexpected in view of the similarities of the two systems. The difference is well outside of statistical uncertainty, and seems unrelated to the details of the assumed interface profile function: a fit to the simple exponential form yields the same ratio of empirical correlation lengths.

4.6 Surface Melting of Ice

In the beginning of Sec. 4 we noted that the freezing and melting behavior of ice and snow prompted the earliest proposals of surface melting. The subject has been controversial for much of the period since then. Numerous experiments have explored the QLL indirectly through macroscopic properties of ice surfaces, but there have been few unequivocal direct observations of the melted layers. One of the most convincing is a very recent study, by means of ellipsometry[71]. Liquid layers, with an index of refraction virtually identical to that of equilibrium liquid water near 0°C, were detected on two faces below the triple point: beginning at -2°C on the (0001) face and between -2 and -4°C on the (1010) face. The temperature range of the liquid film is quite narrow: if its appearance scaled like that of surface melting of rare gas solids, it would begin several tens of degrees below the triple point. It seems that more study of surface melting in ice is needed today, nearly one hundred and fifty years after the question was first raised.

5. TRIPLE POINT WETTING

Wetting, as a formal scientific topic, began at least as far back as 1805, with Young's study of capillarity[72]. There are comprehensive monographs on the subject[73], and several excellent reviews of recent developments are available[74]. Our interest here is on a limited aspect of the field, involving relationships between wetting and surface melting: a recently discovered transition called *triple point wetting.*

The distinction between wetting and nonwetting can be seen in the geometry of contact between a plane wall and two different substances, or phases of the same substance. If a phase β partially wets the interface between the α phase and the wall, the $\alpha\beta$ interface intersects the wall at an angle θ given by

$$\cos\theta = (\gamma_{\alpha W} - \gamma_{\beta W})/\gamma_{\alpha\beta} \; . \tag{14}$$

Complete wetting corresponds to $\theta = 0$; the β phase spreads out completely, forming a uniformly thick layer of β between α and the wall. It is obvious that complete wetting of liquid films is more likely on strongly attractive substrates, where adhesive forces between the liquid and the wall are stronger than the cohesive forces within the liquid. That it should also be the condition for the setting of solid films had been unquestioned, but in 1984 it was shown to be quite wrong. An electron diffraction study[75] of several molecular gases adsorbed on single crystal graphite at low temperature revealed nonwetting for strongly as well as weakly adsorbed gases. Thick equilibrium films could be grown with a few gases in which the cohesive and adhesive forces are comparable: the others could form only two or three molecular layers, after which bulk crystals were nucleated. The authors suggested that the cause of incomplete wetting on strong substrates is the compression of layers near the wall. At about the same time, a model calculation[76] showed such an effect: sufficiently attractive substrates could force the adjacent layers to high density, such that further layers could not grow epitaxially, and bulk growth would ensue. Subsequent theory explored the mechanism in detail.[77-79] An experimental test of the theory was carried out soon after, examining film growth of the same gases on a different substrate.[80] The Au(111) substrate has an attractive potential appreciably greater than graphite; according to the strain mechanism it was expected that the few adsorbates (Ar, Kr, and Xe) that completely wet graphite would not wet gold. This was exactly what was found. As further support, all of the adsorbates crossed over to complete wetting behavior above their triple points. Thus, since no shear distortions could propagate through the liquid, there could be no barrier to the adsorption of unstrained additional layers.[80]

The crossover at T_t could be expected, but the nature of the transition was a surprise. In each system the maximum thickness of uniform film n_c (i.e. at coexistence with the saturated vapor) increased continuously with temperature as a power law, proportional to $[(T_t - T)/T_t]^{-1/3}$. The adsorption isotherms provided a key to explaining the function. Over an interval of temperature below T_t the thickness versus pressure isotherm appears to tend to an infinitely thick film at a virtual P_o' corresponding to the vapor pressure curve of the bulk liquid, extended to $T < T_t$, until abruptly truncated by reaching the vapor pressure of the solid. The interpretation was that the topmost layers of unsaturated film below the triple point are liquid like, extensions of the stable liquid phase to temperatures below the region of true thermodynamic stability. The metastable liquid film could be thickened by adsorption until P reached the actual P_o of the equilibrium phase. Further adsorption produced bulk solid crystal growth. The exponent of the growth law follows from the range dependence of the substrate potential. It is obvious now that triple point wetting is a consequence of surface melting.

Triple point wetting has been observed in other systems, by several experimental techniques. Since wetting at low temperature in equilibrium systems is a rare and accidental phenomenon and surface melting is believed to be typical of most surfaces, triple point wetting should be quite general in all types of film systems.

I am indebted to many colleagues and students for stimulation, advice, and my continuing education during many years of study and research in film phenomena. I am especially grateful to M. Bienfait, J. Krim, M. Schick, J. Suzanne, and D.-M. Zhu, who have done so much to develop the topics of this review. Support by The National Science Foundation, Grant DMR86-11466 is warmly acknowledged.

References

1. R. Pandit, M. Schick, and M. Wortis: In Phys. Rev. B 26, 5112 (1982)
2. J.W. Gibbs: In *Collected Works*, Vol. 1, Thermodynamics (Longmans, Green and Co. 1906); (Dover Reprint 1961) p. 325 footnote
3. P. Curie: Bull. Soc. Fr. Mineral. 8, 145 (1885)
4. G. Wulff: Z. Cryst. 34, 449 (1901)
5. M. Volmer and I. Estermann: Z. Phys. 7, 13 (1921)
6. I.N. Stranski: Z. Phys. Chem. 136, 259 (1928)
7. F.C. Frank: Phil. Mag. 41, 200 (1950)
8. J. Frenkel: J. Phys. USSR 9, 392 (1945)
9. W.K. Burton and N. Cabrera: Disc. Faraday Soc. no. 5, 33, 40 (1949)
10. W.K. Burton, N. Cabrera, and F.C. Frank: Phil. Trans. Roy. Soc. London A 243, 299 (1951)
11. J.D. Weeks: In *Ordering in Strongly Fluctuating Condensed Matter Systems*, ed. by T. Riste (D. Reidel Publ. Co. 1979)pp. 293-318; J.P. van der Eerden, P. Bennema, and T.A. Cherepanova: In *Prog. Cryst. Growth & Characterization*, ed. by B.R. Pamplin, Vol. 3 (Pergamon 1979)p. 219; J.D. Weeks and G.H. Gilmer: Adv. Chem. Phys. 40, 157 (1979)
12. R.H. Swendsen: Phys. Rev. B 17, 5421 (1977); N. Quircke and G. Jacucci: Surf. Sci. 144, 92 (1984)
13. J.Q. Broughton and G.H. Gilmer: J. Chem.Phys. 79, 5119 (1983)
14. V. Rosato, G. Ciccotti, and V. Pontikis: Phys. Rev. B 33, 1860 (1986)
15. Reviews of recent computational results are given by Y. Saito and H. Muller-Krumbhaar (p. 223) and by K. Binder, et al. (p. 299): In *Applications of the Monte Carlo Method in Statistical Physics*, ed. by K. Binder (Springer-Verlag, Berlin, 1987)
16. C. Jayaprakash and W.F. Saam: Phys. Rev. B 30, 3916 (1984); C. Rottman and M. Wortis: Phys. Repts. 103, 59 (1984)
17. A. Pavlovska and D. Nenow: Surf. Sci. 27, 211 (1971)
18. K.A. Jackson and C.E. Miller: J. Cryst. Growth 40, 169 (1977)
19. H.E. Buckley: In *Crystal Growth*, (Wiley 1951) p. 10
20. B.E. Sundquist: Acta Metall. 12, 67, 585 (1964); M. Drechsler: In *Surface Mobilities on Solid Materials*, ed. by V. Binh (Plenum 1983), p. 405; J.C. Heyraud and J.J. Metois: J. Cryst. Growth 50, 571 (1980); Surf. Sci. 128, 334 (1983); C. Rottman, M. Wortis, J.C. Heyraud, and J.J. Metois: Phys. Rev. Lett. 51, 1009 (1984)
21. K.O. Keshishev, A.Ya. Pashin, and A.I. Shalnikov: Phys. Rev. 14, 155 (1984); S. Balibar and B. Castaing: Surf. Sci. Rept. 5, 87 (1985)
22. S.G. Lipson: Contemp. Phys. 28, 117 (1987)
23. A. Thomy and X. Duval: J. Chim. Phys. 66, 1966 (1969); 67, 286 (1970)

24. J.J. Hamilton and D.L. Goodstein: Phys. Rev. B 28, 3838 (1983); D.L. Goodstein, J.J. Hamilton, M.J. Lysek, and G. Vidali: Surf. Sci. 148, 187 (1984)
25. M. Drir and G.B. Hess: Phys. Rev. B 33, 4758 (1986)
26. H.S. Nham, M. Drir and G.B. Hess: Phys. Rev. B 35, 3675 (1987)
27. K. Madih: These de Doctorat, Univ. d'Aix-Marseille 1986 (unpublished)
28. M.J. de Oliveira and R.B. Griffiths: Surf. Sci. 71, 687 (1978)
29. Y. Larher and F. Millot: J. de Phys. 38, C4-189 (1977)
30. W.F. Saam: Surf. Sci. 125, 253 (1983)
31. Da-Ming Zhu and J.G. Dash: Phys. Rev. Lett. 57, 2959 (1986)
32. Da-Ming Zhu and J.G. Dash: to be published
33. V.T. Binh, R. Uzan, and M. Drechsler: J. Phys. Lettres 39, L-385 (1978)
34. J. Lapujoulade, J. Perreau, and A. Kara: Surf. Sci. 129, 59 (1983); J. Villain, D. Grempel, and J. Lapujoulade: J. Phys. F: MetalPhys. 15, 809 (1985); F. Fabre, D. Gorse, B. Salanon, and J. Lapujoulade: J. Physique 48, 1017 (1987)
35. M. den Nijs, E.K. Riedel, E.H. Conrad and T. Engel: Phys. Rev. Lett. 55, 1689 (1985); 56, 4028 (1986)(Erratum); E. Conrad, R. Aten, D. Kaufman, L. Allen, T. Engel, M. den Nijs, and E. Riedel: J. Chem. Phys. 84, 1015 (1986); 85, 4756 (1986)(Erratum)
36. M. Faraday: In *Faraday's Diary*, Vol. IV, (Bell and Sons, London 1933) pp. 79-81. Faraday's first public report of his work on the subject was in a lecture at the Royal Institute in 1850.
37. Brief reviews of Faraday's work and Thomson's opposition are given by W.A. Weyl: J. Colloid Sci. 6, 389 (1951) and P.V. Hobbs: In *Ice Physics*, (Clarendon Press 1974) Ch. 6
38. J. Tyndall: Proc. Roy. Soc. 9, 76 (1856)
39. J. Thomson: Proc. Roy. Soc. 11, 198, 473 (1861)
40. J.K. Roberts: In *Heat and Thermodynamics*, (Blackie & Son, London 1940) p. 322
41. F.P. Bowden and T.P. Hughes: Proc. Roy. Soc. A 172, 280 (1939)
42. W.A. Weyl: J. Colloid Sci. 6, 389 (1951)
43. F.A. Lindemann: Z. Physik 11, 609 (1910)
44. The Lindemann model is nevertheless useful as an interpolation formula linking melting temperatures to vibrational properties of similar solids. It is the basis of a phenomenological model of surface melting by L. Pietronero and E. Tosatti: Sol. State. Commun. 32, 255 (1979)
45. The confusion may arise from the limitations of some theoretical models used to describe surface disorder, having only two possible phases. Burton, et al.[10], for example, equate the roughening transition with surface melting.
46. I.N. Stranski: Z. Physik 119, 22 (1942)
47. J. Frenkel: In *Kinetic Theory of Liquids*, (Clarendon Press 1946) pp. 425-6
48. J.K. Kristensen and R.M.J. Cotterill: Phil. Mag. 36, 437 (1977)
49. T. Kuroda and R. Lacmann: J. Cryst. Growth 56, 189 (1982); D. Nenow and A. Trayanov: J. Cryst. Growth 79, 801 (1986)

50. R. Lipowsky: Phys. Rev. Lett. 49, 1575 (1982)
51. The inequality appears to conflict with *Antonow's rule* for complete wetting (see e.g. J.S. Rowlinson and B. Widom: In *Molecular Theory of Capillarity*, (Oxford Univ. Press 1982) pp. 212-214). If the interface between two phases α and γ is wetted by a third phase β, then Antonow's rule states that the condition on the interfacial coefficients is an equality, $\gamma_{\alpha\gamma} = \gamma_{\alpha\beta} + \gamma_{\beta\gamma}$. However, the rule applies only in the condition that each substance is an equilibrium phase, whereas in surface melting the wetting film is a *quasi liquid layer* at a temperature below the stability range of the equilibrium liquid.
52. J.Q. Broughton and G.H. Gilmer: Acta Met. 31, 31 (1983); J. Chem. Phys. 79, 5119 (1983)
53. J.W.M. Frenken and J.F. van der Veen: Phys. Rev. Lett. 54, 134 (1985); I.W.M. Frenken, P.M. Maree, and J.F. van der Veen: Phys. Rev. B 34, 7506 (1986)
54. Surface relaxation and reconstruction distort the lattice near the surface, introducing a type of disorder distinct from the liquid like melted layer. This may contribute to a slowly varying background at relatively high temperatures. Such a background is easily distinguished from the disorder caused by melting, since its temperature dependence can have no special relation to T_t.
55. I.E. Dzialoshinskii, E.M. Lifshitz, and L.P. Pitaevskii: Adv. in Phys. 10, 165 (1961)
56. See e.g. N.W. Ashcroft and N.D. Mermin: In *Solid State Physics*, (Saunders College, Philadelphia 1976) p. 330 ff
57. R.E. Peierls and E.A. Stern: private communications
58. G. Fritsch, H. Diletti, and E. Luscher: Phil. Mag. A 50, 545 (1984)
59. A. Aharony and B.I. Halperin: Phys. Rev. Lett. 35, 1308 (1975)
60. G. Pfeuty and G. Toulouse: In *Introduction to the Renormalization Group and Critical Phenomena*, (Wiley, N.Y. 1975) Ch. 2
61. J. Als-Nielsen and R.J. Birgeneau: Am. J. Phys. 45, 554 (1977)
62. R. Lipowsky: Phys. Rev. Lett. 52, 1429 (1984); D.M. Kroll, R. Lipowsky, and R.K.P. Zia: Phys. Rev. B 32, 1862 (1985)
63. J. Krim, J.P. Coulomb, and J. Bouzidi: Phys. Rev. Lett. 58, 583 (1987)
64. M. Bienfait: Europhys. Lett. 4, 79 (1987)
65. J.M. Gay and J. Suzanne: private communication
66. A. Trayanov and E. Tosatti: Phys. Rev. Lett. 59, 2207 (1987)
67. D.M. Zhu and J.G. Dash: Phys. Rev. Lett. 60, 432 (1988)
68. A. Bonissent: In *Interfacial Aspects of Phase Transformations*, ed. by B. Mutaftschiev, (Riedel Publ. Co., Holland 1982) pp. 143-182
69. J.Q. Broughton and F.F. Abraham: Chem. Phys. Lett. 71, 456 (1980); A. Bonissent and F.F. Abraham: J. Chem. Phys. 74, 1306 (1981)
70. W.A. Curtin: Phys. Rev. Lett. 59, 1228 (1987)
71. Y. Furukawa, M. Yamamoto, and T. Kuroda: J. Cryst. Growth 82, 665 (1987)
72. T. Young: Phil. Trans. Roy. Soc. 95, 65 (1805)
73. R. Defay, I. Prigogine, and A. Bellemans: In *Surface Tension and*

Adsorption, trans. by D.H. Everett (Longmans Green, London 1966); J.S. Rowlinson and B. Widom: In *Molecular Theory of Capillarity*, (Oxford Univ. Press 1982)
74. Recent reviews on wetting are given by D.E. Sullivan and M.M. Telo da Gama: In *Fluid Interfacial Phenomena*, ed. by C.A. Croxton (Wiley, N.Y. 1986), and S. Dietrich: In *Phase Transitions and Critical Phenomena*, ed. by C. Domb and J. Lebowitz, Vol. 12 (Academic, London 1988)
75. J.L. Seguin, J. Suzanne, M. Bienfait, J.G. Dash, and J.A. Venables: Phys. Rev. Lett. 51, 122 (1983); M. Bienfait, J.L. Seguin, J. Suzanne, E. Lerner, J. Krim, and J.G. Dash: Phys. Rev. 29, 983 (1984)
76. C. Ebner, C. Rottman, and M. Wortis: Phys. Rev. 28, 4186 (1983)
77. R.J. Muirhead, J.G. Dash, and J. Krim: Phys. Rev. B 29, 5074 (1984)
78. D.A. Huse: Phys. Rev. B 29, 6985 (1984)
79. F.T. Gittes and M. Schick: Phys. Rev. 30, 209 (1984)
80. J. Krim, J.G. Dash, and J. Suzanne: Phys. Rev. Lett. 52, 640 (1984); J. Krim, Ph.D. Thesis, University of Washington, 1984 (unpublished)

Discussion

D.R. HAMANN: E.G. MacRae recently discovered a transition on the Ge(111) surface which occurs at $\sim 0.8T_M$, which appears to involve the melting of just a few atomic layers, and not a diverging melt thickness. He has ruled out the possibility that this is a roughening transition. This transition doesn't fit any of the cases you discussed. Can you comment?

J.G. DASH: MacRae's observations are intriguing, and they should stimulate others to study the Ge(111) surface with other probes. I agree that the results do not fit into the patterns of surface melting of Pb(110) or of the molecular films. It is possible that the Ge(111) is an example of "blocked melting" proposed recently by Tosatti.

S. A. RICE: It may be possible to extend the test of "surface melting" versus system dimensionality by using existing data for the helix-coil transition in polypeptides and polynucleotides. This transition is understood to occur because of the entropic difficulty of creating a turn of helix from random coil, which acts as a point edge tension. The statistical mechanics is well worked out (I can provide references) and the data are usually reported as transition temperature ("melting temperature" for helix $\rightarrow$ coil) as a function of number of units in the polymer. A reanalysis of these data, or even new measurements if necessary, could be used to test your ideas in the lowest possible physical dimension.

J.G. DASH: Thank you for pointing out these possible analogues; I had not known about the phenomena, and will look forward to reading about them.

E. TOSATTI: A comment on D. Hamann's question and another question to the speaker. The comment: The melting of only a few layers, as reported by McRae on Ge(111), could represent a case of "blocked" surface melting. As will be discussed later, the growth mode of the quasi-liquid layer as T approaches the triple point is extremely fragile, i.e., sensitive to long-range residual forces, usually deemed to be unimportant. We have found blocked melting in two separate systems: 1) (110) and (100) fcc surfaces with long-range pairwise *repulsive* forces [1]; 2) Short-range many-body "glue" forces modeling of unreconstructed Au(111), as studied by molecular dynamics [2]. In particular there is a striking similarity between these simulations of Au(111) and the Ge(111) results, which suggests a possible attack based on many-body forces, which of course are known to be crucial in a semiconductor, as much as in a metal.

The question to Prof. Dash is what is the present understanding about his reported mean-field-like behavior of edge melting in two dimensions? (References: [1] A. Trayanon and E. Tosatti, *Phys. Rev. Letters* **58**, 2207 (1987); [2] P. Carnevali, F. Ercolessi and E. Tosatti, *Phys. Rev. B(RC)* **36**, 6701 (1987); [3] D-M. Zu, D. Pengra and J.G. Dash, *Phys. Rev. b* **37**, 5586 (1988).

J.G. DASH: (See "General Discussion")

Y. PETROFF: I have two comments.

1. For a long time it was believed that only high index faces should show roughening transitions. In the last two years there have been many experiments showing that low index faces show roughening transition: Ag(110), Cu(110), Pb(100), Pb(110), etc.

2. In the case of Pb(110) and Ge(111) I do not think that proton backscattering or LEED are good to probe short range order. In fact photoemission experiments (*Surface Science* **189**, 605 (1987)) show that there is a strong short range order near melting, in contradiction with the results on Pb(110) (Franken *et al.*) which show fully disordered surface layers.

J.G. DASH: I am not sure that there is a disagreement between the techniques. Thin surface-melted layers are not fully disordered liquids, but retain considerable solid like long range lateral order. One has to be careful in distinguishing whether a particular technique is more or less sensitive to the disorder occurring at different distances from the solid interface.

D.R. HAMANN: Surface melting can be probed by looking at the azimuthal anisotropy with polarization of optical second harmonic generation at a surface. For a crystal with a center of symmetry, second harmonic generation only takes place at the surface (last few atomic layers). C.V. Shank *et al.* used this approach to confirm that Si surfaces are melted for a very short time following irradiation by a femtosecond laser pulse. Perhaps this technique can be used to probe equilibrium surface melting as well.

Theory of Surface Melting and Non-Melting

E. Tosatti

International School for Advanced Studies and International Centre for Theoretical Physics, Strada Costiera 11, I-34100 Trieste, Italy

When the temperature of a crystal approaches the bulk melting point T_M, some low-index surfaces begin to pre-melt below T_M (surface melting), and some do not (surface non-melting). In surface melting, a 'quasi-liquid' layer forms at the crystal-vapour interface. Its thickness diverges characteristically as some power of $1/(T_M - T)$, as $T_M - T$ tends to zero. In surface non-melting, conversely, the quasi-liquid layer does not form at all, or else it may begin to form, but then its growth is 'blocked' to some small thickness, which does not diverge.

Our present theoretical understanding of these phenomena is built upon three different approaches. In order of increasing microscopic contents, they are: (a) phenomenological Cahn-Hilliard theories; (b) computer simulations; (c) a statistical-mechanical lattice model.

The Cahn-Hilliard theories [1-3] are very helpful in setting the broad scenario of possibilities as a function of few simple quantities characteristic of the bulk system.

They are based on the introduction of one [1,2] or more [3] order parameters characterizing the solid, and on the subsequent study of their spatial behaviour at the solid-gas interface. The optimal configuration is obtained by minimizing a phenomenological free energy functional, in the normal Ginzburg-Landau spirit. This approach stresses the analogy of surface melting to the simpler problem of the wetting of a rigid substrate, [4] and shows in fact how many of the results [5] simply carry over.

At the opposite extreme, computer simulations are more likely to stress the specificities of each given surface. They are also harder to carry out and to understand. Particularly in the field of surface melting, simulations have not yet been really crucial. The heart of the problem is simply that finite size fluctuations do not permit a sufficient temperature resolution for the genuine quasi-liquid layer region (which may require control of $(1 - T/T_M)$ down to as little as 10^{-4} or less) to be reached. For example, the otherwise impressive MD study of surface melting by BROUGHTON and GILMER [6], while showing many precursor signs, such as disordering and softening, for all low-index surfaces of the (truncated) LJ crystal, does not really

Springer Series in Surface Sciences Vol. 14
Solvay Conference on Surface Science
Editor: F.W. de Wette

demonstrate - and even less disprove! - surface melting. Paradoxically, there appears to be no such temperature control problem for the case of surface non-melting which is probably well described by MD. This has been recently demonstrated by the MD study of the 'glue model' of Au(111) [7].

The main thrust towards understanding surface melting must now come from solving microscopic statistical mechanical models. We have recently moved one first step in this direction by studying a lattice model. At the cost of severe but well controlled approximations, a first true microscopic treatment of surface melting has been obtained [8]. Numerical implementation of this theory for various modifications of the basic LJ interparticle potential shows directly how surface melting is greatly enhanced by long-range attractive tails, but is blocked by repulsive tails, in agreement with expectations based on the phenomenological models [1,3,5].

1 R Lipowski and W Speth, Phys. Rev. **B28** 3983 (1983)
2 A C Levi and E Tosatti, Surf. Sci. **178** 425 (1986)
3 A C Levi and E Tosatti, Surf. Sci. **189/190** 641 (1987)
4 J W Cahn, J. Chem. Phys. **66** 3667 (1977)
5 S Dietrich, in *Phase Transitions and Critical Phenomena*, edited by C Doumb and J Lebowitz, Academic Press, London, 1987, Vol. 12
6 J Broughton and G H Gilmer, J. Chem. Phys. **79** 5095, 5105, 5119 (1983)
7 P Carnevali, F Ercolessi and E Tosatti, Phys. Rev. **B36** (RC) 6701 (1987)
8 A Trayanov and E Tosatti, Phys. Rev. Lett. **59** 2207 (1987); and to be published

Discussion

W. KOHN: I question whether surface melting (without roughening) is a true thermodynamic phase transition. For, as long as a finite number of atomic layers are "melted," they will have infinite range lateral order due to the solid substrate. Only as the "melted" region becomes thicker does the averaged lateral order continuously tend to zero. Thus I raise the question if surface "melting" should not be regarded as a continuous approach to three-dimensional melting, which certainly is a thermodynamic phase transition.

E. TOSATTI: I have not claimed the actual existence of a surface melting phase transition with thermodynamic singularities anywhere else than $T=T_M$. In particular, the calculated surface free energies in our lattice mean-field theory are certainly continuous all the way to T_M.

However, this does not mean that one or more singularities below T_M [1] could be ruled out in a better theory, or for different interparticle interactions than LJ, or in a true experimental situation. The long range lateral periodicity, certainly present due to the underlying crystal, does not necessarily imply a solid behavior of the quasi-liquid layer. It merely implies a density wave behavior, by which I mean a liquid with a periodic lateral density modulation. If that modulation is weak enough I argue that flow, diffusion and ergodic behavior are still expected in the fluid. In the solid, flow, diffusion and ergodicity are absent (in spite of defects and single action diffusion). I see no way to evolve continuously between the two situations, in spite of the fact that they have the same symmetry. To the extent to which one may neglect surface evaporation and condensation (which, when considered, make the solid surface ergodic at all temperatures), I believe there does exist a genuine question of how ergodicity is lost at a surface which begins to melt. ([1] In addition to roughening, which we argue should *necessarily* occur below T_M if surface melting is under way [A. C. Levi and E. Tosatti, *Surf. Sci.* **178**, 425 (1986)].)

J.G. DASH: I agree with Walter Kohn that surface melting is not a stand–alone transition; it is a surface precursor of a first–order transition, and its transition temperature is the bulk transition temperature. I think part of the issue has to do with whether or not we are to look on the premelted film as a phase—and I think clearly it is *not* a phase; that it cannot be. It is not the same material at all thicknesses; it *is* evolving, it *does* change. I also don't think that it is very profitable to try to pinpoint where it becomes a liquid—that is, how to define 'liquid'. It is an evolving state, and it has a changing order parameter, diffusivity, fluidity, crystallinity, thermal properties...and they're not, I think, one of these (even though I tend to favor thermodynamic properties). I don't think any one of these is to be treated as the criterion for what becomes a liquid. It is the surface melted phase, and as long as it is in contact with the solid, the region near the solid is this intermediate material.

E. TOSATTI: I agree that what we have been dicussing so far does not pinpoint any specific feature that needs to develop in a sharp fashion, so this growth or whatever, the specific heat...anything that we have been describing, both theoretically and experimentally, does not imply any sharp features, sharp transitions in the process. But this simply means that we have not looked carefully enough at fluctuations. For example, there *has* to be a roughening transition, and that is a singularity; if you neglect fluctuations, you do not find it. I think there may very well be other singularities, and my speculation is: if there are going to be any, then they must be connected with the quantity that is either there or not there, like symmetry or what have you, and that quantity, I believe, is ergodicity. The entropy of these two or three or four, five, six melted layers, would you agree, it's hard to distinguish between that and that of the liquid?

L. VAN HOVE: I'm afraid this discussion could go on for the rest of the week, because a thermodynamic limit means infinite lateral dimensions for the same surface layer, and that introduces many questions.

J.B. PENDRY: Concerning identifying a transition: you are thinking of the solid–liquid–gas interface, but what if you had an interface between two commensurate materials and then if you asked for a liquid transition between them? There would be an obvious criterion for whether you *did* have a liquid, and that is the response of the system to a shear. Would that be something that could resolve this question?

E. TOSATTI: Although I am not able to resolve this question, I am very grateful for this opportunity to talk about the open issues. The presence of a well-defined shear constant—or the absence of it—I believe is related to the question of whether you do have a macroscopic diffusion coefficient or not. Therefore, this is the dynamical... If you ask, is there a transition, is there a liquid or not? This is the question you would have to ask if you are doing dynamics, rather than statics, which I am doing so far. So, if there is any sharp transition, it would of course reflect in the shear constant sort of suddenly dropping away. I do not know of any studies that show anything conclusive about that, but is is a very interesting question for the future.

M. CARDONA: Concerning Landau-Ginzburg treatments of surface melting: it seems to me that the phenomenon is rather similar to the existence of surface superconductivity in a magnetic field up to H_{c3} if $H_{c3} > H_c$, and to the surface nucleation of the *ordered* phase if $H_{c3} < H_c$. These phenomena are actually equivalent to the surface solidification and non-melting. They are somewhat richer in phenomenology than the ones you describe since you can vary T as much as the applied H. When varying T in a *constant* H the surface layer thickness does not diverge at the critical T.

E. TOSATTI: Yes, the singularity is very interesting. There are in fact metals such as Cd and Zn, where liquid drops are indeed known to begin crystallization at the surface, while the bulk is still liquid. This would be the exact analogy. In the case of a superconductor, however, I have a side question. I do not understand how long-range phase coherence can be sustained given the absence of anisotropy, and with a thickness of order ξ. Is this not a realization of a 2D x-y system?

Is a Liquid Supported Amphiphile Monolayer Aptly Described as a Two-Dimensional System?

S.A. Rice

Department of Chemistry and the James Franck Institute, The University of Chicago, Chicago, IL 60637, USA

Liquid supported amphiphile monolayers have been studied for many years, yet their properties are surprisingly poorly known and little understood. For example, it is known that surface pressure (the difference between the surface tensions of the covered and clean liquid surfaces)-surface area per molecule isotherms of a liquid supported amphiphile monolayer resemble pressure-volume isotherms for a three dimensional system in that there are well defined regions of each that correspond to different phases [1]. It is generally agreed that when the surface area per molecule is very large, and the adsorbed molecules are independent of each other, the equation of state of the monolayer is like that of a gas. It is also generally agreed that when the surface density is increased a first order transition to a liquid phase occurs, unless the system temperature is less than the triple point temperature. When the surface density is increased further more phase transitions occur, first to a so-called "liquid-condensed" phase and then to a solid phase, but there is disagreement concerning the nature of the transitions [2]. Nothing is known of the intramolecular conformations or the intermolecular distribution functions in the several disordered phases. It is only recently, with the development of new experimental and theoretical techniques, that it is possible to begin to examine the relationships between molecular conformation and the structures of the different surface phases that amphiphile monolayer systems typically exhibit.

Consider the following question: Is a liquid supported amphiphile monolayer best described as a two dimensional system or as a "more than two dimensional" system? Put another way, are the properties of such a monolayer primarily determined by head group interactions or by tail conformation and tail group interactions?

RICE and coworkers [3-6] have carried out several theoretical and experimental studies that yield information relevant to this question. Consider, for example, the conformation of an adsorbed amphiphile molecule at infinite dilution. It is commonly argued that in this limit the tail of the amphiphile lies in the plane of the surface [1], presumably because the attractive interaction with the surface is a more important contribution to the free energy than is the entropy associated with the multitude of chain conformations available when the

Springer Series in Surface Sciences Vol. 14
Solvay Conference on Surface Science
Editor: F.W. de Wette

amphiphile tail projects into the space above the surface. POPIELAWSKI and RICE [3] have described a generalized regular solution model of a liquid supported monolayer of long chain amphiphile molecules. The model, which includes the effects of chain flexibility in the surface and in the space outside the surface layer, leads to selfconsistent equations for the distribution of molecular configurations and the surface density and surface pressure. An examination of available data [2,7] for pentadecanoic acid on water (pH = 2) leads to the strong inference that even at infinite dilution individual molecules take up configurations with a large part of the chain outside the surface layer. This inference is supported by the results of Monte Carlo simulations (Fig. 1) described in the next paragraph.

HARRIS and RICE [5] have carried out Monte Carlo simulations of a lattice model of a supported monolayer of amphiphile molecules. The lattice model allows for the flexibility of the adsorbed molecules and treats their full three dimensional structure, including the effects of nearest neighbor interactions between segments of the amphiphile chains, an attraction between the chain segments and the underlying substrate, and the inter and intrachain excluded volume. It is found that when the surface coverage is very small the conformations of the chains are sensibly independent of the density and, for the parameter set studied, the chains have considerable extension out of the plane of the surface. As the surface coverage increases the interchain interactions compress the chain parallel to the surface and force extension along the normal to the surface (Figs. 2 and 3). It is worthwhile noting that the decrease in

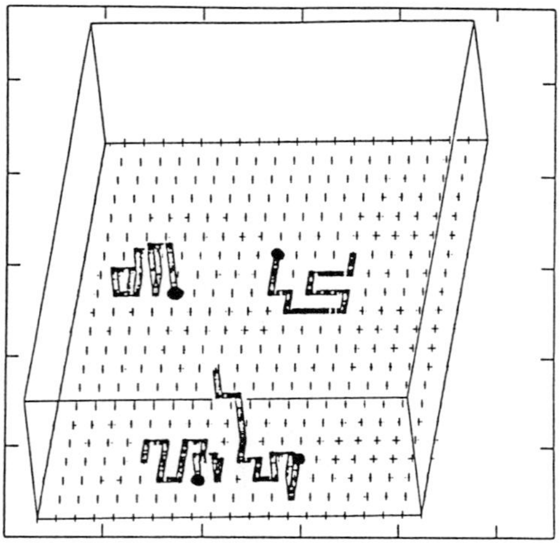

Fig. 1. A sample portion of a monolayer configuration for a chain of 15 segments. The surface coverage is 1% and the parameters used are valid for a system such as pentadecanoic acid on water. T = 200 K.

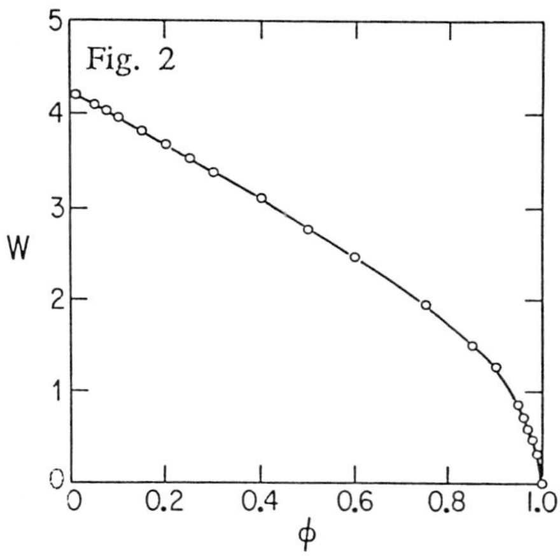

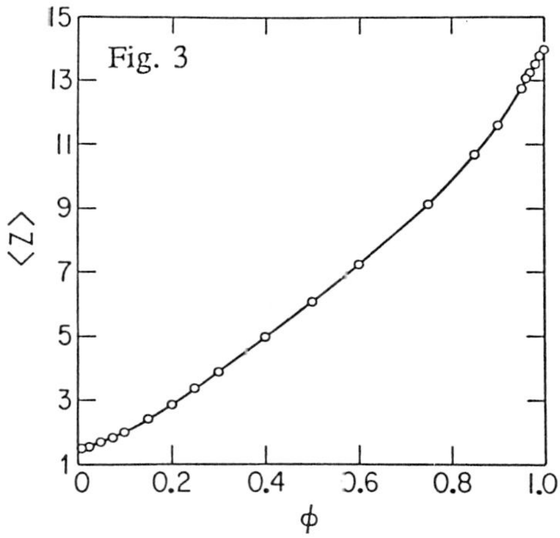

Fig. 2. W, the average end-to-end distance of the projection of 15 segment chains in the surface plane as a function of surface coverage. T = 300 K.

Fig. 3. The average distance from the surface of the last segment of each chain as a function of surface coverage. T = 300 K.

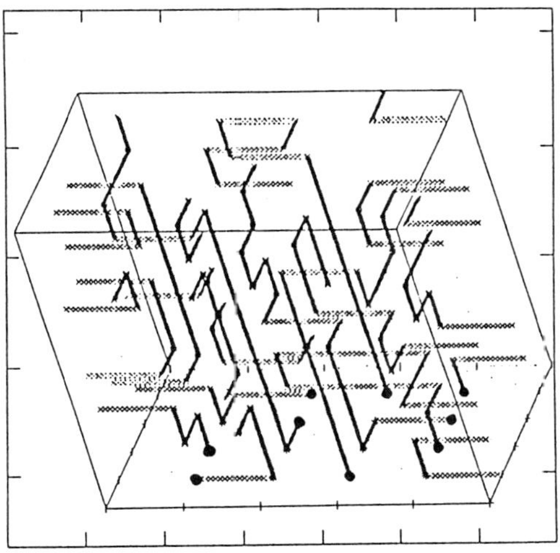

Fig. 4. A sample portion of a monolayer configuration when the surface coverage is 50%. The parameters and temperature are the same as for Fig. 1.

extension of the chain parallel to the surface and the increase in extension normal to the surface are not linear in the surface density. Some feeling for the complexity of the molecular conformations can be obtained from the projection shown in Fig. 4, which is for 50% coverage.

The clear implication of these calculations is that the properties of the monolayer of amphiphile molecules are strongly dependent on the interactions between the chains, hence the monolayer will not be accurately described by only examining the behavior of the head groups. Put another way, the two dimensional character of the ordering induced by head group interactions will influence but not dominate the overall behavior of the monolayer.

Experimental evidence in support of the same conclusion comes from recent grazing incidence x-ray diffraction studies of $C_{21}H_{43}OH$ supported on water (pH = 2) [6]. The surface pressure-surface area isotherms for this system are shown in Fig. 5. Along an isotherm, say 30°C., the observed diffraction pattern consists of a single peak; the temperature dependence of the peak position is shown in Fig. 6. Note that for pressures above the point of change of slope in the isotherm the peak position is independent of pressure, whereas for lower pressures it shifts with pressure commensurate with expansion of the monolayer. Along an isobar, say 30 dyne/cm, the diffraction pattern consists of a single peak for T > 16°C., which splits into two peaks as the temperature is progressively lowered (Fig. 7).

It is not possible to give a definitive interpretation of the x-ray diffraction data without higher order reflections, but such reflections could not be observed with

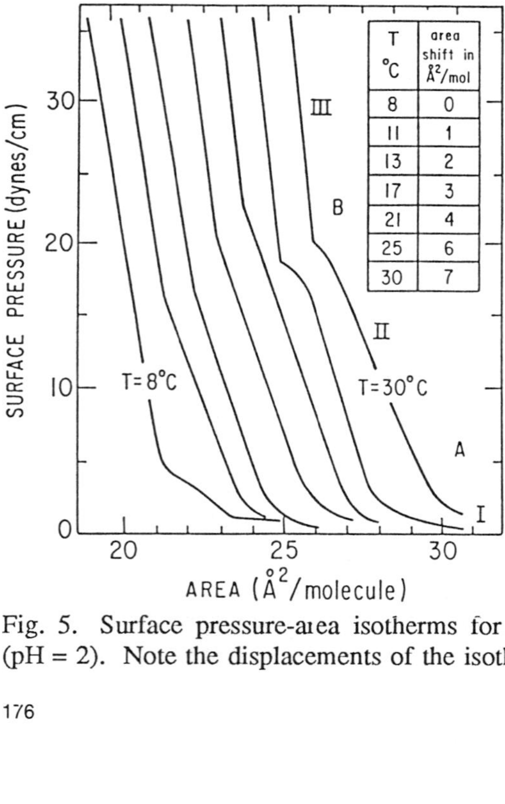

T °C	area shift in Å²/mol
8	0
11	1
13	2
17	3
21	4
25	6
30	7

Fig. 5. Surface pressure-area isotherms for monolayers of $C_{21}H_{43}OH$ on water (pH = 2). Note the displacements of the isotherms as indicated in the box.

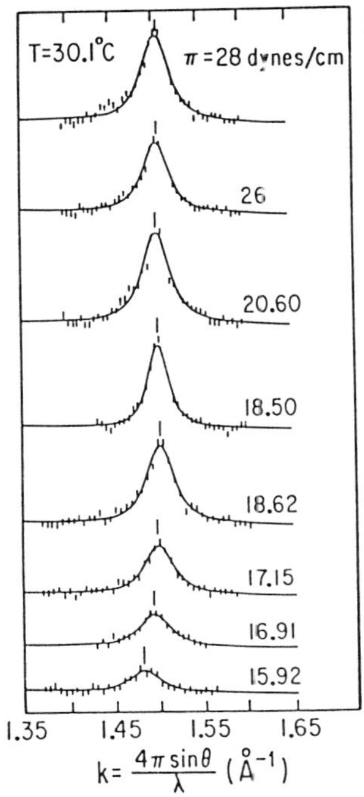

Fig. 6

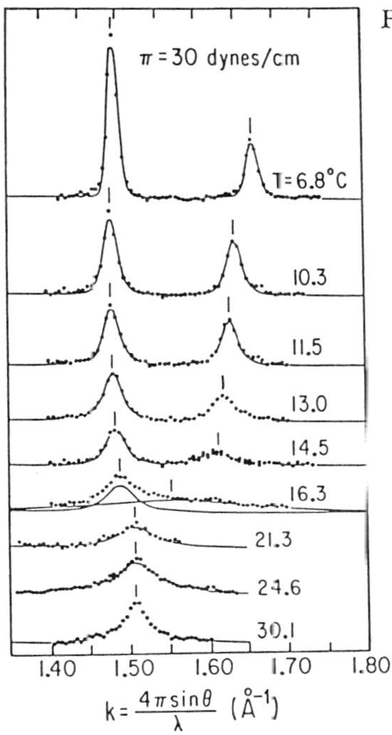

Fig. 7

Fig. 6. Diffraction pattern along the isotherms T = 30°C for $C_{21}H_{43}OH$ on water (pH = 2).

Fig. 7. Diffraction pattern along the isobar π = 30 dyne/cm for $C_{21}H_{43}OH$ on water (pH =2).

the available source intensity (10^{11} photons/sec from the wiggler station at CHESS). The available data are consistent with the following interpretation:

1. The ordered monolayer at high temperature has hexagonal close packing with the observed reflection corresponding to the first reciprocal lattice vector. This inference is consistent with the few other structural data available for amphiphile monolayers [8-10], and the lattice spacing deduced is very similar to that in those monolayers and also in the lamellar crystals of paraffin molecules [11].

2. The kink in the isotherm in the equation of state corresponds to the point of removal of all gauche configurations in the amphiphile tails of the monolayer. Above the kink the monolayer is relatively incompressable and all of the tails are in the all trans conformation. As the surface pressure is lowered below the kink the mean positions of the tails remain ordered but the concentration of gauche configurations increases and the monolayer expands. At the lowest surface pressure for which data have been taken the estimated gauche configuration

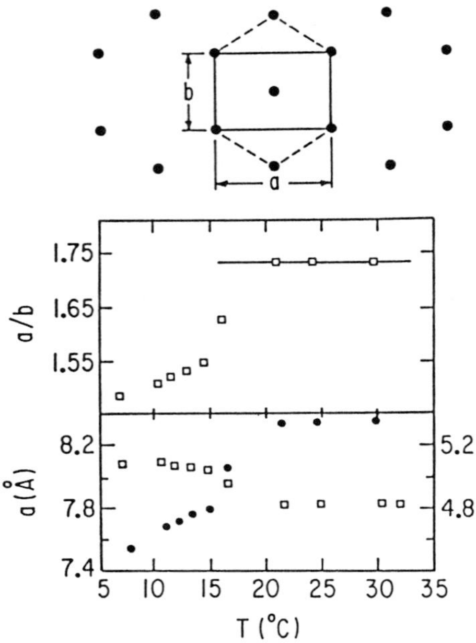

Fig. 8. Sketch of the unit cell and the changes of unit cell parameters in the hexagonal → pseudo hexagonal transition along the isobar π = 30 dyne/cm.

concentration is the same as just below the melting point of the corresponding lamellar crystalline paraffin [12].

3. The splitting of the diffraction peaks as the temperature is lowered along an isobar corresponds to a hexagonal-to-pseudohexagonal transition just as is observed in the rotamer I-to rotamer II transition in the lamellar crystalline paraffins (Fig. 8) [13].

Taken together, the theoretical and experimental evidence supports the contention that the chain-chain interactions in a monolayer of amphiphile are very important and that it is inadvisable to interpret the properties of such monolayers solely in terms of the consequences of two-dimensional ordering transitions of the head groups [14]. Evidence not discussed in this note [9], that the ordered monolayer leads to some ordering of the liquid substrate, also implies that the amphiphile monolayer is a "more than two dimensional" system.

References

1. See, for example, G.L. Gaines, *Insoluble Monolayers at the Liquid-Gas Interface,* Interscience, New York (1966).

2. N.R. Pallas and B.A. Pethica, Langmuir **1**, 509 (1985).
3. J. Popielawski and S.A. Rice, J. Chem. Phys. in press.
4. Z.-G Wang and S.A. Rice, J. Chem. Phys. in press.
5. J. Harris and S.A. Rice, J. Chem. Phys. in press.
6. S. Barton, B. N. Thomas, E. Flom, S.A. Rice, B. Lin, J.B. Peng, J. Ketterson and P. Dutta, to be submitted for publication.
7. M.W. Kim and D.S. Cannell, Phys. Rev. **A13**, 411 (1976).
8. K. Kjaer, J. Als. Nielsen, C.A. Helm, L.A. Laxhuber and H. Möhwald, Phys. Rev. Lett. **58**, 2224 (1987).
9. S.W. Barton, B.N. Thomas, E. Flom, F. Novak and S.A. Rice, Langmuir, in press.
10. P. Dutta, J.B. Peng, B. Lin, J.B. Ketterson, M. Prakash and P. Georgopoulos, Phys. Rev. Lett. **58**, 2228 (1987).
11. J. Nagizadeh, Adv. Chem. Phys. **65**, 45 (1986).
12. M. Maroncelli, Ph.D. Thesis, U. Calif., Berkeley (1983).
13. G. Ungar, J. Phys. Chem. **87**, 689 (1983).
14. See, for example, the multitude of theories discussed in G.M. Bell, L.L. Combs and L.V. Dunne, Chem. Rev. **81**, 15 (1981).

Discussion

E. TOSATTI: The dendritic nature (in the x-y plane) of the islands you show in your MC simulation leads one to suspect that these "island roughening" fluctuations may be responsible for some of the reported or mean-field theory. It is interesting that in the surface melting case the equivalent "disturbing" roughening fluctuations are vertical, i.e., along one direction only. This may constitute the physical reason why they seem to be less important in that case than in yours.

S.A. RICE: You make an interesting observation! As to the amphiphile monolayer, I believe the principal source of failure of the mean field approximation (Z.-G. Wang and S. A. Rice, *J. Chem. Phys.*, in press) is the following: Although the chain connectivity is kept in the theoretical analysis, and thereby the proper distribution of states for an isolated chain, the probability of occupying a configuration in the presence of other molecules is approximated by neglecting the connectivity of those other molecules. Thus, this mean field theory treats the probabilities of configurations in planes parallel to the surface in a different and more approximate way than it does the probabilities of configurations along the chain. It is this inconsistency (or asymmetry) of approximations which I believe dominates the errors in prediction of the mean field model. Clearly, consistent inclusion of all of the chain correlations is needed for an accurate theory.

J. T. YATES, JR.: It has been reported by Stöhr and colleagues that inclined alkyl chains occur in some cases in Langmuir–Blodgett films deposited on Si single crystal substrates. These conclusions are based on NEXAFS results. The inclined structures are postulated to occur as the intersecting alkyl chains connect to each other with a shift of a single CH_2 group between neighboring molecules. Have you observed these configurations in your measurements or in your simulations?

S.A. RICE: First, as to our experimental studies of the transverse structure factor by grazing incidence x-ray diffraction, at present we only measure the intensity as a function of the in-plane azimuthal angle, so that we are only sensitive to in-plane atomic separations. To determine if the chains are tilted we must measure the diffracted intensity as a function of the out of plane angle of scattering. We plan to do this. It is a difficult experiment because the scattered intensity is very low in these monolayer systems. Also note that no claim is made that our limited data uniquely determine the in-plane structure, only that a consistent interpretation can be given drawing on knowledge of the properties of lamellar crystalline long chain alkanes.

Second, as to our own simulations, the calculations completed at this date are for the case that the amphiphile molecule can only move on a simple cubic lattice placed above the surface plane in which the lead group is anchored. We currently have underway both Monte Carlo simulations of amphiphile configurations on a diamond lattice (which permits the amphiphile axis to be tilted relative to the surface plane) and molecular dynamics simulations (which do not restrict the molecular axis relative to the surface plane).

In general, I expect the phase diagrams of systems such as $C_{21}H_{43}OH$ on water to be very rich, and it is likely that there do exist conditions under which an ordered phase with tilted molecules occurs. I also suspect there will be a correspondence between the phases which occur in a monolayer and those that occur in the corresponding bulk phases as argued in my paper and talk.

Measurement of Non-Equilibrium Diffusion from Two-Dimensional Ordering Kinetics

*M.G. Lagally and M.C. Tringides**

University of Wisconsin, Madison, WI53706, USA
*Present Address: Physics Department, Iowa State University, Ames, IA50011, USA

1. Introduction

The thermodynamic behavior of an overlayer on a surface is determined by the microscopic interactions between the adsorbate atoms and between adsorbate and substrate atoms. A determination of these microscopic interactions can be made at equilibrium or also as the system evolves toward equilibrium. The usual equilibrium determination is a measurement of the temperature-coverage phase diagram of the overlayer, for example, with a diffraction or specific-heat experiment. Using a statistical model with a few local interaction parameters and an underlying lattice geometry, the phase boundaries are fitted to determine the magnitude of the interaction parameters. Nonequilibrium determinations of the same interactions can also be made. A system that is initially in an equilibrium state X_i is suddenly forced into a different state (for example, by a temperature or chemical potential quench) where its equilibrium state is X_f. The evolution of the system toward X_f requires mass transport, which can be monitored through a measurement of the change in order of the system with time. The temperature and coverage dependence of the ordering provides information on the microscopic interactions that govern non-equilibrium diffusion in the system.

It is to be noted that this diffusion is termed "non-equilibrium" to distinguish it from equilibrium diffusion. The simplest form of equilibrium diffusion that one can imagine is the random walk of a single particle among lattice gas sites, for which the activation energy for diffusion is simply the corrugation of the substrate potential. For an overlayer, equilibrium diffusion is the random motion of atoms about their equilibrium sites. Measurements of such equilibrium diffusion have been performed by GOMER et al. [1] using the current fluctuation method. The activation energy for diffusion here depends on the interaction energies an adatom sees when it is in an equilibrium environment of neighbors. At temperatures and coverages at which the overlayer is ordered, the activation energy for diffusion will be high. At temperatures and coverages where the overlayer is disordered, the activation energy will be lower, approaching the value for single-particle diffusion as the coverage approaches zero.

Nonequilibrium diffusion depends on the mean environment that exists for the diffusing species at any time in the ordering process. The same adatom interactions that are operative at equilibrium (and are, in fact, driving the system toward equilibrium) also control diffusion here. The activation energy for non-equilibrium diffusion should reflect these interactions as seen by the diffusing atom in the mean environment.

Springer Series in Surface Sciences Vol. 14
Solvay Conference on Surface Science
Editor: F.W. de Wette

In this paper, we briefly summarize the results of ordering kinetics measurements on W(110)p(2x1)-O, compare results of activation energies for equilibrium and nonequilibrium diffusion, and show that a consistent picture of the interaction energies results. More detailed accounts can be found elsewhere [2,3].

2. Nonequilibrium Diffusion Analysis

The system W(110)p(2x1)-O has been investigated in great detail. Relevant here are the determination of the phase diagram [4], fits to the phase diagram to extract microscopic interactions [5], the equilibrium O diffusion measurements mentioned above [1], and ordering kinetics measurements [3,6]. The latter can be used to determine nonequilibrium diffusion. The increase in order in a two-dimensional system is generally found in theoretical analyses [7-9], to follow a power law

$$\langle L \rangle = A(T)\, t^x, \tag{1}$$

where $\langle L \rangle$ is the mean domain dimension and where the growth exponent x depends on the ground state degeneracy of the system and on whether adatom density is conserved (chemisorption) or not (physisorption). Two types of ordering are distinguished: 1) growth of domains for a system that exists in a single-phase region and 2) phase separation, the "coarsening" or growth of the dense (island) phase in the dilute ("sea") phase. In both cases the mechanism for ordering is diffusion of single atoms, but the details of the processes differ (hence different exponents, x, are predicted).

The factor A(T) in Eq. (1) contains the nonequilibrium diffusion coefficient, D. Theoretical determinations of the growth in phase region types 1) [7] and 2) [8] above have found that

$$A(T) \propto D^x , \tag{2}$$

where x is the growth exponent in Eq. (1). This can be generalized on the basis of dimensional arguments to any growth situation (different ground state degeneracies, etc.) with the assumption that the only time dependence on the RHS of Eq. (1) occurs in D and in t^x. There are (as in Refs. 7 and 8) other constants (e.g., boundary energy, molar volumes, etc.) in Eq. (2), but they do not depend on time. The factor A(T) obeys an Arrhenius relationship. The activation energy, E_{meas}, obtained from a plot of lnA vs 1/T can then be related to the activation energy for nonequilibrium diffusion, E'_{diff}, by

$$E_{meas}(\theta) = xE'_{diff}(\theta) . \tag{3}$$

3. Results and Conclusion

The phase diagram for W(110)p(2x1)-O shows [4] both one-phase and two-phase regions, as illustrated schematically in Fig. 1a. Monte-Carlo fits to the phase diagram require (inset, Fig. 1b) a nearest-neighbor attraction ϕ_1, a next-nearest-neighbor repulsion ϕ_2, with $|\phi_1| \simeq |\phi_2|$, and a weaker third-neighbor attraction $\phi_3 \simeq \phi_1/3$. Measurements of equilibrium diffusion [1] have been made over a coverage and temperature range (Fig. 1a) that includes the one-phase region and a disordered lattice gas phase. The activation energy for equilibrium diffusion is

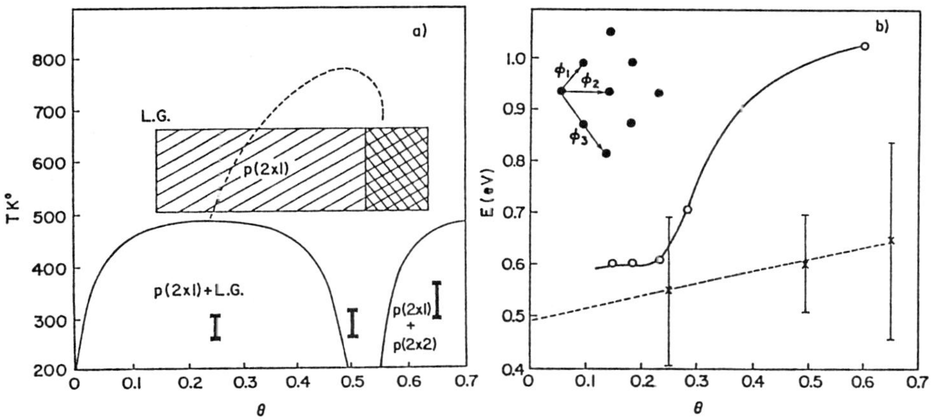

Figure 1. Comparison of the equilibrium and nonequilibrium measurements of activation energies for diffusion of O on W(110). a) Equilibrium phase diagram. The phase boundaries above $\theta = 0.5$ are uncertain. Hatched region indicates T-θ range for equilibrium measurements (Ref. 1). The cross hatching indicates region where these measurements overlap the uncertain phase diagram. Heavy bars indicate T-θ regions of ordering kinetics measurements. b) Activation energy for diffusion. Open circles: equilibrium measurements. Crosses: non-equilibrium measurements. The dashed line shows the coverage dependence of Eq. (4) using the interactions ϕ_1, ϕ_2, ϕ_3 shown in the inset. Their magnitudes [5], determined from a best fit to the phase boundaries shown in a) are -0.09 eV, +0.075 eV, and -0.03 eV. The curve is forced to go through the data point at $\theta = 0.25$. The intercept at $\theta = 0$ gives the activation energy for single-particle diffusion.

shown as a function of coverage in Fig. 1b. It shows a strong coverage dependence, as suggested earlier. The rise in activation energy occurs at the coverage at which the system orders. The magnitude of the rise is consistent with the interaction energies determined from the phase diagram fits. The value at lowest coverage, E_{diff} $(\theta = 0.15) = 0.6$ eV, represents an upper limit of the activation energy, $E_{diff}(0)$, for single-particle diffusion. It is an upper limit because, even though the system is in a disordered phase, fluctuations can still create momentary ordered regions that will raise the activation energy.

Our ordering kinetics measurements [2,3,6] where made at $\theta = 0.25$ and 0.5, coverages that correspond to overlayer equilibrium in two-phase coexistence and single-phase regions respectively. The system was prepared in a disordered state by adsorption at low temperatures where the system shows no order. The evolution to order was subsequently monitored using LEED for anneals at a variety of temperatures. The dependence of the mean size <L> on time was extracted from angular profiles of superlattice diffracted beams, and from it the growth exponent x. The value of E'_{diff} was determined from Arrhenius plots of ln A(T) vs 1/T and using Eqs. (2) and (3). The growth exponent $x = 0.28 \pm 0.03$ at both coverages. Theoretical values, insofar as they exist [8,9], give $x = 0.33$ at both coverages although the mechanisms are quite different. Activation energies are $E_{meas} = .17 \pm 0.03$ eV at $\theta = 0.5$ and 0.16 ± 0.04 eV at $\theta = 0.25$. Hence values for the activation energy for nonequilibrium diffusion are $E'_{Diff} \simeq 0.6$ eV. These values are also shown in Fig. 1b. Measurements were additionally made [3] at $\theta = 0.65$, where the precise nature of the phase diagram is in question. It is likely [3] that two solid phases, p(2x1) and p(2x2), coexist, but there are other possibilities [10]. Two-phase coexistence is analogous to the situation at $\theta = 0.25$, except that now the dilute phase is p(2x1) and the dense one is p(2x2), which has a saturation coverage of $\theta = 0.75$. There are no theoretical analyses of the growth specific to this situation, although it is reasonable that the mechanism for late-stage growth is again coarsening [8]. Experimentally the growth exponent is found to be $x = 0.2 \pm 0.03$, i.e., a slower growth than at lower coverages that may be explained by the existence of a greater number of types of domain walls [3]. The activation energy is $E_{meas} = 0.13 \pm 0.04$ eV. Combining these as in Eq. (3) gives $E'_{diff}(\theta = 0.65) = 0.65 \pm 0.2$ eV.

How are these results to be understood? The value E'_{diff} represents the activation energy for diffusion for an atom in a mean environment. The mean environment of a diffusing atom is disorder - the random occupation of sites at the appropriate coverage. This can be expressed to first order as

$$E'_{diff}(\theta) = E_{diff}(0) + Z\theta(\Delta\phi) \quad (4)$$

where Z is the coordination number of the lattice, θ is the coverage, and $\Delta\phi = \Sigma\phi$ is the net change in diffusion barrier seen by the diffusing atom due to the presence of a second atom in a nearest-neighbor site. As long as the occupation is random, one expects an increase in the activation energy that is linear in coverage and has a slope that depends on the magnitude of $\Delta\phi$. In our case, $\Delta\phi \simeq 0$, and thus (unfortunately) there will be only a very weak dependence of $E'_{diff}(\theta)$ on coverage. In fact, to first order, the values we determine should reflect the activation energy for diffusion for a single particle, because the adatom interactions nearly cancel. The nonequilibrium values are consistent with the values measured at lowest coverage in the equilibrium experiments. We have also shown in Fig. 1b as the solid line the result for the exact value of $\Delta\phi = \phi_1 + \phi_2 + \phi_3$ as a guide only. We do not suggest that the data are precise enough at present to deduce this curve.

This analysis is, of course, only approximately correct. Complete randomness occurs only at zero time. In general, the system is more ordered, which increases the activation energy. In our measurements, the mean domain sizes change from ~ 2.5 atoms wide at early times to ~7 atoms wide at the latest times, and thus there is considerable disorder over the complete time regime. Also, it can occur that the effective coverage for a diffusing species is lower than the mean coverage, e.g., in a two-phase system at a low degree of supersaturation. This latter effect is only important if $\Delta\phi \neq 0$, and then generally it lowers E'_{diff}. A more accurate analysis will require weighting of each configuration with the proper probability factor and the use of Monte Carlo methods. Such investigations are in progress [11]. With regard to the experiment, a wider range of coverages needs to be explored, in particular lower coverages, to determine more precisely any dependence of E'_{diff} on coverage. There is also always the possibility that impurities or defects affect the growth. They would tend to increase the activation energy, in the sense that they act as nucleation sites or sites blocking diffusion, implying that any measurements such as those described here give an upper limit to the activation energy for single-particle diffusion. The fact that they are consistent with the values extracted from equilibrium diffusion measurements that by their nature are very clean suggests that impurities play a negligible role in our experiments.

In conclusion, ordering kinetics measurements provide an opportunity to sample adatom interactions for different configurations. From the coverage dependence and the nature of the phase diagram, it should in principle be possible to verify adatom interactions and determine nonequilibrium diffusion rates under a variety of circumstances. The work reported here is only a small step in this direction.

This research has been supported by NSF, Solid State Chemistry Program, Grant No. DMR 86-15089.

4. References

1. R. Gomer, Surface Sci. 38, 373 (1973); J. R. Chen and R. Gomer, Surface Sci. 79, 413 (1979); M. Tringides and R. Gomer, Surface Sci. 155, 254 (1985).

2. M. C. Tringides, P. K. Wu, and M. G. Lagally, In: Proceedings of the Second Campobello Conference, 1987, ed. by H. J. Kreuzer (Springer, Berlin, Heidelberg 1988) (in press).

3. P. K. Wu, M. C. Tringides, and M. G. Lagally, Phys. Rev. (submitted).

4. G.-C. Wang, T.-M. Lu, and M. G. Lagally, J. Chem. Phys. 69, 479 (1978), M. G. Lagally, T.-M. Lu, and G.-C. Wang, In: Ordering in Two Dimensions, ed. by S. Sinha (Elsevier, New York 1980).

5. E. D. Williams, S. C. Cunningham, and W. H. Weinberg, J. Chem. Phys. 68, 4688 (1978).

6. M. C. Tringides, P. K. Wu, and M. G. Lagally, Phys. Rev. Letters 59, 315 (1987).

7. S. M. Allen and J. W. Cahn, Acta Metall. 27, 1085 (1979).

8. I. M. Lifshitz and V. V. Slyozov, J. Chem. Phys. Solids 19, 35 (1961).

9. A. Sadiq and K. Binder, J. Stat. Phys. 35, 617 (1984).

10. P. A. Rikvold, K. Kaski, and J. D. Gunton, Phys. Rev. B29, 6285 (1984).

11. M. C. Tringides and M. G. Lagally, in preparation.

Discussion

D.A. KING: You compare your diffusion data, described as nonequilibrium measurements, with the data of Gomer, which you describe as equilibrium measurements. However, you do not refer to the nonequilibrium measurements of Butz and Wagner and of Bowker and King, also for O/W(110). Clearly your results are in very poor agreement with the latter set of data. Please comment.

M.G. LAGALLY: Ours are clearly non–equilibrium measurements and Gomer's are clearly equilibrium measurements. I believe the profile evolution measurements you refer to should also be considered a measurement of nonequilibrium diffusion but they are much less microscopic than ours. I do not believe they are as reliable, because the probe integrates over a much larger region.

E. TOSATTI: It seems to me that the main significance to be attached to your finding that the dynamic exponent x is close to the Lifshitz-Slyozov value 0.33 should be that the domains which grow in the coarsening process are compact, as implicitly assumed by Lifshitz-Slyozov, and not fractal, as they may sometimes become. It is not clear to me what this might have to do with the symmetry of the problem.

M.G. LAGALLY: A quick answer to your question is that Monte Carlo studies of coarsening for high density, where the growing domains should become connected and banded, also indicate an exponent of 1/3 [D. Huse, *Phys. Rev.* **B34**, 7845 (1986)]. That, unfortunately, does not provide much insight, and does not address the question of symmetry. The Lifshitz-Slyozov theory is a mean-field theory for the late stages of phase separation (i.e., coarsening) in the limit of vanishingly small droplet volume. The growth mechanism involves 2-D condensation/evaporation and diffusion of monomers from regions of relatively higher vapor pressure (around small islands) to those of relatively lower vapor pressure (around large islands). Ground state symmetry is not considered (in this or extensions of this theory, to my knowledge). But one might see how a high ground state degeneracy could slow the growth. One can imagine a monomer diffusing toward an ordered domain. If this domain has a (1×1) structure, any lattice gas site on which the monomer approaches will be satisfactory for condensation to the domain. If the domain has an $(n\times m)$ superlattice, some of the sites of approach will be "incorrect" for condensation to the domain. Thus, one would expect a slower coarsening for the latter case, i.e., the growth exponent would be lower for a higher ground state degeneracy.

R. GOMER: To what extent do surface imperfections affect your results? With respect to King's comment on profile evolution methods for determining diffusion coefficients, I consider the variation of D with coverage a serious problem, Boltzmann-Matano analysis notwithstanding.

M.G. LAGALLY: Surface imperfections and impurities can clearly affect our results. This can be a serious problem. In general such imperfections act to reduce the rate of ordering and, because they generally act as nucleation sites, would tend to raise the effective activation energy for diffusion. One piece of evidence that this is not a serious problem in our measurement is the agreement at low coverages with Gomer's results, which are obtained on single planes of a field emission tip, and thus are clearly the most perfect surfaces we can obtain.

H. IBACH: With regard to the comment of Gomer, I would like to state that the Boltzmann-Matano analysis used by Butz and Wagner is a relatively straightforward one. The problem with measuring diffusion coefficients by spreading methods is that one integrates over the effect of imperfection of the surface. It has been shown in fact that steps have a strong influence on the diffusion.

M.G. LAGALLY: I would certainly agree with this comment. As I mentioned in my previous answer, any form of structural imperfection will likely increase the measured activation energy for diffusion. The lateral resolution of probes for profile evolution measurements is sufficiently poor so that integration over regions containing many defects must occur.

D. KING: If the results of Butz and Wagner and of Bowker and King are dominated by steps this implies a remarkable coincidence of step direction and density in these two independent studies. I don't believe that you have a good reason for selectively ignoring these two sets of data.

M.G. LAGALLY: I apologize for not having mentioned these papers. They are: R. Butz and H. Wagner, *Surface Sci.* **63**, 448 (1977) and M. Bowker and D.A. King, *Surface Sci.* **94**, 564 (1980). I will give a brief summary of these measurements and attempt to put them into the context of what I have said. Both papers report profile evolution measurements: part of the surface is covered to a particular concentration, the surface is heated, and the evolution of the inital step profile is measured using a probe with high lateral resolution in one dimension. In BW this resolution is $50\mu m$ and in BK it is $100\mu m$. In both cases diffusion measurements were made at temperatures above the highest temperature at which ordered structures exist in O/W(110) [Ref. 4 of paper]. Thus the equilibrium state is disorder, a diffusing atom sees an average environment that is random except for fluctuations, and therefore the activation energy for diffusion should show only the very weak coverage dependence that we have indicated in the paper, and not a large rise, as observed by Gomer, as one passes from disordered to ordered phases. Both BW and BK are consistent with this picture, with BK even showing some indication of a coverage dependence in the appropriate direction.

However, the value of the activation energy in both BW and BK is more characteristic of an ordered state than a disordered state, being about a factor of two too high for the latter. Because there can be no order at these temperatures, it is not adatom interactions but some other interaction that leads to this increase in activation barrier. A likely choice, as already indicated, is surface imperfections. A random configuration of steps and point defects is present on any surface and it is possible that these have sufficient influence, when viewed in a macroscopic scale as in profile evolution measurements, to cause the observed increase in activation energy.

Because there is a concentration gradient in profile evolution, the results should be classified as nonequilibrium diffusion. Locally, the picture is the same as in our measurements: an atom diffuses down a concentration gradient through a mean environment of randomly positioned neighbors.

Metastable Magnetic Epitaxial Thin Films

R.F. Willis

Department of Physics, Cavendish Laboratory,
University of Cambridge, Madingley Road, Cambridge CB3 0HE, UK
Present address: Department of Physics,
The Pennsylvania State University, University Park, PA 16804, USA

Abstract:

We have studied the ferromagnetic behaviour of the metastable phases of f.c.c. cobalt and iron by clamping the lattice spacing in thin films epitaxed to appropriate single-crystal metal substrates/1/. This allows one to effectively "tune" the lattice strain such that changes in lattice volume are reflected in changes in both the magnitude of the magnetic moments and the magnetic exchange coupling between spin moments. This translates into new metastable magnetic systems with unusual properties, i.e., the magnetic moments are predicted to change discontinuously with varying atomic volume and there exist ranges of coexistence of different magnetic phases/2,3,4,5/. However, to date there has been very little experimental evidence to corroborate these views, principally due to the difficulties inherent in measuring the magnitudes of the magnetic moments in ultrathin epitaxial layers.

Recently, we have succeeded in applying spin-polarized neutron reflectance at grazing angle of incidence to study the ferromagnetic properties of thin magnetic films of Co and Fe down to the monolayer thickness level/6/. The technique allows us to determine the magnitude of the magnetic moment per atom, μ, in ferromagnetically ordered layers, as a function of temperature, thickness and applied magnetic field/1/.

The results are unusual in that, whereas the f.c.c. Co(001) layers remain ferromagnetic down to a single monolayer with $\mu \simeq \mu_{bulk} \simeq 1.8\mu_B$ with no sign of a reduced T_C for temperatures up to 450K, the f.c.c. Fe(001) films are ferromagnetically ordered only at temperatures below this value (their T_C values being thickness dependent). Also, the Fe(001) films exhibit strong magnetic anisotropy effects which appear to be strain-related, and dependent on the number of layers. Detailed LEED measurements/7/ have determined the degree of strain as a function of thickness. Perhaps most surprising, whereas the f.c.c. Co(001) films exhibit spontaneous magnetization within the film plane for all layer thicknesses, the f.c.c. Fe(001) films show in-plane magnetization only for the first monolayer, the succeeding layers up to 7 monolayers showing strong spin-anisotropy oriented normal to the interface with remanence strongly dependent on the layer thickness/8/. The 4 monolayer film shows maximum volume expansion and a moment of the order of μ = 1 to $1.5\mu_B$, in agreement with the recently predicted "low-moment" ferromagnetic ground state in the f.c.c. bulk iron phase/3,4,5/.

The unusual uniaxial magnetic anisotropy of the magnetic exchange interaction in these strained lattice films produces behavior different to that which might be expected in a strictly two-dimensional ferromagnetic system. For example, the much higher T_C threshold for the cobalt films

Springer Series in Surface Sciences Vol. 14
Solvay Conference on Surface Science
Editor: F.W. de Wette

would indicate that the spin-wave energy gap is greater than that in the f.c.c. iron films, the latter showing reduced Curie temperatures as low as 150K compared with a bulk phase value, 900 to 1200K /9/. The significance of this latter result for Surface Science is that it offers the opportunity to investigate possible magnetocatalytic behaviour in surface chemical reactions by "tuning" T_C through an appropriate temperature range simply by varying the thin-film thickness.

Acknowledgements

My research students and post-docs in Cambridge; my colleagues in the spin-polarized photoemission group, ETH, Zürich; colleagues at the spin-polarized neutron sources at the Institute Laue-Langevin and Rutherford-Appleton Laboratories; and to the U.K.SERC for financial support for this work.

References:

1. R. F. Willis, Proc. ICOSS II, Amsterdam, June 1987 (Springer-Verlag, in press); also, Proc. 32nd Conf. Mag. Magn. Mater., Chicago, November 1987 (J. Appl. Phys. in press).
2. D. Bagayoko, J. Callaway, Phys. Rev. B28, 5419 (1983).
3. V. L. Moruzzi, P. M. Marcus, K. Schwarz, P. Mohn, Phys. Rev. B34, 1784 (1986).
4. C. S. Wang, B. M. Klein, H. Krakauer, Phys. Rev. Lett. 54, 1852 (1985).
5. F. J. Pinski, J. Staunton, B. L. Gyorffy, D. D. Johnson, G. M. Stocks, Phys. Rev. Lett. 56, 2096 (1986).
6. J. A. C. Bland, D. Pescia, R. F. Willis, Physica Scripta 35, 528 (1987); Phys. Rev. Lett. 58, 1244 (1987).
7. A. Clarke, G. Jennings, R. F. Willis, P. J. Rous, J. B. Pendry, Surf. Sci. 187, 327 (1987).
8. D. Pescia, M. Stampanoni, G. L. Bona, A. Vaterlaus, R. F. Willis, F. Meier, Phys. Rev. Lett. 58, 2126 (1987).
9. W. Kümmerle, U. Gradmann, Sol. State Commun. 24, 33 (1977).

Discussion

H.C. SIEGMANN: The technique of using neutron reflection to determine the materialization is really impressive from a technical point of view. However, would it not really be easier to do the measurements in a SQUID or torque magnetometer? Furthermore, are the theoretical calculations really helpful when they tell us that three phases are extremely close in energy? On the whole, this work marks a major progress in our understanding of ferromagnetism in Fe.

R.F. WILLIS: Magnetization measurements on these film layers can be (have been) made using a SQUID or torque magnetometer, and a value for the magnetic moment μ can be derived. However, I don't believe that it is so easy to do such measurements with single crystal mirrors of a size $5mm^2$ which is very convenient for MBE growth of these films. The advantage of the neutron measurement is two-fold: firstly it tells you immediately the spin-concentration relative to the film plane; secondly, knowing the film thickness, the method is *extremely* sensitive to the magnitude of the spin moment; thirdly, given more time on the neutron source, the accuracy can be very high.

I believe the theoretical calculations of the *bulk* phase ground magnetic state to be a vital starting point in understanding the lattice strain effects in these thin films. The uniaxial strain field (probably) also determines the magnetic anisotropy and hysteresis behavior.

Y. PETROFF: In the case of Fe, do you see the moment perpendicular to the surface only at low temperature or also at higher temperature? How do your data compare with spin-polarized photoemission results? The fact that in the neutron experiment you have a sandwich, do you see any difference?

R.F. WILLIS: The preferred direction of magnetization in the fcc Fe films is perpendicular to the surface for film thicknesses greater than one monolayer, independent of temperature. The preferred direction for one monolayer is in the plane. This is different from the behavior of bcc Fe films on Ag. Here the first monolayer is spin-oriented perpendicular and subsequent layers are parallel to the film plane. The anisotropic exchange coupling in these films appears to be thickness dependent, giving values of T_c as a function of thickness. The sandwich films give similar magnetic anisotropy and T_c behavior to *uncoated* films viewed with spin-polarized photoemission.

P. MARCUS: The magnetization *vs.* lattice constant curve for fcc iron that you show is a continuous line. But part of that curve—those with negative slope—are unstable and should be removed or dashed.

R.F. WILLIS: Thank you—will do!

L. KLEINMAN:

1. I don't think it very useful to consider bulk ferromagnetism when monolayer ferromagnetism is under discussion. After all, there are several metals which have monolayer ferromagnetism but no bulk ferromagnetism.

2. Why were you surprised to find the easy direction of magnetization to be normal to the film? The demagnetizing field is a volume effect and so you should have no *a priori* expectations.

R.F. WILLIS:

1. I avoided the subject of single monolayer magnetism in my talk, due to the well-known interface hybridization effects. However, the choice of Fe and Co on Cu(001) minimizes such effects since the d-bands are not degenerate in these systems. My talk showed that the magnetism of thin films (several monolayers) showed strong magnetic anisotropy effects relating to uniaxial strain in these epitaxial layers. I believe these effects *do* relate to the *bulk* magnetic ground-state-phase.

2. It is true that the demagnetizing field is a volume effect in general which gives a shape anisotropy with the magnetization *uniphase* as the film thickness is reduced. However, the argument ignores uniaxial strain effects. These effects can produce an easy direction either in plane or normal to the surface. The fcc iron films are surprising in their behavior since the spins flip direction from parallel to normal for thicknesses *greater* than a single monolayer. Current theoretical expectations predict the opposite behavior.

General Discussion

L. VAN HOVE: In opening the general discussion of this session, which unfortunately has to be brief, I would like to raise two questions which may become of some interest in the long range.

Our session was entitled *"Two-Dimensional Physics"*. Are the surfaces and interfaces which were discussed really two-dimensional? Remember geography and fractals! Is a coastline between land and sea one-dimensional? We now know that the answer is no; a careful definition of dimension gives a very interesting number between 1 and 2. Surely there is a similar question for physical surfaces and interfaces. Can we define a useful concept of dimension for them? By how much does it exceed two?

In his beautiful lecture, Prof. Dash took a long-range view, going as far back as Faraday. What about the long-range future? I believe that space experiments will confront you and/or your successors with a qualitatively new challenge, to wit, surface and interface phenomena under conditions of microgravity. I suspect that a vast and rich field of investigation lies ahead in that direction.

G.A. SOMORJAI: Surface melting is a fascinating phenomenon that, in my view, is in search of experimental verification. I am impressed by Professor Petroff's studies of photoemission that appear to be very promising. I would like to suggest another possible experimental approach that was alluded to by Faraday, that is, to measure changes of mechanical properties of the surface; friction or plastic deformation onset upon contact. Recent studies indicate that the scanning tunneling microscope can be used to measure friction coefficients on the atomic scale. Is there a possibility to develop a theory or model that predicts changes of friction coefficient or onset of plastic deformation as surface melting commences?

J.G. DASH: One extremely promising development in this direction is the recent *Physical Review Letter* that reports a correlation between stick–slip motion and surface structure. That would seem to be rather different from the friction that involves surface melting, as on ice. But both types of friction are prime targets for fundamental study on an atomic scale.

Y. PETROFF: I think that the question of surface melting of Ge(111) could be solved by Takayanagi by VHV election microscopy.

D.R. HAMANN: A liquid metal surface is layered, according to S.A. Rice's talk yesterday. This suggests that variation with polar angle of photoemission spectra does not rule out liquid surface layers.

Y. PETROFF: No—The angle-resolved photoemission of liquid lead does not show any difference for $\theta = 0$ and $\theta = 20°$. I should mention that photoemission of liquids like Ge, Aℓ, Pb and many others have been done by Oelhafhen and his group.

J.T. YATES, JR.: The comment on the relationship between the melting of a surface layer and the coefficient of friction reminds me of a classic experiment on single monolayer Langmuir–Blodgett films by Zisman in the mid 1950's. Zisman showed that carboxylic acid monolayers exhibited a *higher* coefficient of friction when they were above the melting point of the organic molecule (in bulk), and the coefficient was *lower* for the films below their bulk melting point. It would seem, on this basis, that a disordered interfacial layer might exhibit a *higher* coefficient of friction.

E. TOSATTI: I have three separate comments, which address different problems touched on this morning.

1. A nice way to probe surface melting would be using the atomic force microscope. Due to the extreme fragility of the actual thickness of the surface melted layer, I would expect that this thickness could be modulated by the van der Waals interaction of a closely approaching mechanical tip. Slow tip vibrations could be followed by the substrate where local melting and crystallization would occur in phase. Fast tip vibrations could not be followed by the liquid layer at all, leading to a change in the mechanical response.

2. I would like to suggest that a possible explanation of the unexpected mean-field-like behavior reported for edges of 2D crystals by Prof. Dash could be the fact that roughening fluctuations (present at all temperatures) cannot affect the two interfaces liquid-vapor and solid-liquid in an independent, i.e., uncorrelated manner. Rather, in line with our previous discussion of the 3D case [1], I would expect a joint roughening of the two surfaces, with the actual thickness of the liquid layer remaining roughly constant everywhere. This seems required by the necessity to select only fluctuations of small free energy cost. If this were the case, then roughening could *not* exert any renormalization of the liquid-layer-related specific heat, which in turn would remain mean-field-like, as seen. ([1] A.C. Levi and E. Tosatti, *Surface Sci.* **178**, 425 (1986).)

3. It would be nice to hear more from the people who study thin magnetic films about the critical behavior of their systems with temperature. In

particular, I wonder if the case of a very thin Fe film, when the magnetization is about to switch from vertical to in-plane, is a Kosterlitz-Thouless case as one would expect.

J.G. DASH:

1. There are several suitable probes of surface melting in addition to those which have been used in the experiments I mentioned in my talk. For example, nuclear magnetic resonance and ellipsometry have been employed in studies of the surface melting of ice, and thermal luminescence was the technique in one of the first indications of the facet dependence of surface melting. It is in principle possible to use any surface sensitive technique to study surface melting in particular systems, but some are much less convenient. For example, LEED cannot be used at pressures typical of the saturation pressure of most materials at their melting points. STM might be used, as you suggest, but it too seems to be quite limited in application. This is because the devices are extremely sensitive to temperature variation, except at very low T, and require typically very long equilibration time after the temperature is changed.

2. Dr. Tosatti refers to a study of "edge melting" [1] mentioned in my talk but not described in the written version. In a recent experiment on the lower dimensional analogue of surface melting, performed on fractional monolayers of Ne on graphite, we found calorimetric signals at the 2D triple point similar to the power law behavior seen in Ne and Ar multilayers at their 3D triple points. The power law exponent in the monolayer study is close to 5/4, which corresponds to mean field theory, i.e., to edges without fluctuations. Dr. Tosatti's suggestion for the discrepancy is intriguing. Certainly, one expects some degree of correlation between the fluctuations of the two interfaces. It seems improbable, however, that the correlation should extend over all frequencies, given the difference between the shear constants of the solid and liquid. Other possible explanations, which we consider in our paper, involve the finite thickness and lateral extent of the interfaces. ([1] D.M. Zhu, D. Pengra, and J.G. Dash, *Phys. Rev. B***37**, 5586 (1988).)

C.B. DUKE: An important contribution to "friction" forces comes from the electrostatics associated with contact charge transfer between two dissimilar surfaces. This phenomenon can mask the absence of shear created by a liquid layer between two solid interfaces, so it should be considered when evaluating mechanical experiments designed to detect and/or evaluate such layers.

Part IV

Scanning Tunneling Microscopy

Tunneling Microscopy

J.A. Golovchenko

Harvard University, Physics Department and
Division of Applied Science, Cambridge, MA 02138, USA, and
The Rowland Institute of Science, 100 Cambridge Parkway,
Cambridge, MA 02142, USA

The tunneling microscope, a new instrument capable of imaging surfaces in real space with atomic resolution, has been available for about six years. The ultimate impact of this tool on the whole of surface science must still await the critical judgment from future generations with a broad perspective on what will hopefully one day be considered a mature field. Nevertheless it is clear today that the microscope will certainly play a role in the determination of surface crystal structures in cases where traditional diffraction methods have been used in the past as well as in situations where such methods have not been applicable.

The microscope is sensitive to atomic structure through both spatial and energy dependence of the electronic states connected with the tunneling process. Viewed from one perspective the possibility of obtaining both spatial and electronic information might seem as much as a surface scientist could hope for. Yet it is the position of the *atomic cores* that has traditionally been the cornerstone upon which we build our understanding of the electronic and chemical properties of surfaces either as the basis for calculations or analogies with related structural systems which are understood.

Thus the main focus in the following will be on the tunneling microscope as a tool for determining surface structure. By keeping to a particular case, semiconductor surfaces, admittedly the richest one that has yet been studied with the instrument, I hope to demonstrate the varied and sometimes surprising ways in which the instrument can contribute to our knowledge and understanding of surfaces.

While the rudiments of obtaining images of surfaces using tunneling processes can be traced back over a decade it was the fundamental progress by BINNIG et. al. [1] that led to the attainment of atomic scale resolution on crystal surfaces by performing experiments in ultrahigh vacuum, establishing fine control of mechanical motions, and the elimination of vibration problems.

A schematic illustration of the instrument [2] in Fig. 1 shows a probe tip, often composed of sharpened tungsten wire, mounted on a

Springer Series in Surface Sciences Vol. 14
Solvay Conference on Surface Science
Editor: F.W. de Wette

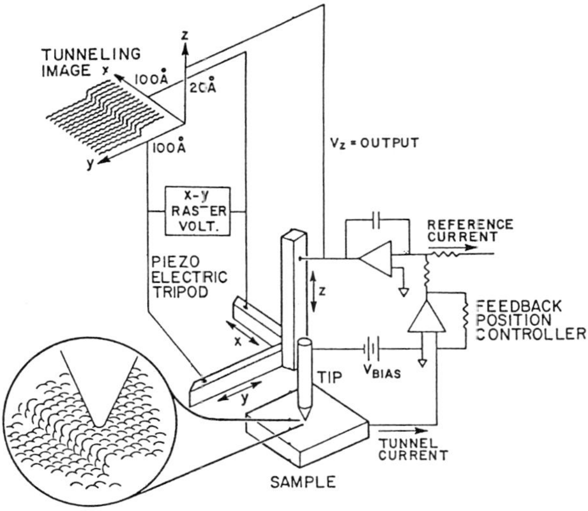

Figure 1. Shematic illustration of the tunneling microscope. The tunnel current induced from tip to sample by V_{bias}, is maintained constant by an electronic feedback system (right), which controls the tip position normal to the sample surface via the "z" leg of a piezo electric tripod. A record of the piezoelectic feedback voltage, as the x-y tripod legs raster scan the tip laterally, constitutes a tunneling image, which is a kind of replica of the sample surface (inset at bottom. The sample tip and tripod are maintained in a u.h.v. vacuum chamber. From ref [2].

piezo-electric "robot" capable of moving the tip around over the surface with positional control of order 0.1 Angstrom. The motion of the tip in the plane of the surface is usually rastered or scanned over some rectangular region of the surface which defines the area displayed in a tunneling image. The tip position normal to the surface may be controlled in several different ways. The most commonly used method attempts to make the tip hover at a constant distance above any given surface feature by adjusting the tip height, with the help of a feedback circuit, so that a constant tunnel current flows regardless of the lateral position of the tip during the scan. That tip height recorded as a function of lateral position constitutes a tunneling image of the type that has proved the most useful to date in the comparison of theory and data.

Alternatively for surfaces that are flat enough (which currently usually means graphite) many workers [3] have had success with scanning the tip parallel to the average surface plane and record rather than changing tip height the changing tunnel current in this so called current imaging mode of operation.

Finally an important area where there has been much recent effort [4,5,6] has been in recording electrical information about the microscopic tunnel junction. Here current voltage characteristics in one form or another are obtained some times at each pixel element of a tunneling image (for those with the time and memory storage for such a feat). The information gleaned from such studies can sometimes be of direct relevance in the deduction of geometrical surface structure as we shall show shortly.

As regards order of magnitudes of the the basic physical quantities involved in the operation of the tunneling microscope it is easy to frame a simple model on a planar metal vacuum metal tunnel junction from which one can base an intuitive understanding of the instrument. For example if the metal is taken to be free electron-like and each eigenstate is considered as a superposition of current eigenstates with a component going towards and away from the surface in question one easily deduces that within ten millivolts of the Fermi level there are 10^{12} electron/Angstrom^2/sec. impinging on the surface. A barrier penetration factor for these electrons to tunnel across the vacuum gap assuming a work function of 4.5 eV and a distance of 5 Angstrom is or order 10^{-4}. Hence there will be a current flow of about a nanoampere associated with each circular cross-section of the junction of order 5 Angstroms in radius. A tip of the same radius would be expected to provide about the same current which is indeed the typical current at which the tunneling microscope is usually operated for the other parameters indicated in the calculation, ie.tip spacing above metal 5 Angstroms, bias ten millivolts, and workfunction 4.5 eV. Because of band gaps and density of states problems the bias used to study semiconductors is usually closer to one volt.

At any rate the basic physical picture that emerges is of an evanescent electronic wavefunction in the vacuum gap region, whose origin is on the tip, which interacts with the surface. The interaction strength is measured by the magnitude of the tunnel current injected into the sample. Alternatively with the bias reversed the current can be made to tunnel out of the surface with the evanescent waves, whose origin is the sample, being probed by the tip which accepts the electron current.

It is of interest to compare this type of surface structural probe with others that take advantage of wave mechanical principals. With low energy electron diffraction, x-ray diffraction , and high energy transmission electron diffraction, a diffracted itinerant wave in the elastic channel yields information about the magnitude of the fourier components of the density distribution of atoms near the surface. Alternatively surface extended x-ray absorption fine structure (SEXAFS) uses an inelastic channel to detect electron wave interference effects and X-ray standing wave methods use x- ray interference effects in an inelastic channel (ie fluorescence scattering)to detect atom positions. Both of these take advantage of interference between traveling waves. It is only the tunneling microscope that takes advantage of evanescent waves to probe the surface and in this sense it is quite unique in being able to provide highly local information about even a single atom rather than an ensemble of atoms (hopefully) in equivalent sites on the surface.

Figure 2 shows the so called line scans that typically come in on a computer screen or x-y chart recorder when a tunneling image is being taken. This image is one of the silicon (111) surface obtained in the standard mode, ie the tip follows the surface contours of constant tunnel current as it raster scans along the region of interest. Aside from the constant offset of each scan line to indicate the advance of tip across the surface the vertical displacements on the plot indicate atomic structure in the form of atoms protruding above the surface and an atomic step. If that data is digitally recorded it can be redisplayed on a computer in different formats, one of which is shown in Fig. 3, where the individual pixels of data are connected by lines in such a way to indicate the presence of structure in the data even at the atomic step edge. Figure 3 also shows the length scales which have been calibrated from the known reconstruction periodicity of this surface and electron interferometry methods to be discussed shortly.

Data of this type is now commonly being displayed in a grey scale format which in fact enhances greatly ones ability to quickly

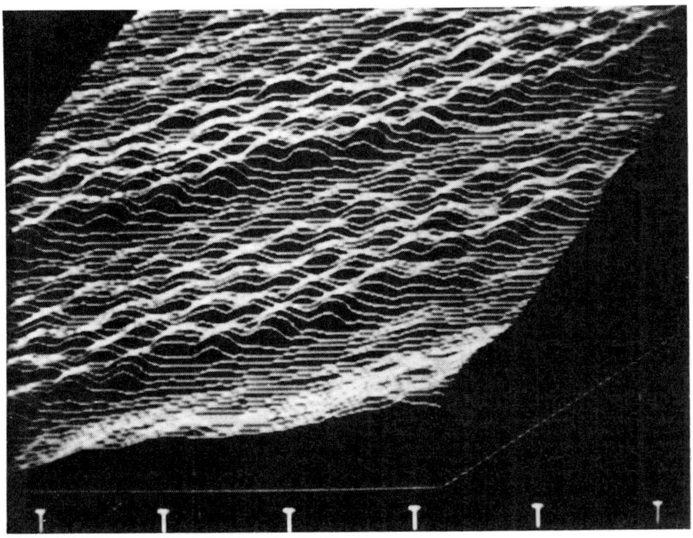

Figure 2. Photograph of computer screen showing the initial view one has of the atomic structure of a crystal surface just after the data comes in. Sample is a (111) oriented silicon crystal surface.

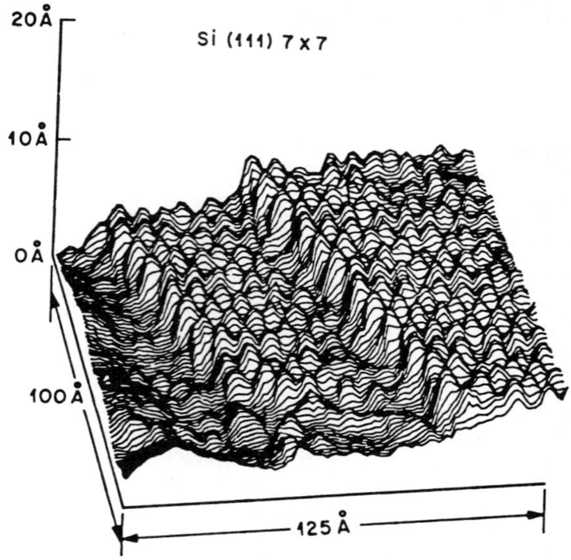

Figure 3. Same data as Fig. 2 but redisplayed to more clearly show atomic step details with the length scale calibrated. From ref. [2].

recognize geometrical structure in the tunneling images. Figure 4 shows the data displayed in this way. High regions of the surface are colored more lightly than lower regions. The periodicity of the lattice reconstruction of this surface as well as the (112) atomic step now appear in a striking way.

Finally having seen the global picture of a surface it is often desirable to take an appropriate cut through the data showing specific features of interest in a quantitative way. The rhombic unit cells seen in Fig. 4 may be viewed as being bounded by four deep corner holes each of which has an apparent sixfold ring around it. A single cut through the long diagonal of this unit cell shown in Fig. 5. The deep corner hole is seen as one proceeds from the left of the plot to the right. Then one encounters a major protrusion associated with a single adatom and then a small protrusion associated with the saddle point between two adatoms. Finally proceeding up to the center of the cell another adatom is directly passed over. The second half of the cell then looks like the mirror image of the first half with adatom modulations of order one Angstrom. While we shall see shortly that some of the apparent symmetry in pictures of the type described above is illusory, data of this type already provided major clues to the atomic explanation of the silicon 7x7 reconstruction.

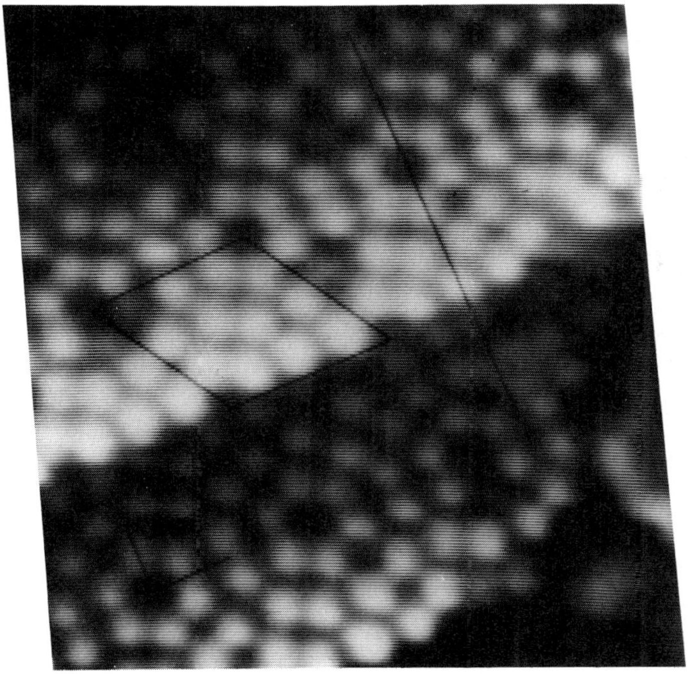

Figure 4. Gray scale plot of the data in Fig 2. The rhombic unit cell of the 7x7 reconstruction is indicated. The cell edge is 27 Angstroms.

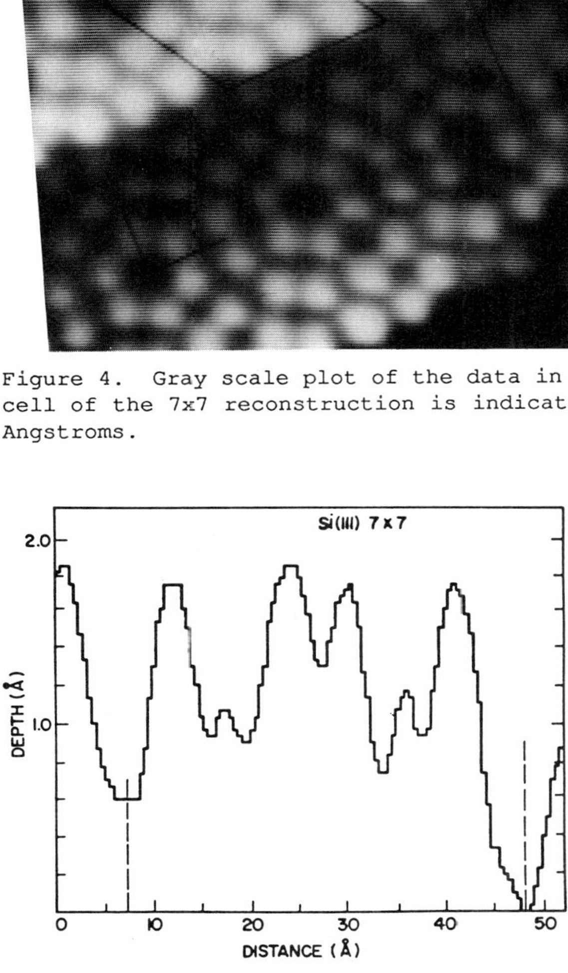

Figure 5. "Cut" showing the changing tip height as one passes along a long diagonal of the 7x7 unit cell.

Early calculations on adatom positions as well as the interpretation of pictures of the type shown above proved to be inadequate in explaining the atomic structure of this surface. Additional information was necessary and part of it came directly from the tunneling microscope. It is at this stage that it proved to be useful to investigate more fully the electrical behavior of the junction, which led rather directly to more information which has helped confirm the currently accepted atomic model for this surface.

There are several different and inequivalent ways to characterize the electrical behavior of the tunneling microscope tunnel junction because the spatial gap may or may not vary with the electrical characteristic being treated as an independent variable. A very early experimental approach to the problem involves keeping the feedback circuit that controls the tip position operable at low frequencies and to apply a "dither" a.c. voltage signal at high frequencies in addition to the junction bias. As a result of this an a.c. component will appear in the tunnel current and the ratio of these two a.c. components yields the differential conductivity of the tunnel junction at the applied d.c. bias and the tip sample spacing. (Note that precautions must often be taken to eliminate the effect of displacement current from stray capacitance.) The differential conductivity may be measured as a function of applied bias and a great deal of structure may often be observed in such spectra [4].

Figure 6 shows the results of this procedure at two different positions within the unit cell of a silicon 7x7 reconstructed (111) surface. The actual positions within the cell where the data was taken will be shown shortly. The important thing to note is that there appear to be two types of structure in the curves. Above about 4.5 volts (which is about the work function of the surface) the differential conductivity undergoes an oscillatory behavior whose origin lies in the operation of the junction as a Fabry-Perot interferometer for electrons. Basically the after tunneling through a barrier which exists only on the side of the junction near the tip, the electrons gain positive kinetic energy already within the gap on the sample side. Before they enter the sample there is an amplitude for reflection at the sample surface which redirects some of the amplitude back towards the classical turning point on the tip side of the junction etc. The resulting interference curves can be calculated using this fairly simple model and quantitative agreement can be obtained as shown in the figure. The results are sensitive to the scale of increasing gap spacing as the d.c. bias is raised and this effect can be used to calibrate the vertical tip position vs. bias to high accuracy.

At lower biases the kinetic energy of the electrons never goes positive in the gap and the transferred electrons wavefunction has an evanescent behavior through the whole gap region. Electrons

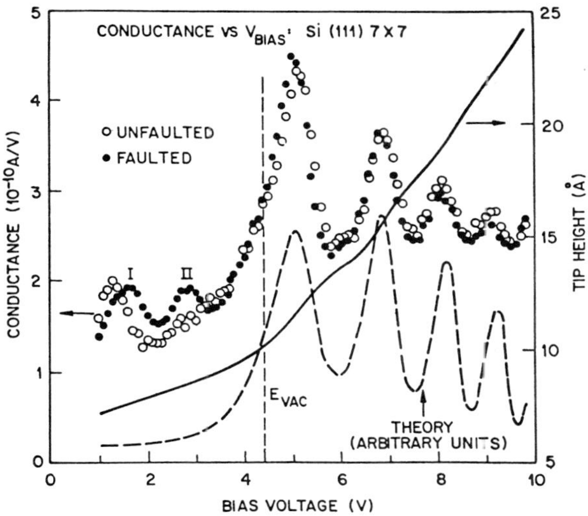

Figure 6. Mechanical and electrical of the tip-vacuum-sample tunnel junction for a silicon (111) 7x7 surface. Curve **a** shows the tip height above the surface versus bias voltage for a tunnel current of 1 nA. Curves **b1** and **b2** show the electrical conductance of the junction (dI/dV_{bias}) versus bias voltage for two different places in the unit cell indicated in Fig. 7. The dashed curve shows the theoretical behavior expected of an electron Fabry-Perot interferometer. From ref. [4].

are then injected into the sample at energies where they are normally bound to the solid and one may expect to observe features in differential conductivity measurements that reflect the eigenvalue spectrum of excited single particle-like states of surface terminated solid. If the bias is chosen so as to extract rather than inject electrons into the solid the results will reflect the distribution of occupied states. There are two caveats here that must be born in mind. First the results of such measurements will also generally reflect not only the available states of the sample but also those of the invariably (at least in all current experiments of this type) uncharacterized tip. Secondly it is not always true that the proximity of the tip leaves the states being probed unperturbed, particularly at high biases and or high injection or extraction currents. Hence great care must be exercised in the interpretation of such spectra in many cases.

Nevertheless it is possible to draw strong conclusions about surface structure from such electrical data particularly when the surface is scanned not only in terms of the tip height for constant tip current but also for the differential conductivity at some interesting bias. An early example [4] of such a study was based on the appearance of structure in the differential

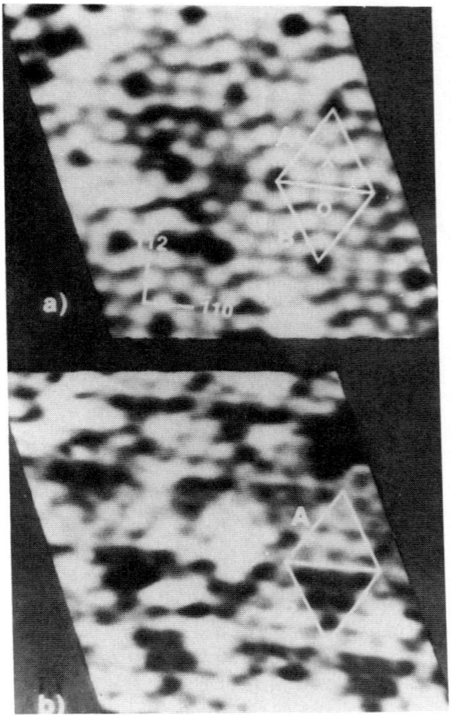

Figure 7. Tip height tunneling image in **a** and differential conductance image in **b** taken at V_{bias} = 2.0 volt. From ref. [4]

conductivity curves on silicon (111) surfaces. As can be seen in Fig. 6 very strong modulations appear at approximately 2.5 volts of bias. The figure actually contains two curves that were obtained at two different places within the reconstructed surface unit cell. Figure 7 shows a tunneling image of the standard tip height in a) and the differential conductivity in b) for exactly the same region of the surface. The two curves in figure 6 were taken at the positions marked "o" and "x" in figure 7a.

The reader will observe a clear asymmetry across the short diagonal of the unit cell in the spatial plot of differential conductivity, which is not observable in in the standard tunneling image. This feature led to the real space identification of a sub-surface stacking fault occupying half of the 7x7 unit cell which confirmed the dimer adatom stacking fault model proposed by TAKAYANAGI et. al [7] based on their beautiful analysis of high energy electron diffraction experiments. Figure 8 shows a calculation of the surface density of allowed states available for tunneling above the fermi level on such a silicon surface for a totally faulted and unfaulted surface both with out adatoms or dimers [4]. The strong modulation and its sensitivity to the presence of a fault supports the interpretation presented above. One must refrain from attempting a very ambitious quantitative

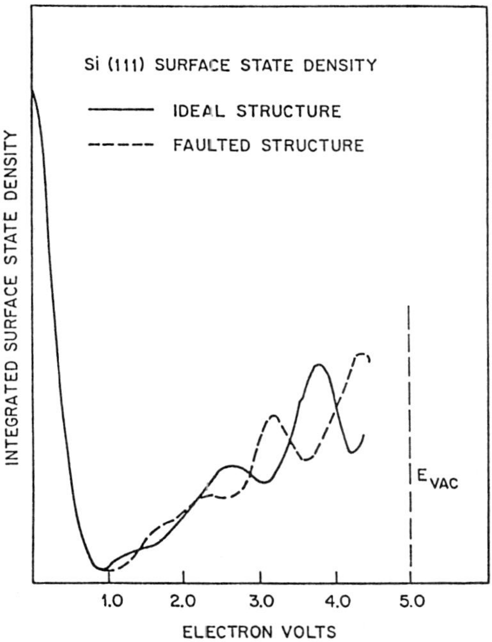

Figure 8. Theoretical surface density of states versus energy for a faulted and unfaulted Si(111) plane slabs. The curves were calculated by the linear augmented plane wave density functional method without the contribution from adatoms or dimers [4].

comparison with such calculations however because they do not contain either adatoms or dimers (computers are not yet large and fast enough to make such calculations practical yet) and the caveats mentioned earlier must be born in mind as well. Dr. Demuth will present further evidence from the study of extracted electrons from this surface which within the same limits also support the model.

A top view showing all the essential features of a unit cell of the reconstruction as it is currently understood is shown in Fig. 9 for the reader to compare with the various tunneling images presented so far. It may also be mentioned that study of the most fundamental defect, the atomic step using the tunneling microscope also can be simply understood on the basis of this model [8].

Tunneling images of related semiconductor surfaces have lead to interesting new insights into both the surfaces being studied and new ways in which the microscope may be used to unravel surface structure problems in general. For example a completely new adatom model of the c2x8 reconstruction on the germanium surface resulted the tunneling microscope study of epitaxially grown films as well as bulk samples [9]. In that case not only was a structure observed which had the symmetry required to explain low energy electron diffraction observations but new reconstructions 2x2 and c4x2, whose unit cells serve as building blocks for the more complicated c2x8, were also discovered. Figure 10 shows an image of this surface with all the related related reconstructions.

Figure 9. Computer generated top view of the dimer adatom stacking fault model for the silicon (111) 7x7 surface.

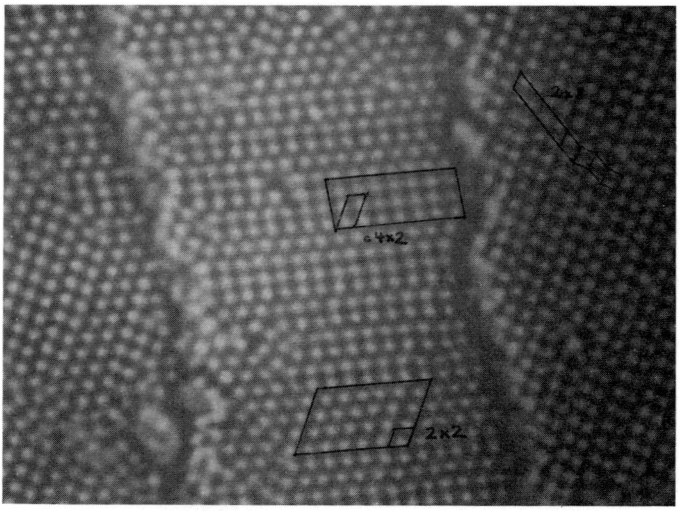

Figure 10. Reconstructions on a germanium (111) surface. Note the disorder in the atomic steps perhaps caused by contamination.

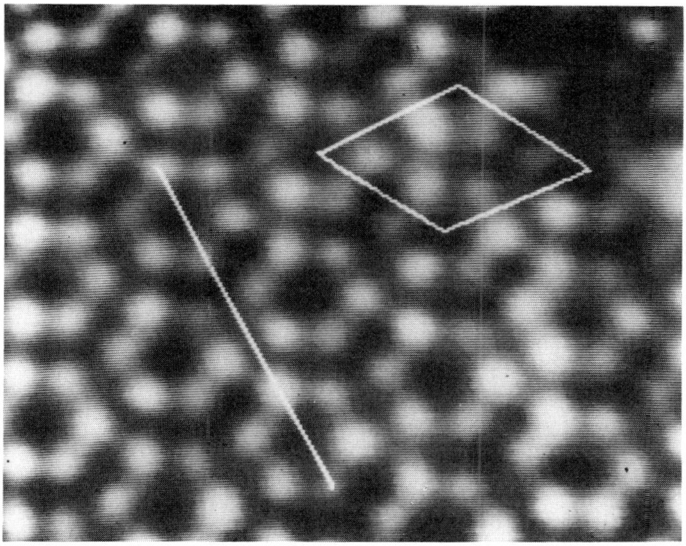

Figure 11. Reconstruction of equal component germanium silicon alloy showing 5x5 surface superlattice [10].

Images (Fig. 11) from equal component silicon germanium alloys of this surface orientation have revealed the structure of the indigenous 5x5 reconstruction to be a shrunken version of the 7x7 with rather strong and as yet unexplained modulations in observed adatom peak heights.

The physics of this family of surface reconstructions possible on this surface has be further studied by using the laser as a tool to prepare nonequilibrium but nevertheless metastable surfaces that could subsequently be observed with the tunneling microscope [11]. The laser melts the near surface region and the subsequent rapid thermal quench associated with the microsecond time scale cooling of the surface leaves the surface in a stable but clearly nonequilibrium state. Small regions of 2x2 and c2x4 are observed first. Partial thermal anneals then show the growth of a defect ridden surface containing 5x5 and 9x9 structures in addition to the equilibrium 7x7.

An ambitious recent attempt at understanding the energetics of these structures has been reported by Vanderbilt [12] who used a combination of density functional theory to estimate total energy contributions from adatom and stacking fault features and Keating model for dimer features. A phase diagram from that work which indicates the relationship of (2n-1)x(2n-1) structures to one another as well as 1x1 structures is shown in Fig. 12. The energy differences between 2x2, c2x4 and 2x8 are assumed small and cannot be calculated within the framework of this model assuming

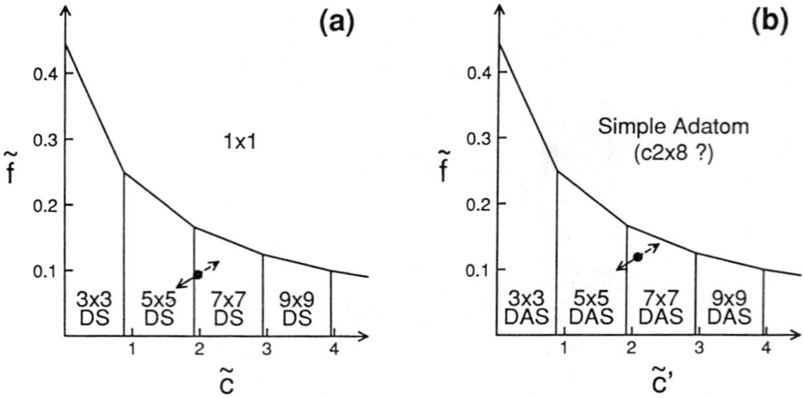

Figure 12. (**a**) Ground state structure as a function of dimensionless corner hole energy c and stacking fault energy f in the absence of adatoms. Filled circle shows estimated position of the parameters for silicon; solid (dashed) arrows show expected variation of the parameters with a o.5% compression (expansion) of the surface lattice constant. (**b**) same as (**a**) but after decoration by adatoms [12].

only adatom reconstructions. Anyway they are all lumped into the 1x1 region of the phase diagram. Vanderbilt's best estimates of the various contributing energies shows that the presence of the dimer rows on the surface is a main driving force in minimizing the number of high energy surface bonds and hence the total energy of the system. His calculations encouragingly show the 7x7 to be the low temperature equilibrium state for silicon (111).

Aside from the observations of atomic steps referred to earlier on silicon an metallic surfaces a number of stable defects such as surface twin boundaries and line defects have also been observed on germanium (111) surfaces where the structure can be seen in great detail as Fig. 13 shows.

In addition this surface has yielded images of unstable defects whose structure appears to change with each new picture of exactly the same region. Figure 14 shows a region of the germanium (111) c2x8 reconstruction that appears to have a single adatom missing with a neighboring adatom that jumps back and forth between two equivalent but not surface substitutional sites from one image to the next. The time to acquire an image is about a minute which is also the time between images. The last image in the figure shows the transition occurring as the tip passes over the unstable atom. It is not clear as of this writing whether this transition is thermally stimulated or stimulated by the tip.

Finally one of the most interesting phenomena to be observed on this class of surface is the possibility to modify it on what appears to be the single atom scale in a purposeful way [13].

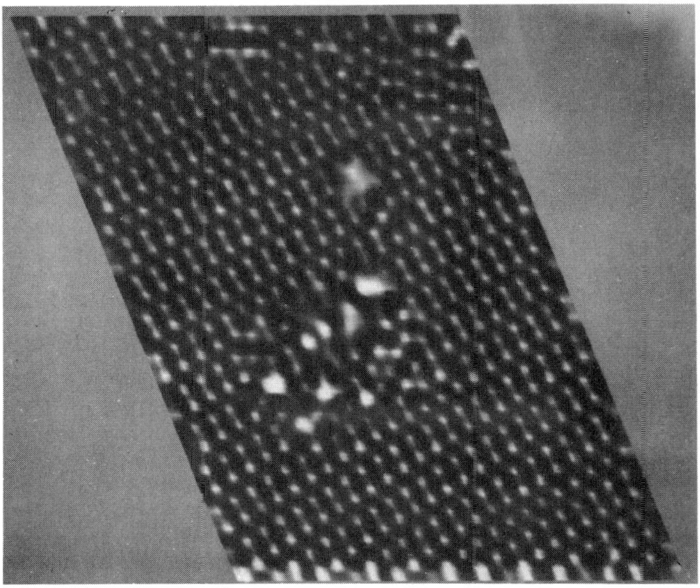

Figure 13. Tunneling image showing variety of defects on a germanium (111) c2x8 reconstructed surface. A twin boundary is seen in the upper left hand corner from which line defects can be seen extending into the left hand side of the image.

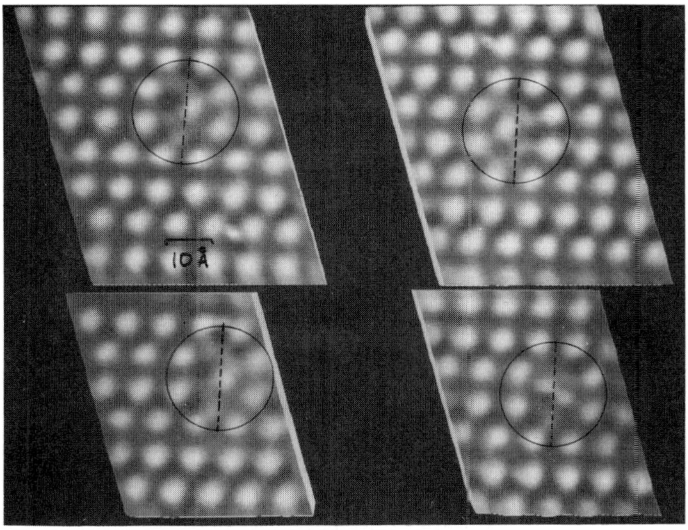

Figure 14. Time dependent defect structure on germanium (111) c2x8.

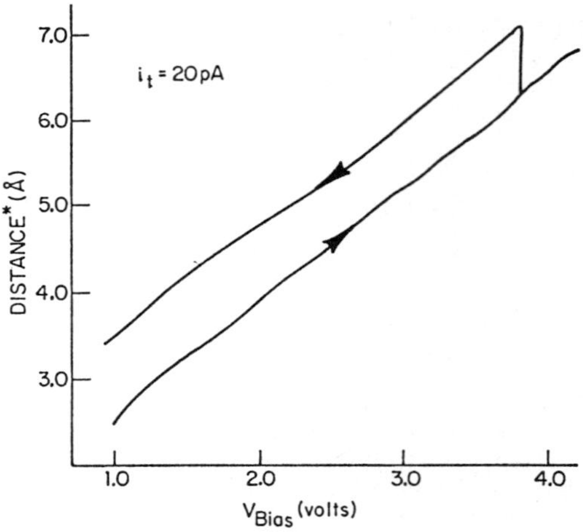

Figure 15. Electromechanical junction response [13].

Figure 15 shows a plot of the tip distance above a germanium (111) surface for a constant current of 20 picoamps as the tunneling bias is increased to several volts and then decreased. On this surface it is common to see the irreversible behavior of the tip retracting rapidly and going back over a path suggesting that the surface under the tip has extended about an angstrom. Indeed tunneling images of the region of the surface before and after this exercise in Fig. 16 show that the surface contains a new induced atomic structure of the same order of size as a germanium adatom. Most likely an atom from the tip has been transferer to the surface below by some mechanism controlled by the bias conditions during the transfer. While this mechanism is hardly understood as yet it is clear that the structures available for study with the tunneling microscope will surely be not only those provided to us by nature but may well include a new generation in which we have control over the atomic features on top of crystalline surfaces.

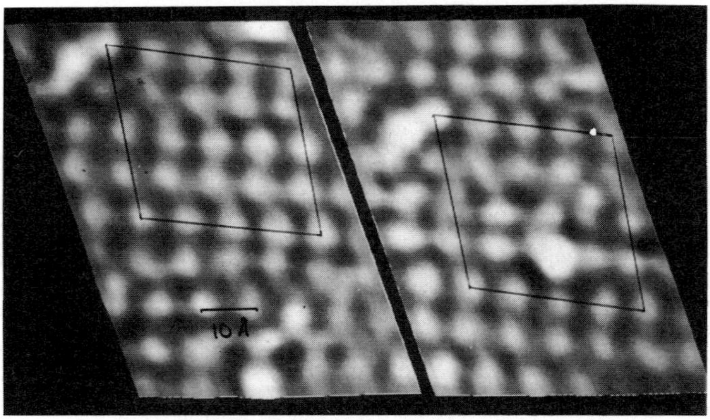

Figure 16. Tunneling images before (**a**) and after (**b**) atomic scale modification procedure has been applied [13].

References

1. G. Binnig, H. Rorher, C. Gerber, E. Weibel: Phys. Rev. Lett. **50**,120 (1983), see also R. Young and F. Scire: Rev. Sci. Instrum. **43**, 999 (1972)

2. J. A. Golovchenko: Science, **232**, 48 (1986)

3. S. Park and C.F. Quate: Rev. Sci. Instrum. **58**, 2010 (1987)

4. R.S. Becker, J. A. Golovchenko, D.R. Hamman, and B.S. Swartzentruber: Phys. Rev. Lett. **55**, 2032 (1985)

5. R.M. Feenstra, W.A. Thompson and A.P. Fein: **56**, 608 (1986)

6. R.J. Hamers, R. M. Tromp and J.E. Demuth: Phys. Rev. Lett. **56**, 1972 (1986)

7. K. Takayanagi, Y. Tanishiro, M. Takahashi and S. Takahashi, J. Vac. Sci. Technol. **A 3**, 1502 (1985)

8. R.S. Becker, J.A. Golovchenko, E.G. McRae, B.S. Swartzentruber: Phys. Rev. Lett. **55**, 2028 (1985)

9. R.S. Becker, J.A. Golovchenko and B.S. Swartzentruber: Phys. Rev. Lett. **54**, 2578 (1985)

10. R.S. Becker, J.A.Golovchenko and B.S. Swartzentruber: Phys. Rev. B **32**, 8455 (1985)

11. R.S. Becker, J.A. Golovchenko, G.S. Higashi and B.S. Swartzentruber : Phys. Rev. Lett. **57**, 1020 (1986)

12. D. Vanderbilt : Phys. Rev. B **36**, 6209 (1987)

13 R.S. Becker, J.A. Golovchenko and B.S. Swartzentruber : Nature **325**, 419 (1987)

Discussion

R.F. WILLIS: Si(001) surface is easier as regards symmetry properties. The energy balance appears to be as follows: (1) the surface wishes to reduce number dangling bonds by forming (2×1) dimers , *but* (2) this costs strain energy; (3) this strain energy is released to some extent by creating missing dimer defects along the dimer chains to produce $(2 \times n,\ n = 8, 9, 10)$. Our He diffraction results suggests the most stable structure is n=8 with *pairs* of missing dimer defects. This also *raises* the configurational entropy of the system. Your Ge(111) results suggest that strained dimer arrays dictate similar behavior.

J.A. GOLOVCHENKO: Our studies of the Ge c(2 × 8) reconstruction with the tunneling microscope have shown no indication of dimers at the surface. Such structures do appear in the images of Si and Ge(001) surfaces but without periodic missing dimers along the rows.

M. CARDONA: You just briefly touched upon the fact that the electrons have to tunnel from the surface states into the bulk through the space charge barrier. What are the consequences of this? Typical depths of semiconductor–barriers are 1000Å but they can be reduced down to as little as 20Å in very heavily doped materials (~ 10^{20} or more in n– and p–Si and p–Ge, 10^{21} in laser annealed ion implanted in n–Si). Have such heavy doped materials been measured by STM? Would the STM images be better for them? Would the high dopings affect the structure of the surfaces? Could one get information about the structure of the impurities (bond length changes, clustering, etc.)?

J.A. GOLOVCHENKO: Carriers injected into surface states in the gap will also have to penetrate the space charge layer at the surface. This penetration will generally require that part of the applied bias voltage to the junction appear across the sample. The interpretation of I-V curves in terms of surface state locations in the band gap can be complicated by not knowing what part of the applied voltage appears across the vacuum gap. The type samples you suggest with extremely high doping have, to my knowledge, never been used in tunneling microscopy studies. They would certainly significantly reduce the possible problems with the space charge layer by insuring that tunneling through, rather than thermal activation over it, is the dominant charge transport mechanism.

As to your other question, I suspect that the quality of images could not be affected by these considerations at all as long as reliable methods of carrier transport exist from the surface into the bulk. The tunneling microscope is not a particularly accurate tool for measuring bond lengths and I would tend to leave that job to diffraction methods. Finally, one might anticipate surface segregation effects from such highly doped samples and the structure on the surface induced would certainly be capable of study in the microscope.

C.B. DUKE: Following up Prof. Cardona's question, I might recall that the total voltage across the tip is divided between the vacuum and the space charge region in a semiconductor such as the Si samples which you discussed. The absolute energies of structure in the dynamic conductance characteristics requires the specifications of this voltage division. How was this done in the integration of the "surface state" peaks observed for voltages below the Fowler-Nordheim region in your data? Even the standing wave resonances in the Fowler–Nordheim region depend on the boundary conditions at the surfaces as demonstrated for the analogous resonances in field ion microscopy many years ago by M.E. Alferiell and myself. Therefore, even they do not determine the absolute energy normalization.)

J.A. GOLOVCHENKO: I do not believe any significant problem exists as far as voltage drops across the surface space charge region is concerned, as long as carriers are injected above the conduction band edge at the surface. The interference effects observed in this Fowler-Nordheim region are calculated with surface boundary conditions taken into account and agreement with experiment is quite good.

D.R. HAMANN: Following up Roy Willis' comment about strain, I'd like to point out that $\sim 1\%$ strain drives Ge(111) from $c(2 \times 8)$ to 7×7. The Ge is strained through epitaxial growth on Si(111).

K. TAKAYANAGI: You mentioned relative stability of the family of 7×7 structures. We discussed (in *JVST*) the relative stability by estimating the elastic energy of dimer walls using the Keating model. It is concluded: (1) the dimer chain works as the dominant mechanism, which determines the relative stability; (2) if the elastic constant of the bulk Si crystal is assumed for the Keating parameter, the 7×7 structure (DAS model) does not stabilize, but the 1×1 structure instead; (3) the 2×8 structure with stacking–fault (*Phys. Rev. B.*) becomes stable only for the very soft bonding, which might suggest that the model is unlikely.

Question: You mentioned that you believe the $c(2 \times 8)$ structure of the Ge(111) surface is a simple adatom structure. Do you have any evidence supporting your idea?

J.A. GOLOVCHENKO: The calculations of Vanderbilt I mentioned in my talk use Keating model parameters for the dimer walls and corner hole structures on the Si 7×7 but evaluates adatom and stacking fault contribution from LDA methods.

The main evidence supporting the simple adatom model of the Ge $c(2 \times 8)$ is the common observation of 2×2 and $c(4 \times 2)$ structure on the surface which can coherently form to make the 2×8 structure without the need for dimers or stacking faults.

Theory of Single-Atom Imaging in the Scanning Tunnelling Microscope

N.D. Lang

IBM Research Division, Thomas J. Watson Research Center, P.O. Box 218, Yorktown Heights, NY 10598, USA

We discuss first the current flow in the scanning tunnelling microscope./1/ The vacuum tunnelling current between two planar metallic electrodes with a small bias voltage between them was studied in the instance in which there is an adsorbed atom on each electrode, with one electrode corresponding to the tip, and the other to the sample (a surface with a chemisorbed atom)./2/ The tip and the sample are thus treated on an equal footing.

The method developed in Ref. 3 (equivalent to a spatial decomposition of the BARDEEN tunnelling-Hamiltonian formalism/4/) was used to calculate the current density distribution in terms of the wave functions determined separately for each electrode in the absence of the other. These wave functions were calculated starting with the results on atomic chemisorption of LANG and WILLIAMS/5/, as described in Ref. 3. The jellium model for the metal surface was used (r_s=2); it can be expected to be adequate for a discussion of many of the properties that depend on the wave functions well outside the surface.

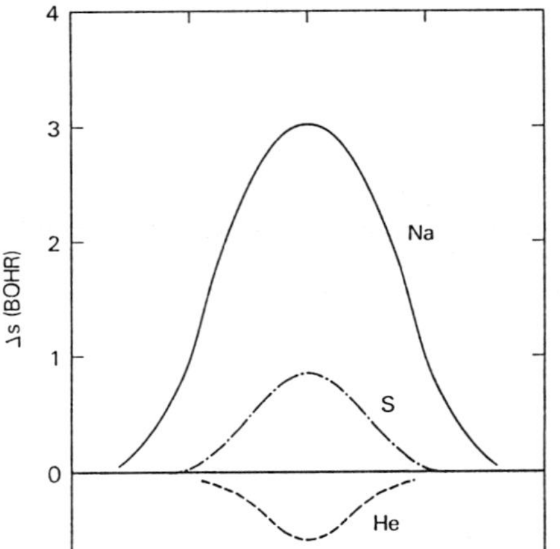

Fig. 1 Plots of change in tip distance $\Delta s = s(Y) - s(\infty)$ versus lateral separation Y for constant current $I_{ST} - I_S$, where I_{ST} is the total current with sample and tip atoms both in place, and I_S is that with just the sample atom in place. $s(\infty)$ = 16 bohr (1 bohr = 0.529 Å.) Tip atom is Na; sample adatoms are Na, S, and He.

We first discuss results obtained when the same atom (Na) on one electrode, taken as the tip, is scanned past several *different* atoms on the other electrode: Na, S and He. These chemically very different atoms produce characteristically different tunnelling-current behaviors. Curves of tip displacement vs. lateral separation for constant current are shown in Fig. 1.

We note first in Fig. 1 that the maximum change in tip distance Δs is much smaller for S than for Na. In part this is due to the fact that S sits closer to the surface than Na, and in part to the fact that the Fermi-level state density for S is appreciably smaller than that for Na. (For the case of low bias considered here, only states near the Fermi level participate in the tunnelling process.) Even more striking, however, is the fact that the change in tip distance for He is slightly negative. The He atom sits rather far from the surface, and so would cause a large enhancement of the tunnelling current if there were to be even a small increase of the Fermi-level state density due to broadened levels of the atom. The closed valence shell is very far down in energy, however, and its only effect is to polarize metal states away from the Fermi energy, producing a decrease in Fermi-level state density, which leads to the negative tip displacement.

Springer Series in Surface Sciences Vol. 14
Solvay Conference on Surface Science
Editor: F.W. de Wette

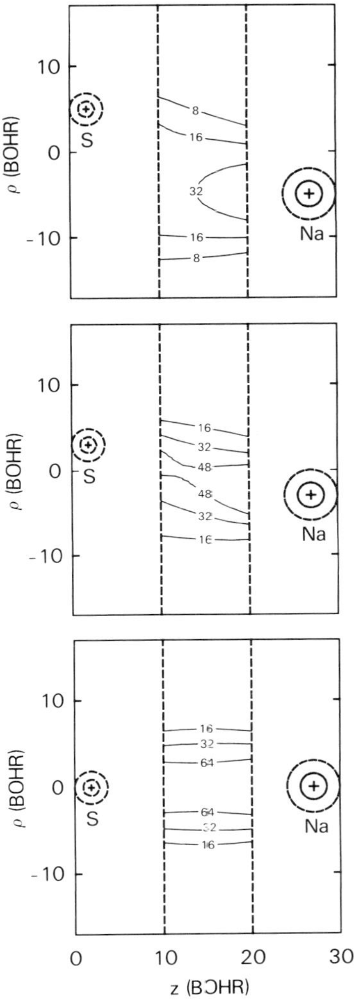

Fig. 2 Contour maps of j_z/j_0 for S adatom on left electrode and Na adatom on right electrode for three values of lateral separation Y (10 bohr, 6 bohr and 0), in the plane normal to the surfaces that includes both atomic nuclei.

We now discuss the spatial distribution of the current in our problem. Figure 2 shows, at several values of lateral separation Y, contour maps of j_z/j_0 for Na on the right and S on the left, with j_z the z-component of the current density and j_0 the current density that would be present without the adsorbed atoms. Results are given only in a strip in the center of the vacuum barrier; much closer to the surfaces, the equation used is not adequate (see Ref. 3). In this figure, the left and right edges of the box correspond to the positive background edges of the two electrodes. The presence of the adsorbed atom is indicated schematically by two dashed circles with a cross at the equilibrium distance of the nucleus, with the larger of the two circles given a radius equal to this distance. At the largest Y shown, the map is very similar to that found for Na alone, in the absence of the S atom; obviously the presence of S will lead to a rather small z-axis displacement of a Na tip at this lateral separation. As Y is decreased, the fact that the z-axis displacement of the tip would have to grow to maintain constant current is clear.

We now consider the way in which topographic images obtained with the scanning tunneling microscope are affected by the sample density of states./6/ The vacuum tunneling current was calculated for the tip atom/sample atom configuration described above, as a function of the bias between the two electrodes. Take s to be the z-axis separation (i.e., that along the surface normal) between the center of the tip atom and the center of the sample atom. Let us hold the current fixed; then the z-axis separation will be a function of the lateral separation Y and the applied bias voltage: $s = s(Y,V)$. For any given Y, the separation s will increase substantially as V is increased from the millivolt range to several volts, in order to maintain the current at its fixed value. The measure of z-axis displacement due to scanning laterally over the atom on which we focus our attention is $\Delta s(V) = s(0,V)-s(\infty,V)$. We can call this quantity the apparent vertical size (or displacement above the surface) of the adatom. It is the most important piece of information in the topograph. The calculations of the currents are done in the same way as those described in Ref. 7, with the same approximation, in which the stationary-state wave functions are computed in the absence of the electric field; this approximation leads to a rather small error for biases (typically ~1 eV) that are not too large on the scale of the work function.

Three different sample atoms were considered: Na, Mo, and S, as prototype alkali, transition metal, and non-metal. A Na atom was used for the tip in all calculations, but the results are rather independent of the tip atom used. The curves of $\Delta s(V)$ for these atoms are shown in

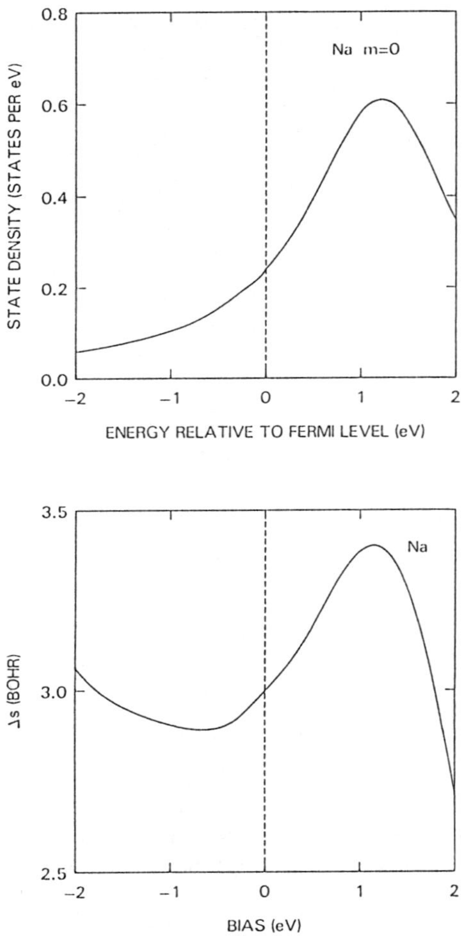

Fig. 3 Top: Na eigenstate density difference. The 3s resonance is clearly evident. Bottom: Δs(V) for Na sample adatom.

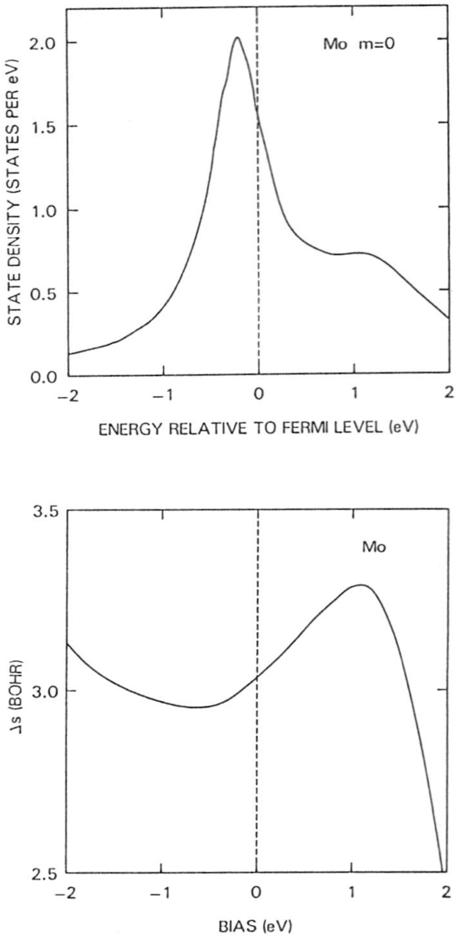

Fig. 4 Top: Mo eigenstate density difference. The lower-energy Mo peak corresponds to 4d, the upper, smaller peak to 5s. Bottom: Δs(V) for Mo sample adatom.

Figs. 3-5, for V in the range -2 to +2 eV, together with the additional eigenstate density in this energy range due to the presence of each of the three sample atoms on a metal surface (only the m=0 component is shown). Positive V corresponds to the polarity in which electrons tunnel into *empty* states of the sample.

Figure 3 shows the computed Δs(V) for the case of the Na sample atom; it clearly reflects the mostly empty 3s resonance. The graph of Δs(V) for Mo in Fig. 4 is very similar to that for Na, with the large peak at positive bias associated with the mostly empty 5s state of the adsorbed Mo atom. There is very little evidence of the 4d state in this graph. The reason for this is that the valence d orbitals in the transition elements are in general quite localized relative to the valence s orbitals, and will thus have a much smaller amplitude at the tip.

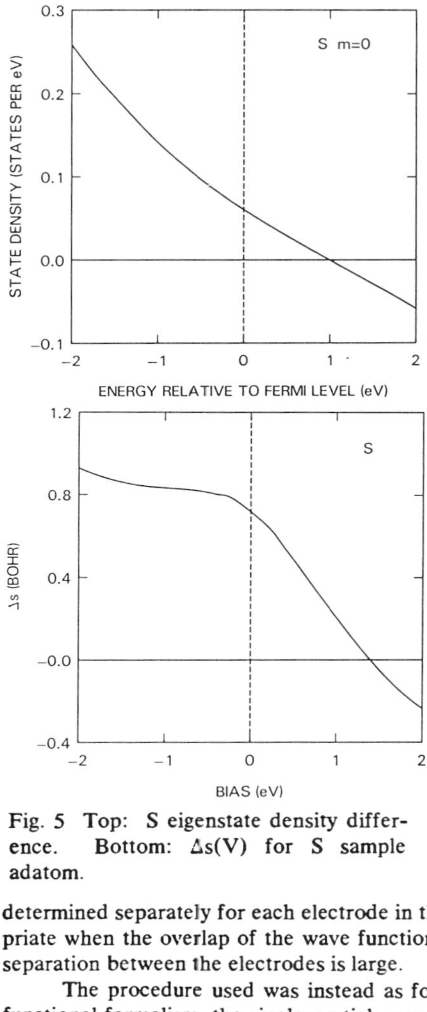

Fig. 5 Top: S eigenstate density difference. Bottom: Δs(V) for S sample adatom.

Figure 5 gives Δs(V) for S. We note the rather striking fact that for a bias just above +1 eV, the S atom is essentially invisible, that is it causes no displacement of the tip, and for biases much above this, the tip moves *closer* to the atom as it passes over it (negative displacement). The negative displacement is a direct consequence of the fact that the additional state density due to the presence of the S atom is negative in the relevant energy region. (Note that the piece of the state density curve shown is just the tail of the 3p resonance, whose center lies ~3 eV below the Fermi level.) Since the total (metal plus atom) state density at these energies is thus lower than it is for the bare metal, the tip must move closer to the surface when it is over the atom in order to maintain a constant tunneling current.

We now turn to consider the region of transition from the tunneling to the point-contact regimes in the scanning tunneling microscope, where the initial contact takes place between a single tip atom and the sample surface. The current that flows between two planar parallel metallic electrodes, one of which has an adsorbed atom (this electrode represents the tip), was studied as a function of distance between the electrodes, for the case of a small applied bias voltage. In this initial calculation, the atom was not allowed to move relative to its electrode as the tip-sample separation was decreased, but was kept at its large-separation equilibrium distance *d*. In contrast to the calculations described above, however, a tunneling-Hamiltonian formalism/4/ can no longer be used to compute the current in terms of the wave functions determined separately for each electrode in the absence of the other, because this is only appropriate when the overlap of the wave functions of the two electrodes is small, that is, when the separation between the electrodes is large.

The procedure used was instead as follows: First, within the framework of the density functional formalism, the single-particle wave functions (only the standing-wave solutions are needed initially) and self-consistent density distribution were found for the pair of bare metallic electrodes, assuming them for simplicity to be identical (r_s = 2 jellium model). This is the previously solved problem of the bimetallic junction./8/

Next the procedure of LANG and WILLIAMS/5/ was used to find the self-consistent density distribution and single-particle wave functions (again just standing-wave solutions) for the total system consisting of the two bare electrodes plus the atom. It will be recalled that this method had originally been used to study an atom adsorbed on a single bare metallic surface, and proceeded by solving a Lippmann-Schwinger equation that involved a Green's function for the bare metal. The only significant difference in the present case is that the Green's function is that appropriate to the bimetallic junction, instead of the single surface. This calculation provides the potential in which the electrons in the system will move in the zero-bias limit.

Consider now current-carrying states of the bimetallic junction without the atom. Deep in the left electrode (to make a particular choice) each such state consists of a plane wave moving to the right plus a reflected wave; deep in the right electrode it consists of a transmitted plane wave moving to the right. All such waves were considered whose energy is equal to the Fermi energy, but whose other quantum labels are different (the azimuthal quantum number m and parallel-momentum label κ defined in Ref. 5). All of these states (continuum-normalized as in Ref. 5) were weighted equally, and an overall normalization factor was chosen such that all of the plane waves moving to the right deep in the left electrode, taken together, yield a unit current density (atomic units).

The Lippman-Schwinger equation was solved again, using the procedure of Lang and Williams, but this time to obtain the current-carrying states at the Fermi level in the presence of the atom, starting with the current-carrying states of the bimetallic junction without the atom. The potential that goes into this equation is the difference between the self-consistently determined potentials in the bimetallic junction with and without the atom./9/ These wave functions are then used to compute the total additional current δI due to the presence of the atom. We can define the additional conductance due to the presence of the atom as $\delta G = \delta I/V$, and it is convenient to define an associated resistance $R \equiv 1/\delta G$.

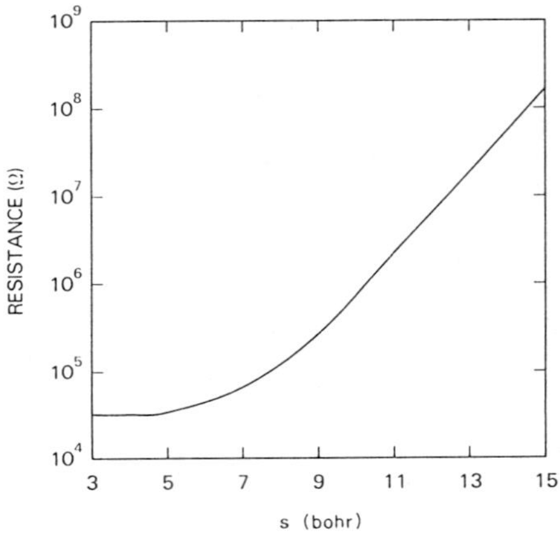

Fig. 6 Resistance $R \equiv 1/\delta G$ as a function of tip-sample separation s for a Na tip atom.

In Fig. 6, the resistance R is shown as a function of tip-sample separation s for the simple case of a Na tip atom. At large separations, R changes exponentially with s. As s is decreased toward d, the resistance levels out at a value of 32,000 Ω. (For $s=d$, the atom is midway between the two metal surfaces, so this can in some sense be taken to define contact between the tip atom and the sample surface). If the same calculation is done for a Ca atom, R is found to level out at 18,000 Ω. We can write these resistances as $R_{lim} = A\pi\hbar/e^2$, with A=2.5 for Na and 1.4 for Ca. A physical discussion of this result and a comparison with recent experimental data of GIMZEWSKI and MÖLLER/10/ are given in Ref. 11.

References

1. For reviews of scanning tunneling microscopy, see G. Binnig and H. Rohrer: IBM J. Res. Develop. **30**, 355 (1986); R. J. Behm and W. Hösler: In *Chemistry and Physics of Solid Surfaces VI*, ed. by R. Vanselow and R. Howe (Springer, Berlin, 1986), p. 361; P. K. Hansma and J. Tersoff: J. Appl. Phys. **61**, R1 (1987).
2. N. D. Lang: Phys. Rev. Lett. **56**, 1164 (1986).
3. N. D. Lang: Phys. Rev. Lett. **55**, 230 (1985); **55**, 2925(E) (1985); IBM J. Res. Dev. **30**, 374 (1986).
4. J. Bardeen: Phys. Rev. Lett. **6**, 57 (1961).
5. N. D. Lang and A. R. Williams: Phys. Rev. B **18**, 616 (1978); Phys. Rev. Lett. **37**, 212 (1976).
6. N. D. Lang: Phys. Rev. Lett. **58**, 45 (1987).
7. N. D. Lang: Phys. Rev. B **34**, 5947 (1986).
8. J. Ferrante and J. R. Smith: Phys. Rev. B **19**, 3911 (1979); **31**, 3427 (1985); A. J. Bennett and C. B. Duke: Phys. Rev. **160**, 541 (1967).

9. All image effects are omitted, as in the previous calculations.
10. J. K. Gimzewski and R. Möller: Phys. Rev. B **36**, 1284 (1987).
11. N. D. Lang: Phys. Rev. B **36**, 8173 (1987).

Discussion

A. DE LOZANNE: You showed a graph of resistance *vs.* spacing which had an exponential region and a saturation region. Does the tunneling Hamiltonian formalism stop being valid as soon as one leaves the exponential region at small spacing?

N.D. LANG: It is certainly invalid in the saturation region and is probably invalid part way into the exponential region. The tunneling-Hamiltonian formalism is only appropriate when the overlap between the wave functions of the two electrodes is small, i.e. only for relatively large separations.

S. G. LOUIE: The quasiparticle energy spectrum of adsorbates can be quite different from the LDA Kohn–Sham spectrum. The position of peak features in the DOS can shift by as much as 1 eV or more. Care should therefore be taken in using Kohn–Sham state densities in interpreting STM spectroscopy/image quantitatively.

N.D. LANG: The problem of a shift in peak positions between the LDA spectrum and the true quasiparticle energy spectrum is most important for very narrow resonances. It can be expected to be less significant for the broader resonances that I have discussed here, but I certainly agree in principle that care ought to be taken when using LDA eigenstate densities in a quantitative interpretation of experimental results. It should of course be noted that the calculation of the quasiparticle spectrum of an adsorbate is a difficult and as-yet unsolved problem.

R.F. WALLIS: What was the electron density in the jellium electrodes?

N.D. LANG: The electron density n deep in the electrodes was taken to be that corresponding to $r_s = 2$ bohr, where $\frac{4}{3}\pi r_s^3 \equiv n^{-1}$. This is roughly the density appropriate to aluminum.

G.A. SOMORJAI: Would your model and calculation be applicable if you had an atomic vacancy instead of an adsorbed atom on the surface?

N.D. LANG: To the extent to which a vacancy could be modelled by, e.g., a spherical hole cut out of the jellium positive background (with the electron density calculated self-consistently with this new positive charge distribution), the calculation I have described would be entirely applicable.

Electronic Processes at Disordered Surfaces

J.B. Pendry

The Blackett Laboratory, Imperial College, London SW7 2BZ, UK

The calculation of electronic properties for infinite systems that are also disordered has been severely hampered by the absence of a Bloch theorem. Recent theoretical work [1,2] has begun to remove these restrictions to enable analytic and computational studies to be extended to this relatively unexplored class of matter.

The general case of a completely disordered surface in the strong disorder limit has been a major challenge with all the complications that Anderson localisation involves [3,4]. New methods based on symmetry principles have been evolved to make analytic studies of the qualitative properties of these systems. Two particular phenomena will be discussed.

Firstly the localisation produces trapping states at varying depths below the surface [5]. The widths, d, of these localised states are much too narrow to be observed by conventional spectroscopy whose resolution can rarely exceed one milli- electron volt [6]. So small is d in general that $\hbar/d$ is of the order of seconds. We have developed a formalism [5,7,8] that enables us to calculate the statistical distribution of level widths. Analytic solutions show that:

i) most of the levels have a 1/d distribution of widths
ii) this distribution implies 1/f noise in the total surface charge
iii) saturation of the widths occurs at very small values of d with $\hbar/d$ being many hours or even days
iv) the noise [9,10] arising in surface charge and its deviation from a 1/f law can be used as a spectroscopy of the level widths
v) the cut off is related to the diffusion limited frequency response of the disordered material in transmission
vi) the cut off has implications for the approach to thermodynamic electronic equilibrium in disordered systems which may take many days and invites comparison with spin glasses
vii) scanning tunneling microscopy [11] has the potential to examine very small areas and explore the cut off in the noise spectrum

We have treated the mathematical side of the problem by solving a simplified model of the surface exactly using the technique of the symmetric group [12]. Our conclusions can be summarised as follows: we are interested in the reflection coefficient of a chain spanning a thick layer of disordered material. It will have the form:

$$R = |R| \exp(i\Theta) \tag{1}$$

Springer Series in Surface Sciences Vol. 14
Solvay Conference on Surface Science
Editor: F.W. de Wette

$$|R| \approx 1 \tag{2}$$

If the disordered region is much thicker than a localisation length, $|R|$ will be unity.

The phase, Θ, has a second interpretation: it is proportional to the density of states inside the surface,

$$\sigma(E) = (2\pi)^{-1} \, d\Theta/dE. \tag{3}$$

In localised systems, where an electron deep inside the surface stands little chance of tunnelling out, the states have a long lifetime and are in consequence very narrow in energy. Thus σ will consist of a series of sharp resonances as the phase jumps through 2π. This effect has also been observed by Azbel [11] in numerical calculations.

There is a third interpretation of the phase which is the crucial one for our purposes: suppose a packet is constructed from waves of closely related energies so that the energy spread is small and $d\Theta/dE$ is essentially constant for all the waves. If this packet is incident on the surface it will not immediately be reflected back into the vacuum, but will be delayed by a time proportional to $\hbar d\Theta/dE$.

How do these ideas relate to noise? Suppose we have a random electron wave incident on the surface. Its noise spectrum is not important provided that there is some time dependence and the phases of successive energies are random. If Θ were not a function of energy, then the delay in reflecting the randomly fluctuating currents from the surface would be zero and the in going and out going fluxes would cancel at each point in time. In contrast when there are localised states, some of the fluctuating current is not immediately reflected, but is delayed as we discussed above. Hence in localised systems there is not instantaneous cancellation of current at the surface. The net current shows fluctuations on a time scale dictated by the trapping times, and the net charge on the surface will fluctuate.

Application of our symmetrised theory enables us to reconstruct the probability distribution of the widths of the localised states,

$$P(d) = \text{const}/d. \tag{4}$$

From this follows the distribution of the reflection coefficients, and of the fluctuations in charge on the surface:

$$<|Q(w)|^2> \approx \text{const}'/w. \tag{5}$$

Hence the "1/f" noise. The value of the constant can depend on temperature in a variety of ways depending on whether the traps are actually at the Fermi energy.

It is important to note that the charge accumulating within the disordered region does not produce a voltage <u>across</u> the device, rather a fluctuation in the electrostatic potential of the device as a whole and therefore no thermodynamic principles are violated by the noise being coloured. Van der Ziel [9] has discussed how intrinsic fluctuations of trapped charge can produce mobility fluctuations in MOSFETS.

Regions of noisy surface are seen in the STM [14]. Selection of sufficiently thin regions of disorder would enable the cut- off to be reached at reasonable frequencies giving access to some of the parameters of the disorder. In fact measurement of the noise spectrum provides us with an extremely high resolution tool with which the widths of the Anderson localised levels can be measured to an accuracy of $\hbar$/(measuring time). The STM has the potential to become a unique probe of localisation at surfaces.

The second phenomenon discussed is the nature of electron states dominating the transport process. It has been suggested that in a strongly disordered material, states localised near the centre of the system dominate the transport properties because only these states have a high transmission coefficient. This argument is in fact incorrect: being far from either metallic contact, such states are very long lived and contribute to the conductivity over a restricted range of energies. Our conclusions can be derived in a rigorous fashion [15], but here we shall give the intuitive argument as it is shorter and more easily understood.

The new idea is quite simple: instead of considering only a single state we imagine a whole necklace of n quasi- extended states stretching from one side of the disordered sample to the other. If they all have approximately equal energies they will hybridise to form a sub- band. The band width is determined by the overlap integrals between the n individual states. We argue that this is maximised by requiring the n states to be approximately equally spaced across the sample so that the overlap integrals are all of the order,

$$\Omega(n) \approx \exp(-\delta'^2 L/(4n)). \tag{6}$$

The transmission coefficient for such a state will rise to unity in the centre of the broadened sub- band and, contributing as it does over a wide range of energies, it will be an effective contributor to the transport process as well as having a wider band width for communication. Of course, as n grows larger, it will become increasingly unlikely that we shall find such a fortuitous coincidence of energies required for the hybridisation of n states. There will be an optimum value of n determined by a trade off between the width of the sub- band given by (6) and the probability of its occurrence,

$$P(n) \approx [c\ \Omega]^n = \exp(-\delta'^2 L/4 + n \ln c), \tag{7}$$

where $c<1$ but is of order unity. We can see that the most probable sub-bands are those containing only a single state in the centre of the sample. However, contrary to our assumption in section 1, these states do not dominate the average band width which is given by:

$$w_{1/2} \approx \sum_{n=1}^{L} \Omega(n)P(n)/[\sum_{n=1}^{L} P(n)]$$

$$= \sum \exp(-\delta'^2L/(4n)-\delta'^2L/4+n \ln c)/[\sum P(n)] \quad (8)$$

The summand is a maximum when

$$+\delta'^2L/(4n^2_{max}) + \ln c = 0,$$

$$n_{max} = \sqrt{[\delta'^2L/|4 \ln c|]}. \quad (9)$$

Hence to logarithmic accuracy,

$$w_{1/2} \approx \exp(-\delta'\sqrt{L}\sqrt{|\ln c|})\ (1/c - 1). \quad (10)$$

Our new theory has successfully reproduced the curious dependence of the band width on $\sqrt{L}$ in the exponent predicted by more rigorous theory [13]. To obtain the pre- exponential would require a slightly more subtle theory with which we shall not concern ourselves here.

The quasi- extended states which dominate the conductivity form a sub-band containing of the order of $\sqrt{L}$ states strung out more or less evenly over the sample, like beads on a necklace. The dimensionality of the sub-band is fractal and equal to 1/2 because it occupies a fraction $\sqrt{L}/L$ of the sample.

We stress that these states occur extremely rarely. From (6) we deduce the probability of finding an $n=\sqrt{L}$ state in a given sample to be,

$$\approx \exp(-\delta'\sqrt{L}\sqrt{|\ln c|/4})\ (1/c - 1). \quad (11)$$

For large L this is a very small number. In a numerical simulation it would be necessary to study a large ensemble in order to be sure of finding a quasi- extended state.

There are two sorts of states in the system and each has its own properties. The quasi- extended ones discussed above have a remnant of the character of the delocalised states of an ordered system. They form a relatively broad band which enables them to dominate the conductivity despite their being extremely uncommon. In addition this broad band width means that they are "fast" states: they can carry signals rapidly across the sample, and they quickly adjust their population to changes in Fermi energy. Their typical width is given by $w_{1/2}$ and is of the order of 10^6 Hz. In contrast the most numerous states tend to be localised on a single site and have a narrow width of the order of w_{min} which can be of the order of 10^{-8} Hz. These are "slow" states that may take many hours to adjust to a change of Fermi energy. They dominate the density of states and constitute the trapping states observed in strongly disordered materials.

References

1. PD Kirkman and JB Pendry: J. Phys. C 17 4327 (1984)
2. PD Kirkman and JB Pendry: J. Phys. C 17 5707 (1984)
3. PW Anderson: Phys. Rev. 109 1492 (1958)
4. R Landauer: Phil. Mag. 21 863 (1970)
5. JB Pendry, PD Kirkman, and E Castaño: Phys. Rev. Lett. 57 2983 (1986)
6. H Ibach and DL Mills: "Electron Energy Loss Spectroscopy and Surface Vibrations" (Academic, New York, 1982)
7. JB Pendry: J. Phys. C 15 3493 (1982)
8. JB Pendry: J. Phys. C 17 5317 (1984)
9. A Van der Ziel: Advances in Electronics and Electron Physics 49 225 (1979)
10. P Dutta and PM Horn: Rev. Mod. Phys. 53 497 (1981)
11. G Binnig, H Rohrer, Ch Gerber, and E Weibel: Appl. Phys. Lett. 40 178 (1982); and Phys. Rev. Lett. 49 57 (1982)
12. JB Pendry: J. Phys. C 15 4821 (1982)
13. M Ya Azbel: Phys. Lett. 78A 410 (1980)
14. ME Welland and RH Koch: Appl. Phys. Lett. 48 724 (1986)

Discussion

C.B. DUKE: There is an extensive literature on tunneling through disordered films (e.g., thermally grown oxides, plasma deposited insulators). These junctions occasionally exhibit hysteresis phenomena associated with trapping, but rarely anomalous noise (i.e., $< \Delta i^2(\omega) >$, $\Delta i = i(\omega) - \bar{i}$) characteristics. Could you indicate the experimental circumstances in which one would observe the resonances which you predict? Specifically what form of $< \Delta i^2(\omega) >$ do you predict for thin insulating films. Is this curve a universal one for all static dc bias values eV or is it a function of eV?

J.B. PENDRY: The fluctuations predicted are in charge trapped within the dissolved layer. Observation of this fluctuating charge is via second order effects. For example, if the insulating film were an SiO_2 layer in a MOSFET, fluctuations in the MOSFET characteristics would be induced by the charge. This gives rise to the characteristic feature of 1/f noise: observation as a fluctuation in the *conductivity* of a device such that $< \Delta G^2(\omega) > \approx 1/\omega$.

D.R. HAMANN: For very narrow resonances, won't the phonon fluctuations dominate the charge transport, bringing nearby traps in and out of resonance and leading to so-called variable range hopping?

J.B. PENDRY: Yes: at higher temperatures, in thick samples, variable range hopping dominates. However at very low temperatures in their disordered layers the elastic mechanisms involved here dominate.

Recent Results in Scanning Tunneling Microscopy

A. de Lozanne

Department of Physics, University of Texas, Austin, TX 78712, USA

1. INTRODUCTION

We review two of our recent developments in scanning tunneling microscopy (STM). The first is our use of a unique low-temperature STM to perform spectroscopic measurements of the new oxide superconductors. A second recent development is the direct writing of nanometer features with the STM.

2. HIGH T_c SUPERCONDUCTORS

The recent discoveries by Bednorz and Muller[1] of superconductivity in the 30 K range and by Wu et al.[2] in the 90 K range have generated a tremendous interest because of their scientific and technological importance. We have performed spectroscopic measurements on both classes of materials with a newly developed scanning tunneling microscope (STM). This instrument operates at low temperature in an ultra-high-vacuum chamber and has a number of unique capabilities:

* Temperature range: 10K to 400K
* Topographic imaging and I-V characteristics
* In-situ Auger analysis, capable of scanning
* In-situ LEED
* In-situ ion milling, annealing and thermal evaporation
* Load-lock exchange of samples and tips

The last feature is very important in this experiment since it has allowed us to measure a large number of combinations of tips and samples in a short time. Figure 1 shows the differential conductance (dI/dV) of a $La_{1.85}Sr_{0.15}CuO_{4-y}$ bulk sample with a tungsten tip[3]. A gap-like structure is clearly seen with a value of approximately Δ=12 mV. The superconducting transition temperature, T_c, was measured resistively on a sample from the same batch, giving an onset of 40 K, a midpoint of 38.5 K and a width (10%-90%) of 1.1 K. Therefore the ratio $2\Delta/kT_c$=7 is twice as large as the BCS prediction.

Figure 2 shows similar data for a $YBa_2Cu_3O_{7-y}$ bulk sample[4]. The three different curves illustrate the improvement achieved by using softer tips. A tungsten tip was used for the top curve, while an aluminum tip was used for the lower two. Sharper characteristics are obtained with the softer aluminum tip. The most common value of $2\Delta/kT_c$ for this material is approximately 11, with values even twice as large sometimes observed.

One must ask whether these large gaps are real and whether they are related to superconductivity. The experimental setup has been calibrated with resistors to check the voltage scale. Furthermore, we do not believe that there is a series voltage drop anywhere because this would reduce the

Springer Series in Surface Sciences Vol. 14
Solvay Conference on Surface Science
Editor: F.W. de Wette

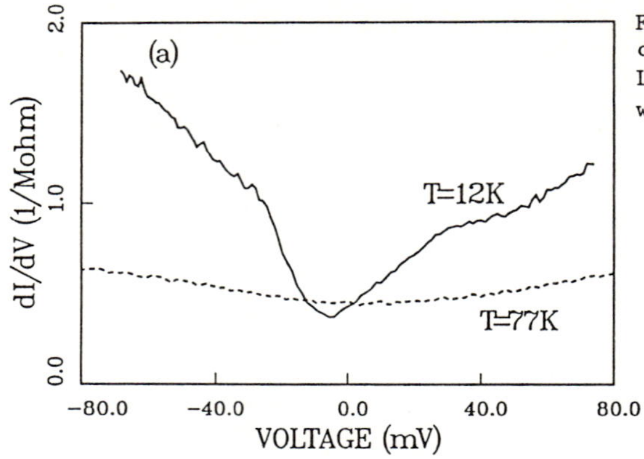

Figure 1. Differential conductance of a $La_{1.85}Sr_{0.15}CuO_{4-y}$ sample with a tungsten tip

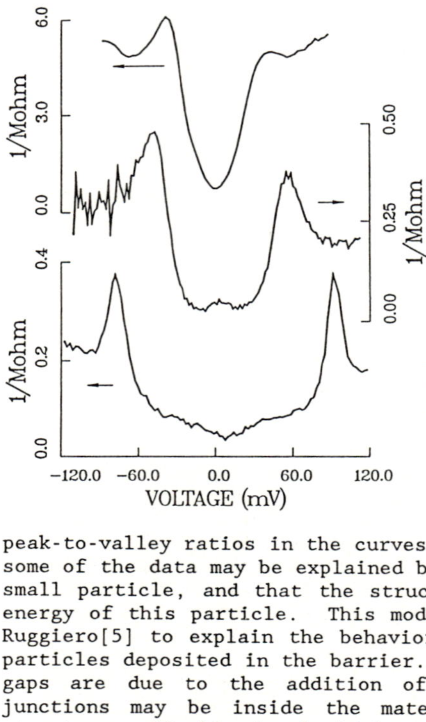

Figure 2. Differential conductance of a $YBa_2Cu_3O_{7-y}$ bulk sample and a tungsten tip (top) or aluminum tip (center and bottom)

peak-to-valley ratios in the curves. A strong possibility is that at least some of the data may be explained by assuming that the tunneling involves a small particle, and that the structure we observe is due to the charging energy of this particle. This model has been used recently by Barner and Ruggiero[5] to explain the behavior of a macroscopic tunnel junction with particles deposited in the barrier. One last possibility is that the large gaps are due to the addition of several junctions in series. These junctions may be inside the material due to its granular and twinned structure. Finally it should be pointed out that given the current state of the theoretical understanding of these materials, the BCS value for $2\Delta/kT_c$ need not be appropriate.

3. DIRECT WRITING OF NANOMETER STRUCTURES

A second recent development is the direct writing of nanometer features with the STM[6]. We allow organometallic gases into a dedicated STM chamber and use the electron current between tip and sample to break the organometallic molecules and deposit the appropriate metal. So far we have written cadmium dots and lines with sizes down to 40 nm. There are several mechanisms that may explain these results, with inelastic tunneling being one of the more interesting possibilities.

Figure 3 shows a schematic of the STM used for this experiment. The scanner in this STM has a range of 10 microns on all axes, which is longer than usual. The organometallic gas is fed through a needle valve. So far we have used mostly dymethylcadmium as the source gas. Some of the deposits have been characterized with Auger spectroscopy, which shows mostly cadmium. A typical writing procedure is to image the surface at a low voltage (tens of mV) to ensure flatness, bring the tip to the desired location on the surface, increase the voltage to a few volts (which breaks the organometallic molecules), and image the surface again. Examples of the deposits obtained are shown in fig. 4. Both images, about 100 nm on a side, are obtained with the STM immediately after writing. In figure 4(a) the tip was moved across the surface to write a short line while in fig. 4(b) the tip was stationary at the center. The short dimension of these

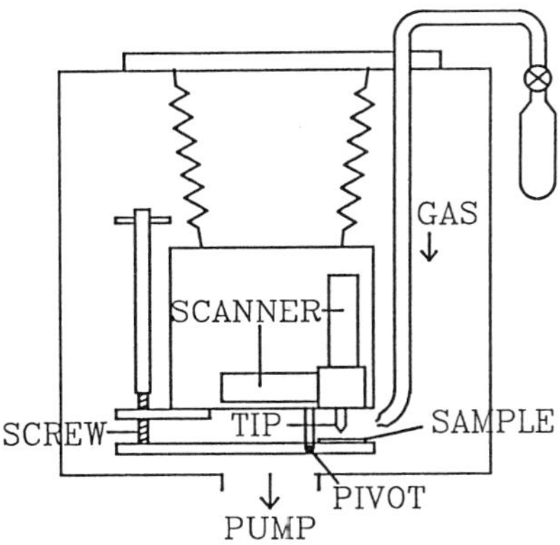

Figure 3. Schematic of the STM for direct writing

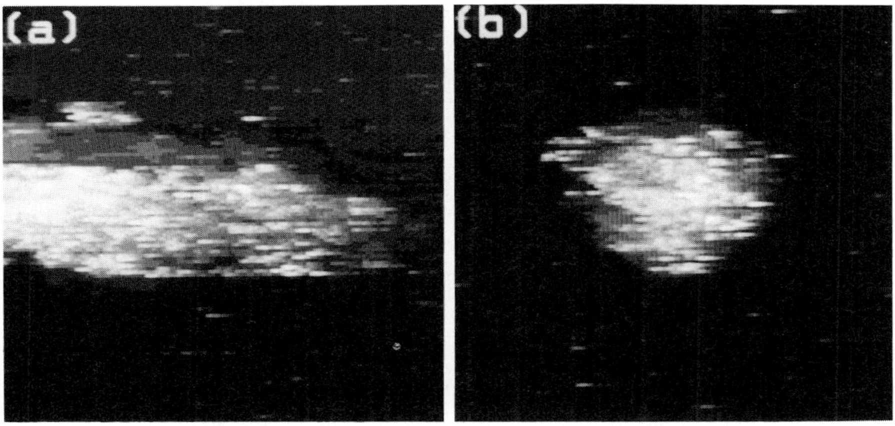

Figure 4. STM images of a line (a) and a dot (b) written by the process described in the text. Each image is 100 nm on a side

features is about 40 nm , which is already one order of magnitude smaller than the smallest features that can be made by laser photodeposition.

Possible applications of this new technique include mask repair, custom wiring of a few devices on a chip, fabrication of sub-tenth-micron devices, and archival memory with densities up to $2x10^{13}$ bits/cm^2 (assuming 1 nm dots with 2 nm center-to-center spacing). A possible scientific application is the fabrication of fine (<50 nm) metallic lines for the study of electron localization in one dimension.

We are currently studying the physical mechanisms responsible for this process and attempting to write smaller features and complex patterns.

4. ACKNOWLEDGEMENTS

This work has been done in collaboration with K.W. Ng, S. Pan, R.M. Silver, and E.E. Ehrichs in my laboratory, J.M. Tarascon and L.H. Greene at Bell Communications Research, and J. Talvacchio and A.J. Panson at Westinghouse R&D Center. Support from NSF and AFOSR is gratefully acknowledged.

5. REFERENCES

1. J.G. Bednorz and K.A. Muller B 64, 189 (1986)
2. M.K. Wu, J.R. Ashburn, C.T. Torng, P.H. Hor, R.L. Meng, L. Gao, Z.J. Huang, Y.Q. Wang, and C.W. Chu: Phys. Rev. Lett. 58, 908 (1987)
3. S. Pan, K.W. Ng, A.L. de Lozanne, J.M. Tarascon, and L.H. Greene: Phys. Rev. B 35, 7220 (1987)
4. K. W. Ng, S. Pan, and A. L. de Lozanne: Jap. J. Appl. Phys. 26-3, 993 (1987)
5. J.B. Barner and S.T. Ruggiero: Phys. Rev. Lett. 59, 807 (1987)
6. R.M. Silver, E.E. Ehrichs, and A.L. de Lozanne: Appl. Phys. Lett. 51, 247 (1987)

Discussion

Y. PETROFF: It seems that the value that you obtain for $2\Delta/kT_c = 11$ is 2 or 3 times larger than values reported by various groups using different techniques. Could you comment on that?

A. DE LOZANNE: The Stanford group also reported $(2\Delta/kT_c) \approx 20$, and Schlesinger *et al.* (IBM) have recent IR measureements with a ratio up to 8. It is possible that lower values are obtained due to surface damage. The BCS ratio of 3.5 is not likely to hold for the new superconductors and should not be used as a standard.

M. CARDONA: What is the best estimate $2\Delta/kT_c$ you can give from your data? Most optical measurements (Raman and ir, including our own) give 3–4 for ceramic materials although one set of ir data reports ~ 8 for single crystals (IBM). NQR data from Bell suggest the existence of two gaps, one giving $2\Delta/kT_c$ around 3, the other ~ 8. MPI data on phonon softening at T_c (MPI, IBM) suggest the existence of an additional gap with $2\Delta/kT_c \geq 5$. The possible existence of two gaps is a rather intriguing question at this time. A value of $2\Delta/kT_c \simeq 11$, as suggested by your data, seems, however, too high as compared with all other determinations.

A. DE LOZANNE: Our best estimate is $(2\Delta/kT_c) \approx 11$. A ratio of 11 is not much higher than 8, particularly when we have no solid theoretical prediction for this ratio. Stanford has also reported values around 11. Our data consists of the best I–V curves observed by any group.

J. T. YATES, JR.: What excitation processes can you envision which will lead to the decomposition of organometallic molecules under the tip of the STM, given that your electron energy is only $\sim 1eV$?

A. DE LOZANNE: The simplest mechanism is localized heating of the surface, which causes pyrolysis. The fact that we observe deposition on both copper and silicon substrates makes this unlikely because these two materials have very different resistivities and thermal conductivities. A second possibility is the dissociation of a single molecule by a single electron. The dissociation energy for a free dymethyl cadmium molecule is 3.13 eV, but this may be substantially lower when the molecule is placed in the strong ($\sim 10^7$ to 10^8 V/m) electric field between tip and sample. This process may be able to dissociate single molecules and thus give the highest resolution. A third possibility is the formation of a microscopic plasma between the tip and the sample, due to high electric fields and current densities. Given that our minimum feature size is currently $\sim 300\mathring{A}$ in diameter, we believe that the microscopic plasma is the dominating effect.

Z.Y. HUA: (1) Have you ever obtained a topographical map of an oxide superconductor by STM? (2) In your Auger spectra there exists a large signal of carbon. Where is this coming from? Is there any influence on your result? (3) The carbon might come from incomplete decomposition of the starting material barium carbonate ($BaCO_3$). If barium peroxide is used the residual carbon content should be decreased or even none.

A. DE LOZANNE: (1) No. These materials have a very rough surface, so that images obtained so far are not useful. (2) The Auger measurements could not be done *in–situ*. The sample was exposed to the atmosphere for several days before the Auger spectra was taken. Thus the sample may have been contaminated during this time. It is also possible that incomplete dissociation of the organometallic molecule may leave carbon residues. More recent measurements indicate that the deposit is mostly cadmium. Furthermore, the fact that we can image the deposit with the STM means that it is conductive. (3) The Auger spectrum was not for a superconductor, it was for the material deposited by the STM using an organometallic gas.

Preliminary Studies on the Interface Structures of CaF_2/Si(111)

Zhang Kaiming, Tian Zengju, and Xie Xide

The Institute of Modern Physics, Fudan University, Shanghai, People's Republic of China

1. Introduction

As a prototype of insulator-semiconductor interfaces, the epitaxial growth of CaF_2 on silicon is of much interest from both theoretical and experimental point of view. The high degree of lattice matching between the fluorite structure CaF_2 and the diamond structure Si crystal makes the growth of high quality crystalline epilayers possible. Recently a number of experimental studies [1-5] have been devoted to properties of the interface, however, theoretical studies on microscopic structure of the interface CaF_2/Si(111) are still relatively scarce.

It is shown by Rutherford backscattering data [1] that the interface is sharp, and the type B epitaxial triple layer F^-–Ca^{2+}–F^- is rotated by 180^0 with respect to the substrate Si(111). There exist different models concerning the interface structure of CaF_2/Si(111). It is found by the Xerox group [6] that the Si(111) surface may serve as catalyst for the dissociation reaction of CaF_2, at the interface one fluorine atom is dissociated. Their model is given in Fig.1A, where only Ca interacts with Si. However, a model with stacking fault on the top double layer(see Fig.1B), and the Ca and F atoms interact with the first and second layer Si atoms respectively at the interface has been suggested by the IBM group [7].

In the present paper the interface structure and the electronic states of CaF_2/Si(111) system are investigated by the tight binding extended Huckel band calculations. The slab models adopted consist of three triple layers of CaF_2 and three double layers of Si (see Fig.1 A, B and C). In addition to the model A and B a new model (Fig.1C) is also considered. As shown by experiments, at about two triple layers of CaF_2 the interface seems to be completely formed. Therefore, all three models can reasonablely represent a complete interface.The parameters for Si and Ca, F included in the extended

Springer Series in Surface Sciences Vol. 14
Solvay Conference on Surface Science
Editor: F.W. de Wette

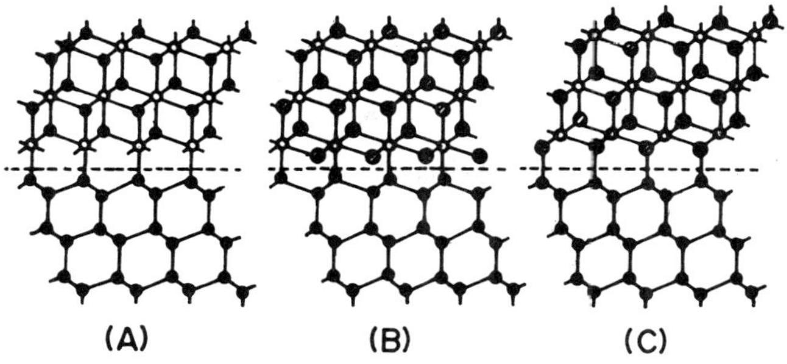

Fig.1 B type epitaxial models of CaF_2/Si(1 1 1)

Huckel approximation are selected to fit the bulk energy bands of Si [8] and CaF_2 [9] respectively.

2. Results

For considering the stability of different models, a total valance band energy per unit cell is introduced by the formula

$$E^{tot} = \int^{E_F} ED(E)dE \qquad (1)$$

where E_F is the Fermi level, D(E) is the density of states of the valence energy bands.

It is interesting to find out that the energy of model B and C are very close. Except for the energy of the fluorine layer at the interface in models B and C, the total energy of model A is relatively lower. This result indicates that the model A is the most stable for CaF_2/Si(1 1 1), however, the new model C suggested by the present work should not be ruled out either.

The layer densities of states for model A is shown in Fig.2, the ith layer from the interface labeled by X_i, where X can be Si, Ca, or F. It can be seen that by losing a fluorine atom one 4s electron of Ca atom at the interface is released from CaF_2, and interacts with the dangling bond of Si to form a covalent bond. The peak near the E_F is contributed by this Si-Ca interaction. In the densities of states for model B two peaks contributed by Si-F and Si-Ca are found. While for model C the strong interaction between Si and F forms a peak located at about 9.1 eV below the Fermi level, and the configuration of Ca atom is almost the same as that in the bulk of CaF_2

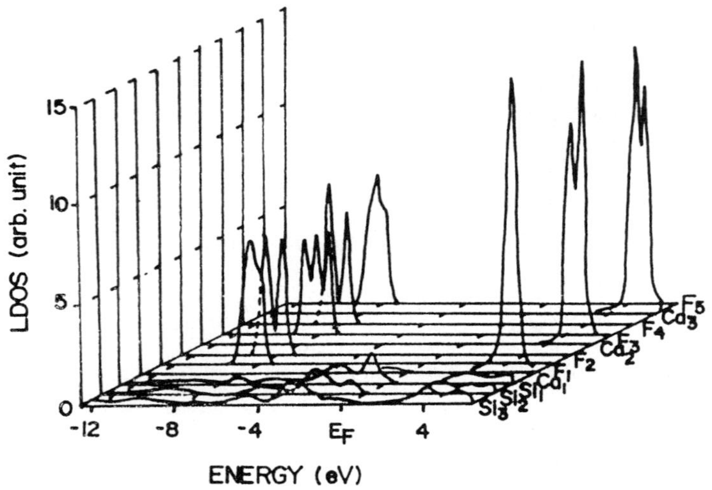

Fig. 2 Local density of states of model A

In the above densities of states the peak near the Fermi level has a tail extending into the gap, it means that the interface is semimetallic, just as observed by experiments [6,7].

3. Discussion

The above results indicate that all three models seem to be possible for the structure of the interface CaF_2/Si(111). In other words, three different structure might coexist at the interface.

In both model A and B the Ca atom sits on the top of the Si atom, in the structure B due to the stacking fault of the first layer Si the fluorine is located on the top of the second layer Si. While in the structure C the fluorine is located at the top site of the substrate. For further understanding of those structures the initial stage of Ca (or F) atom chemisorption on Si(111) is considered by a cluster calculation and charge self-consistent extended Huckel approximation. The calculations show that the top site is more favorable for both Ca and F atom. The binding energy for Ca sitting on the threefold hollow site differs only a little with that on the top site. However, the isolated F atom on the threefold hollow site prefers to penetrate into surface. This possibility might imply that dissociated F atoms at the interface would diffuse into the substrate along the threefold hollow site in the structure A [6].

1. T. Asano, H. Ishiwara: Appl. Phys. Lett. 42, 517 (1983)
2. R. W. Fathauer, L. J. Schowalter: Appl. Phys. Lett. 45, 519 (1984)
3. F. J. Himspel, F. U. Hillobrecht, G. Hughes, J. L. Jordan, U. O. Karlsson, F. R. McFeely, J. F. Morar, D. Rieger: Appl. Phys. Lett., 48, 596 (1984)
4. L. J. Schowalter, R. W. Fathauer, R. P. Goehner, L. G. Turner, R. W. DeBlois, S. Hashimoto, J. L. Peng, W. M. Gibson, J. P. Krusius: J. Appl. Phys. 58(1), 302 (1985)
5. M. A. Olmstead, R. I. G. Uhrberg, R. D. Birngans, R. Z. Bachrach: J. Vac. Sci. Technol., B4(4), 1123 (1986)
6. M. A. Olmstead, R. I. G. Uhrberg, R. D. Bringans, R. Z. Bachrach: Phys. Rev. B35(14), 7526 (1987)
7. D. Rieger, F. J. Himpsel, U. O. Karlsson, F. R. McFeely, J. F. Morar, J. A. Yarmaff: Phys. Rev. B34(10), 7295 (1986)
8. J. R. Chelikowsky, M. L. Cohen: Phys. Rev. B10, 5095 (1974)
9. R. A. Heaton, C. C. Lin: Phys. Rev. B22, 3629 (1980)

Spectroscoping Imaging of Surfaces with Atomic Resolution

J.E. Demuth, R.J. Hamers, and R.M. Tromp

IBM Research Division, Thomas J. Watson Research Center,
P.O. Box 218, Yorktown Heights, NY 10598, USA

With its ability to probe surfaces with atomic resolution, the scanning tunneling microscope (STM) has attracted wide attention. Since tunneling depends on the overlap of charge densities of sample and tip, the 'topograph' determined with the STM also reflects the density of filled and empty states of the tip and surface, respectively, which are involved in tunneling [1,2]. For biases, of 1-2 eV several different surface states can be involved, particularly for semiconductor surfaces. Since the tip is presumably a point which one attempts to maintain the same while scanning, the 'topograph' also reflects variations associated with the local surface density of states (DOS) of the sample. The surface states on semiconductors can dominate tunneling and produce variations in 'topographic' images up to 1Å for homogeneous surfaces, e.g. Si [3], ~ 2Å for inhomogeneous systems, e.g. Ag/Si [4], and up to 10Å where strong charge transfer can occur, e.g. 0/GaAs [5]. Understanding the role of these electronic states and delineating DOS effects from geometric features on the surface is important for using the STM. Furthermore, delineating DOS features from geometric features can allow the spatial imaging of surface states [3,6]. Relating the locations and energies of these states to the different atoms on the surface can provide new insight to the bonding interaction arising at semiconductor surfaces. This paper summarizes some recent methods and issues regarding spatially resolved surface spectroscopy with the STM as well as some recent findings.

I. EXPERIMENTAL CONSIDERATIONS IN SPECTROSCOPIC IMAGING

Tunneling spectroscopy for rigid, metal oxide junctions is widely practiced and both electronic and inelastic (i.e. vibrational) features have been observed in the I-V, dI/dV or d^2I/dV^2 characteristics of these junctions [7]. Spatially resolved spectroscopy with the STM is more complicated due to (1) the variable tip-sample separation, d, (2) identifying and maintaining one position on the surface as well as (3) maintaining a chemically and structural constant tip. As discussed by BINNIG et al. [8], performing dI/dV in the constant current mode does not produce spectroscopic images since tunneling depends on both d and V. This exponential dependence on two variables introduces an inverted topograph into this type of dI/dV image. The speed of acquiring I-V information with the STM is also important as the tip can unexpectedly change in time. There also is the usual hysteretic and/or thermal drift of the piezo scanner during the measurement which places uncertainties in where one is on the surface. (At low temperatures, this problem seems to disappear [8]). HAMERS et al. [6] introduced a method called CITS, to obtain I-V data at a constant sample-tip separation for each pixel element of a topographic scan [6]. In this way, for example, the I-V spectra at any pixel element of the

Springer Series in Surface Sciences Vol. 14
Solvay Conference on Surface Science
Editor: F.W. de Wette

topograph can be later examined in detail. Also, any change in the tip during the scan can be identified from the topograph. The greatest problem of this method is rapidly acquiring and storing this large amount of I-V data at an appropriate signal to noise level needed for later analysis. FEENSTRA et al.[10] have used interlaced scanning at two different energies to obtain topographs of the filled and empty states at the same time. Using a blunt tip, FEENSTRA et al.[10] have found the quantity dI/dV/I/V removes the strong dependence of I on V and d, and shows the overall features of the surface DOS. LANG [11] has also confirmed this from energy-dependent tunneling calculations for different adsorbed atoms.

II. SPECTROSCOPIC IMAGING OF SEMICONDUCTOR SURFACES

Here, we briefly discuss some of our recent examples to illustrate spectroscopic imaging results with the STM. In applying the CITS method to obtain I-V measurements, it has been shown that the I-V features can be extremely sharp and localized on the atomic level. In Fig. 1a is shown I/V versus V for various atomic positions in the Si(111) 7x7 surfaces. The onsets and intensity variations of these I-V features at different positions are sufficient to identify the location at the photoemission surface states shown in panel b. In panel c is also shown an area averaged dI/dV/I/V measurement taken after averaging many pixels together. Here the averaging is numerically needed to improve the signal to noise so that this derivative/ratio produces a continuous function. All the features in the photoemission spectra are seen in the tunneling spectra. Extra features in the tunneling near E_F are seen in low temperature photoemission measurements [13]. While this quantity produces a tunneling 'spectra' having features common to

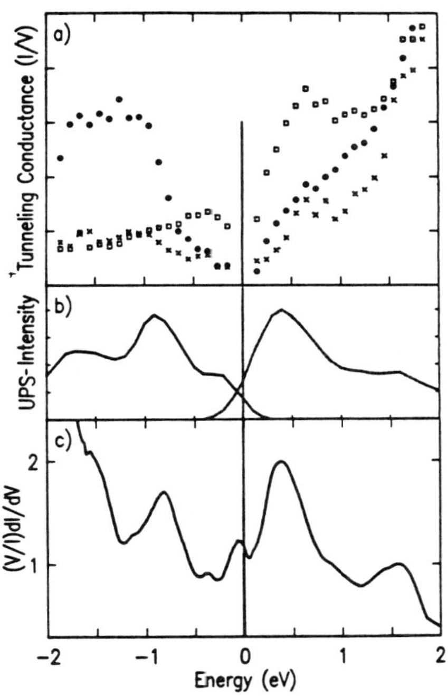

Figure 1. (a) Tunneling conductance at different atomic positions on Si(111) 7x7: over an adatom (squares), over a rest atom (dots) and in between both on the same side of the unit cell (X's). The surface state DOS determined by photoemission [12] is shown (b) together with the area averaged tunneling spectra (c) obtained by averaging many atomically resolved I-V spectra in the unit cell.

photoemission, it is not a DOS spectra, since for example, the normalization requires this quantity to equal unity at V = 0.

In the case of Si(111) 7x7 where the structure is known and a contour of constant barrier height could be established at one particular bias voltage [14], the tunnel current variations at different bias energies can be used to create images of these surface states [3,6,16]. However, such a contour is not known apriori for an unknown structure. Controlling the feedback at higher biases to 'integrate' the contributions of several surface states and looking at differences in tunneling for lower lying surface states does provide a quick way to understand the symmetries and relative locations of several surface states [4]. In situations where primarily one filled or empty surface state arises, the locations of these surface states are reflected by the topographs themselves. Shown in Fig. 2 are STM images of the filled and empty states of an ideal region of Si(100) together with the corresponding dimer model. Such images have also been taken with the CITS method with qualitatively the same spatial information resulting [4]. For the case of GaAs(110), Stroscio et al. have provided images of the Ga and As derived states, essentially locating these atoms on the surface [17].

Tunneling spectroscopy becomes an essential complement to surface spectroscopic methods when unknown inhomogeneities exist or defects are important. For semiconductor surfaces or heterjunctions small concentrations of these can markedly affect the the overall electrical properties such as band bending. Many of these features have only recently been directly observed and identified with the STM. For example, clean Si(100) typically contains numerous missing dimer defects that localize charge near the top of the valence band [18]; submonolayers of Ag [4] or Pd [21] on Si(111) produce small (~ 10Å) metallic islands [4]; Al [19], As [20] and Pd [21] on Si(111) each show several coexisting phases that are intermixed and, in some cases, form small islands due to the lattice mismatch. In Figure 3a we illustrate this local aspect of tunneling spectroscopy by showing the photoemission spectra for $\sqrt{3}$x$\sqrt{3}$ Al on Si(111) obtained by KINOSHITA et al. [22]. The state labeled as D could not be accounted for on the basis of theoretical calculations [19] and was attributed to a surface state associated with islanding [22]. In our STM studies of ideal regions of the $\sqrt{3}$x$\sqrt{3}$ covered surface, we find the tunneling

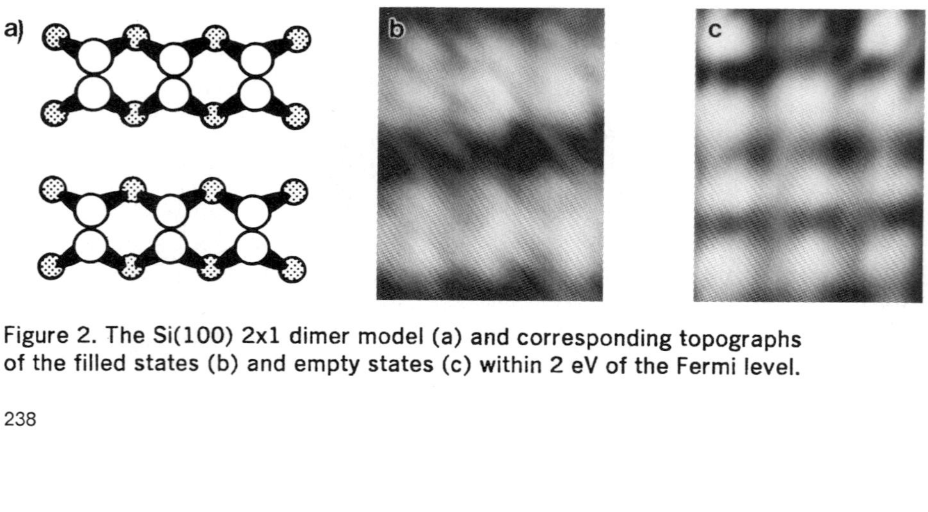

Figure 2. The Si(100) 2x1 dimer model (a) and corresponding topographs of the filled states (b) and empty states (c) within 2 eV of the Fermi level.

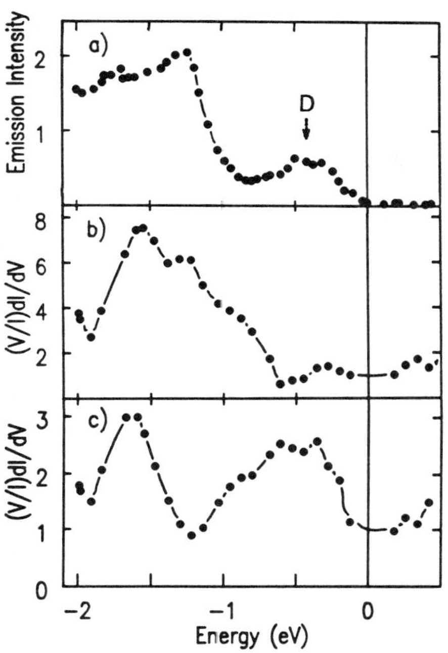

Figure 3. (a) Area averaged photoemission spectra of $\sqrt{3} \times \sqrt{3}$ Al on Si(111) and STM spectra for (b) ideal areas of $\sqrt{3}$ Al on Si(111) and (c) Si substitutional defects in this $\sqrt{3}$ structure [19].

spectra shown in Fig. 3b which agrees with the calculated energy bands [23]. In some regions of the $\sqrt{3}$ structure, two different point-like defects can be observed. A small fraction of these defects appear as missing atoms in the $\sqrt{3}$ structure while the majority of them as something else which is strongly bias-dependent. I/V spectra taken at these latter defects reveal the tunneling spectra shown in Fig. 3c. These defect states agree well with the surface states seen previously by photoemission, and they are interpreted as Si atoms substituted for Al in the $\sqrt{3}$ structure. These Si surface states correspond to isolated dangling bonds which are sufficiently localized that electron correlation effects in the tunneling process reduce their binding energy and make them appear below E_F[19].

III. THE ROLE OF THE TIP IN STM SPECTROSCOPY

The electronic structure of the tip is clearly important in spectroscopy with the STM. Fortunately, the tunneling spectra of the clean W-tip has been determined from studies on Ni films [10] and on a metallic $\sqrt{3}$ Co/Si(111) layer [24] and appears to be broad and structureless. The electronic effects of an electronegative adatom on the tip have also been observed and correlated with theory [25]. Here, the electronegative atom on the tip produces negligible tunneling from the empty states, and markedly influences the observed topography at positive bias. Another electronic effect of the tip can arise to displace the apparent relative locations of the filled and empty surface states of the sample. Here the filled and empty states on the tip can be physically displaced by $\sim 1-2$Å so as to produce an additional offset between the filled and empty states of the sample [25]. Such electronic tip effects argue that tips should also be spectroscopically characterized by scanning a well defined reference surface.

REFERENCES

1. J. Tersoff and D. R. Hamann, Phys. Rev. B31, 2 (1985).
2. N. D. Lang, Phys. Rev. Lett. 55, 230 (1985); 56, 1164 (1986), 58, 45 (1987).
3. R. M. Tromp, R. J. Hamers and J. E. Demuth, Science 234, 304 (1986).
4. E. J. Von Loenen, J. E. Demuth and R. M. Tromp, Phys. Rev. Lett. 58, 373 (1987); to be published in Jour. Vac. Sci. and Technol.
5. J. A Stroscio, R. M. Feestra and A. P. Fein, Phys. Rev. Lett. 58, 1668 (1987).
6. R. J. Hamers, R. M. Tromp and J. E. Demuth
7. E. L. Wolf, Principles of Electron Tunneling Spectroscopy, Oxford University Press, (1985).
8. G. Binnig and H. Rohrer, IBMJournal of Res. and Dev. 30, 355 (1986).
9. D. Eigler, private communication.
10. J. Stroscio, R. Feenstra and A. P. Fein, Phys. Rev. Lett. 57, 2579 (1986).
11. N. D. Lang, Phys. Rev. B34, 5947 (1986).
12. F. J. Himpsel and Th. Fauster, Jour. Vac. Sci. and Technol. A2, 815 (1984), A1, 111 (1983).
13. J. E. Demuth, B.N.J. Persson and A. J. Schell-Sorokin, Phys. Rev. Lett. 51, 2214 (1983).
14. R. M. Tromp, R. J. Hamers and J. E. Demuth, Phys. Rev. B34, 1388 (1986).
15. J. E. Demuth, R. J. Hamers and R. M. Tromp, in *Proceedings 18th International Conference on Physics of Semiconductors*, Olof Engstrom, Ed. World Scientific Publishers, Vol. 1, 49 (1987).
16. R. J. Hamers, to be published.
17. R. M. Feenstra, J. A. Stroscio, J. Tersoff and A. P. Fein, Phys. Rev. Lett. 59, 1192 (1987).
18. J. E. Demuth, R. J. Hamers and R. M. Tromp, MRS Symposium Proceedings, Vol. 77, 1 (1987).
19. R. J. Hamers and J. E. Demuth, to be published.
20. M. Copel, U. Koehler and R. M. Tromp, to be published.
21. U. Koehler, J. E. Demuth and R. J. Hamers, to be published.
22. T. Kinoshita, S. Kono and T. Sagawa, Phys. Rev. B32, 2714 (1985).
23. J. Northrup. Phys. Rev. Lett. 53, 683 (1984); 57, 154 (1986).
24. B. J. Orr, R. J. Hamers and J. E. Demuth, to be published.
25. R. M. Tromp, E. van Loenen, J. E. Demuth and N. D. Lang, to be published.

Discussion

C. B. DUKE: How do you account for the voltage division between the tip–surface spacing and the space–charge within the semiconductor? Your curves look remarkably like little or none of the space–charge drop occurs inside the semiconductor. Do you believe that such is really the case? Would this still be the case for less heavily doped samples?

J.E. DEMUTH: For all our spectroscopic measurements we have used heavily doped samples to minimize such voltage drops. (Our Si samples are typically 6 milli-ohm/cm.) This doping seems to minimize these effects as all our spectroscopic measurements to date on both n and p–doped Si(100) and Si(111) seem to agree with photoemission to within ~ 0.2 eV.

For lightly doped samples we always see several problematical effects associated with the space-charge layer, as first discussed many years ago by Binnig and Rohrer. It is for these problems that we, as well as most other people now doing STM on semiconductors, have avoided lightly doped samples in favor of heavily doped samples.

M.G. LAGALLY: Considering that the STM measures a combination of topology and electronic structure, is it possible that the observed alternating dimer tilts at defects in Si(100) that subsequently damps out along the dimer chain are only to a small degree a topological effect and largely an electronic effect associated with charge at the defect?

J.E.DEMUTH: Well, I think Golovchenko referred to this earlier in his talk. Basically, tunneling (and Norton Lang showed that very nicely in his calculations), doesn't just look at geometry—there's a very strong electronic component. I cannot change where the atoms are; nature gives that to us, and what we measure is a combination of where the densities and where the atoms are. What you really want to do is look at, and integrate is as much of the charge density as you can to get something that's closer to a real topographic image.

M.G. LAGALLY: More specifically, what I was addressing is when you have these tilted dimers near a defect they seem to tilt the other way and then they anneal themselves out again. Presumably that's a combination of a topographic and a charge defect. You *could*, if you resolved in bias, presumably sort *some* of that out, couldn't you?

J.E. DEMUTH: Well we've done that, and you still can't sort it out because what you see here. . .you can't quantitatively sort it out. Qualitatively, you know that the filled state—this is an image of the filled states. The filled state is localized on this atom in the asymmetric dimer. And then you can go switch bias and look at the other. The empty state is really localized way over here. And so you know that the reason this looks low is because there are no filled states there, but also, as Charlie will point out, is that if you have this charge distribution you have to *flip* those atoms. And that's what's partly happening here; we can't distinguish the geometric part from the charge density part of it.

J. T. YATES, JR.: Could you comment on the possibility suggested previously, that the symmetric Si(100) dimers are in fact averages over a dynamic situation in which buckled dimers are interconverting over a time scale which is rapid compared to the sampling time of the STM?

J.E. DEMUTH: The images we observe for Si(100)-2 × 1 show time–averaged features appearing as symmetric and asymmetric dimers. As we have discussed in our previous papers, the symmetric–looking dimers could in fact be time–averaged asymmetric dimers which are thermally fluctuating. We have also performed tunneling spectroscopy on both structures and compared them to theoretical calculations to assess this question. Unfortunately, the experimental differences are too small to draw a firm conclusion. One might also theoretically calculate the energy barrier for these fluctuations to see whether such transitions could be accessible at room temperature.

A. DE LOZANNE: An easy way to determine if voltage drops in the sample are affecting the features observed in the spectra may be to change the tip–sample spacing. If there is a resistive drop in the sample, the voltages of these features will shift.

J.E. DEMUTH: To see significant energy shifts requires large changes in tunneling current which, in most cases, is hard to realize experimentally. In cases where we have varied the tunnel current by a factor of five we cannot distinguish any meaningful shifts. One problem in seeing such voltage shifts experimentally is that most of the features one sees in dI/dV are rather broad and riding on a "background."

K. TAKAYANAGI: You mentioned that the symmetric dimer is a statistical average of a Si(100)2 × 1 structure. In TED analysis, which should give a statistical average, the results support a rather asymmetric structure (temperature of the specimen is $\simeq 600°C$).

C.B. DUKE: Are the STM and TEM microscopy samples prepared the same? Some of the differences between surface typographical features could be due to sample preparation differences. What is the current status of understanding of this issue?

J.E. DEMUTH: Yes, the STM samples are prepared differently from those used for TEM. However, both samples are cleaned by thermally removing the oxide layers. Thus one might expect the resulting Si surfaces to be similar.

One of the major differences in the appearance of STM and TEM surfaces can arise from the imaging process. In TEM one typically reconstructs the image from specific diffraction features (i.e., Fourier components). This might make various point defects look different, but I expect the steps to still be equally visible. Differences in step density distributions would most likely be related to sample preparation methods.

General Discussion

D.R. HAMANN: (Question to J. Demuth) At higher biases, like 2–4 eV, the electric field may significantly distort the states of the isolated surface. How can you be sure that features in scanning tunneling spectroscopy aren't joint states of the surface–biased tip system?

J.E. DEMUTH: It is clear that the electric field will modify the boundary conditions and hence the wave functions. The real question is how much. Consider the field strength between the surface and the tip of $\sim 0.2 V/Å$, for typical conditions. This is about one tenth of that arising in the surface dipole layer $\sim 2 V/Å$. Thus, one might imagine that the external fields act only as a perturbation on the surface states.

C.B. DUKE: (Question for Alex de Lozanne.) Have you calibrated your tunneling spectrometer with a superconductor of known energy gap?

A. DE LOZANNE: No. Our typical base temperature is 15K. Most of the known superconductors are normal or have a small gap at this temperature. However, we have carefully calibrated our electronics with precision resistors.

J.L. ERSKINE: (Addressed to de Lozanne.) What are the prospects of expanding the capabilities of STM to include spin–sensitive detection (ferromagnetic surfaces) or sensitivity to local vibrational modes (inelastic tunneling)?

J.E. DEMUTH: The prospects are very good, and should materialize within the next five years.

A. DE LOZANNE: Atomic force microscopes have been developed at IBM by C. Williams *et al.* which can measure magnetic forces between the tip and surface. As far as spin detection, one possibility is to develop a ferromagnetic tip with a single domain at the end, so that the electrons tunneling out or into it are spin polarized. The tunneling probability would be high only if the spin domain in the sample aligns with that of the tip. As far as local vibrational modes are concerned, some work has been done by Quate *et al.* and Hansma *et al.* I believe these are adsorbate/substrate combinations that have strong signatures in inelastic tunneling measurements. This could yield atom–specific information with the STM, although much ground work needs to be done to identify and characterize these spectroscopic features.

P. MARCUS: I should like to comment on the structure of the Si(001)2 × 1 surface. In our LEED study of this surface, impurities were deliberately added to the surface to suppress $c(4 \times 2)$ and other reconstructions, leaving just the 2×1. The analysis consistently gave asymmetric dimers. It might be useful for the scanning tunneling microscope to look at a surface with the same amount of impurity, which may determine the extent to which the surface is made up of asymmetric dimers.

M. G. LAGALLY: (Question for Paul Marcus.) You made the statement, I believe, that F. Jona stabilized the Si(100) surface that gave reproducible I-V LEED spectra that could be fitted with calculations requiring a dimer tilt by introducing a controlled amount of impurities. Are we to conclude that you believe: (1) that clean surfaces do not produce I-V spectra that require tilting of the dimer to be fit (i.e., that contamination is necessary to produce tilted dimers) and, (2) conversely, that surfaces that exhibit dimer tilting are of necessity contaminated or contain defects?

P. MARCUS: Jona used impurities on the Si(001) surface to prevent the formation of larger cell reconstructions such as the c(4 × 2), thereby cleaning up the LEED spectra, which showed only beams attributable to the 2 × 1 reconstruction. I suggest in the light of the evidence from Demuth's STM study of Si(001)2 × 1, which showed tilted dimers concentrated near defects, that perhaps the tilt found in Jona's work was induced by inpurities. This question might be settled by an STM study of the Si(001) surface with the same impurities used in the LEED study.

D.R. HAMANN: The time scale for the Si(100) dimer to flip may be short on the time scale relevant to STM, but long for LEED or TED, so that those techniques see instantaneously buckled dimers, while STM only sees buckled dimers pinned in a time–averaged sense.

Part V

Surface Scattering and Surface Dynamics

The Study of the Forces Between Atoms of Single Crystal Surfaces from Experimental Phonon Dispersion Curves

J.P. Toennies

Max-Planck-Institut für Strömungsforschung, Bunsenstr. 10, D-3400 Göttingen, Fed. Rep. of Germany

The observation of relaxation and reconstruction suggests that the forces between individual atoms at the surface of single crystals are different from those in the bulk. With the recent development of new techniques for high resolution inelastic helium- and electron-impact scattering it is now possible to measure phonon dispersion curves on insulators, semiconductors and metals. In all three types of materials a number of unexpected phenomena are observed many of which have been analyzed in terms of interatomic force constants describing the lateral interactions between atoms in the first layer. For example on the fcc(111) metal surfaces only the nearly free electron metal Al shows a behavior consistent with bulk force constants, whereas all the noble metals and the transition metals Pd and Pt show a strong anomaly in their surface lateral force constants even though they do not reconstruct. In W(001), a metal which is known to reconstruct below room temperature, the dispersion curves show a normal behavior at 1100 K, an anomalous behavior at 450 K and a folded dispersion curve characteristic of a reconstructed surface at 220 K. These and other examples of unexpected behavior in the insulator LiF and semiconductors such as Si, GaAs, GaSe and $TaSe_2$ are discussed in terms of recent theoretical models of force constant modifications at the surface.

I. Introduction

A. Historical Survey

In most crystals the structure at the surface is different than in the bulk as indicated by the observation of structural changes such as "relaxation" (changes in interlayer spacing near the surface) and "reconstruction" (different two-dimensional geometries in the surface plane). Although such changes could be due simply to the significant reduction in the number of nearest neighbors there is now considerable evidence that the forces between surface atoms are different than in the bulk. As in all physical studies of the microscopic behavior of matter a knowledge of these forces is fundamental for understanding not only the structure and the electronic states but also dynamic processes such as diffusion at surfaces. Although theory cannot at the present time predict these forces it is evident that there is a subtle interplay between the forces, structures and electronic states which because of the loss of 3D symmetry at the surface are more delicately coupled than in the bulk. At present experimental data on dispersion curves of surface phonons provide the most direct information on the interactions between individual surface atoms.

The history of surface phonons is now exactly 100 years old and dates back to Rayleigh [1] who showed from continuum theory that the macroscopic waves, which now bear his name, can exist at an interface. Born and von

Springer Series in Surface Sciences Vol. 14
Solvay Conference on Surface Science
Editor: F.W. de Wette

Karman [2] in 1912 provided the first microscopic theory of bulk phonons but it was only in 1948 that Lifshitz and Rosenzweig [3] carried out the first microscopic calculation of surface modes using a Green's function theory. This work was continued by Wallis and coworkers [4] in the 1960's and culminated in the first realistic calculations of surface phonon dispersion curves by de Wette and coworkers [5] in 1971. Benedek [6] then applied the Green's function method to the same alkali crystals studied by de Wette and obtained similar results.

The analogous bulk phonons were first measured by neutron scattering in 1955 [7]. The same technique was later attempted for surface phonons but with little success largely because of the very small cross sections of neutrons [8]. The first attempts to study surface vibrations by low energy electrons goes back to 1965 when thermal diffuse scattering was attributed to inelastic scattering from Rayleigh modes [9]. Then in 1967 Probst and Piper [10] demonstrated the ability of electrons to resolve adsorbate vibrations and launched the field of electron energy loss spectroscopy (EELS). In these experiments the energies are low (E = 2-10 eV) and the scattering is restricted to near specular directions. The advantages of going over to the impact regime at higher collision energies of 100-200 eV for measurements of surface phonon dispersion curves was pointed out in a theoretical study by Li, Tong and Mills [11]. The first such measurements were reported by Szeftel, Lehwald, Ibach and coworkers [12] in 1983 who succeeded in measuring a full dispersion curve for one direction in Ni(100) using an electron beam of energy halfwidth 7 meV.

The suggestion to use helium atom scattering for measuring surface phonons goes back to two theoretical studies by Cabrerra, Celli and Manson [13] (1969) and Manson and Celli [14] (1971). These authors adapted the theory for scattering of neutrons from bulk phonons to the surface problem. This work immediately stimulated several groups to carry out experiments. S.S. Fisher and coworkers [15] at the same institution as the theoreticians were the first to clearly resolve energy losses in time-of-flight scattering of He from LiF(001). The groups of D.R. Miller [16] at La Jolla and B. Williams [17] at Ottawa continued this work in the 1970's and succeeded in resolving structures which they were able to attribute to single phonons near the zone origin. These early experiments were, however, hampered by inadequate velocity resolution ($\Delta v/v \cong 5\%$) and detector sensitivity. The first successful measurements of surface phonon dispersion curves out to the zone boundary were carried out by Brusdeylins, Doak and Toennies on LiF(001) [18] in 1981 and on Ag(111) [19] in 1983. In He atom scattering measurements energy transfers of up to about 30-50 meV with a resolution of 0.5 to 1 meV are achieved. These experiments were made possible by the introduction of a new mode of operation of nozzle beam sources which provides a relative velocity resolution of 1%, which is almost a factor 5 to 10 better than in previous experiments, and the use of a carefully designed system of differential pumping to increase the sensitivity of the detector. Since these initial experiments many groups have entered the field.

The present review is organized as follows. In the next section we review concepts of force constants between atoms in metal, insulator and semiconductor lattices. Then we survey briefly the dynamical theory of surface phonons. Next we discuss the kinematics common to both electron and helium atom scattering and the experimental techniques and data analysis. In the second half of the review we provide an up to date survey of experimental results on phonon dispersion curves on the clean surfaces of insulators, semi-conductors, layered components and metals. These results are discussed in terms of current models for the modifications of forces at surfaces for the different types of materials. The review closes with a brief summary.

B. Interatomic Forces in Lattices and at Surfaces

The interactions between atoms in a solid are usually much more complex than those between free atoms or molecules or between the different constituents of large molecules. Whereas, in the latter two extremes, theory has provided us with good quantitative descriptions, at least for certain simple model systems; in the case of solids reliable information from theory is largely lacking. The best results on solids are obtained from density functional theory which provides a method for calculating the total energy of the lattice also for frozen phonon distortions [20]. Unfortunately, this method does not provide us with much physical insight into the physics of the interactions nor has it been extensively applied to surfaces. The only density functional calculations reported so far for a metal are for the Al(110) surface where the theory is able to predict the observed large relaxation and to a certain extent the measured surface phonon frequencies (see Section IIIC) [21].

In view of these difficulties it is presently customary to assume additive two body pair potentials and to make the harmonic approximation. The validity of both approximations is strongly dependent on the nature of the crystals. For example, the addition of pair potentials is a much better approximation for a rare gas solid or an alkali halide than for a metal. Among the metals large differences are observed. For instance for Ni an additive model with only nearest neighbor interactions is surprisingly successful in describing the bulk dynamics, whereas for other metals such as Al many more force constants involving second nearest and more distant neighbors are required for a satisfactory description. In the following very brief summary of the extensive field of lattice dynamics we follow the notation and convention of the Bortolani group [22,23].

Within these approximations the total energy of a monatomic crystal is given by

$$\phi = \frac{1}{2} \sum_{ij} v\,(R_{ij}) \quad , \tag{1}$$

where $v(R_{ij})$ is the potential between two atoms and $R_{ij} = |\vec{R}_i - \vec{R}_j|$, where $\vec{R}_i = \vec{r}_{io} + \vec{u}_i$ represents the actual position of the atoms with $\vec{u}_i$ the dynamical displacements. Expanding in a Taylor series we get

$$\phi = \phi_0 + \sum_{\ell\alpha} \phi_\alpha\,(\vec{r}_{\ell_\circ})\, u_{\ell\alpha} + \frac{1}{2} \sum_{\substack{\ell,\alpha \\ \ell',\beta}} \phi_{\alpha\beta}\,(\vec{r}_{\ell_\circ}, \vec{r}_{\ell'_\circ})\, u_{\ell\alpha}\, u_{\ell'\beta} \quad , \tag{2}$$

where the sums are over the ℓ atoms of the crystal and the three Cartesian coordinates α, β and γ. The first and second derivatives of the total energy ϕ_α and $\phi_{\alpha\beta}$, respectively, are evaluated at the equilibrium positions $\vec{r}_{\ell_o}$.

In the pairwise interaction model the ϕ_α and $\phi_{a\beta}$ can be expressed in terms of the first and second derivatives of the interatomic potentials taken at the equilibrium positions

$$\phi_\alpha(r_\ell) = \frac{1}{2} \sum_{ij} \frac{\partial v(R_{ij})}{\partial R_{\ell\alpha}}$$

$$= \frac{1}{2} \sum_{ij} \frac{\partial v(R_{ij})}{\partial R_{ij}} \frac{\partial R_{ij}}{\partial R_{\ell\alpha}} , \tag{3}$$

and

$$\phi_{\alpha\beta}(r_\ell, r_{\ell'}) = \frac{1}{2} \frac{\partial^2 v(R_{ij})}{\partial R_{\ell\alpha}\, \partial R_{\ell'\beta}}$$

$$= \frac{1}{2} \sum_{ij} \left\{ \frac{\partial^2 v}{\partial R_{ij}^2} \frac{\partial R_{ij}}{\partial R_{i\alpha}} \frac{\partial R_{ij}}{\partial R_{\ell'\beta}} + \frac{\partial v}{\partial R_{ij}} \frac{\partial^2 R_{ij}}{\partial R_{\ell\beta}\, \partial R_{\ell'\beta}} \right\} . \tag{4}$$

Thus only first and second derivations of the two-body potential enter into the total energy of the lattice and ultimately into the classical Hamiltonian.

For these derivatives it is customary to use the following notation:

$$\alpha_{ij} = \frac{1}{R_{ij}} \frac{\partial v}{\partial R_{ij}} \Big|_0 , \tag{5}$$

$$\beta_{ij} = \frac{\partial^2 v}{\partial R_{ij}^2} \Big|_0 . \tag{6}$$

After some computation one obtains an expression for the central force acting on the ℓth atom by some other atom i in the lattice with the displacements of all other atoms set equal to zero [22]:

$$\vec{F}_{\ell i} = - \nabla_{\vec{u}_\ell} \Phi = - \left[\alpha_{\ell i} \vec{u}_{\ell i} + (\beta_{\ell i} - \alpha_{\ell i}) \frac{\vec{R}_{\ell i} \cdot \vec{u}_{\ell i}}{R_{\ell i}} \frac{\vec{R}_{\ell i}}{R_{\ell i}} \right] \tag{7}$$

where $\vec{u}_{\ell i}$ $(= \vec{u}_\ell - \vec{u}_i)$ is the relative displacement of the two atoms. With the aid of Eq. (7) one can obtain a physical interpretation of α and β [24]. Consider first the case $u_{\ell i} \parallel R_{\ell i}$ then

$$\vec{F}_{\ell i} = - \beta_{\ell i} \cdot \vec{u}_{\ell i} . \tag{8}$$

If on the other hand $\vec{u}_{\ell i} \perp \vec{R}_{\ell i}$, then one obtains

$$\vec{F}_{\ell i} = - \alpha_{\ell i} \cdot \vec{u}_{\ell i} . \tag{9}$$

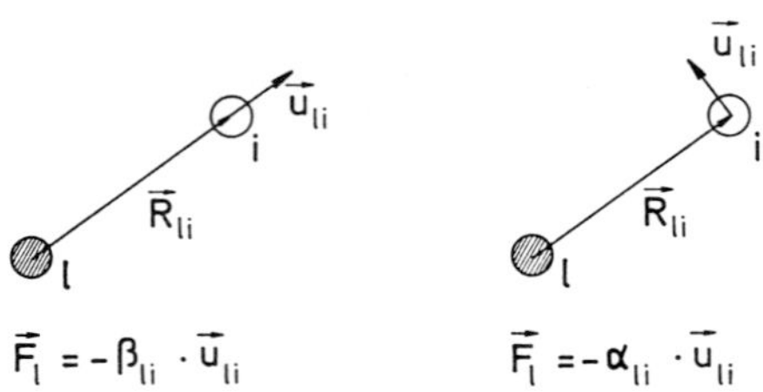

Fig. 1. Schematic diagram illustrating the physical interpretation of the force constants α and β in a coordinate system which follows the dynamic displacements of atom ℓ.

These two cases are illustrated in Fig. 1. This interpretation justifies the notation of Eq. (5) and (6) and explains why α is called the tangential force constant and β the radial force constant. From here on α_n and β_n will be used to denote these force constants where now n indicates the order of the nearest neighbor "shell" involved. Thus n=1 describe nearest neighbor interactions, n=2 second nearest neighbor interactions etc.

For most crystals the Cauchy relationship $C_{12} = C_{44}$ relating two of the three elastic constants C_{11}, C_{12} and C_{44} for a cubic crystal is violated. This indicates that the assumption of central forces is no longer valid. The Cauchy relationship is rather well satisfied for ionic crystals and the alkali metals. However, for other metals and the semiconductors the Cauchy condition fails and this is attributed to the many-body effects arising from the binding electrons. In their calculations of phonon dynamics of metals Bortolani and coworkers account for these by introducing an additional three-body potential [22]

$$\Phi^{(3)} = \frac{1}{3!} \sum_{ijk} w(\cos\vartheta_{ijk}) \quad , \tag{10}$$

where ϑ_{ijk} is the angle formed by the vectors $\vec{R}_i-\vec{R}_j$ and $\vec{R}_i-\vec{R}_k$. The first order angular force constant is defined by

$$\gamma_{ijk} = \frac{\partial w}{\partial \cos\vartheta_{ijk}} \quad . \tag{11}$$

Figure 2 shows the first and second nearest neighbors in an fcc crystal and illustrates the coordinates used to define the three-body potential.

The requirement of stability imposes important constraints on the force constants. In the bulk the stability condition requires that

$$\sum_{ij} \alpha_{ij} R_{ij}^2 = 0 \quad , \tag{12}$$

provided the solid is not under a net force. Since the R_{ij}^2 are all positive, this condition requires for a model with only nearest neighbors that all $\alpha_{ij} = 0$ or in the case of several shells of nearest neighbors, the signs of the α_{ij} corresponding to different shells must be different. For an harmonic interatomic potential there are no similar constraints on β since for such a potential β is independent of R_{ij}.

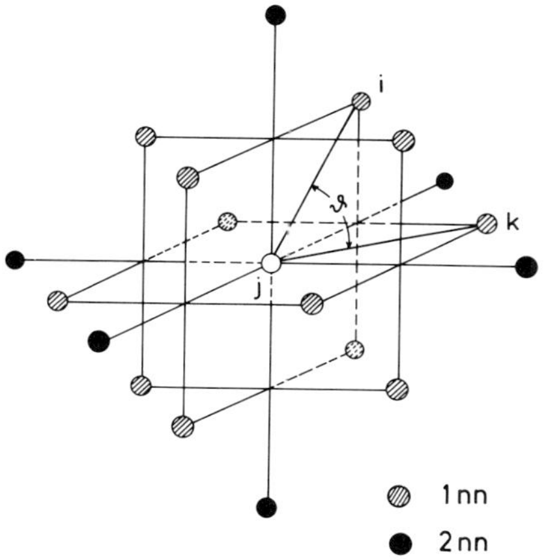

Fig. 2. Schematic diagram showing the first (1 nn) and second (2 nn) nearest neighbors in an fcc crystal. The meaning of the angle ϑ_{ijk} is illustrated.

Up to this point we have not made a distinction between interatomic force constants at the surfaces and in the bulk. Early attempts to explain the inward relaxation of the first atomic layer observed on many metal surfaces in terms of additive potentials failed because additive potentials with bulk parameters invariably predict an expansion of the layer spacing. It is now generally accepted that the observed contraction is due to a rearrangement of electrons in the topmost layer, which leads to a change in the force constants. The frequent occurrence of reconstruction is additional evidence that the potentials at the surface are substantially modified. As we shall see later the experimental surface phonon data suggests that even for surfaces without significant relaxation and without reconstruction the force constants are considerably changed at the surface. In the central forces model discussed above these changes have been accounted for by modifying α and β at the surface. The superscript s will be used to denote surface force constants and the subsripts $\perp$ and $\parallel$ denote the force constants betweeen atoms in the first and second layers and the force constants within the first layer, respectively. As noted above, these force constants can usually be changed to some extent without affecting the crystal stability. The $\alpha_{\parallel}^{s}$ are related to surface stress within the surface layer.

The concept of surface stress is quite old and goes back to early work of Shuttelworth [25] and Herring [26]. Recently this concept has been revived in connection with the driving forces of adsorbate induced reconstruction [27] and at least one ab initio calculation has been reported [30]. Franchini [23] has investigated the effect of changing force constants on relaxation for a model with 10 nearest neighbors and finds that the relaxation is directly proportional to the surface tangential force constant.

$$\Delta d_{12} \propto \alpha_{1\perp}^{s} \quad . \tag{13}$$

The tangential stress is defined as the force per unit peripheral length or the increase in total energy with area and is equivalent to the surface tension which in turn is defined by [25]

$$\gamma = \frac{d(AF)}{dA} = F + A\,\frac{dF}{dA} \quad , \tag{14}$$

where F is the surface Helmholtz free energy and A is the area of the surface. Because of the high mobility of molecules in a liquid $dF/dA = 0$, but for a solid both terms in Eq. (14) make nearly equal contributions [25]. For a rare gas solid at 0° K Shuttelworth [25] has calculated the surface tension assuming central additive forces. If we apply his formulae to a nearest neighbor model we find simply

$$\gamma_0 = \frac{1}{2} \alpha_{1_{\parallel}}^{s} \tag{15}$$

Thus the concept of surface stress provides another way of explaining the changes in α in the surface layer.

The models used for describing the force constants in insulators and semiconductors are basically similar. In these crystals the electrons are more localized in the lattice and thus they can be accounted for explicitly in the models. On the other hand the charges of the ions are not completely locally screened as in metals and long-range Coulomb forces enter into the force constants. We briefly review two current models since we will have occasion to refer to them in later sections. Fig. 3(a) displays a schematic diagram of the shell model successfully used to describe the alkali halides [28]. The outer electrons on both ions are located in massless shells surrounding the charged cores. The shells are allowed to move rigidly with respect to the cores and the shells and ion cores in one site interact with each other and also with shells and ion cores on adjacent sites. Thus in the Taylor expansion of the total energy (Eq. 2) four different interactions appear in place of the single term for metals [29]. These describe the core(1)-core(2), core(1)-shell(2), core(2)-shell(1) and the shell(1)-shell(2) interactions. The core-shell interactions within the atoms are usually described by simple harmonic forces. The four interatomic potentials are decomposed into long-range Coulomb terms and short-range repulsive potentials. For relaxation studies the latter are described by the Born-Mayer formula A exp(-bR). In the lattice dynamical calculation of the alkali halides by de Wette and coworkers [31] only the shell-shell short range forces between nearest- and second-nearest neighbor interactions are included. They are each described by force constants for the radial and tangential forces. Thus in addition to the shell charges and the shell spring constants there are altogether two nearest neighbor force constants β^{+-}, α^{+-} and four second nearest neighbor interactions β^{--}, α^{--}, β^{++} and α^{++} where the indices + and - denote the charges on the ions. Non-central interactions are only included indirectly in α since they are expected to be small and reliable models are not available.

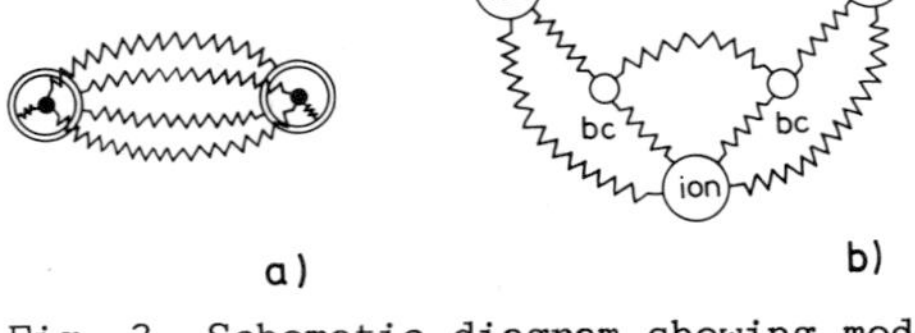

Fig. 3 Schematic diagram showing models used to describe the interatomic forces in insulators (a) and semiconductors (b). In the shell model shown in (a) the outer electrons are concentrated in shells which interact with the cores as well as with neighboring cores and shells. In the bond charge model the electrons are localized in positions between the ion cores. Bending forces are included between the bond charges to provide for shear stability of the lattice.

In the semiconductors the so-called bond charge model of Weber [32] has proved to be particularly successful. In this model (Fig. 3b) the local bonding character of the strong covalent bonds in C, Si and Ge is accounted for by localizing the bonding electrons in positions half-way between the cores. Shear stability of these lattices is accounted for by a bond-bending potential of the Keating type [33] between the bond-charges and other tangential terms are not included. Additional parameters account for the nearest neighbor stretching, Coulomb interactions and the ion bond-charge stretching force constants. This model is successful in describing the bulk phonons in diamond, Si, Ge and α-Sn with only four adjustable parameters [32]. A recent comparison of interplanar and interatomic force constants between the bond charge model and ab initio calculations has shown very good agreement in the case of Ge [34]. As we shall see in Section IIIB this model has been applied successfully in a modified form to the surface phonons of Si(111) + H (1x1) by Benedek and coworkers [35].

C. Lattice Dynamics at Surfaces

The calculation of surface phonon dispersion curves has been extensively described in the literature [36,37]. Of the two available methods involving either slab calculations [5] or the use of Green's functions [6], the former is gaining in popularity with the advent of large fast computers. The slab method is basically an extension of the standard methods developed to treat the lattice dynamics in the bulk. Since at the surface the symmetry is reduced from 3 dimensions to 2 dimensions the dynamical matrix has dimensions 3N x 3N, where N is the number of layers in the slab compared to a 3x3 dynamical matrix in the bulk. The eigenvalue equation is thus given by

$$\sum_{\ell'\beta} D_{\alpha\beta}(\ell,\ell',\vec{Q})\,\xi_\beta(\ell',Q,p) = \omega_p^2(Q)\xi_\alpha(\ell,\vec{Q},p) \quad , \tag{16}$$

where ℓ = (1 ...N) labels the planes in the slab, and p labels the 3N modes for a monatomic slab for a given parallel momentum $\vec{Q}$. The dynamical matrix in Eq. (16) is given by

$$D_{\alpha\beta}(\ell,\ell',\vec{Q}) = \frac{1}{M}\sum_{\ell'}\left[\frac{\partial^2\Phi}{\partial u_\alpha(\vec{\ell}')\,\partial u_\beta(\vec{\ell}')}\right]\Bigg|_{r_{\ell_0}}\left[\exp i\vec{Q}\cdot(\vec{r}_{\ell_0}-\vec{r}_{\ell'_0})\right] . \tag{17}$$

For a given Q each of the $3N_3$ modes is characterized not only by its frequency ω_p but also by the eigenvector components ξ_α. They are normalized following the condition

$$\sum_{\ell\alpha} |\xi_\alpha(\ell,Q,p)|^2 = 1 \quad . \tag{18}$$

From an examination of the eigenvector components as a function of penetration ℓ we can distinguish between the <u>surface modes</u> which exhibit a strong decay of ξ with ℓ, the <u>bulk modes</u> with little change of ξ with ℓ, and the <u>mixed modes</u> (also called <u>resonances</u>), which show a weak decay of ξ going over to a nearly constant small value of ξ in the bulk [5]. The polarization of surface modes is not as clearly defined as in the bulk. If the direction of $\vec{Q}$ lies in a plane normal to the crystal surface with respect to which the lattice has reflection symmetry then the eigenvector components of a single layer lie either in this plane or are normal to it. In the former case they are called <u>sagittal plane</u> (SP) modes and in the latter case shear horizontal (SH) modes. The sagittal plane modes may either be predominantly vertically (T) polarized (z-direction) or longitudinally (L) polarized (x-direction). Only at the zone boundary are surface modes either pure vertical or pure longitudinal. Along directions in the surface plane in which the condition of reflection symmetry is not fulfilled such as along the <110> direction of the densely packed (111) surface of an fcc lattice this classification breaks down and the modes may have eigenvector components in all three directions x, y and z.

By way of illustration consider an ideal two dimensional slab of densely packed atoms corresponding to the top most layer of a (111) surface of an fcc lattice. Fig. 4 shows the calculated dispersion curves for this simple model system [38]. The two-dimensional Brillouin zones and the labelling of the symmetry directions is given in Fig. 4a. The dispersion curves in Fig. 4b were calculated for a nearest neighbor pairwise interaction characterized by a single radial force constant β_1 with $\alpha_1 = 0$ corresponding to a surface without stress. The two modes show a transverse (T) and longitudinal (L) shear horizontal polarization in the plane. If one now adds α_1 then an additional mode with transverse polarization out of the plane appears. Thus this simple model illustrates the importance of the tangential force constant for out-of-plane vibrations. The effect of adding additional layers to the two dimensional slab while retaining the same two force constants β_1 and α_1 is illustrated in Fig. 5 [36]. Already the addition of one additional layer shows the appearance of new modes with finite frequencies at the origin $\bar{\Gamma}$. These modes can be simply attributed to vibrations of atoms with momentum vectors along directions normal to the planes and correspond to optical modes in a lattice with more than one atom in a unit cell. As one adds more and more layers these modes increase in number. Beyond about 10 layers a definite structure in the dispersion curves becomes apparent with single modes splitting off from dense bands. In the limit of $\ell = \infty$ the former are the surface modes, while the bands are attributed to bulk modes. Recall that in the infinitely extended bulk the dispersion curves are few in number and sharp. However, when projected onto the surface they have different components depending on their direction with respect to the surface plane and appear as broad bands.

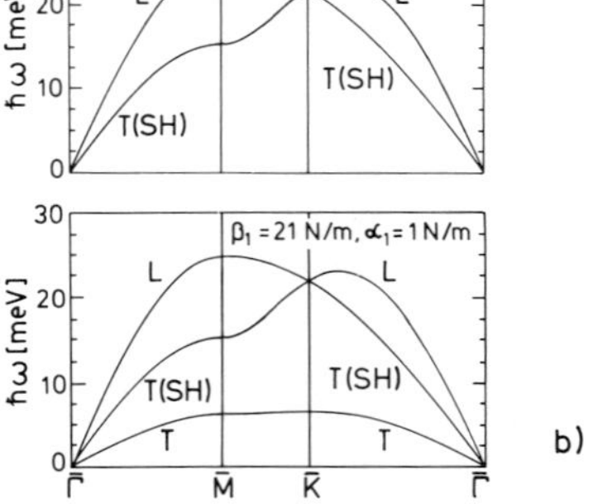

Fig. 4. Calculated dispersion curves for a single (111) layer equivalent to the surface layer of an fcc crystal [38]. Part a) shows the surface geometry of the closely packed hexagonal structure of atoms and the Brillouin zone and the irreducible part of the zone boundary. Part b) shows the dispersion curve for nearest neighbors interaction with only radial forces. Part c) shows the effect of adding the nearest neighbor tangential force constant. L denotes longitudinal, T transverse out-of-plane and T(SH) transverse shear horizontal (in-plane) polarization.

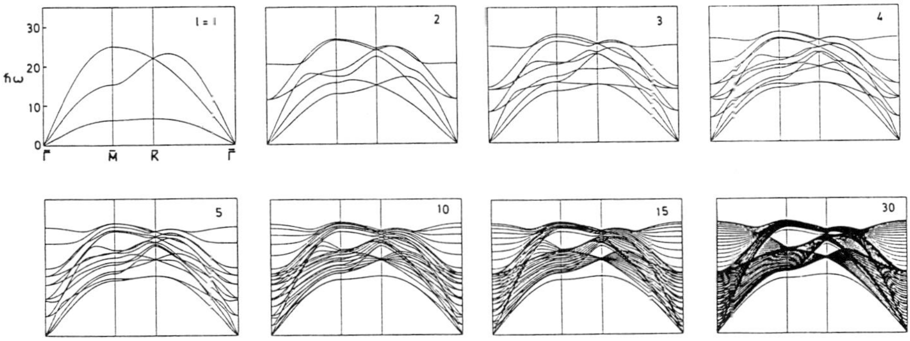

Fig. 5. Schematic diagram showing the evolution of the surface phonon dispersion curves as additional layers ℓ are added to the single layer also shown in Fig. 4c [38]. At ℓ = 30 the behavior for an infinitely thick slab is approached and the distinction between surface modes and bulk bands becomes apparent.

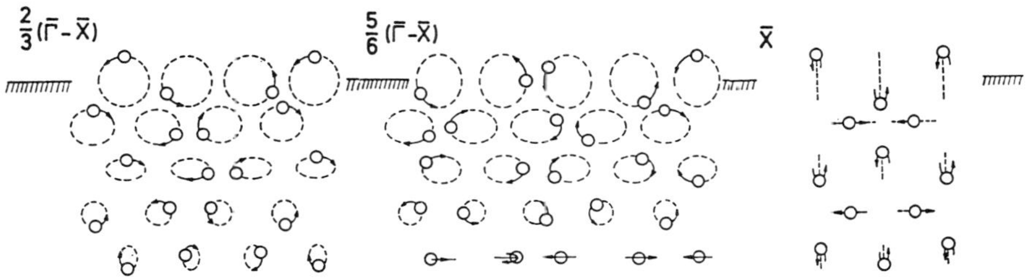

Fig. 6. Calculated displacement trajectories of the atoms for three different wave vectors of the Rayleigh mode with Q = 2/3, 5/6, 1 times Q_{max} (Q_{max} is the value at the $\bar{X}$ point) along the $\bar{\Gamma}$-$\bar{X}$(<110>) direction of the (100) surface and several layers below for an fcc crystal [39]. The amplitudes are plotted in a logarithmic scale. For example at the zone boundary ($\bar{X}$) the value of $|\xi|^2$ at ℓ = 4 is $8 \cdot 10^{-3}$ compared to 0.8 at the surface layer ℓ = 1. Five force constants were used in the calculation.

As discussed above, the true nature of the modes is only apparent from an analysis of the polarization vectors. Calculated results for the polarization vectors along the $\bar{\Gamma}$-$\bar{X}$ direction of the (100) surface of an fcc crystal are presented in Fig. 6 [39]. The calculations are based on a second nearest neighbor model with two radial and two tangential force constants and one for angle bending. For simplicity we have chosen a sagittal-plane polarized Rayleigh wave moving through the fcc crystal along a direction with inversion symmetry. The displacement ellipses, showing the momentary positions of the atoms, are plotted on a logarithmic relative scale for the first few layers near the surface. These pictures illustrate the complicated normal mode motion involving all the atoms near the surface.

Of particular interest for studying surface force constants is the effect of changing force constants at the surface on the frequencies of surface modes. Fig. 7 presents the results from recent model calculations

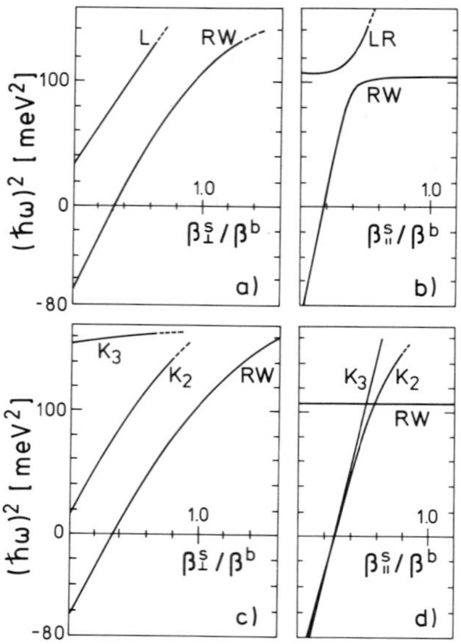

Fig. 7. The effect of changing the lateral in-plane force constant $\beta_\parallel$ and the force constant between the first two layers $\beta_\perp$ on the zone boundary surface mode frequencies is plotted for the $\bar{M}$-point in a) and b) and for the $\bar{K}$-point in c) and d) [40]. The calculations are for Pt(111) where the other force constants are held fixed at the values obtained for the bulk dispersion curve with interactions to 6 nearest neighbors.

[40] of the frequencies of the surface modes at the $\bar{M}$ and $\bar{K}$ points for a Pt(111) surface, which has surface dispersion curves similar to those shown in Fig. 5 for ℓ = 30. Only the surface frequencies are shown since the bulk bands are not significantly affected by the force constants at the surface. At both $\bar{M}$ and $\bar{K}$ the reduction of $\beta_\perp$ leads to a linear reduction in ω^2 as expected for a simple harmonic oscillator model $\omega^2 = \beta/m_{eff}$. The mode L which is shown in Fig. 7a) is a new mode involving vertically polarized motion localized in the second plane and horizontal motions in the first and third planes. Its appearance can be explained by the increasingly loose coupling of the topmost layer which leads to a transfer of the motion of the Rayleigh wave from the first to the second layer. At the $\bar{K}$ point (Fig. 7c and d) similar new modes designated by K_2 and K_3 appear. K_2 is similar to L discussed above while K_3 is a mode in which the transverse vertical vibrational motion has been transferred to the third layer.

Fig. 7d) shows that $\beta^s_\parallel$ has no effect on the Rayleigh mode at the zone boundary as one might expect on the assumption that transverse and longitudinal motions are not coupled. The strong effect of changing $\beta_\parallel$ at $\bar{K}$ on the new modes K_2 and K_3 is attributed to the large longitudinal motions in the surface layer of these modes. At the $\bar{M}$ point changing $\beta_\parallel$ leads to a new surface mode, which since it is imbedded in the transverse polarized bulk bands is called a resonance. This mode is designated LR since it has a predominant longitudinal polarization in the surface plane. This surface mode peels off from the lower edge of the longitudinal bulk bands at a frequency of about 20 meV as a result of the perturbation produced by changing the in-plane surface force constant. Since this mode and the Rayleigh mode are in the sagittal plane they can interact with each other and hybridize when their frequencies become equal at $\beta^s_\parallel/\beta^b_\perp$ = 0.4. At smaller values of $\beta^s_\parallel/\beta^b_\perp$ the lower mode has mainly longitudinal character and the upper nearly horizontal mode mainly vertical character.

II. Experimental Methods

A. Kinematics of He Atom and Electron Scattering

From the point of view of kinematics both experimental techniques are in principle very similar [41]. For phonon studies the electrons are accelerated to energies of 100-300 eV at which the electron wave vector is about equal to that of a reciprocal lattice vector. This mode of operation is called impact scattering in contrast to the more usual dipole scattering with energies of less than about 10 eV [42]. By virtue of their factor 10^4 greater mass He atoms have the same wave vectors as electrons at energies which are 10^{-4} smaller or about 20 meV. It is a fortunate circumstance that at energies of 200-300 eV the electron penetration is only two or three layers. At the much lower energies of He atom scattering there is of course no penetration at all. Thus He atom scattering is one of the few surface techniques which is sensitive only to the outermost layer and is absolutely non-destructive.

Although the energies are very different the kinematics are nevertheless the same in both cases and are determined by the conservation of total energy of probe particle and lattice and the conservation of parallel momentum of particle and phonon:

$$\hbar\omega(Q) = \frac{\hbar^2}{2m} (k_f{}^2 - k_i{}^2) \quad , \qquad (19)$$

and

$$\Delta\vec{K} = \vec{G} + \vec{Q} = k_f \,\vec{\sin}\, \theta_f - k_i \,\vec{\sin}\, \theta_i \quad , \qquad (20)$$

where k_i and k_f are the initial and final wave vectors of the probe particle with mass m, $\vec{G}$ is a reciprocal lattice vector and θ_i and θ_f are the incident and final polar angles measured with respect to the surface normal. The coordinates and conservation equations are illustrated in Fig. 8. It is customary to refer to $\hbar\omega > 0$ as annihilation and $\hbar\omega < 0$ as creation scattering and processes with $Q > 0$ as forward and $Q < 0$ as backward-directed phonons.

For simplifying the interpretation of the experiments $\vec{k}_i$ and $\vec{k}_f$ are located in the sagittal plane (see Fig. 8). From conservation of parallel momentum it follows that the wave vector transfer $\Delta\vec{K}$ also lies in the sagittal plane. Experimentally it is also found that even in cases where

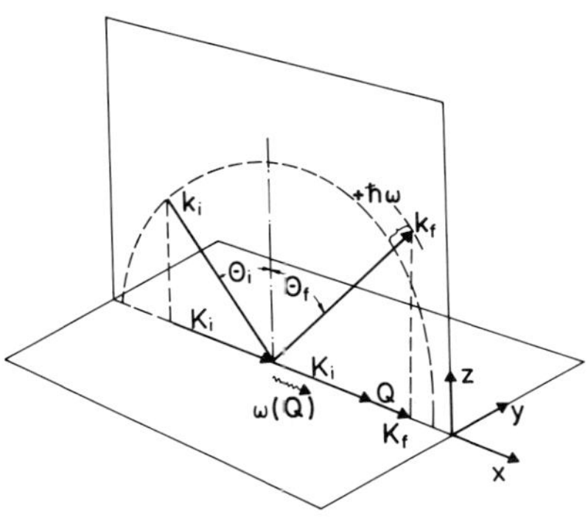

Fig. 8. Schematic diagram illustrating the conservation equations for in-plane scattering. The xz-plane is called the sagittal plane.

substantial out-of-plane diffraction occurs that $\vec{Q}$ is in the sagittal plane as well. In order to study shear horizontal polarized phonons the crystal would have to be tilted so that the scattering and sagittal planes are no longer coincident. Such experiments have, however, so far not been reported.

In the experiments it is convenient to select a certain fixed incident energy as well as the incident and final scattering angles. Then the energy distribution of the scattered particles is measured. From the observed peaks the energies and wave vectors of the phonons are determined via the conservation equations. The curve connecting all values of $\hbar\omega$ and ΔK accessible for a given incident energy and set of scattering angles is called a scan curve [41]. An expression for the scan curve for $\vec{G} = 0$ can simply be obtained by substituting Eq. (19) into Eq. (20) with the result

$$Q = k_i \left[(\sin \theta_i - \left(1 + \frac{\hbar\omega}{E_i}\right)^{1/2} \sin \theta_f \right] \quad . \tag{21}$$

This expression illustrates an important difference between electrons and He atoms. For electrons $\hbar\omega/E_i$ is typically 10^{-3} - 10^{-4} so that Q is nearly independent of $\hbar\omega$. This is not true for He atoms where $\hbar\omega$ is comparable to E_i. Another way of understanding this difference is in terms of the much larger group velocity dE/dk of electrons compared to He atoms. Thus for electrons the measurements are carried out under conditions where Q is independent of $\hbar\omega$. In neutron inelastic scattering this mode of operation is called a constant Q scan. In He atom scattering experiments the scan curves intersect the dispersion curves at nearly right angles for either annihilation-backward or creation-forward phonons. Because of the increased resolution this is the preferred mode of operation in He scattering experiments.

B. He Atom and Electron Scattering Experiments

Fig. 9 presents a schematic diagram of an apparatus used for He atom inelastic surface scattering [43] and also an apparatus for inelastic electron scattering [44]. In the He atom machine a nearly monoenergetic beam of atoms is formed by expanding the gas from a high pressure (≈ 100 atm) through a small orifice (≈ 10 μm) into a vacuum. In the expansion the gas acquires a sharply peaked velocity distribution with a most probable velocity 30% greater than in the source chamber [45]. The most probable velocity can be varied anywhere from 2000 m/s (≈ 60 meV, k_i = 11 Å, λ = 0.5 Å) down to ~ 700 m/s (≈ 8meV, k_i = 4 $Å^{-1}$, λ = 1.5 Å) by simply changing the temperature of the source from 300 to 40 K, respectively. For this range of incident beam energies the energy half-width is between 1 meV and 0.2 meV. The energy distribution of the scattered atoms is usually measured by the time-of-flight technique. For this purpose the incident beam is mechanically chopped by a rotating disk chopper and the time of arrival of the scattered atoms at the mass spectrometer detector is analyzed by a multichannel scaler. Extensive differential pumping (not shown in Fig. 9), is required to prevent helium gas from diffusing into the detector chamber, and in practice it has been possible to keep the diffuse helium background partial pressure down to $<10^{-15}$ Torr (< 10 helium atoms/cm^3). This background places a lower limit on the smallest detectable beam of about 10 counts/s. The straight-through incident beam intensity is estimated to be about 10^7-10^8 counts/s illustrating the large dynamical range in sensitivity.

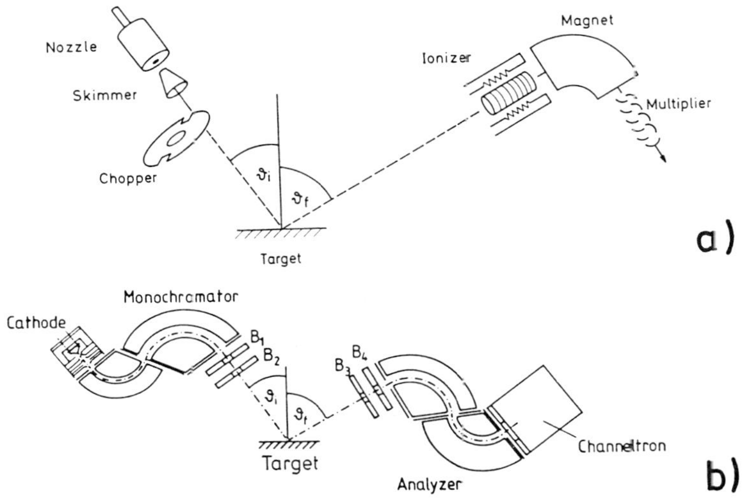

Fig. 9. Schematic diagram showing (a) the time-of-flight apparatus used in He atom scattering and (b) the energy selector-analyzer apparatus used in electron scattering.

In the electron scattering experiments the electrons are first accelerated from a relatively cold lanthanum hexaboride cathode with an energy halfwidth of less than 0.1 eV. They are then decelerated to energies of a few eV and passed through a 127° sector field energy selector. In most machines used for phonon studies two energy selectors are mounted in series in order to achieve a high energy resolution of the incident beam. With such an apparatus an energy resolution of between 7 and 3 meV can be achieved with reasonable intensities [46]. The electrons are then accelerated to the desired incident energy. After scattering the electrons are collected by a lens system in which they are also decelerated before passing into another 127° sector field which serves as an analyzer. The detector is usually a Channeltron multiplier. The count rates in electron scattering are much the same for zone boundary phonons as in the helium atom experiments. The available electron currents, which are limited by space charge, are smaller than the He atom flux by about a factor 10^5-10^6. This is compensated for, however, by the 100% detection efficiency for electrons compared to the 10^{-5} - 10^{-6} efficiency of the mass spectrometer used to detect He atoms.

In presenting the He atom scattering data it is customary to transform the time-of-flight of spectra into an energy transfer spectrum [41]. Fig. 10 shows a comparison of energy loss spectra measured by He atom and electron scattering both on the Ni(100) surface [47]. The spectra from the two different experiments were chosen to correspond to about the same points on the dispersion curve. Note that as expected for the different scan curves the electron energy loss distributions are for a fixed value of Q whereas the He atom spectra are for a fixed incident angle. For each He energy loss spectrum the values of Q are indicated next to the corresponding inelastic peaks. This side-by-side comparison of the two sets of spectra reveals the much greater resolution of the He atom scattering experiments.

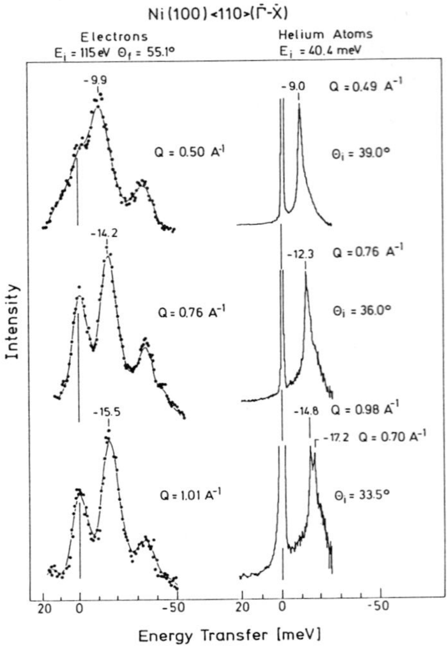

Fig. 10. Comparison between electron scattering energy transfer spectrum and helium time-of-flight spectra showing energy loss peaks due to surface phonons of clean Ni(100) along the $\bar{\Gamma}$-$\bar{X}$ direction.

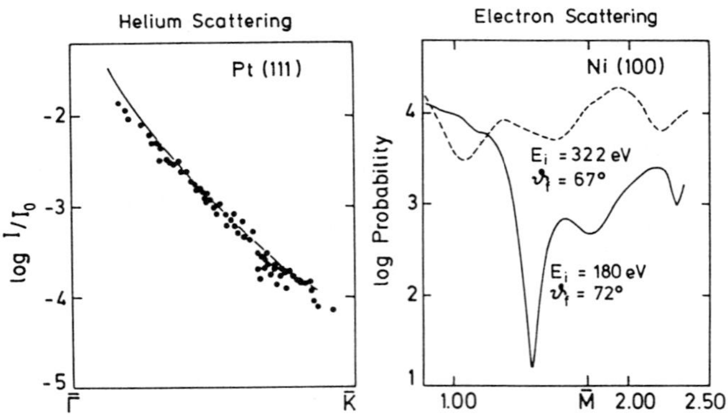

Fig. 11. Comparison of inelastic intensities measured in electron scattering for two different energies and angles [44] and He atom scattering [49] as a function of the phonon wave vector. In the case of electron scattering the intensities show an erratic behavior which is due to interference associated with the multiple scattering [48]. The drop-off in the He scattering intensities is largely due to the rapidly decreasing inelastic reflection coefficient with increasing wave vector [50].

Fig. 11 shows a comparison of inelastic intensities as a function of phonon wave vector for the two techniques. In the case of electron scattering the intensities show an erratic behavior with wave vector which strongly depends on the incident energy and scattering angles [44]. The strong dip of the E_i = 180 eV curve is attributed to destructive interference accompanying the multiple scattering of the electrons from the surface layers [48]. The He scattering intensities are characterized by a smooth but sharp drop-off by several orders of magnitude with increasing energy transfer [49]. As mentioned previously the actual count rates for zone boundary phonons are in fact about the same for both techniques.

C. Data Analysis

Considerable care must be exercised in evaluating the phonon energy transfer spectra. In some cases it is possible to determine the phonon dispersion curves with the aid of the conservation equations (Eqs. 19 and 20) directly from the experimental spectra. This procedure can, however, lead to erroneous results in some special situations. For one the apparatus resolution may not be sufficient to resolve the peaks actually present. By theoretically simulating the spectra for a given lattice dynamical model a set of closely spaced peaks may, on the other hand, be predicted which, after averaging over the apparatus resolution, are consistent with the measurements. In other cases interference phenomena may suppress or enhance certain modes making them difficult to detect or resolve. Finally it is sometimes not easy to distinguish an observed peak due to a true surface mode or a resonance from a structure which may only be due to a high density of states of bulk phonons. For these reasons it is customary to carry out direct comparisons of the measurements with either calculated densities of states or preferably with a complete simulation of the measured spectra. The procedures are basically the same in the case of He atom scattering and electron scattering, despite the fact that the theories used for calculating the inelastic intensities are quite different. These procedures are schematically illustrated in the case of experimental He atom time-of-flight spectra by the flow chart in Fig. 12. An analogous diagram with slightly different labels for the different steps would also apply to electron scattering. Starting from the upper left a set of force constants which provides a best fit of bulk dispersion curves and the elastic constants is used in a first trial calculation. From the solution of the dynamical matrix for the surface phonons the dispersion curves and polarization vectors for this set of force constants are determined. Then the distorted wave Born approximation is used to calculate the inelastic reflection coefficient for one phonon events [50]. In this theory the reflection coefficient is proportional to the following product of dynamical factors [48]:

$$|iQ\xi_x - \beta\xi_z|^2 |\langle f|U_0|i\rangle|^2 \exp\left[-(Q/Q_c)^2\right], \qquad (22)$$

where ξ_x is the longitudinal and ξ_z is the normal component of the polarization vector of atoms in the surface plane. The $\langle f|U_0|i\rangle$ are the matrix elements of the laterally averaged effective He-surface potential U_0 and $\exp[-(Q/Q_c)^2]$ is the dynamic cut-off function which accounts for the simultaneous interaction of the He atom with several surface atoms [15]. The He atom-surface potential has been successfully estimated in a number of different ways. One procedure involves the determination of an effective two-body potential between the He atom and an atom of the solid surface, of which the parameters are determined in part from theory and in part by fitting other experimental data. This procedure has been shown to work quite well for both alkali-halide [51] and metal surfaces [50]. The cut-off parameter Q_c is in good approximation directly proportional to the range of the repulsive potential responsible for the phonon excitation. Soft potentials with a long range lead to small values for Q_c and a steeper fall-off in intensity than hard potentials. In the final step in

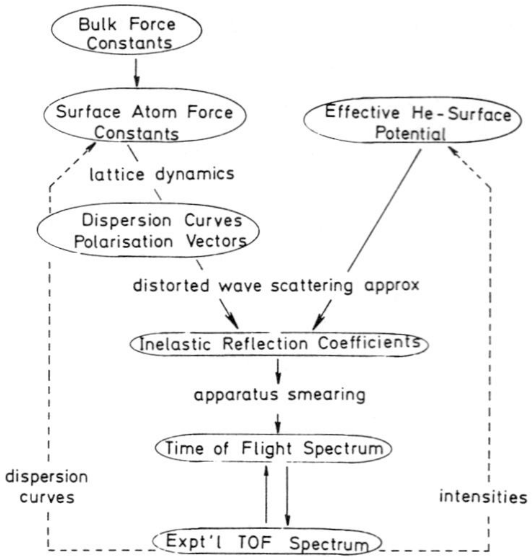

Fig. 12. Schematic diagram showing the theoretical steps and computational pathways in the best fit simulation of experimental He atom time-of-flight spectra.

Fig. 12 the inelastic reflection coefficients are smeared out to take account of the apparatus resolution. These calculations are then repeated with modified force constants for the surface layers until a best fit of the time-of-flight distributions is achieved.

In the case of electron scattering the theory is more complicated because the electrons may undergo multiple scattering with several atoms in the first few layers before being finally inelastically scattered. The procedures, which borrow heavily from LEED theory and the theory for photoelectron emission, are, however, now well developed [11]. Thus for example the theory is able to predict in agreement with experiment the large oscillation in intensities with beam energy shown in Fig. 11 [49]. Because of these interferences the inelastic intensity for certain modes may be enhanced and this phenomenon can be helpful in identifying the origin and assignment of the experimentally observed modes. Under certain kinematic conditions the theory also indicates that the apparent phonon energies of surface modes may be significantly shifted away from their true positions [49].

III. Experimental Results

A. Insulators, especially Alkali Halides

The alkali halides were the first crystal surfaces for which full dispersion curves out to the zone boundary could be measured by He atom scattering [18]. The relative inertness of the LiF and NaF surfaces to residual gas contamination, which was already exploited in the first He diffraction studies in the 1930's [52], made these crystals an ideal choice for the first experiments. The experiments were also stimulated by the availability of theoretical calculations of surface phonons which were published as early as 1971 by Allen et al. [5]. In the early 1970's Chen et al. calculated full dispersion curves for LiF, NaF, NaCl, NaI, RbF, RbCl and MgO using the slab method with 15 layers and this work is summarized in Reference [31]. These extensive computational studies were based on the shell model parameters for the dynamical interactions in bulk ionic crystals. In 1976 Benedek reported on extensive calculations for LiF, NaCl and NaI based on the Green's function method [6].

Fig. 13 shows the experimental results for the Rayleigh mode measured by He scattering along the $\bar{\Gamma}$-$\bar{M}$ direction (<100>) of LiF as reported by Brusdeylins, Doak and Toennies in 1981 [18]. The two dashed line curves display the results from the calculations of Benedek [6] and Chen et al. [31] using shell model parameters determined from a best fit of bulk dispersion curves. It is clearly seen that these theoretical curves lie above the experimental points at the zone boundary by about 11%. A similar discrepancy was not observed in the other crystals NaF, KCl [18] and NaCl [53], which were subsequently studied. Benedek employed the breathing shell model [54] in which the electron shell of the negative ion is allowed to adjust its radius to the vibrations of the surrounding cage of ions, whereas the calculations of Chen et al. are based on a rigid shell model [31]. As seen in Fig. 13 both predict essentially the same dispersion curves. This is also found for other modes (including the optical modes), as well as for other crystals, so that the 11% discrepancy between the experiment and these theories does not appear to be due to any shortcomings in the interaction model used. Recent extensive calculations of the dynamics for the relaxed crystal show that the discrepancy also can not be attributed to the small relaxation $\Delta d_{12}/d_{12}$ = - 0.8% nor to the relative displacements of adjacent ions (rumpling) which amounts to about 3% or less [55]. Very recently two possible explanations have been proposed by Kress et al. [56]. One of the proposed models derives from the relatively large breakdown of the Cauchy relationship (C_{12} = 0.420 versus C_{44} = $0.628 \cdot 10^{12}$ dynes/cm^2 at T = 300 K) for LiF compared to the other alkali halides. Thus the nearest neighbor value of B, which in our notation is proportional to α_1, is expected to have a rather large non-central part B^b_{ncent} in the bulk (superscript b) to indirectly account for many body forces. At the surface one might expect that many-body effects will be more strongly modified than the two-body central interactions between the closed shell ions. Thus by substantially reducing the non-central force constant at the surface to B^s_{ncent} = - 0.5 B^b_{ncent} Kress et al. find that the Rayleigh frequency is suppressed at the zone boundary. The solid line in Fig. 13, which is in good agreement with the experimental results was obtained after this modification. An alternative model leading to the same curve is based on the fact that in the bulk the valence charges of ionic crystals have to be corrected to account for the overlap from neighboring charge distributions. This overlap is expected to be disturbed at the surface and Kress et al. found that by slightly reducing the shell charge

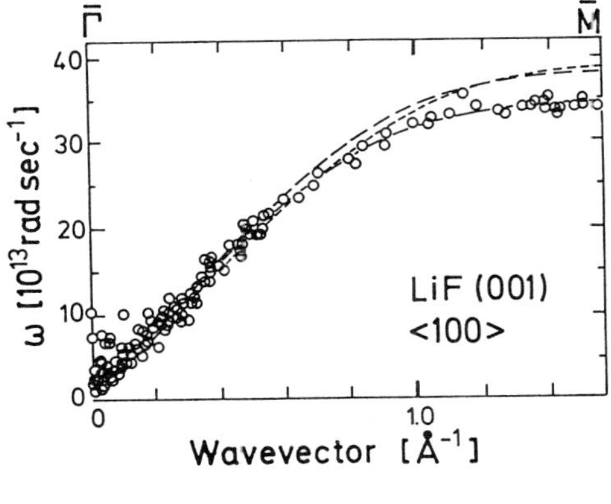

Fig. 13. Comparison of the experimental dispersion curve [18] for the Rayleigh phonon measured by helium atom scattering on LiF(100) with three theoretical calculations. The dashed line is from the calculations of Chen et al. [31], the short dashed lines from Benedek [6] and the solid line from Kress et al. [56].

of Li^+ from 1.376 to 1.316 and that of F^- by the same amount from -1.834 to - 1.774 at the surface the same good agreement with the experiments could be achieved. The strong effect of these subtle changes in charges compared to the large modifications of B_{ncent} illustrate the overriding influence of electrostatic forces in the ionic crystals. Kress et al. have no preference for one of the above models nor do they provide an explanation for the apparent absence of charge transfer on the other alkali halide surfaces.

Although the early He scattering experiments were not able to resolve additional modes Brusdeylins et al. in 1985 [57] finally succeeded in resolving a number of optical modes with energies of the order of 30 meV on NaF (100). NaF was chosen over LiF since the optical modes of this surface should have larger cross sections. In both crystals the fluorine ion interacts more strongly with He because of its larger size, but in NaF the fluorine ion has a much larger amplitude in the optical modes since it is somewhat lighter than Na^+. In LiF, on the other hand, the fluorine ion amplitude is greatest in the acoustical modes. Thus the excitation of optical modes is more favorable in NaF than in LiF. Nevertheless Bracco et al. [58] have recently succeeded in observing a series of optical modes up to 50 meV in LiF as well. Fig. 14 shows the experimental dispersion curves for both directions for NaF as well as for LiF and compares them with the shell model calculations of Benedek et al. [59] and Kress et al. [56], respectively. The NaF measurements along the $\bar{\Gamma}$-$\bar{X}$ directions clearly show the S_6 longitudinal mode for the first time. Since the Green's function calculation did not have sufficient resolution to distin-

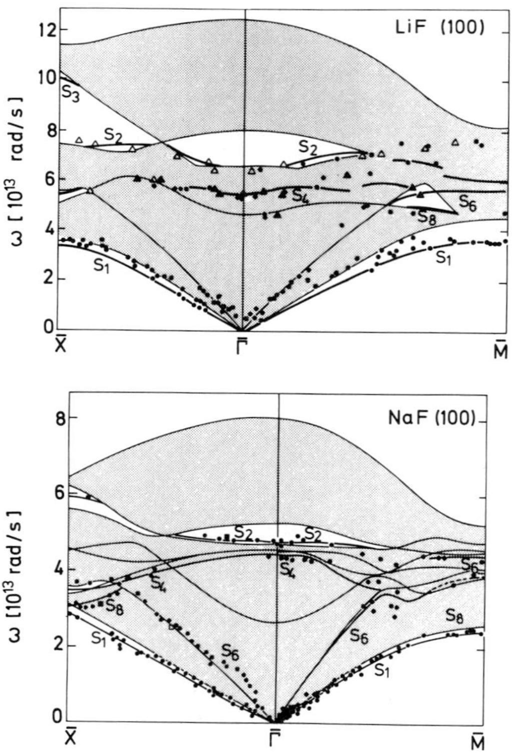

Fig. 14. Comparison of He atom scattering measurements and theoretical dispersion curves for the alkali halides. In a) the experimental results of the Genoa group [58] for LiF(100) are compared with calculations [59] of sagittal polarized modes based on the breathing shell model. In b) the experimental results of the Göttingen group [57] for NaF(100) are compared with similar calculations [59].

guish the S_6 mode from the lower edge of the longitudinal acoustical band it is not clear if the S_6 mode is simply due to the bulk band edge or due to a new surface resonance produced by the surface. It is of interest that this surface resonance has been observed in a much more extreme way in several metals (see Section IIIc). The shape and location of the S_8 and S_4 modes, which are nicely resolved along the same direction, have a simple explanation which derives from the near equality of the masses of the Na^+ and F^- ions [60]. If they were indistinguishable and had equal masses the transverse polarized Rayleigh mode S_1 and the longitudinal mode would have their zone boundaries at twice the value of Q. But because the ions in the NaF crystal are distinguishable they show up in the underlying bulk band continuum as surface resonances which are folded back from the NaF Brillouin zone boundary. In analogy to the diatomic linear chain it is expected that the folded mode pairs (S_1, S_8) and (S_4, S_6) have equal polarisations, but complementary displacement patterns. These folded modes do not show up nearly as clearly in LiF because of the large differences in masses of the individual ions. In both LiF and NaF the optical surface modes S_2 and S_4, however, show up nicely especially in the vicinity of the zone origin. Here the S_2 mode has a predominantly transverse character while the S_4 mode, also commonly called the Lucas mode [61], has a stronger longitudinal character. With the present experimental resolution it is not possible to identify any serious discrepancies with the available theoretical calculations.

Finally we note that electron scattering experiments have also been attempted on the alkali halides [62]. Serious problems are encountered because of charge build-up at the surfaces and the high electron energies required for this reason may lead to substantial damage of the surface.

B. Semiconductors

At the present time the semiconductor surfaces Si(111) (2x1) [63], GaAs (110) [64] and the layered components GaSe [65], 2H-$TaSe_2$ [66] have been studied by He scattering. Electron energy loss spectroscopy has only been applied to the Si(111) (2x1) reconstructed surface where a 56 meV loss was observed [67]. These results have recently been confirmed and extended with respect to the phonon polarization and defect dependence [68]. The lack of electron energy loss investigations at lower energy transfers on these crystals appears to be due to special experimental problems arising from a large background signal possibly due to excitation of electron-hole pairs. On the other hand there is a great interest in surface phonon measurements because of the remaining uncertainties concerning the detailed atomic structure of especially Si(111) in its naturally occurring (2x1) and (7x7) reconstructed phases and the forces which drive these structural phase transitions.

Fig. 15 shows the measured dispersion curves for Si(111) (2x1) [63] and GaAs(110) [64] measurd by He atom scattering. There are rather striking similarities between the two sets of measurements despite the differences in the structures. In both crystals there is clear evidence for the Rayleigh mode in the two directions studied. The frequency of this mode in Si(111) at the $\bar{X}$ point (11.5 meV) is significantly below the value of 14 meV calculated by Allerhand et al. [69]. These authors used a model in which the total energy is partitioned into two contributions. One term accounts for the sum of independent particle energies from the filled valence bands, and is calculated in the tight binding approximation. The second term accounts for the interatomic elastic potentials of all the bonds. The model qualitatively predicts the observed Rayleigh mode, but does not explain the 10 meV mode.

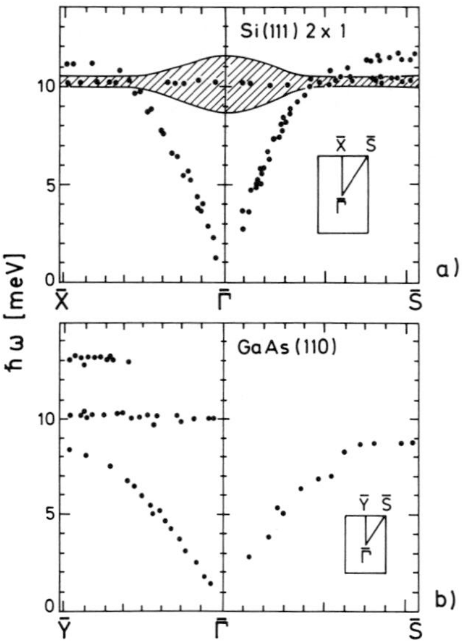

Fig. 15. Surface phonon dispersion curves measured by helium atom scattering on a) reconstructed Si(111) 2x1 [63] and b) on GaAs(111) [64]. The shaded area in a) indicates the width of the observed phonon peaks.

At the present time no lattice dynamical calculation of low frequency surface phonons are available for GaAs and thus only a comparison with the measured bulk phonon frequencies can be made. The energy of the expected surface modes at the Brillouin zone boundary can be estimated from the energy of the bulk modes at the bulk zone boundary, which is the K point for the <110> direction. The lowest-lying bulk transverse mode has an energy of about 9 meV at the K point. This compares well with the maximum frequency of 8.3 meV measured for the Rayleigh mode. The surface mode at 13 meV seen near $\bar{Y}$ is conceivably related to a mixed longitudinal-transverse bulk mode which has an energy of 14 meV at K. If this surface resonance is situated in a gap of the bulk phonon bands near the zone boundary, then it would probably hybridize with the bulk phonons when entering the band as it approaches the zone center. All other bulk phonon modes have an energy greater than 25 meV at the K point and thus do not provide an explanation of the observed 10 meV mode.

Very recently Allerhand and Mele [70] have proposed an interesting mechanism to explain the 10 meV mode on Si(111) (2x1). They note that the π-bonded chains can undergo a vibrational motion which at $\bar{\Gamma}$ consists of an up and down "bobbing" of all of the atoms in the chain with a frequency of about 10 meV. Ordinarily this mode is expected to increase in frequency with increasing wave vector as the motion changes over to a twisting motion along the chains at the zone boundary $\bar{X}$ (in our notation). There, however, it is folded back and goes over into a "see-saw" rocking motion of the entire chain as Q decreases again with an expected frequency of about 31 meV at the zone origin. In order to explain the lack of dispersion of the lower branch of this mode these authors invoke an additional mechanism consisting of an electron-phonon coupling at the zone boundary. This coupling is attributed to an indirect resonant electron transition with large ΔK into the depressed empty electronic state at $\bar{J}$. It is reminiscent of Kohn anomalies [71] which, however, lead to disturbances of the phonon curves at well defined values of $\vec{Q}$. In the present situation the coupling distorts the lower branch of the chain vibrations to make it have a constant energy independent of Q. Calculations for the shift in frequency due to the electron-phonon coupling are consistent with this mode, although they do not appear detailed enough to precisely explain the extremely flat curve observed. If this model is correct it would provide the first evidence for the importance of electron-phonon coupling at the surfaces of semiconductors and may provide new insight into the old question of the driving mechanism of the surface reconstruction of semiconductors [72]. A Kohn anomaly has been observed on the Pt(111) surface [73] for which there is now substantial evidence that the anomaly is merely a projection of the bulk anomaly onto the surface [47]. Obviously more experiments and theory are called for to further clarify the situation.

Our understanding of the surface phonons on the Si(111) + H (1x1) surface is in a more satisfactory situation largely due to the theoretical work of Benedek and Miglio and colleagues [33]. The adsorption of H on Si(111) (2x1) has the effect of restoring the Si lattice back to the unreconstructed geometry which otherwise is not stable at T < 900 K. This phenomenon is attributed to the saturation of the surface dangling bonds by the hydrogen atoms and thus it is expected that the hydrogenated surface will exhibit dynamical properties similar to a hypothetical unreconstructed surface. Fig.16 shows the calculated dispersion curves based on the bond charge model of Weber [30] and compares them with the He scattering measurements [33]. In this model the hydrogen atom at the surface is introduced in the same way as a bond charge with, however, a finite mass and a weakly polarizable shell. This is coupled to its first nearest neighbor by a central force constant as well as by the Coulomb interaction. When fitted to the measurements, this model yields a force constant to the hydrogen atoms which is roughly the same as that of the Si-Si bond in the bulk confirming the assumed similarity between the hydrogenated surface and the ideal termination. The agreement between experiment and theory in Fig. 16 is very good. The discrepancy at small Q values is attributed to a preferential coupling of the He atoms to the longitudinal band edge instead of to the Rayleigh mode. The theory can also explain an energy loss peak of about 60 meV found by an off-specular EELS experiment [74]. Since the EELS experiment was performed at low collision energies (≅5 eV) the dominant scattering mechanism is dipole scattering and only the phonons at the $\bar{\Gamma}$ point are probed.

The experimental studies of the layered components GaSe [65] and $TaSe_2$ [66] were undertaken to test for possible differences between their sur-

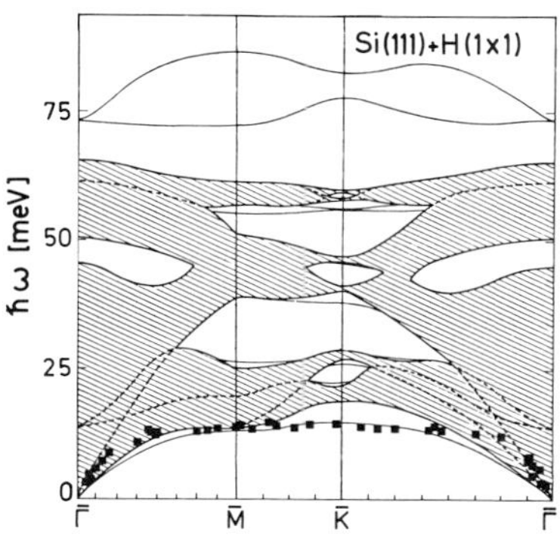

Fig. 16. Comparison of surface phonon dispersion curves measured by helium atom scattering with calculations based on the bond-charge model [33].

face and bulk dynamics. Since the layers are only very weakly coupled by van der Waals forces which are negligibly small compared to the strong covalent bonds within the layers one might expect insignificant differences between the surface dynamics and the bulk dynamics studied by neutrons. In fact, however, in both GaSe and $TaSe_2$ very sizable differences are observed both with respect to the overall shape and frequencies of the dispersion curves and in the case of $TaSe_2$ also with respect to the location of incipient mode softening accompanying the charge density wave transitions. Some theoretical models have been developed to explain these unexpected results but all of the phenomena are not well understood at the present time.

C. Metals

The surfaces of metals have been extensively studied by both He atom scattering and electron scattering. In He scattering rather weak diffraction peaks with intensities of the order of 10^{-3} of the specular intensity are observed for all but the highly corrugated fcc (110) and bcc (111) surfaces. Since atom scattering is sensitive to the electron density profiles far from the surface this is attributed to the smoothing of the corrugation of these surfaces by the free electrons at the surface (Smoluchowski smoothing) [75]. For similar reasons the He inelastic phonon scattering intensities are somewhat smaller for metals and the peaks are usually somewhat broader than for the alkali halides. In the following we review some of the recent results for nearly free electron metals, noble metals and the transition metals.

1. Nearly Free Electron Metals

Of the nearly free electron metals such as the alkalis, Be, Mg, Al and Pb only Al has been studied extensively largely because of the experimental difficulties in the preparation of the other metals. The simple electronic structure of Al which consists of a Ne core with two 3s and a single 3p electron makes it readily accessible to ab initio calculations. On the other hand because of the nearly free electrons the interactions are particularly long-ranged. This manifests itself by the fact that coupling parameters up to at least 8 nearest neighbors are needed to fit the bulk phonon dispersion curves. For the surface at least two ab initio calculations of the strongly relaxed ($\Delta d_{12}/d_{12} = -6.8\%$) (110) face have recently been reported [76,77]. Ho and Bohnen [78] have also recently calculated the surface phonon dispersion curves by an ab initio method. For comparison with the ab initio calculations He atom scattering experiments have recently been reported [79]. Unfortunately the (110) surface shows extensive disorder and this may effect the results especially at small wave vectors. At large wave vectors near the zone boundary where the wavelength and penetration are small it is felt, however, that the experiments are reliable since the immediate environment of an individual surface atom is not expected to be affected. Table I shows a comparison of the experimental and theoretical results for the zone boundary phonons. The largest discrepancy is found at the zone boundary at $\bar{X}$ where the experiment is nearly 3 meV ($\approx 20\%$) smaller than the theory. This difference is surprising in view of the good success of theory in predicting the relaxation. In this connection it should be recalled that relaxation depends on the tangential force constants and is thus related to the first derivatives of the two-body potential whereas the phonons are also sensitive to the second derivatives, which may well be more difficult to predict accurately.

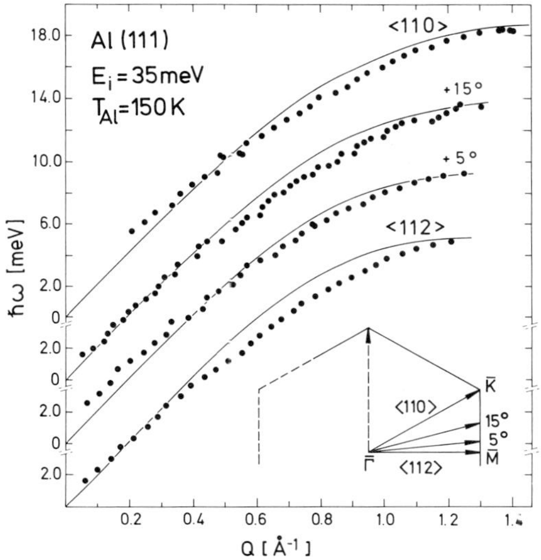

Fig. 17. Comparison of experimental and theoretical Rayleigh surface dispersion curves for Al(111) along the two symmetry directions and two directions in between [80]. The smooth line curves have been calculated with force constants fitted to the bulk phonon dispersion curves.

Table I Comparison of ab initio frozen phonon calculations [78] with the results from a helium scattering experiment [79] for Al(110) for relaxed surface, all values in meV.

		$\bar{X}$				$\bar{Y}$	
Mode	Pol.	Theory[a)]	Expt.[b)]	Mode	Pol.	Theory[a)]	Expt.[b)]
-	L	16.96	-	S_1	L	7.86	8.9 ± 0.3
S_1	T	17.37	14.6 ± 0.2	S_3	T	14.06	13.5 ± 0.2
S_2	SH	22.75	-	S_2	SH	14.89	-
-	T	24.82	-				

The phonons on the Al(111) surface, where relaxation amounts to only about -0.9%, have also been measured [80]. The dispersion curves are shown in Fig. 17 for the two symmetry directions and for two azimuthal directions in between. These results have been carefully analyzed by Bortolani and collaborators [80] following the procedure outlined in Section II. They first refitted the bulk neutron data with a 10 nearest neighbor model and determined altogether 23 force constants. In view of the very long range nature of the interaction a total of 90 layers was required in the lattice dynamical calculations to obtain complete convergence. The good agreement shown for the continuous curves in Fig. 16 with the experimental data could be achieved for all four directions with only a very small 1.5% increase in the central force constant between the first and second layer. This good agreement is gratifying since as we shall see later on this is in fact the only metal surface without a striking anomalous behavior. Even here on closer examination we notice that especially along the <112> direction there appear to be some small yet significant oscillations around the theoretical curves, which are not understood [81].

2. Noble Metals

Several faces of all three of the noble metals Cu, Ag and Au have now been studied. Although in the noble metals the d-shells are filled they nevertheless play an important role in the bonding through (sp)-d hybridization. The (111) surfaces were chosen for the first high resolution He atom scattering experiments since, with the exception of Au, they do not reconstruct and the relaxation is small. The reconstruction in Au has a 23 x 1 super-structure which is consistent with a small 4.6% contraction along the <110> direction. This weak reconstruction could be clearly resolved in a He atom diffraction experiment [82] and the observed soliton walls have recently been explained by molecular dynamics calculations for a Double-Sine-Gordon potential describing the interactions of the first layer atoms with the substrate [83].

Fig. 18 presents a comparison between the He atom dispersion curves [84] and calculations based on a simple central forces nearest neighbor model with one force constant. This single force constant was obtained by a simple fit to the maximum bulk frequency [85]. The comparison shows a reasonable agreement of the experimental Rayleigh mode with the predictions except in the case of Au where the experimental points are lower by about 10%. A much better agreement with the Rayleigh zone boundary frequencies is found if the single force constant is adjusted to provide a best fit of seven characteristic bulk phonon frequencies [86]. With a force constant determined in this way the largest deviation reduces to 5% for Au at the $\bar{K}$ point.

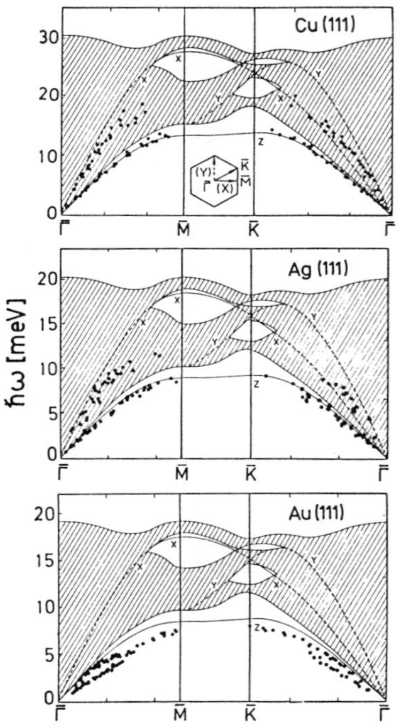

Fig. 18. Comparison of dispersion curves measured by He atom scattering for Cu(111), Ag(111) and Au(111) with calculations for a simple central-forces nearest-neighbor model [84]. Note that the energy scales and wave-vector scales are different for the three systems.

The He atom scattering experiments also reveal an additional mode at higher frequencies in the region of the transverse polarized bulk bands where a sagittal plane polarized surface mode is not predicted by the simple theory. This additional mode, which was first reported for Ag(111) [19], has been attributed to an unexpected softness of the lateral force constants in the surface layer. The group of Bortolani et al. [87] were able to fit this anomalous mode only after they had reduced $\beta^{s}_{\shortparallel}/\beta^{b}$ by 70, 52 and 70% in Cu, Ag and Au, respectively. They used typically 15 force constants with interactions up to 6 nearest neighbors in order to first fit the bulk phonons. Recently the same anomalous mode could be detected along one direction in Cu(111) in electron scattering experiments by Kesmodel and coworkers [88]. Using a single central force constant model Mills and collaborators were able to fit these results by reducing the surface lateral force constant by only 15% [89]. So far, however, the one force constant model has not been tested for the other symmetry directions and it remains to be seen if this simple model is able to fit both symmetry directions and also predict the correct intensities seen in He scattering.

These He scattering experiments on the noble metals have stimulated a new interaction model for describing the interatomic forces in metals which was recently introduced by Jayanthi, Bilz, Kress and Benedek [90]. The model attempts to account for many-body effects in a more realistic fashion than by a mere expansion of two-body potentials discussed previously. In the new model the many-body effects are attributed to sp electrons located in the interstitial regions. They are accounted for by introducing massless deformable pseudoparticles between the ion cores. The total energy of the lattice now depends not only on the dynamic displacements but also on the distortion fluctuations of the charges at the ionic and pseudo-particle sites. The model is found to be quite successful in fitting the bulk dispersion curves and elastic constants of Cu and Ag with only two radial force constants for the ion-ion and ion-pseudocharge Coulomb forces (tangential force constants disappear in the bulk by the requirements of translational and rotational invariance) and a dipolar and quadrupolar deformability of the pseudoparticle. In the case of Au an additional small quadrupolar and octupolar deformabilities of the ions are required in addition to these four parameters.

At the surface the model is modified to account for the significant charge redistribution. For example charge density calculations for Cu(111) indicate an increase of the sp charge per atom by 20% with a relaxation of sp charges from the top layer into the underlying layers [91]. These electronic charges are accounted for in the model by a reduction in the ion-pseudoparticle radial force constant. This is estimated for Cu via a simple model with estimated values for the work function, pseudoparticle-ion separation distances and the magnitudes of the sp charges in the surface layer and in the bulk. Unfortunately data on the charge redistribution is not available for Ag(111) and Au(111) and for these systems it is necessary to rely on rough estimates.

Fig. 19 presents a comparison between the prediction of this new theory and the surface phonon measurements. For Cu(111) and Ag(111) the agreement with experiment is very good. Thus the anomalous longitudinal mode can be explained simply by the reduction in ion-ion repulsion between ions in the top layer produced by charge redistribution. However, in the case of Au(111) the theory is not in agreement with the analysis of Bortolani et al. [92,93] in that it predicts a longitudinal mode at much higher energies where the experiments do not reveal a surface phonon mode. It is

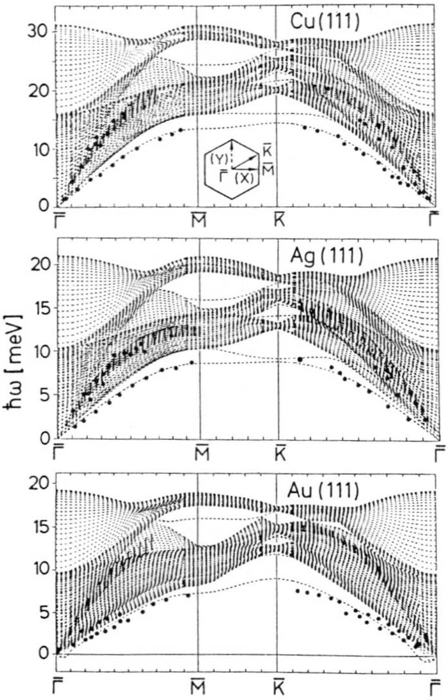

Fig. 19. Comparison of calculated surface-phonon dispersion curves of Cu(111), Ag(111) and Au(111) based on the model of Jayanthi et al. and helium atom scattering data. The spectral width of the surface-resonance mode S_6 is indicated by the lengths of the vertical bars. For Au(111) there is a lattice instability along the $\bar{\Gamma}$-$\bar{K}$ near $\bar{\Gamma}$. For the sake of clarity not all experimental points, also shown in Fig. 18, are indicated.

interesting, however, to see that along the $\bar{\Gamma}$-$\bar{K}$ direction the new theory predicts a surface instability consistent with the observed reconstruction along this direction. These authors propose that the observed splitting in the low lying surface mode in Au(111) is not due to the longitudinal anomaly as discussed by Bortolani et al. but rather due to reconstruction. In the lattice dynamical calculations of Bortolani et al. reconstruction is found to have only a minor effect on the peak positions of the Rayleigh phonons [93].

Cu(100) has been studied both by EELS [94,95] and in an early He atom scattering experiment [96]. The EELS data has been compared with a single force constant model calculation and indicates a stiffening of the transverse force constants by about 20%. Since the Cu(100) surface has only a weak relaxation of about -1%, the stiffening can not be directly explained by the presence of relaxation. As we shall see in the next subsection the stiffening may be an artifact coming from the oversimplified force constant model used to describe the bulk. Cu(110) has also been studied by He atom energy loss spectroscopy but the data did not extend to wavevectors beyond half way to the zone boundary.

Very recently the Au(110) surface, which spontaneously reconstructs to the missing row structure, has also been studied by He atom scattering [98]. An ab inito calculation is in progress for this system [99] and when available it will be possible to carry out another direct comparison of experiment with theory. Ag(110) has recently been studied by the Genoa group of Bracco et al. [100] in the direction along the rows. In addition to the Rayleigh mode another longitudinal mode is observed. The Ag(110) results have also been simulated by Bortolani et al. [101] and they find that the same best fit force constants determined from the fit of the Ag(111) measurements including the lateral softening could also explain these results.

3. The Transition Metals Ni, Pd and Pt

The surface phonons in the transition metals are of particular interest because of the strong catalytic activity which is thought to be related to the unfilled d-shells. The transition metals with nearly filled d shells such as Ni, Pd and Pt are in many respects very similar to the noble metals. So far, only the phonons on a few selected faces of Ni, Pd, Pt and W have been studied experimentally.

Ni(100) was the first system to be studied by electron scattering [12] and recently the system has also been studied by He atom scattering [47]. Fig. 20 shows the good agreement of the two sets of data. The lowest observed mode in both experiments is attributed to the transverse polarized Rayleigh mode S_4 whereas the other modes are surface resonances embedded in the continuum of bulk vibrations projected onto the surface. Note that helium atom scattering reveals a mode not seen in electron scattering and is sensitive to wave vector regions of the same mode not seen by the electrons. These differences are related to the more complicated multiple scattering interaction of electrons which also probe the second and third layer vibrations, whereas helium atoms are sensitive only to vibrations at the surface.

The electron scattering data for Ni(100) were analyzed by several authors [102,103] in the framework of a lattice dynamical model with a single radial force constant between nearest neighbors. This model was justified by the fact that with one force constant the bulk dispersion curves can be fitted to within about 4%. The EELS surface phonon frequencies at the zone boundary were found to be 15% greater than these theoretical results, which was attributed to a stiffening of the surface inter-

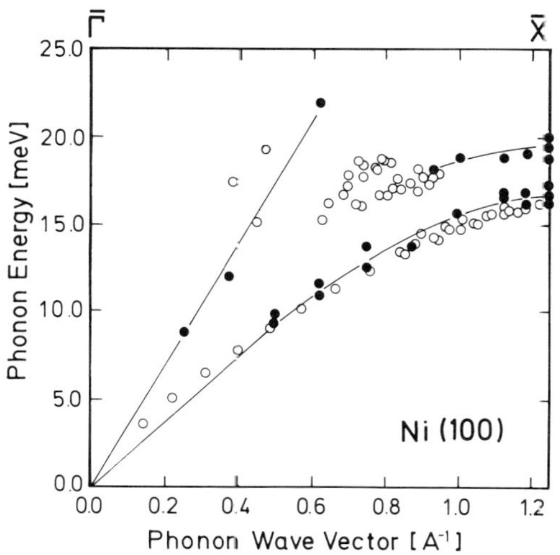

Fig. 20. Comparison between surface phonon dispersion curves of clean Ni(100) measured by electron scattering [44] (•) and by helium atom scattering [47] (o) along the Γ̄-X̄ direction.

layer force constant $\beta_{\perp}^{s}$ by 10% [12]. This stiffening was explained by a -4% inward relaxation of the distance between the first and second layers at the surface as measured by ion backscattering [104] as suggested by the well-known dependence of force constants on bond lengths in diatomic molecules (Badger's Rule). This explanation must be reexamined in view of the observation of a similar increase of frequencies in Cu(100) [94] where the relaxation is known to be negligibly small. Also He atom experiments on Pd(110) [105], which has a large inward relaxation of -6%, and on Pb(110) [106], which has a large relaxation of -15.8%, both reveal no significant increase in the transverse force constant $\beta_{\perp}^{s}$. There is also some evidence that the apparent stiffening in Ni(100) is not due to an inadequate force constant description of the bulk dynamics. This is surprising since a best fit set of five force constants by Black et al.[86] (up to second nearest neighbors) appears to provide a better fit of the experiments. However, an even larger set of force constants involving central interactions up to the 6th nearest neighbor and angle-bend interactions up to the 4th nearest neighbor give nearly the same frequencies as obtained by the first nearest neighbors model [107]. Thus there appears to be an oscillatory dependence of the Rayleigh mode frequency on the number of shells of nearest neighbors to which interactions are included and consequently for Ni a nearest neighbor force constant model is adequate. If this is the case the only way to explain the observed stiffening is by increasing α^{s} with respect to the bulk. There is strong evidence supporting this explanation coming from calculations by Franchini [23]. She has shown that for both the Cu(111) and Pt(111) surfaces the amount of relative relaxation of the first layer and the square of the zone boundary frequency are both linearly proportional to the lowering in $\alpha_{\perp}^{s}$.

Some additional evidence on the nature of surface force constant changes comes from studying the effect of hydrogen absorption at the surface. Hydrogen adsorption provides an ideal way to change the electronic density of states at the surface since it forms a 1x1 overlayer and therefore does not change the diffraction pattern. Moreover because of its small mass it does not appreciably change the effective mass loading of the surface oscillators. In the hydrogenation of Ni(100) the He atom experiments show that the highest frequency mode on the clean surface (see Fig. 20), which had been assigned to a longitudinal mode, disappears almost completely and that the Rayleigh mode is slightly shifted downwards by 0.7 meV at $\bar{X}$ and by 1.5 meV at $\bar{M}$ [45]. It is especially noteworthy that the hydrogen adsorption restores the phonon frequencies to the values predicted from calculations using bulk force constants. This is in accord with similar observations on Pt(111) [49] and Si(111) + H (1x1). We attribute this phenomenon to a stiffening of the lateral forces in the surface layer at the expense of the transverse bonding between the first two layers resulting from a saturation of the dangling d-orbitals at the surface. A similar phenomenon was also found on Si(111) 2x1.

Pt(111) has been extensively studied by He atom scattering and results have now been reported by 3 groups [73, 108, 109]. These experiments appear to show a similar two peak energy loss structure as found on the other (111) surfaces. In fact the true nature of the energy loss spectrum along the $\bar{\Gamma}$-$\bar{K}$ <110> direction could only be correctly understood after a careful lattice dynamical analysis had been carried out by the Modena group [110]. These calculations revealed that actually three surface modes are present along this azimuthal direction while along the $\bar{\Gamma}$-$\bar{M}$ direction only two clearly resolved modes as in the noble metals are present. Fig. 21 shows the dispersion curves obtained by the lattice dynamical calculations which provide the best fit of the time-of-flight spectra [49]. The

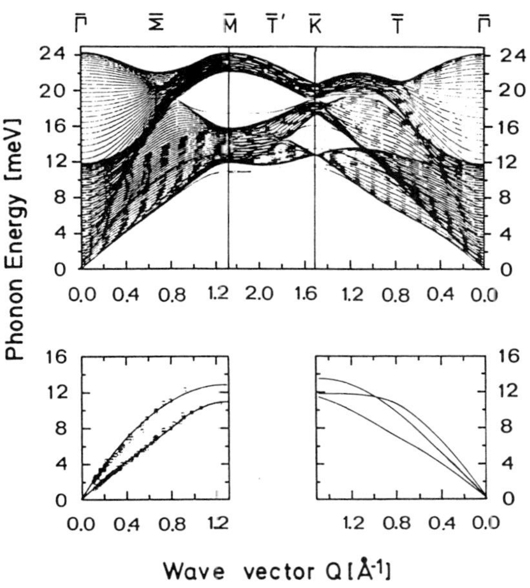

Fig. 21. Calculated dispersion curves for Pt(111) based on a six nearest neighbor force constant model with surface force constants modified to fit the measured surface phonon dispersion curves [49]. The bottom panels show the surface modes which fit the experimental data.

third mode in Pt(111) along $\bar{\Gamma}$-$\bar{K}$ is attributed to a pseudo-Rayleigh mode which has a sizable coupling to the bulk modes. This mode hybridizes with the anomalous longitudinal mode at about 2/3 of the way from the zone origin leading to a complicated change in polarization as a function of Q. The best fit calculations for Pt(111) reveal once more that the lateral force constants are appreciably reduced compared to the bulk, whereas the transverse force constants again appear to be unchanged. The best fit nearest neighbor force constants in the surface plane were found to be $\alpha_{\parallel}^{s} = 1.7\ \alpha^{b}$, $\beta_{\perp}^{s} = 0.39\ \beta^{b}$ and $\delta^{s} = 0.045\ \delta^{b}$ [49], where the index b denotes bulk force constants.

Through several improvements in technique it was possible to greatly improve the precision of the Pt(111) measurements compared to the earlier noble metal data [73]. This improved precision has made it possible to determine the shape of the Rayleigh dispersion curve with root mean square deviations of about 0.09 meV (90 μeV!). The group velocity obtained from a nine-term Fourier expansion fit reveals an oscillatory behavior with Q. The experimental surface mode exhibits two weak anomalies at about 0.5 $Å^{-1}$ and 1.0 $Å^{-1}$ which appear to be much stronger and sharper than in the bulk. This is consistent with theoretical studies of electron phonon coupling in other systems such as TiN [111]. A theory based on 16 bulk force constants [49] can nicely account for the anomaly at Q = 0.5 $Å^{-1}$ suggesting that this anomaly is in fact a projection of the bulk behavior onto the surface and is probably not associated with surface electronic states.

In addition to the clean surface the Pt(111) + H (1x1) surface was also investigated by helium atom scattering [49]. The longitudinal and the pseudosurface mode both disappear and the Rayleigh mode is lowered by 10 and 15% at the zone boundary $\bar{M}$ and $\bar{K}$ points, respectively. The observed frequencies could be nicely explained by a lattice dynamical calculation in which the force constants at the surface are identical to those in the bulk. This observation suggests that the addition of one electron per unit cell by the hydrogen atom is sufficient to remove the longitudinal softening.

Both directions of the Pd(110) surface have also been studied [105] using He atoms. The experimental results have been interpreted in terms of a dynamical model with first and second nearest neighbor interactions. The (110) surface is especially interesting because of its open structure with alternating densely packed rows and furrows resulting in a large corrugation along a scattering direction perpendicular to the rows. Pd(110) was chosen for the phonon studies since it is one of the few (110) surfaces without a missing row reconstruction. The relaxation of the first two layers of Pd(110) is -5.7% and is comparable to that of the Al(110) surface. Another reason for studying Pd(110) was to see if the lateral softening seen in Pt(111) and the transverse stiffening predicted by the ab initio calculations for Al(110) can both occur simultaneously. The best fit lattice dynamical calculations reveal a lateral softening of about 30% for the longitudinal mode directed along the close packed rows of the top layer. In addition, the transverse force constant is, surprisingly, not significantly changed despite the strong contraction. Thus in a first approximation it appears that in this crystal relaxation has in fact little effect on the transverse forces and on the longitudinal softening.

Table II summarizes the results on the reduction of radial force constants in the surface plane for the noble metals and transition metals adjacent in the periodic table. The results seem to indicate with the possible exception of Cu that for both the noble and transition metals the amount of softening increases with the increasing size of the d-shells. Moreover perhaps surprisingly for all three sets of systems there is definite evidence that the effect is smaller for the transition metals

Table II Electronic structures and lateral softening in the noble and transition metals determined from measured surface phonon dispersion curves

Transition Metals			Noble Metals		
Surface	outer shell	$\beta^s_{\parallel}/\beta^b$	Surface	outer shell	$\beta^s_{\parallel}/\beta^b$
Ni(100)	$3d^8 4s^2$	> 0.9 a)	Cu(111)	$3d^{10} 4s^1$ (0.30) b)	0.70-0.85 c)
Ni(100) + H (1x1)	-	1.00	-		
Pd(110)	$4d^9 6s^1$	0.7	Ag(111)	$4d^{10} 5s^1$	0.48
Pt(111)	$5d^9 6s^1$	0.39	Au(111)	$5d^{10} 6s^1$	0.30
Pt(111) + H (1x1)	-	1.00		-	

a) Preliminary evaluation of new data [39]
b) Preliminary
c) Single force constant fit of electron scattering experiments [88,89]

than for the noble metals. The sequence of decreasing $\beta_{\parallel}^{S}/\beta^{b}$ follows the trend of expected increase in sp-d hybridization. It now appears on the basis of the experimental results on W(100) to be discussed next that longitudinal softening becomes much greater as the d-shells become less filled.

The (100) surface of the refractory bcc transition metal W is probably the best known system showing what appears to be a nearly perfect two-dimensional phase transition. At temperatures above about 280°K the clean surface appears to have an ideal (1x1) structure although the amount of disorder is still open to question. As the temperature drops below 280 K a ($\sqrt{2}$ x $\sqrt{2}$) R 45° structure is observed in LEED [112]. This has been attributed to the formation of zig-zag chains along the diagonal direction shown in Fig. 22).

Two mechanisms have been proposed for describing the atomic interactions driving the reconstruction on the clean surface. The one mechanism is called the charge density wave mechanism which is currently used to explain the reconstruction in the layered compounds such as $TaSe_2$ [113]. According to this model the reconstruction is driven by an electron-phonon coupling which is enhanced by the occurrence of large nearly parallel "nesting" regions of electronic states at the Fermi surface in an extended zone scheme. Thus at sufficiently low temperatures a Peierls instability sets in at the corresponding wave vector Q_c connecting these regions and a new structure with a periodicity related to Q_c appears. In this picture electronic states sharply localized in K-space are involved implying a delocalization in real space. Photoemission experiments [114,115] show, however, that the projection of the Fermi level on the surface is in fact very curved and that the expected "nesting" is unlikely. The other model is called the Jahn-Teller model and is suggested by tight-binding calculations which reveal an inherent geometrical instability in which the density of electronic states shifts to lower energies for the reconstructed surface [116]. The latter model is currently favored, but the issue is by no means settled as indicated by the appearance of two recent publications each supporting one of these mechanisms [117,118].

Fig. 23 summarizes the results of a recent He atom study [119] of the changes in diffraction structures and surface phonon dispersion curves accompanying temperature induced reconstruction [120] and hydrogen adsorption on the unreconstructed surface. In panel c) we display the diffraction pattern taken in the high temperature phase. The narrow sharp specular peak is interpreted as indicating a large degree of order. This is supported by a detailed study of the incoherent elastic intensity measured

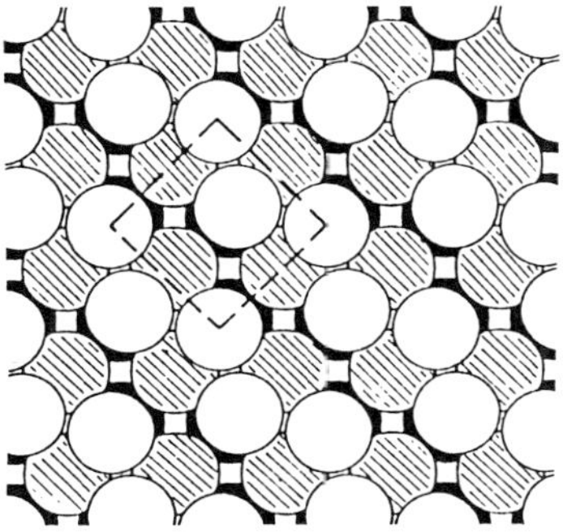

Fig. 22. Schematic diagram showing the reconstructed ($\sqrt{2}$ x $\sqrt{2}$) R 45° surface of W(100) [112]. Note that the atoms of the top layer (white circles) are displaced from their equilibrium positions (x) to form diagonal chains.

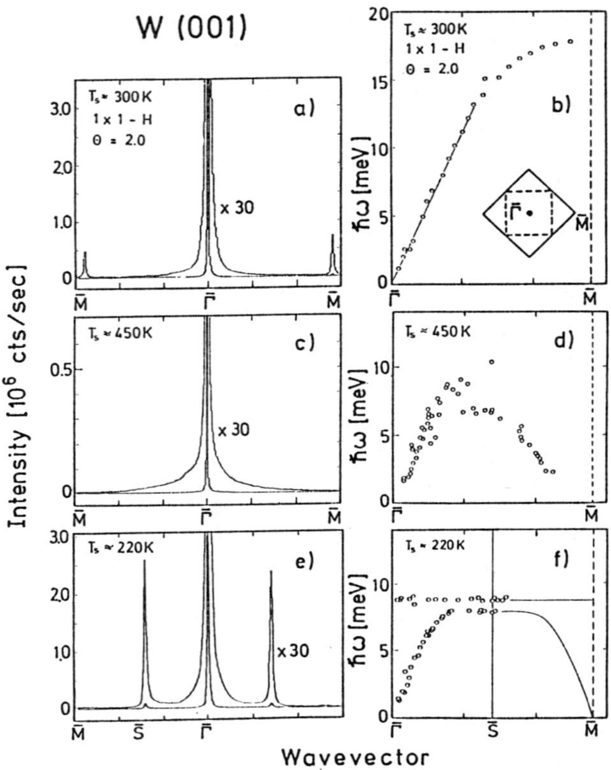

Fig. 23. Three sequences in the changes in the structure and dynamics of the W(001) surface studied by He atom scattering [119]. In panels a) and b) the diffraction pattern and phonon dispersion curves are shown for a (1x1) H atom covered surface at T = 300 K [121]. In panels c) and d) the diffraction pattern and phonon dispersion curves for the clean surface are shown at T = 450 K. Finally panels e) and f) show the changes which set in when the surface has been cooled down to 220 K revealing a reconstructed surface.

by time-of-flight spectroscopy between the diffraction peaks which also reveal a relatively small intensity consistent with a high degree of order. In panel d) we show the two distinct phonon dispersion curves obtained from time-of-flight spectra measured under the same conditions. The upper mode is attributed to a transverse Rayleigh mode in accordance with several recent calculations [118]. It was not possible to observe sharp peaks near the zone boundary because of a broadening of the peaks possibly due to a significant amount of disorder on an atomic scale. The lower dispersion curve is attributed to a longitudinal mode and its behavior is expected to be closely linked to the shear horizontal mode which is expected to be responsible for the reconstruction. This mode shows a typical soft phonon behavior with a frequency approaching zero near the Brillouin zone boundary. The close similarity of this dispersion curve in the high temperature phase with the general pattern of the softening of longitudinal modes observed in the noble metals and in Pd and Pt is remarkable. The extent of the longitudinal softening in W was also not anticipated by theory and indicates that the surface is in fact nearly metastable in the high temperature phase. Recent unpublished experiments reveal as the temperature is further increased up to 1100° K that the frequency of the anomalous mode rises proportionally and appears to converge towards the normal behavior of a dispersion curve with a positive slope [121]. Lattice dynamical calculations fit to the measured anomalous surface phonon dispersion curve indicate that both $\alpha^S_{\parallel 2}$ and $\beta^S_{\parallel 2}$ are greatly effected at the surface compared to their bulk values [122].

In Fig. 23 the panels above and below show the effect of H-chemisorption and a temperature induced reconstruction, respectively. With the addition of H the first order peaks are clearly observed (panel e) and are in agreement with the expected 1x1 structure. The anomalous soft mode disappears altogether and the Rayleigh mode stiffens considerably. The stiffening of the Rayleigh mode and disappearance of the longitudinal anomaly suggest that the presence of the H atoms tends to increase both the longitudinal and transverse forces on the W atoms of the first layer. It is recalled that this effect is qualitatively similar to that observed with hydrogen adsorption on Ni(100) and Pt(111). On the clean surface a further reduction of the temperature shifts the half order peaks outward until they assume the correct positions in reciprocal lattice space appropriate to the ($\sqrt{2}$ x $\sqrt{2}$) 45° reconstructed structure at about 220 K. The surface phonon dispersion curves now end at the new Brillouin zone boundary. In panel f) a new "optical" mode is observed, which can be explained by the fact that after reconstruction there are now two W atoms per unit cell.

The observed temperature dependent softening in phonon frequencies provides the first direct confirmation of the soft mode mechanism suggested by Tosatti and coworkers [123]. Their mechanism assumes that the surface atoms move in a potential determined by the atoms of the second layer. This potential has two local minima displaced by about 0.18 Å so that the system "locks" into one of these at low temperatures. The temperature dependent shift of the 1/2 order diffraction peaks as well as the broadened phonon peaks at $Q > 1.0$ Å^{-1} all suggest that the transition does not, however, proceed in a simple way from the (1x1) to the ($\sqrt{2}$ x $\sqrt{2}$) R 45 structure. This might be interpreted as indicating a competition of longitudinal and shear horizontal displacive forces with a complicated chaotic dynamics leading possibly to a succession of complex intermediate incommensurate phases.

IV. Summary

We have briefly surveyed the present understanding of atomic forces at the surface of clean single crystal surfaces obtained from experimental surface phonon dispersion curves. With the data on a fairly large number of systems it is tempting to make a few generalizations. In doing so we realize that as new data and theory becomes available these will have to be revised or even totally discarded. One reason for extreme caution is the lack of consensus among theoreticians concerning the use of models which account for many nearest neighbor interactions as opposed to simple models with only nearest neighbor central forces. Obviously with one force constant it is usually not possible to adequately fit the bulk phonons. On the other hand there is certainly no problem of uniqueness which is inherent when a large number of force constants is used [124]. As experimentalists we favor the use of the more elaborate models since they guarantee that the bulk forces are simulated in the most optimal way. For this reason the modified force constants at the surface should be more reliable even though the fit may not be unique. It will be recalled that in order to explain the changes in force constants at the surface only nearest neighbor interactions were invoked. In cases where there are no nearest neighbors in the surface plane the interactions to the second nearest neighbors had, of course, to be considered as with Pd(110) and W(100). Thus there is at least unanimity in how to describe the surface.

We can therefore concentrate our discussion on the three nearest neighbor surface force constants $\beta_\perp^S$, $\beta_\parallel^S$ and $\alpha_\parallel^S$. We recall from the discussion of stress and stability in Section IB that the tangential force constant $\alpha_\perp$ is largely responsible for relaxation. Its effect on vibrational frequencies is secondary to that of $\beta_\perp^S$, $\beta_\parallel^S$ and $\alpha_\parallel^S$. Also we recall from Fig. 7 that the effect of $\beta_\perp^S$ and $\beta_\parallel^S$ on zone boundary frequencies is largely decoupled, with the former effecting only the vertical transverse polarized Rayleigh modes and the latter, the longitudinal resonances. The experiments seem to tell us with only a few exceptions that $\beta_\perp^S$ is not greatly effected by the surface truncation nor by the presence of a large amount of relaxation as in the case of Pd(110). Notable exceptions are LiF(100), Ni(100) and Cu(100) but for most other crystals it is found that the Rayleigh mode can be well reproduced by bulk force constants. As noted in Section III A in LiF(100) the discrepancy could be due to a modification of α at the surface and the author suspects this might also be the case for Ni(100) and Cu(100).

On the other hand the radial force constant in the surface plane $\beta_\parallel$ appears to be significantly affected on all metal surfaces with the exception of the nearly free electron metal aluminum. Recent data on Pb(110) seem to bear this out but has not yet been fully analyzed [125]. The softening of $\beta_\parallel$ appears to be quite strong on all the noble metals as well as on the transition metals with the possible exception of Ni (see Table II). The only transition metal with half-filled d-shells studied, W(100), shows a dramatic temperature dependent anomaly which could be shown experimentally to be responsible for the reconstruction. In this case it appears that both $\alpha_{\parallel 2}$ and $\beta_{\parallel 2}$ are affected. Recently experiments have become available on Fe(110) which appear also to show an anomaly but have not been fully analyzed [126].

The longitudinal anomaly has been attributed to a reduction of sp-d hybridization at the surface [87], but could also be explained by a simple model in which the electrons are represented by pseudocharges [90]. This latter model still has to be tested for nearly free electron systems like Al. In this connection it is worth noting that the lack of substantial changes of force constants in Al is consistent with the sp-d hybridization explanation. In the case of Ni(100), Pt(111), Si(111) (2x1) and to a certain extent on W(100) it was observed that adsorbed H atoms restored the forces at the surface to those of an undisturbed truncation. At present it is not clear which of the two models is preferentially favored by this remarkable effect.

Another surprising anomaly is that found in the layered compounds notably GaSe and $TaSe_2$. Despite the weak van der Waals bonding between layers the observed phonon dispersion curves in GaSe show significant deviations from those expected for the bulk. The wavevector of the charge density wave phonon anomaly in TaSe is significantly shifted away from the bulk value. These results are not yet completely understood.

Thus we have seen that the many observed surface phonon anomalies have provided us with a new look at forces in crystal lattices. In the bulk the higher three dimensional symmetry apparently hides many of the details of the competing forces by a fortuitous cancellation of components. At the two dimensional surface, however, the electron distributions are sufficiently altered to change the balance of repulsive and attractive forces. Moreover we have demonstrated, in the case of hydrogen, that with a controlled addition of adsorbates we can study their effect on surface interaction forces.

Acknowledgements:

The author thanks G. Benedek, K.H. Bohnen, V. Bortolani, A. Eguiluz, H.-J. Ernst, D.L. Mills, G. Santoro and Ch. Wöll for many fruitful discussions concerning forces at surfaces. He is grateful to R. Berndt, V. Bortolani, H.-J. Ernst, E. Hulpke, Ch. Wöll and G. Zhang for permission to present unpublished data and to F.W. de Wette, H.-J. Ernst, J. Hinch, D.-M. Smilgies and Ch. Wöll for reading and commenting on the manuscript.

List of References

[1] Lord Rayleigh, Proc. London Math. Soc. 17, 4 (1887)
[2] M. Born and Th. von Karman, Phys. Zeit. 13, 297 (1912)
[3] I.M. Lifshitz and L.M. Rosenzweig, Zh. Eksp. Teor. Fiz. 18, 1012 (1948)
[4] D.C. Gazis, R. Herman and R.F. Wallis, Phys. Rev. 119, 533 (1960)
[5] R.E. Allen, G.P. Alldredge and F.W. de Wette, Phys. Rev. B4, 1648 (1971); B4, 1661 (1971); G.P. Alldredge, R.E. Allen and F.W. de Wette, Phys. Rev. B4, 1682 (1971)
[6] G. Bendedek, Surf. Sci. 61, 603 (1976)
[7] B.N. Brockhouse and A.T. Stewart, Phys. Rev. 100, 756 (1955)
[8] K.H. Rieder and E.M. Hörl, Phys. Rev. Lett. 20, 209 (1968); K.H. Rieder and W. Drexel, Phys. Rev. Lett. 34, 148 (1975)
[9] J. Aldag and R.M. Stern, Phys. Rev. Lett. 14, 857 (1965); R.F. Wallis and A.A. Maradudin, Phys. Rev. 148, 962 (1966); D.L. Huber, Phys. Rev. 153, 772 (1967)
[10] F.M. Probst and T.C. Piper, J. Vac. Sci. Technol. 4, 53 (1967); H. Ibach, Phys. Rev. Lett. 24, 1416 (1970); 27, 253 (1971); J. Vac. Sci. Technol. 9, 713 (1971)
[11] C.H. Li, S.Y. Tong and D.L. Mills, Phys. Rev. B21, 3057 (1980)
[12] S. Lehwald, J.W. Szeftel, H. Ibach, T.S. Rahman and D.L. Mills, Phys. Rev. Lett. 50, 518 (1983); J.M. Szeftel, S. Lehwald, H. Ibach, T.S. Rahman, J.E. Black and D.L. Mills, Phys.Rev. Lett. 51, 268 (1983)
[13] N. Cabrera, V. Celli and R. Manson, Phys. Rev. Lett. 22, 346 (1969)
[14] R. Manson and V. Celli, Surf. Sci. 24, 495 (1971)
[15] S.S. Fisher and J.R. Bledsoe, J. Vac. Sci. Technol. 9, 814 (1971); Surf. Sci. 46, 129 (1974)
[16] R.E. Subbarao, D.R. Miller, J. Vac. Sci. Technol. 9, 808 (1972); J.M. Horne and D.R. Miller, Phys. Rev. Lett. 41, 511 (1978)
[17] B.E. Mason and B.R. Williams, J. Chem. Phys. 55, 3220 (1971); 61, 2765 (1974)
[18] G. Brusdeylins, R.B. Doak and J.P. Toennies, Phys. Rev. Lett. 44, 1417 (1980); G. Brusdeylins, R.B. Doak and J.P. Toennies, Phys. Rev. Lett. 46, 437 (1981)

[19] R.B. Doak, U. Harten and J.P. Toennies, Phys. Rev. Lett. 51, 578 (1983)
[20] See for example K.M. Ho, C.L. Fu, B.N. Harmon, Phys. Rev. B29, 575 (1984) and "Dynamical Properties of Solids", G.K. Horton and A.A. Maradudin (Eds.), Vol. 3, North Holland, Amsterdam 1980.
[21] K.M. Ho and K.P. Bohnen, Phys. Rev. B32, 3446 (1985); Phys. Rev. Lett. 56, 934 (1986)
[22] V. Bortolani, A. Franchini and G. Santoro in "Electronic Structure, Dynamics and Quantum Structural Properties of Condensed Matter", J.T. Devreese and P. van Lamp (Eds.), Plenum Press, New York 1984, p. 401.
[23] A. Franchini, Dissertation, University of Modena, 1985/86
[24] J. deLaunay in "Solid State Physics", F. Seitz and D. Turnball (Eds.), Academic Press, New York 1956, p. 219.
[25] R. Shuttelworth, Proc. Phys. Soc. London A63, 444 (1950)
[26] C. Herring in "The Physics of Powder Metallurgy", W.E. Kingston (Ed.), McGraw-Hill, New York 1951, p. 162 ff
[27] J.E. Müller, M. Wuttig and H. Ibach, Phys. Rev. Lett. 56, 1583 (1986)
[28] B.G. Dick Jr., and A.W. Overhauser, Phys. Rev. 112, 90 (1958)
[29] A.D.B. Woods, W. Cochran and B.N. Brockhouse, Phys. Rev. 119, 980 (1960)
[30] R.J. Needs, Phys. Rev. Lett. 58, 53 (1987)
[31] T.S. Chen, F.W. de Wette and G.P. Alldredge, Phys. Rev. B15, 1167 (1977)
[32] W. Weber, Phys. Rev. B15, 4789 (1977)
[33] P.N. Keating, Phys. Rev. 145, 637 (1966)
[34] A. Fleszar and R. Resta, Phys. Rev. B34, 7140 (1986)
[35] U. Harten, J.P. Toennies, Ch. Wöll, L. Miglio, P. Ruggerone, L. Colombo and G. Benedek, submitted to Phys. Rev. B
[36] G. Benedek in "Collected Excitation in Solids", B. di Bartolo (Ed.), Plenum Press, New York 1983, p. 523
[37] G. Benedek in Springer Series in Chemical Physics, Vol. 21, G. Benedek and U. Valbusa (Eds.), Springer Verlag, Berlin 1982, p. 227.
[38] Ch. Wöll, Dissertation, University of Göttingen, 1987
[39] R. Berndt, Diplom thesis, University of Göttingen, 1987
[40] V. Bortolani, A. Franchini and G. Santoro, private communication
[41] J.P. Toennies in "Surface Phonons", W. Kress (Ed.), Springer Verlag Berlin 1988, to be published
[42] H. Ibach and D.L. Mills "Electron Energy Loss Spectroscopy and Surface Vibrations", Academic Press, New York 1982
[43] G. Lilienkamp and J.P. Toennies, J. Chem. Phys. 78, 5210 (1983)
[44] M. Rocca, Berichte der Kernforschungsanlage Jülich Nr. 2000, 1985. See also M. Rocca, H. Ibach, S. Lehwald and T.S. Rahman in "Structure and Dynamics of Surfaces I", W. Schommers and P. von Blanckenhagen (Eds.), Springer Verlag Berlin 198., p. 245
[45] J.P. Toennies and K. Winkelmann, J. Chem. Phys. 66, 3965 (1977); G. Brusdeylins, H.-D. Meyer, J.P. Toennies and K. Winkelmann, Progress in Astronautics and Aeronautics 51, J.L. Poetter (Ed.), AIAA New York 1977, p. 1047
[46] H. Ibach, J. Vac. Sci. Technol. A5, 419 (1987)
[47] R. Berndt, J.P. Toennies and Ch. Wöll, J. Electron. Spectrosc., 44, 183 (1987)
[48] B.M. Hall and D.L. Mills, Phys. Rev. B34, 8318 (1986)
[49] V. Bortolani, A. Franchini, G. Santoro, J.P. Toennies, Ch. Wöll and G. Zhang, to be published
[50] D. Eichenauer, U. Harten, J.P. Toennies and V. Celli, J. Chem. Phys. 86, 3693 (1987)

[51] V. Celli, D. Eichenauer, A. Kaufhold and J.P.Toennies, J. Chem. Phys. 83, 2504 (1985)
[52] I. Estermann, R. Frisch and O. Stern, Z. Physik 73, 348 (1932)
[53] G. Benedek, G. Brusdeylins, R.B. Doak, J.G. Skofronick and J.P. Toennies, Phys. Rev. B28, 2104 (1983)
[54] U. Schröder, Solid State Commun. 4, 347 (1966); V. Nüsslein and U. Schröder, Phys. Stat. Sol. 21, 309 (1967)
[55] W. Kress, F.W. de Wette, A.D. Kulkarni and U. Schröder, Phys. Rev. B35, 5783 (1987)
[56] W. Kress, F.W. de Wette, A.D. Kulkarni and U. Schröder, Phys. Rev. B35, 2467 (1987)
[57] G. Brusdeylins, R. Rechsteiner, J.G. Skofronick, J.P. Toennies, G. Benedek and L. Miglio, Phys. Rev. Lett. 54, 466 (1985)
[58] G. Bracco, R. Tatarek, S. Terreni and F. Tommasini, Phys. Rev. B34, 9045 (1986); G. Bracco, M. D'Aranzo, C. Salvo, R. Tatarek, S. Terreni and F. Tommasini, Surf. Sci. 189/190, 684 (1987)
[59] G. Benedek, G.P. Brivio, L. Miglio and V.R. Velasco, Phys. Rev. B26, 497 (1982)
[60] G. Benedek, L. Miglio, G. Brusdeylins, J.G. Skofronick and J.P. Toennies, Phys. Rev. B35, 6593 (1987)
[61] A. Lucas, J. Chem. Phys. 48, 3156 (1968)
[62] Thiry et al., J. Electron Spectroscopy, to be published
[63] U. Harten, J.P. Toennies and Ch. Wöll, Phys. Rev. Lett. 57, 2947 (1986)
[64] U. Harten and J.P. Toennies, Europhys. Lett. 4, 833 (1987)
[65] G. Brusdeylins, R. Rechsteiner, J.G. Skofronick, J.P. Toennies, G. Benedek and L. Milio, Phys. Rev. B34, 902 (1986)
[66] G. Benedek and L. Miglio, J.G. Skofronick, G. Brusdeylins, C. Heimlich and J.P. Toennies, J. Vac. Sci. Technol. A5, 1093 (1987)
[67] H. Ibach, Phys. Rev. Lett. 27, 253 (1971)
[68] J.A. Schaefer, J. Anderson and G.J. Lapeyre, J. Electron Spectroscopy, 44, 373 (1987)
[69] O.L. Allerhand, D.C. Allan and E.J. Mele, Phys. Rev. Lett. 55, 2700 (1985)
[70] O.L. Allerhand and E.J. Mele, Phys. Rev. Lett. 59, 657 (1987)
[71] W. Kohn, Phys. Rev. Lett. 2, 393 (1959); E.J. Woll Jr. and W. Kohn, Phys. Rev. 126, 1693 (1962)
[72] E. Tosatti in Festkörperprobleme XV, 113 (1975)
[73] U. Harten, J.P. Toennies, Ch. Wöll and G. Zhang, Phys. Rev. Lett. 55, 2308 (1985)
[74] H. Froitzheim, H. Lammering and H.-L. Günter, Phys. Rev. B27, 2278 (1983)
[75] R. Smoluchowski, Phys. Rev. 60, 661 (1941)
[76] K.M. Ho and K.P. Bohnen, Phys. Rev. B32, 3446 (1985)
[77] A.G. Eguiluz, to be published
[78] K.M. Ho and K.P. Bohnen, Phys. Rev. Lett. 56, 934 (1986)
[79] J.P. Toennies and Ch. Wöll, Phys. Rev. B36, 4475 (1987-I)
[80] A. Lock, J.P. Toennies, Ch. Wöll, V. Bortolani, A. Franchini and G. Santoro, to be published
[81] A. Lock, Diplom thesis, University of Göttingen, 1986
[82] U. Harten, A.M. Lahee, J.P. Toennies and Ch. Wöll, Phys. Rev. Lett. 54, 2629 (1985)
[83] M. El-Batanouny, S.Burdick, K.M. Martini and P. Stanciaff, Phys. Rev. Lett., to be published
[84] U. Harten, J.P. Toennies and Ch. Wöll, Faraday Discuss. Chem. Soc. 80, 137 (1985)
[85] G. Armand, Solid State Commun. 48, 261 (1983)
[86] J.E. Black, F.C. Shanes and R.F. Wallis, Surf. Sci. 133, 199 (1983)

[87] V. Bortolani, A. Franchini, F. Nizzoli and G. Santoro, Phys. Rev. Lett. 52, 429 (1984)
[88] L. Kesmodel et al., as communicated to the author by D.L. Mills
[89] D.L. Mills et al., to be published
[90] C.S. Jayanthi, H. Bilz, W. Kress and G. Benedek, Phys. Rev. Lett. 59, 795 (1987)
[91] A. Euceda, D.M. Bylander and L. Kleinman, Phys. Rev. B28, 528 (1983)
[92] V. Bortolani and G. Santoro, U. Harten and J.P. Toennies, Surf. Sci. 148, 82 (1984)
[93] G. Santoro, A. Franchini, V. Bortolani, U. Harten, J.P. Toennies and Ch. Wöll, Surf. Sci. 183, 180 (1987)
[94] M. Wuttig, R. Franchy and H. Ibach, Solid State Commun. 57, 445 (1986)
[95] K.-L. Kesmodel, M.L. Xu and S.Y. Tong, Phys,. Rev. B34, 2010 (1986)
[96] B.F. Mason and B.R. Williams, J. Chem. Phys. 75, 2199 (1981), Phys. Rev. Lett. 46, 1138 (1981)
[97] B.F. Mason, K. McGreer and B.R. Williams, Surf. Sci. 130, 282 (1983)
[98] J.W.M. Frenken, A.M. Lahee, J.P. Toennies and Ch. Wöll, to be published
[99] K.M. Ho and K.P. Bohnen, Europhys. Lett. 4, 345 (1987)
[100] G. Bracco, R. Tatarek, F. Tommasini, U. Linke and M. Persson, Phys. Rev. 36, 2928 (1987)
[101] V. Bortolani, A. Franchini and G. Santoro, private communication
[102] S. Lehwald, J.M. Szeftel, H. Ibach, T.S. Rahman, D.L. Mills, Phys. Rev. Lett. 50, 518 (1983)
[103] M.-L. Xu, B.M. Hall, S.Y. Tong, M. Rocca, H. Ibach, S. Lehwald and J.E. Black, Phys. Rev. Lett. 54, 1171 (1985)
[104] J.W.M. Frenken, J.F. van der Veen and G. Allen, Phys. Rev. Lett. 51, 1876 (1983)
[105] A.M. Lahee, J.P. Toennies and Ch. Wöll, Surf. Sci. 191, 529 (1987)
[106] J.W.M. Frenken and Ch. Wöll, unpublished
[107] V. Bortolani, A. Franchini and G. Santoro, private communication
[108] K. Kern, R. David, R.L. Palmer, G. Comsa and T.S. Rahman, Phys. Rev. Lett. B33, 4334 (1986)
[109] D. Neuhaus, F. Joo and B. Feuerbacher, Surf. Sci. Lett. 165, L90 (1986)
[110] V. Bortolani, A. Franchini and G. Santoro, Surf. Sci. 189/190, 675 (1987)
[111] G. Benedek, M. Miura, W.Kress and H. Bilz, Phys. Rev. Lett. 52, 1907 (1984)
[112] D.A. King, Physica Scripta T4, 34 (1983)
[113] E. Tosatti and P.W. Anderson, Solid State Commun. 14, 773 (1974)
[114] J.C. Campuzano, D.A. King, C. Somerton and J.E. Inglesfield, Phys. Rev. Lett. 45, 1649 (1980)
[115] M.I. Holmes and T. Gustafsson, Phys. Rev. Lett. 47, 443 (1981)
[116] C.L. Fu, A.J. Freeman, E. Wimmer and M. Weinert, Phys. Rev. Lett. 54, 2261 (1985)
[117] D. Singh, S.H. Wei and H. Krakauer, Phys. Rev. Lett. 57, 3292 (1986)
[118] X.W. Wang and W. Weber, Phys. Rev. Lett. 58, 1452 (1987)
[119] H.-J. Ernst, E. Hulpke and J.P. Toennies, Phys. Rev. Lett. 58, 1941 (1987)
[120] H.-J. Ernst and E. Hulpke, J. Vac. Sci. Technol. A5, 460 (1987)
[121] H.-J. Ernst and E. Hulpke, unpublished data
[122] A. Kulkarni, F.W. de Wette and J. Prade, Phys. Rev. B36, 6190 (1987)

[123] E. Tosatti, Solid State Commun. 25, 637 (1978); A. Fasolino, G. Santoro and E. Tosatti, Phys. Rev. Lett. 44, 1684 (1980)
[124] R.L. Leigh, B. Szigetti and V.K. Tewary, Proc. Roy. Soc. A320, 505 (1971)
[125] J.W.M. Frenken and Ch. Wöll, unpublished data
[126] Ch. Wöll and G. Zhang, unpublished data

Discussion

L. KLEINMAN:

(1) I believe you misspoke and said Ho and Bohnen made a frozen phonon calculation, which is not possible for surface phonons because you don't know the displacements of the atoms. I think they made an *ab initio* interplanar force constant calculation.

(2) You said the s-d hybridization at transition metal surfaces is decreased. Actually it is increased because of the lower symmetry.

(3) You mentioned our projected Cu(111) charges. Although these are correct as far as I know, they should be taken with a small grain of salt. The projected d surface charge, for example, represents the sum of the squares of the coefficients of all the d basis functions centered on surface atoms after these basis functions have been symmetrically orthogonalized to all the other basis functions. Because these orthogonalized functions can be very long range, this only approximates the actual d-like charge on the surface.

J. P. TOENNIES: With regard to (1), I agree with you concerning the nomenclature. What I meant to imply is that their calculations are essentially equivalent to a "frozen phonon" calculation at the zone boundary.

Concerning (2), I wish to repeat that Bortolani and coworkers feel that the decrease in the pseudopotential operator ΔV coming from the reduction in the number of nearest neighbors and the reduction in the sp wave function by spill-out will override the symmetry effect. Thank you for calling our attention to the proper interpretation of your calculations. It certainly would be very desirable if you could repeat your calculations to obtain the additional electronic charge in the surface plane as accurately as possible since this is needed in the theory of Jayanthi *et al.* referred to in my talk.

J.E. DEMUTH: In regards to interpreting unusual surface phonon bands in your results, can one rule out possible defects on the surface? For example, the dispersionless band seen on Si(111)2×1 may also be a result of vibrations of the chain defects seen in STM. Also, could the unusually "softened" Rayleigh modes on the noble and transition metals be a result of native random point defects (i.e., vacancies) on these surface formed by the way we prepare them.

J. P. TOENNIES: We considered the possibility that defects could be affecting our results. However, essentially because of the very good reproducibility from one cleave to another over about 10-20 cleaves, we do not feel that defects were of major importance. With regard to the metals I wish to point out that our results on Ag(111) and Pt(111) have since been reproduced by several groups in different laboratories. Also, I would like to call your attention to our ability to measure the concentration of single steps and point defects by observing the incoherent diffraction of He atoms (M. Lahee, J. R. Manson, J.P. Toennies and Ch. Wöll, *Phys. Rev. Lett.* **57**, 471 (1986)). These and other techniques indicate that our Pt(111) surface had undisturbed regions of the order of 500Å or greater.

D. A. KING: I have a question and a comment. My question is: you compare your data from W(100) at high temperatures with calculations from the group of de Wette *et al.*, which presumably refer to an *ordered* (1×1) surface lattice; since you see appreciable dispersion, also shown in the calculations, do you conclude that the high temperature phase is an ordered surface structure?

My comment is that we have recently looked at ion channeling from a "flat" W(100) surface through the phase transition, and find that the number of surface atoms displaced from bulk sites per string is unchanged through the transition. The implication is either that the high temperature phase is disordered, with random $< 11 >$ displacement vectors; or the high temperature phase is an ordered structure involving a uniform $< 10 >$ displacement of the surface lattice with respect to the second layer.

J. P. TOENNIES: In reply to your question we do indeed conclude that our surface is *ordered* in the sense that we were able to observe sharp (1×1) diffraction peaks over the entire range of temperatures from 220K to 1000K. Also, since the single phonon process is a coherent diffraction-like process, the observation of sharp inelastic peaks is consistent with this observation. Concerning your comment, we like the explanation of a uniform displacement of the surface lattice. A careful lattice dynamical calculation for this geometry would be of interest in this connection.

R. F. WILLIS: A speculative comment/question!

On the curious W(100) results at 1000K: Although you have shown that electron scattering from ion cores dynamics basically agrees with the atom scattering from the associated charge density dynamics, is it possible that in materials with very strong electron-phonon coupling (the charge density wave being extreme) you could be seeing additional effects other than simple–ion lattice dynamics? For example, you have already demonstrated that "charged shell" models involving charge rearrangement in simple metals can explain the more "simple anomalies."

J.P. TOENNIES: So far, in all systems studied to date, the results appear to indicate that we are only exciting the phonons at the surface. The only indication for electron-phonon coupling is in Pt(111) where we have observed two Kohn anomalies [U. Harten, J.P. Toennies, Ch. Wöll, and G. Zhang, *Phys. Rev. Lett.* **55**, 2308 (1985)], in Si(111) 2×1 according to the theory of Allerhand and Mele [*Phys. Rev. Lett.* **59**, 657 (1987)], and in 2H–$TaSe_2$. In the latter system the Kohn-like anomaly is strongly temperature-dependent [G. Benedek, G. Brusdeylins, C. Heimlich, L. Miglio, J.G. Skofronick, J.P. Toennies, and R. Vollmer, *Phys. Rev. Lett* **60**, 1037 (1988)]. Otherwise there is no indication for a coupling to electrons. In the shell model theories the charged shells have no mass and therefore cannot have any kinetic energy; they are only introduced to mediate the interactions.

J.B. PENDRY: It is not published yet, but we have done a structure determination of the high–temperature phase with the Erlangen group and there is definite and strong evidence that the high–temperature phase does consist of disordered displacements. I don't think this needs to contradict your measurements at all, but what it *does* mean is that if you are trying to identify the mode which *you* see as a mode which goes soft in the phase transition, the disordered model would not agree with that. But it's perfectly possible for your mode to be some other sort of mode which isn't affected so strongly by the disorder.

E. TOSATTI: The surface phonon temperature behavior you show for W(001) agrees very nicely with the theoretical results of our molecular dynamics simulations [C.Z. Wang, A. Fasolino and E. Tosatti, *Phys. Rev. B* (January 1988)], based on the same effective Hamiltonian used to explain the $T = 0$ properties [A. Fasolino and E. Tosatti, *Phys. Rev.* **B35**, 4264 (1987)] and the phase transition static properties at $T \neq 0$ [C.Z. Wang, A. Fasolino and E. Tosatti, *to be published.*]

Upon heating, the zone-boundary M-point (x,y) surface phonons undergo partial softening, then broadening around T_c, then again hardening much above T_c, very much like in a standard bulk structural, or ferroelectric phase transition. This lends us confidence that our overall understanding of the $c(2 \times 2) \rightarrow 1 \times 1$ transition of W(001) is sound. The transition occurs with the surface displacement magnitude remaining essentially constant, but with the relative phase angle of displacements of different atoms developing first long-range disorder, and then progressively short-range disorder. This picture also seems to agree with that suggested by John Pendry's remarks on structural analysis.

J.P. TOENNIES: With the exception of the question of zone boundary peak widths (see my question to Tosatti) the overall agreement between theoretical and experimental phonon frequencies and widths is very good. However, the shift in the location of the 1/2 order diffraction peaks with temperature seen only in the scattering experiments still needs to be explained.

C. B. DUKE: The surface force constants for GaAs(110) are outputs of a calculation of the surface atomic geometry by C. Mailhiot *et al., Surf.*

Sci. **149**, 366 (1985). This calculation suggests the identification of the flat "optical" modes which you observe as vibrations of the top-layer Ga-As chains relative to the substrate. It also suggests the existence of a soft mode associated with bond–length conserving rotational top-layer displacements oriented normal to the chain direction. Use of the predicted force constants in a lattice dynamics calculation should lead to a detailed description of the new surface modes associated with the changes in effective force constants caused by the electronic rehybridization effects at zincblende (110) surfaces. The scaling of these surface phonon spectra from one material to another also should be of interest, given the linear scaling of the equilibrium relaxed geometries with bulk lattice constant. Another interesting topic would be the comparison of the surface phonon spectra for zincblende *versus* wurtzite cleavage faces of a material like ZnSe which occurs in both crystal structures.

J.P. TOENNIES: We are grateful to you for calling our attention to the work of Mailhiot and the suggestion concerning ZnSe.

M. CARDONA: I notice the lack of reference in your talk to the Fuchs and Kliewer type of electrostatic surface modes. Can one see them with your type of spectroscopy? These should be at $\sim$ 32 meV for GaAs but may be lowered to $\sim$ 20 meV for InSb.

J. P. TOENNIES: Thank you for this interesting suggestion. For reasons of resolution we used beam energies of 20-30 meV in these experiments so that the Fuchs-Kliewer modes would have only very small annihilation cross sections. At any rate in this region of energy losses nothing was observed.

Self-Consistent Theory of Surface Phonon Dispersion Curves at the (110) Surface of Aluminum

A.G. Eguiluz [1], *A.A. Maradudin* [2], *and R.F. Wallis* [2]

[1] Department of Physics, Montana State University, Bozeman, MT 59717, USA

[2] Department of Physics, University of California, Irvine, CA 92717, USA

1. Introduction

Considerable progress has been made in recent years in the experimental determination of surface phonon dispersion curves for metals by inelastic helium atom scattering [1] and by electron energy loss spectroscopy [2]. Relatively little progress has been made, however, in the first-principles calculation of surface phonon dispersion curves for metals. In the pseudopotential perturbation-theoretic calculations of CALANDRA et al.[3] for alkali metal surfaces, the ground state and response properties of the electronic subsystem were calculated using the infinite-barrier model (IBM) for the electron wave functions. Surface phonon dispersion curves were then calculated from the response function. Although this procedure is relatively simple, it does not provide a self-consistent treatment of the metal surface. In a different approach HO and BOHNEN [4] used the local density approximation (LDA) to density functional theory and the frozen-phonon method to calculate the surface phonon frequencies for Al(110) at a few high symmetry points on the boundary of the surface Brillouin zone. They then employed a force constant model to interpolate between these points and the origin and thus obtain surface phonon dispersion curves. Clearly, such a calculation is a hybrid and not a bona fide first-principles calculation of surface phonon dispersion curves.

2. Theoretical Development

In the present work we calculate surface phonon dispersion curves for Al(110) using a self-consistent pseudopotential perturbation-theoretic approach. The one-electron states of jellium slabs with thicknesses up to that corresponding to 27 atomic layers were calculated by the self-consistent solution of the Kohn-Sham equations including exchange and correlation in the LDA. These states were then used to calculate the response function for non-interacting electrons $\chi^{(0)}$. The response function χ for interacting electrons was then calculated by solving the integral equation

$$\chi = \chi^{(0)} + \chi^{(0)} V\chi \tag{1}$$

where V contains both Coulomb and exchange-correlation contributions. Defining $V_1(\vec{x})$ to be the difference between the pseudopotential of the ions and the jellium potential, we obtain the ground state energy E_{GS} of the conduction electrons to second order in $V_1(\vec{x})$,

$$E_{GS} = E_{GS}^{(0)} + \int d^3x n_o(\vec{x}) V_1(\vec{x}) + \frac{1}{2} \int d^3x \int d^3x' V_1(\vec{x}) \chi(\vec{x}\vec{x}') V_1(\vec{x}'), \tag{2}$$

Springer Series in Surface Sciences Vol. 14
Solvay Conference on Surface Science
Editor: F.W. de Wette

where $n_o(\vec{x})$ is the zero-order electron number density. Differentiating Eq. (2) twice with respect to the displacements of the ions from their equilibrium positions gives the electronic contribution to the dynamical matrix. The ion-ion interaction contribution is obtained by standard Ewald summation techniques. Diagonalization of the dynamical matrix then yields the eigenvectors and frequencies of both surface and bulk phonon modes.

The pseudopotential was taken to have the local Heine-Abarenkov form whose two parameters were determined by requiring that the total bulk energy be a minimum at the experimental value of the lattice constant and that the square of the bulk phonon frequency averaged over the Brillouin zone agree with the value extracted from neutron scattering data [5,6]. The bulk phonon dispersion curves calculated with this pseudopotential are in excellent agreement with the neutron data.

3. Results

In Fig. 1 we show the results of a calculation of the phonon spectrum for a 27-layer ideal, unrelaxed slab bounded by a pair of (110) surfaces. (The zone-boundary frequencies are independent of the slab thickness for slabs thicker than about 23 atomic layers.) We emphasize that the surface calculation has no adjustable parameters; it has been carried out from first-principles, according to the outline given above, with pseudopotential parameters determined from bulk properties only.

The presence of surface modes below and in the gaps of the surface-projected bulk branches is clearly seen in Fig. 1. In fact, the overall nature of the surface phonon spectrum is qualitatively similar to the results obtained by ALLEN et al.[7] in their seminal work on surface phonons for model fcc crystals with Lennard-Jones interatomic interactions.

In Fig. 2 we show the phonon dispersion curves in the presence of multilayer relaxation [8]. The values obtained for the percentage change of the first three interplanar spacings are Δ_{12} = − 5.4, Δ_{23} = +0.8, and Δ_{34} = −2.6, where a negative (positive) value denotes a contraction (expansion). The main effect of the relaxation process is to raise the frequencies of the surface modes.

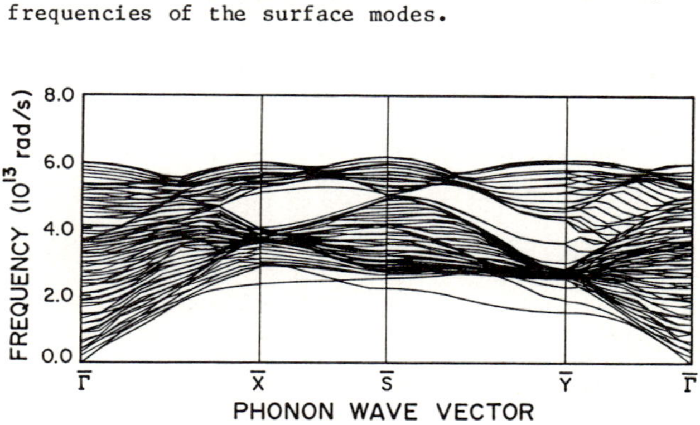

Fig. 1. Calculated surface phonon dispersion relations for an unrelaxed 27-layer aluminum slab with (110) surfaces

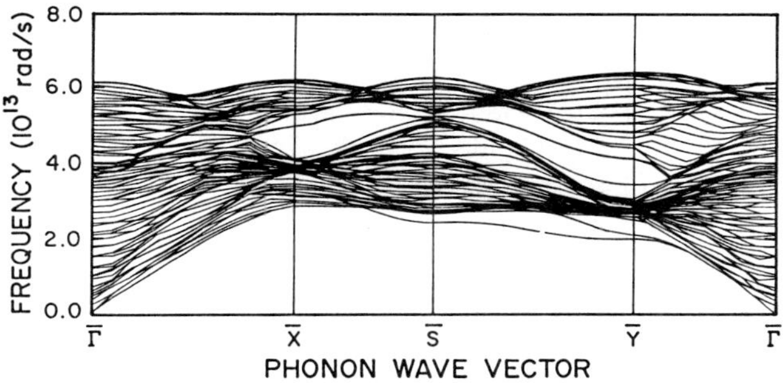

Fig. 2. Calculated surface phonon dispersion relations for a relaxed 27-layer aluminum slab with (110) surfaces

For both the relaxed and unrelaxed crystals we find that the xx- and yy-elements of the dynamical matrix are reduced substantially for the outermost plane in relation to their values for the inner planes, a result favoring the creation of surface modes. For the case of the zz-element, the smallest value is obtained for the second atomic plane. This also favors the creation of surface modes.

In Table I we compare the values of the frequencies of the surface phonons measured by TOENNIES and WÖLL [9] at the $\bar{X}$ and $\bar{Y}$ points in the SBZ, with the corresponding theoretical values obtained in the present work, and by HO and BOHNEN [4]. We have that: (i) at the $\bar{X}$ point both theoretical values are close to the measured surface phonon frequency when the calculation is done for the unrelaxed surface, but not for the relaxed one; (ii) at the $\bar{Y}$ point the same conclusion applies for the lower-frequency mode. On the other hand, our calculated value of the frequency of the higher surface mode agrees very well with experiment, in the presence of relaxation.

Table 1. Comparison of the values of the energies of the surface phonons measured by Toennies and Wöll for Aℓ(110) at the $\bar{X}$ and $\bar{Y}$ points in the SBZ with the theoretical values obtained in this work by Ho and Bohnen. (Energies are in meV.)

	$\bar{X}$		$\bar{Y}$	
	Unrelaxed	Relaxed	Unrelaxed	Relaxed
Present work	15.8	19.0	10.0 12.4	13.0 13.8
Ho and Bohnen [4]	15.3	17.4	8.7 13.2	7.9 14.1
Toennies and Wöll [9]		14.6		8.9 13.5

The discrepancy between theory and experiment may be due to a number of factors, including among others the neglect of third and higher order terms in the expansion of E_{GS} in powers of the pseudopotential, inadequacy of the local density approximation and local exchange and correlation in the surface region, and experimental error.

4. Conclusion

We have presented a calculation of surface phonon dispersion for Al(110) whose salient features are: (i) the underlying bulk calculation gives dispersion curves which agree very well with experiment along symmetry lines; (ii) the surface calculation has no adjustable parameters; (iii) both relaxed and unrelaxed geometries are treated on the same footing; and (iv) the full set of interatomic force constants for the slab has been explicitly obtained in the evaluation of the dynamical matrix.

Acknowledgement

The work of A.G.E. was supported in part by MONTS under Grant No. 16210882, and by the National Science foundation under Grant No. DMR 86-03820. The work of A.A.M. and R.F.W. was supported by NSF Grant No. DMR 85-17634. The work of all authors was supported by the San Diego Supercomputer Center. We wish to thank Professor J. P. Toennies for communicating the experimental results contained in Ref. 9 prior to publication.

References

1. R. B. Doak, U. Harten, and J. P. Toennies, Phys. Rev. Lett. 51, 578 (1983)
2. J. M. Szeftel, S. Lehwald, H. Ibach, T. S. Rahman, J. E. Black and D. L. Mills, Phys. Rev. Lett. 50, 518 (1983)
3. C. Beatrice and C. Calandra, Phys. Rev. B10, 6130 (1983); C. Calandra, A. Catellani and C. Beatrice, Surf. Sci. 148, 90 (1984)
4. K. M. Ho and K. P. Bohnen, Phys. Rev. Lett. 56, 934 (1986)
5. G. Gilat and R. M. Nicklow, Phys. Rev. 143, 487 (1966)
6. D. C. Wallace, Phys. Rev. 187, 991 (1969); ibid B1, 3963 (1970)
7. R. E. Allen, G. P. Alldredge and F. W. DeWette, Phys. Rev. B4, 1661 (1971)
8. A. G. Eguiluz, Phys. Rev. B35, 5473 (1987)
9. J. P. Toennies and Ch. Wöll, Phys. Rev. B36, 4475 (1987)

Discussion

D. R. HAMANN: In your approach, using the fixed jellium electron density as a zero-order reference, should create the following problem: If all the slab atoms are moved uniformly perpendicular to the slab, the first-order term in your energy expression should create a restoring force, which is non-physical. Is there some subtlety in the formalism which cancels this effect?

R. F. WALLIS: The total energy of the crystal consists of the ground state energy of the conduction electrons given by Eq.(2), the energy of the neutral

Madelung subsystem, and the energy of interaction between the Madelung subsystem and the positive uniform background of the electronic subsystem (also a neutral subsystem). This last energy plays a crucial role in eliminating the non–physical restoring force to which you make reference. This is discussed in detail in Sections 2 and 3 of Reference 8. As a consequence, the force constants that we calculate satisfy very well the condition of infinitesimal translation invariance perpendicular to the slab.

S. A. RICE: Our experience has been that the simple Heine-Abarenkov pseudopotential often leads to quantitative errors in the prediction of the compressibility and heat of sublimation of simple metals; the former error influences the accuracy of the calculated phonon dispersion relations. It is found that a simple semi-empirical modification to the pseudopotential—addition of a simple two parameter function of r_s (the radius of a sphere occupied by one electron at the system electron density) with the two parameters fit to the zero frequency compressibility and the heat of sublimation—leads to very good agreement between calculated and observed bulk phonon dispersion relations. Your suggestion for improving the approximation you have used goes in the same direction but is likely very tedious. Have you considered using the simple semi-empirical correction I mention, or one like it? I suspect its use would more easily lead to accurate predictions of the surface phonon dispersion relations.

R. F. WALLIS: Our calculations using the Heine-Abarenkov pseudopotential yield good results for the phonon dispersion curves and elastic constants of bulk aluminum. I am not familiar with your procedure and would like to learn the details.

W. PLUMMER: My comment is a general one concerning the validity of the various theoretical models in the region of the surface, where there is a large charge variation. First, what proof is there that the Lang-Kohn charge density is correct? Secondly, what proof is there that LDA works in the surface region? And finally, what proof is there that RPA works in the surface region? As an example we now know that if an angle-resolved photoemission band mapping measurement agrees with theory based on LDA then there is something wrong with the theory. It is possible that the same thing will happen with phonon calculations, i.e., the better the theory the worse the agreement with experiment.

R. F. WALLIS: It would certainly be nice to solve the problem under consideration exactly. Unfortunately, that is not possible at the present time. One either does nothing or one uses approximations. We have chosen the latter course. The density functional procedure in the local density approximation has given valuable results concerning the surface physics of metals. The random phase approximation is a standard method for handling electron correlation effects. Time will tell how appropriate these procedures are for the surface phonon problem.

W. KOHN: Dr. Plummer expressed understandable unease about the use of local density functional theory (LDA) for surfaces, where the density varies rapidly. I offer three comments.

1. Several workers (Langreth and others) have gone beyond the LDA and concluded that corrections are fairly small.

2. Krotschek, Qian and Kohn calculated surface energies by the hypernetted chain theory, in which slow variation of the density is *not* assumed, and found semiquantitative agreement with LDA.

3. For surface vibrations, errors due to rapid density variation perpendicular to the surface would be expected to drop out in large measure.

R. F. WALLIS: I concur in Professor Kohn's comments.

Carrier Excitation Yield from Atomic Collisions at Semiconductor Surfaces

P.S. Weiss, A. Amirav, P.L. Trevor, and M.J. Cardillo*

AT&T Bell Laboratories, Surface Physics Research,
600 Mountain Avenue, Murray Hill, NJ 07974, USA

*Permanent Address: Chemistry Dept., Tel Aviv Univ., Tel Aviv, Israel

We have furthered our understanding of the collisional excitation of carriers at semiconductor surfaces by hyperthermal neutral rare gas atoms. We have introduced optical carrier excitation in order to characterize the samples, to determine carrier recombination rates, and to calibrate the collisional e^-h^+ excitation probability for Xe on the InP(110) surface.

I. Introduction

Recent experiments in the hyperthermal scattering of rare gas atoms ($1<E(eV)<15$) from single crystal surfaces have revealed a variety of interesting phenomena including the excitation of electron-hole pairs (e^-h^+) [1-3]. In this paper we examine the excitation yields for Xe and Kr for InP(100) and for Xe for an ordered InP(110) surface. We describe an optical excitation experiment designed to characterize these samples and to derive a value of the absolute collision-induced e^-h^+ excitation probability.

II. Experimental

The ultrahigh vacuum molecular beam surface scattering apparatus used in these studies has been described elsewhere [3,4]. A Xe or Kr beam seeded in H_2 was expanded through a heated platinum nozzle to give incident energies of up to 15.5 eV for Xe, and 9.2 eV for Kr. The crystals used were Fe-compensated semi-insulating InP(100) and InP(110). The e^-h^+ creation efficiency was measured by applying a voltage bias across two ohmic contacts on the front surface of the crystal as described in Ref. 1. All experiments discussed here were performed at room temperature.

Carriers were also optically excited using either a HeNe laser at 632 nm or a HeCd laser at 442 nm. The incident laser powers were varied from 1 nW to 10 mW using a series of neutral density filters. The laser was opened to the crystal for a few msec at a repetition rate of less than 1 Hz. The photoconductance was recorded as a complete waveform on a multichannel signal averager. From the steady state number of excess carriers and the generation rate due to the known laser power absorbed, the minority carrier

Springer Series in Surface Sciences Vol. 14
Solvay Conference on Surface Science
Editor: F.W. de Wette

lifetime was determined as a function of excess carrier density [3,5]. These lifetimes were then used to calibrate the collisional e^-h^+ yield as described below.

III. Results and Discussion

The collisional excitation of carriers was measured as a change in the conductivity of a semi-insulating InP(100) or InP(110) crystal as in Ref. 1. This efficiency was studied as a function of incident energy, angle, and atom. The excitation yield was found to increase with increasing incident energy [1-3]. Xe was more efficient than Kr at exciting electron-hole pairs at each incident energy [2,3]. The measured yields could be well fit as being exponential in the inverse of the energy deposited in the collisions (ΔE). We have previously discussed the excitation of e^-h^+ in terms of a local "hot spot" with rapid electronic equilibration due to the energy transfer of the projectile to the surface which could lead to this dependence [1-3]. Assuming the high temperature heat capacity for a solid $C_v = 3nk$ is appropriate, the expected yield (Y) would then be:

$$Y = A \exp\{-3nE_{bg}/\Delta E\} \qquad (1)$$

where $E_{bg} = 1.35$ eV is the InP band gap, and n is the effective number of lattice atoms involved in the electronic equilibration. The effective number of atoms, n=8, was derived from the slope of a Boltzmann plot of the excitation yield vs. ΔE.

In the optical excitation determination of the recombination lifetimes it was assumed that each absorbed laser photon excited an electron-hole pair with unit efficiency. The carrier generation rate was then the measured number of photons absorbed by the crystal per second. The carrier lifetime was determined as a function of laser power, and was defined to be the number of excess carriers excited at steady state (after a few msec) divided by the generation rate, $\tau = \Delta N/G$. Only the lifetime at low excess carrier density (where the lifetime remains nearly constant with increasing carrier density) were considered, because the carrier densities obtained in the collisional excitation experiment were all in this regime.

For an ordered InP(110) surface, both collision-induced e^-h^+ excitations and optically excited carrier lifetimes were measured. For this surface the bulk-averaged minority carrier (hole) lifetime at low laser power was $\tau = 50$ nsec. We assume that the only difference in yield between the two types of excitation is due to the carriers collisionally excited and immediately trapped at the surface. Because of this, the *measured yield* provides a lower limit on the *excitation probability*.

The yield of the carrier excitation per incident Xe atom is defined to be: $Y = G/F_{Xe}$, where G is the e^-h^+ generation rate, and F_{Xe} is the Xe atom flux. At steady state, the generation rate and recombination rate are equal, so that the excess carrier density (ΔN) is determined by the generation rate and the carrier lifetime (τ):

$$\Delta N = G\tau. \qquad (2)$$

From the optical excitation experiment the excess carrier density and generation rate determine the lifetime. We assume this lifetime can be used to derive the collision-induced generation rate. The excess number of carriers has been measured as a change in conductance, which is related by a geometrical factor to the change in conductivity ($\Delta\sigma$). The change in conductivity is:

$$\Delta\sigma = q\mu\Delta N, \qquad (3)$$

where q is the charge of a carrier, and μ is the mobility of a carrier. For μ, the sum of the electron and hole mobilities for InP, 5000 cm^2/V-sec, will be taken. Then, the yield so defined is:

$$Y = \frac{\Delta\sigma}{q\mu\tau F_{Xe}}. \qquad (4)$$

Using the measured values for InP(110), for Xe atoms at an incident energy of 9.8 eV normal to the surface, from equation (4), the experimentally derived yield is $Y = 0.002$. This value is smaller than the order of magnitude estimate of Ref. 1, despite the comparable flux-normalized e^-h^+ yields, because the value of the carrier lifetime in the previous work had to be assumed.

The effective local temperature change from which the electronic equilibration was established was determined by simply dividing the energy deposited by the high temperature heat capacity $\Delta T = \Delta E_{Xe}/3nk$, which for an incident Xe energy of 9.8 eV, and an assumed energy transfer $\Delta E = 8$ eV, gives a temperature rise of 3900K. From the effective number of lattice atoms, $n = 8$, derived from the energy dependence of the InP(100) data, and the calibrated yield at 9.8 eV for InP(110), the value of the preexponential is $A = 0.002 \exp\{3nE_{bg}/8.\} = 0.11$. As the measured excitation yields were approximately the same for the two crystal surfaces at 9.8 eV at normal incidence this should be a valid estimate of the value of the preexponential.

The preexponential, A of equation (1) can be approximated by the intrinsic carrier concentration at this temperature for a volume of 8 atoms, for which the preexponential that would be expected is:

$$A = 2\left(\frac{2\pi\sqrt{m_e^* m_h^*}kT}{h^2}\right)^{3/2} V_n \qquad (5)$$

where m_e^* and m_h^* are the effective masses of the electron and hole, respectively, T is the (effective) temperature, and V_n is the volume taken up by the n atoms involved. Taking $m_e^* = 0.077m_e$ and $m_h^* = 0.8m_e$, [7] $T = 4200K$ from above, and the volume of $n = 8$ atoms for InP ($\rho = 4.787$ g/cm^3) [7] to be $V_8 = 4.635 \times 10^{-21}$ cm^3 gives $A = 0.66$.

The preexponential, A of equation (1) can also be approximated by analogy to the thermodynamic equilibrium responsible for the ionization of atoms, for which the preexponential that would be expected is:

$$A = 2\,\frac{g_i^+}{g_k}\left(\frac{2\pi m_e^* kT}{h^2}\right)^{3/2} V_n \qquad (6)$$

where g_k and g_i^+ are the degeneracies of the ground and ionized states, respectively, and the other terms are as defined above. Taking the same values used above, and assuming 1 for the ratio of the degeneracies gives $A = 0.11$.

The reasonable agreement between these estimates ($A = 0.66$, 0.11) and the value derived from the experiments ($A = 0.11$) serves to show the consistency of a local thermal hot spot description with the experimental results.

IV. Summary

The excitation of carriers at the surface of InP(110) was measured to be comparable to that of InP(100). Using an optical technique to determine effective carrier lifetimes the yield was found to be 0.002 e^-h^+ per incident Xe atom at Xe energies of 9.8 eV at normal incidence on InP(110). Based on the exponential dependence of the yield on ΔE^{-1} and the agreement of the preexponentials derived form experiments and from statistical arguments, we conclude that the excitation occured in a manner consistent with electronic equilibration with the initial local lattice energy in the immediate vicinity of the impact.

Acknowledgements

The authors would like to thank Ami Appelbaum, Vince Lambrecht, Murray Robbins, and Peter Thomas for evaporating contacts on the InP crystals, Ed Chaban for assistance in setting up the experiments, and Greg Blonder, Michel Lannoo, Mark Stiles, and John Tully for helpful discussions.

REFERENCES

1. A. Amirav and M. J. Cardillo, Phys. Rev. Lett. 57, 2299 (1986).
2. P. S. Weiss, A. Amirav, P. L. Trevor, and M. J. Cardillo, J. Vac. Sci. Technol. A, in press.
3. P. S. Weiss, P. L. Trevor, and M. J. Cardillo, in preparation.
4. M. J. Cardillo, C. C. Ching, E. F. Greene, and G. E. Becker, J. Vac. Sci. Technol. 15, 423 (1978).
5. S. M. Sze, Semiconductor Devices, Physics and Technology, Wiley, New York, pg. 47 (1985).
6. A. Amirav, P. L. Trevor, M. J. Cardillo, C. Lim, and J. C. Tully, J. Chem. Phys. 87, 1796 (1987).
7. M. Neuberger, Handbook of Electronic Materials, Volume II: III–V Semiconducting Compounds, IFI/Plenum, New York (1971).

Discussion

C. B. DUKE: In the analysis of your absolute yields how did you treat the balance between bulk and surface recombination? In particular, what are the magnitudes of the optical absorption depth, the electron mean free path, the hole mean free path, the surface recombination velocity and the bulk recombination lifetime?

M.J. CARDILLO: We don't separate the two. The lifetimes are always surface dominated. In both collisional and optical excitation experiments we measure the bulk-averaged change in conductance. The optical experiment gives us an effective lifetime at low laser powers which we think is close to what is appropriate for collisional excited carriers, perhaps missing only those carriers excited at the surface and immediately trapped at the surface.

Recalling that these low carrier density materials are n-type so that we measure the hole properties, some rough numbers for InP(110) are as follows: optical penetration depth $\sim 10^{-4}$cm, surface recombination velocity $> 6 \times 10^{4}$cm/sec, bulk (hole) lifetime $\sim 3\mu$sec (from literature), and hole diffusion length $\sim 30\mu$.

E. TOSATTI: I would like to understand something you didn't have the time to touch upon. Is it correct to say that, when you have surface in different conditions, this conductivity or whatever it is changes very much—is it true to say that, as you change the band bending, you'd be changing the localization of the carrier's wave function underneath the surface—if they're very localized against the surface they will scatter a lot and therefore they will decay very fast; if they extend a lot, then they scatter less? I don't understand when you try to relate to the band bending these enormous changes in mobility or whatever that was. What's the explanation for that?

M.J. CARDILLO: Well we have some experts in the audience on that–perhaps they'd like to give the explanation. The classical model of band bending *does* work to explain this. I'm not sure what you're actually asking me. How that works, in a sense? Or, what's the relation between lifetime and resistance?

E. TOSATTI: Once you have the band bending, what does one do with it? Calculate the resistivity? What is the physics underneath?

M.J. CARDILLO: This is attributable to the change in the surface potential, and from a change in the surface potential there is a simple, integral relationship between that and the excess surface conductivity, which is essentially what we're measuring. So one can just work that out numerically—what the conductivity is, how it changes with the surface potental. That gives you a plot of what the conductivity is — *should* be — *versus* surface potential. And the surface potential is defined as whatever range this effect has. As one example that I know of, Tom Buck worked these out and they're in text books. One works out one of these plots of how the excess surface conductivity changes the surface potential; you know the resistivity of your sample, you watch to see what happens as you make a controlled change on the surface—if you sputter, for example, you know what the conductivity does and you know that you... For example, when you create defects that you've put yourself up on the n-branch so the defects are acting like donors. There's a regular, well–defined procedure for working yourself through that. How you would like to interpret the detailed mechanism is a physics problem which I'm not up to yet.

W. KOHN: Can I ask a question in sightly different terms? You're producing a time–dependent perturbation that causes the electrons to be excited from the valence band into the conduction band. We have a frequency associated with that gap, right? What are the characteristic frequencies of your time-dependent perturbation? They would seem to me to be very very much smaller, and so is there enough, so to speak, in the tail of that frequency distribution to account for what is a fairly sizable effect?

M.J. CARDILLO: Well, we do see electrons excited—so in some sense, Nature has answered that, and all I can do is conjecture about how this might happen, but I don't have much of a basis other than my limited knowledge of chemistry. The fact is, we are compressing this atom rapidly, we are doing something analogous to curve crossing in molecule collisons. Or, it must be the same in solids. And so, we've distorted the local states. When this target atom comes rocketing back, some electrons are going to be left on the states which have crossed, and the question is how many of those are lying sufficiently high to go out as carriers. That's my simple picture of this. The time–dependence must say how much of this will be adiabatic; how much of it will be left, so there's no question that when we distort the bonding scheme with this tremendous amount of energy that is given up to essentially one atom, that the interaction potential looks like curve crossing in the simplest sense. The question is, the time scale must determine how many of them get left up there. What's disconcerting is to see that krypton and xenon give the same result, which suggests that we're not measuring a direct result of that process, we're measuring something afterwards. Perhaps hundreds of femtoseconds afterwards—I don't want to get into the details here, it's between many tens or many hundreds of femtoseconds, but it's of that order. I don't know how to think about that except that I am a little concerned that what we're measuring is the result of the carriers trying to get out after the excitation. They all start with zero mobility; some drop down, some don't. I don't know how to think about how they divide themselves up other than the fact that because of the slow speed at which it happens—the excitation ought to be proportional to the speed, and yet the result isn't—and so I think it's happening afterwards.

Simulation Studies of Structure and Vibrations of Clean Transition Metal (001) Surfaces Across the Reconstruction Phase Transition

*C.Z. Wang, A. Fasolino, and E. Tosatti**

International School for Advanced Studies, Strada Costiera 11, I-34100 Trieste, Italy
*also at International Centre for Theoretical Physics, Strada Costiera 11, I-34100 Trieste, Italy

The interest in displacive surface reconstruction phase transitions lies both in their nature of genuine two-dimensional critical phenomena and in their being driven by soft surface vibrations. This then is particularly pertinent to this meeting, where surface phonons are being widely discussed, their full temperature-dependent spectrum being now measurable and measured [1].

The prototype systems chosen for our theoretical studies are the clean (001) surfaces of tungsten and molybdenum. A rich literature is available on the phenomenology of these surfaces [2], which undergo a continuous reversible displacive transition between a reconstructed phase - c(2x2) commensurate for W, 'c(2.2x2.2)' incommensurate for Mo - below a critical temperature T_c (slightly below room temperature for both) and a nominally 1x1, unreconstructed high temperature phase above T_c. One particularly intriguing issue has been in the past the question of whether these high temperature phases are statically 'ordered' (order-order transition), or 'disordered' (order-disorder transition).

In order to be able to perform computer simulations of these surfaces, it is necessary to deal first with the modelling of electronic forces. While Car-Parrinello methods [3] seem to represent a desirable method for the future, we have observed that an effective ion-ion force model is not a priori unjustified for these transitions. The crucial point is, whether the transition is driven by *lattice* entropy or by *electron* entropy. There are good reasons to believe that electron entropy is irrelevant to this problem. A consequence of this is that electron dynamics can be removed from the problem, and replaced by effective, *temperature-independent* interactive forces, i.e., by an effective 'lattice' hamiltonian.

Based on a carefully optimized two-body effective hamiltonian, whose main feature is to be vastly different for atoms which are near-neighbours on the same surface [4], we have carried out extensive molecular dynamics (MD) simulations of the displacive surface reconstructions of W(001)

Springer Series in Surface Sciences Vol. 14
Solvay Conference on Surface Science
Editor: F.W. de Wette

and Mo(001). The MD approach gives the possibility of studying the temperature behaviour not only of the surface lattice structure across the reconstruction phase transition but also of the reconstruction-related surface vibrations which strongly characterize, in an experimentally accessible manner, the high and low temperature phases.

The MD simulations are conducted on slabs with in-plane periodic boundary conditions and low-temperature structures are obtained by careful annealing cycles. The temperature-dependent evolution of all reconstructions can be followed in great detail, including reconstruction phase transitions. Moreover, velocity-velocity correlation functions of first-layer atoms provide a very simple and direct access to dynamical properties and in particular to surface phonons, during a surface molecular dynamics simulation.

The phase transition of W(001) is found to be driven by loss of long-range coherence of the in-plane displacement *direction* [5]. This direction is (110) below T_C, and becomes disordered above T_c, with only *short-range* order being preserved well above T_c. The distortion *magnitude*, of about 0.2 Å, hardly changes at all. This picture fits quite naturally in the general phenomenology of structural displacive phase transitions [6], and in our view does provide the basic answer: the high-temperature surface is dynamically disordered, and is 1x1 ordered only in a statistical sense.

The situation of Mo(001) is rather similar, however, with the non-trivial complication of incommensurability. The $T = 0$ phase has soliton-like walls between stripes with 'vertical' and 'horizontal' c(2x2) reconstructions. The 'wiggling' of these solitons increases with temperature, finally wiping out all phase coherence [7], and leading to a high-temperature state quite close to that of W(001).

With this understanding of the static properties, we can proceed further to consider dynamic properties, i.e., surface phonons. Our results from velocity-velocity correlations show well developed surface phonon peaks, their k-dispersion depending on temperature to different degrees. The strongest T-dependence is that of the reconstruction-related in-plane modes at the $\bar{M}$-point, $k_M = (\frac{1}{2}, \frac{1}{2})$ for W(001), and both in-plane and vertical modes at the incommensurate superstructure k-point $k_I = (0.44, 0.44)$ for Mo(001).

For W(001), the k_M surface phonons exhibit both softening and broadening near T_c, in agreement with recent experiments [8]. Also other non-critical features, particularly those connected with shear-vertical modes, agree well with experiments. Another feature we discovered is a high-frequency mode, i.e., a mode well above the bulk continuum. This mode is strongly broadened by temperature, and broadens out completely just below T_c. By a detailed analysis we find that this mode is due to a reconstruction-related stiffening of the first-second layer bonds, and appears to explain very well [9] recent EELS results of WOODS and ERSKINE [10].

For Mo(001), the k_I phonons soften on cooling, down to $\omega = 0$ for $T = T_c$, and they become acoustical phasons originating out of $k = k_I$ for $T < T_c$. There are as yet no surface phonon measurements on this surface, but we hope they will soon become available, to check these predictions of our model.

REFERENCES

1 J P Toennies, this conference; H Ibach, this conference
2 For a review see, for example, D A King, Phys. Scripta **T4** 34 (1983); L D Roelofs and P J Estrup, Surf. Sci. **125** 51 (1985); J E Inglesfield, Rep. Prog. Phys. **45** 223 (1982)
3 R Car and M Parrinello, Phys. Rev. Lett. **55** 2471 (1985)
4 A Fasolino, G Santoro and E Tosatti, Phys. Rev. Lett. **44** 1648 (1980); A Fasolino and E Tosatti, Phys. Rev. **B35** 4264 (1987)
5 C Z Wang, M Parrinello, E Tosatti and A Fasolino, Europhys. Lett. to be published
6 see e.g. A D Bruce and R A Cowley, *Structural Phase Transitions*, Taylor and Francis, London, 1981
7 C Z Wang, E Tosatti and A Fasolino to be published
8 H J Ernst, E Hulpke and J P Toennies, Phys. Rev. Lett. **58** 1941 (1987)
9 C Z Wang, A Fasolino and E Tosatti, Phys. Rev. Lett. **59** 1845 (1987)
10 J P Woods and J L Erskine, J. Vac. Sci. Technol. **A4** 1414 (1986)

Discussion

J. P. TOENNIES:

1. One of the salient differences between your molecular dynamical simulations and the recent phonon dispersion case measurements by J. Ernst and E. Hulpke from our laboratory is that whereas you predict reasonably sharp energy loss peaks at the zone boundary at temperatures above the transition temperature, the experiments are unable to detect sharp peaks beyond about 1.2Å^{-1} (80% of the zone boundary). Is this discrepancy due to some lack of short–range order in the crystal or an artefact of the theory?

2. I note that in your recent molecular dynamics simulation that you use a repulsive potential to describe the interactions of the atoms in the surface plane. Earlier (Fasolino *et al., Phys. Rev. Lett.* (1980)) you described those

forces by two force constants, α_s and β_s. Is the new potential consistent with those earlier force constants?

E. TOSATTI: These α_s and β_s are the same as in the 1980 paper. We had a misspelling there—it said "attractive" but that was a mistake. The first and second derivatives of these full potentials at the equilibrium point are exactly the same as on that phase diagram, except that this one has a lot more to it: not only the harmonic path—there's a lot more, and you need that for molecular dynamics.

J.P. TOENNIES: To get back to my first question, is this now an error in the theory?

E. TOSATTI: Well, I don't really know; it is a very good open question. I would stress the following: First, the width of the modes as we go are around the critical temperature region, the width of the modes is maximum at the zone boundary, so it takes some fantasy to draw a point there, even for us. Maybe I can put this slide up again. In these curves you should disregard these "zig–zags"; they have to do with the fact that we can't do infinitely many layers—we have something like seven layers. But here it takes some fantasy to put a frequency as your phonon freqency. The width is larger than the frequency, so one might as well have put nothing in this region. That's the first comment. The second comment is that, as you said before and stress now, not only do you have less intensity *at* the zone boundary, but somehow your dispersion, as measured, seems to want to have a minimum slightly away from the $\frac{1}{2}-\frac{1}{2}$ point, unlike what is seen in LEED. I think there is a consensus that we do not understand this. It would be very welcome to come up with ideas as to why two experiments should disagree in this respect. On the theoretical side, we get pictures that look very much like the LEED, and we have honestly not tried to calculate helium scattering spectra, which is quite a heavier job than what we have done. We should have potentials and all that....

D. R. HAMANN: In a recent preprint, Henry Krakauer reports *ab initio* local-density results for W(100) in which he considers a variety of displacements besides the Debe-King mode, including the effects of interlayer relaxation. He claims that his results are not consistent with the repulsive potential between surface atoms in your model. Have you seen this preprint, and what is your opinion?

E. TOSATTI: Yes, I have seen the preprint, and I agree. First of all, that is a first–principle calculation and there is a lot to learn by studying it; however, when it comes to branding something 'repulsive' or 'attractive', there may be a slight, how should I say, *confusion.* It is not clear until you cast your first–principle results in terms of pair-wise, three-body, many-body, what have you, what actually does it mean to say attractive or repulsive? All you can do is to qualify 'what does the energy do, when you move the atoms in a certain way?' And, what we call a 'repulsive' interaction does certainly lower the energy if you move the atoms the same way as Henry moves them. Furthermore, we get anisotropies very much of the same type; of course in our case it's not surprising—we *chose* it that way. In his case, it's more of a remarkable result that he can use to explain a certain type of thing. But, calling it attractive or repulsive—I find that rather subjective.

D.R. HAMANN: I thought he claimed that in some motions his results contradicted your model; that's the way I read it.

E. TOSATTI: I haven't understood it that way.

F.W. DEWETTE: It is known that bcc crystals are unstable under a wide class of central potentials. It is also known that deviation from the Cauchy relation is evidence of the presence of non-central forces in the material. Can you comment on how the absence of non-central forces in your molecular dynamics calculations might affect the results?

E. TOSATTI: At the harmonic level, we use Spanjaard *et al.*'s parametrization of first and second neighbor force constants of bulk W as well as Mo. This provides a rather decent, though not perfect, fit of T=0 bulk phonons, including Cauchy relations between elastic constants. Our full anharmonic potentials, built consistently with those force constants, have been checked by bulk molecular dynamics to yield a stable bcc structure with the correct lattice constant and thermal expansion coefficient up to 1500K. Thus, we believe our description of the background bulk crystal is totally adequate as it is.

D. A. KING: My question relates to the uniqueness of the particular set of potentials for v_S, v_1, and v_2 that you choose to use for W(100). Presumably these are chosen to describe the low temperature surface structure. If they are not unique, presumably a different set of potentials would lead to different conclusions regarding the nature of the phase transition.

E. TOSATTI: First of all, there is nothing such as a unique choice in this game. However, I would claim that if the only thing you want to change in this scheme, with respect to the bulk, is interaction between surface atoms, then I would claim there's no other way you can do it. In particular, if you had an attraction it would go 2×1 and not 2×2 and that's very simple to see just by counting neighbors and doing little things like this. You could do a better job if you had more parameters: suppose you change the interaction between the first and the second layer; suppose you include glue effects to account for the relaxation on the first and the second; you can build on this, or you can change your approach to make it more adherent to the reality. What I would claim is that this is a minimal sort of model that seems to have basically everything in it. This is the critical behavior that we get from the molecular dynamics. The order parameter has an amplitude and a phase—the amplitude is ρ and doesn't budge at the transition, the phase ϕ is the angle between the displacements. It is this order at low temperature (well, this is a finite-size effect), and it is this order at high temperature. So this shows that the transition occurs in this model because the atoms are still displaced but they lose long-range order in the relative orientations of their displacements. Now, I believe that everything that this picture has to offer does fit with what one knows and I'm rather gratified to hear John Pendry's remark—I don't know whether I'm interpreting it right—there may be some Freudian correction to it. But, the high temperature phase is disordered in that sense.

Soluble Models in Surface Collision Dynamics

J.W. Gadzuk

National Bureau of Standards, Gaithersburg, MD 20899, USA

1. INTRODUCTION

Selective control of the rate and outcome of chemical events, based on molecular-level manipulations, is a major goal in chemical physics. Laser assisted chemistry is one area which has offered such hope [1]. Although "surface-assisted chemistry" is certainly capable of statistically controlling rates and outcomes [2], it is only recently that bond selectivity has been considered within the context of surface dynamics [3]. The physical basis for many proposed scenarios in selectivity can be summarized as follows. At some time $t = 0$, the molecular system is placed in an initially prepared, non-stationary state (say by electronnic excitation) and is then allowed to time-evolve. At a later time $t = \tau_R$, the time-evolution of the initially excited state is interrupted. By choosing the time delay τ_R appropriately (depending upon system specifics), it is possible to achieve non-statistical control of the distribution of final quantum states and/or products. Ideally, the time delay τ_R should be comparable with intra-molecular vibrational timescales for the following reason. In order for selectivity to be realized, the interruption at time $t = \tau_R$ must occur before anharmonic forces both erase the memory of the birth process and "locally thermalize" the initially prepared state. Thus τ_R must be on a sub-picosecond time scale. This requirement has hindered proposals based on two-photon, pump-probe configurations using current laser technology [4].

It seems timely at the first joint Physics and Chemistry Solvay Conference to address this important issue in chemical physics, one that has some interesting realizations in surface physics. We will consider several examples of molecule-surface collision phenomena which i) achieve some degree of selectivity; ii) depend upon the "realistic" use of sub-picosecond time control; and iii) are exactly soluble in certain simple limiting cases. Specifically, we will consider aspects of electron-hole pair, phonon, and intra-molecular vibrational contributions to the problem of energy redistribution and sticking at metal surfaces.

2. MODEL

Consider a generic "two-body" collision (incident atom or molecule and target/surface) in which each partner has its own set of internal quantum states. Further suppose that the molecule is given some center-of-mass translational energy $\equiv \epsilon_p$ and that both "particles" are prepared in some prescribed initial state (or distribution of states). The internal state specifies a potential energy function which enters

Springer Series in Surface Sciences Vol. 14
Solvay Conference on Surface Science
Editor: F.W. de Wette

into the center-of-mass equation of motion. Curve crossings are experienced along the scattering trajectory, where charge transfer or intra-molecular electronic transitions occur [5-8]. In a full scattering event, the curve crossing is traversed not only on the inward but also on the outward branch of the trajectory where the reverse electronic transition can occur. From the point of view of both the translational as well as the internal degrees of freedom, these (hopefully coherent) transitions produce sudden switches in the potential energy functions operating within each subsystem; i.e., time-dependent Hamiltonians. This time-dependence permits energy redistribution which can lead to controlled selectivity in molecule-surface collisions.

The preceding discussion suggests a generic "two-body" collision involving double curve crossing and hence the formation of an intermediate electronic state for the controllable duration τ_R between crossings. For the piece-wise continuous "top-hat" time structure, the time dependence of the (molecule and/or target=surface) internal Hamiltonian is $H(t) = H_{int}$ for $0 \le t \le \tau_R$; $= H_o$ otherwise. It is then elementary to demonstrate [6,9] that the probability for the system to emerge in some final state $|f\rangle$, is

$$P_f(\tau_R) = |\langle 0;f|e^{-iH_{int}\tau_R/\hbar}|0;0\rangle|^2. \tag{2.1}$$

Inserting a complete set of eigenstates of H_{int}, (2.1) becomes

$$P_f(\tau_R) = |\sum_{\tilde{m}} \langle 0;f|int;\tilde{m}\rangle e^{-i\epsilon_{\tilde{m}}\tau_R/\hbar}\langle int;\tilde{m}|0;0\rangle|^2. \tag{2.2}$$

This coherent sum on intermediate states clearly displays the dependence of the final-state excitation probability on τ_R, the intermediate state lifetime or time delay between curve crossings and the generalized internal-state overlap integrals between eigenstates of H_o and of H_{int}.

The no-loss probability obtained from (2.2) is

$$P_o(\tau_R) = |\sum_{\tilde{m}} |\langle 0|\tilde{m}\rangle|^2 e^{-i\epsilon_{\tilde{m}}\tau_R/\hbar}|^2. \tag{2.3}$$

Note that $P_o(\tau_R)$ is given in terms of $|\langle 0|\tilde{m}\rangle|^2$ the "generalized Franck-Condon factors", which are known analytically for some model systems.

The next general point concerns a prescription for establishing the value of τ_R, as this is the source of the proposed selectivity. Within the classical limit, this is expressed as

$$\tau_R(\epsilon_p) = \oint_{z_{in}}^{z_{out}} \frac{dz}{v_z(z)} \tag{2.4}$$

where the integral is performed along the trajectory, z is the molecule-surface separation, and z_{in}/z_{out} are the locations of the curve crossings on the inward and outward legs [9]. The velocity as a function of z is obtained from $\epsilon_p = \frac{1}{2} M v_z^2(z) + V_{int}(z)$ which requires a functional expression for $V_{int}(z)$. In order for (2.2)-(2.4) to be most useful, a given value of ϵ_p should imply a unique value of τ_R or at least a spread in which $\Delta\tau_R/\tau_R \ll 1$. This requirement is in many ways equivalent to adopting the socalled no-recoil trajectory approximation [9-11] which, when used cautiously, provides a meaningful framework for continued analytic development.

Finally we acknowledge the so-called "zero-energy sticking problem" [12,13]. The basic premise is that in the limit of a zero kinetic energy incident beam, any inelasticity in the molecule-target collision is sufficient to cause the molecule to stick to the target/surface. Thus within the present picture, the zero-energy sticking probability is

$$S \approx \lim_{\epsilon_p \to 0} \{1 - P_o(\tau_R(\epsilon_p))\} \tag{2.5}$$

which demonstrates that "selectivity" in sticking can, in principle, be controlled by varying τ_R.

3. APPLICATIONS

3.1 Electron-Hole Pairs/X-ray Edge

The Anderson orthogonality theorem (AOT) says that the many-body ground state of an infinite Fermi system (electron gas) is orthogonal to the ground state of the same system, within which a localized potential has been introduced [14]. One consequence of the AOT is that the permanent switching on of a potential in an electron gas always results in excitations of electron-hole pairs with zero probability for the elastic or no-loss line, independent of the details of the switching process. Another consequence is that the strength of the no-loss line is always less than unity when a localized potential is turned on and then off after a finite duration [14]. In other words, introducing and then withdrawing a scattering potential into an infinite Fermi system, as in particle scattering from metallic surfaces, cannot be done without some finite probability for leaving the Fermi system in an excited state. Thus truly adiabatic processes in such systems are rigorously impossible, no matter how slowly they proceed. There is thus the possibility that sufficient energy could be transferred to the electron-hole pair excitations (from the incident-particle kinetic energy) that the particle would remain trapped at the surface. Since the potential is a localized atomic-like potential, the electron-hole pair excitation probability resulting from this permanently switched-on potential (i.e., $H(t) = H_o + H_{int}\,\theta(t)$) is precisely of the form of the infrared-divergent X-ray edge [15]. Thus

$$P(\epsilon_*) = |\langle 0;0|int;*\rangle|^2 \simeq \eta \, \frac{e^{-\epsilon_*/\epsilon_c}}{\epsilon_*^{1-\alpha}} \tag{3.1}$$

where $\eta \equiv 1/(\Gamma(\alpha)\epsilon_c^\alpha)$ is a normalization factor, $\epsilon_c \equiv \hbar\omega_c$ a cutoff energy of order the conduction bandwidth, and $\alpha \approx (\delta/\pi)^2 < 0.25$, where δ is the dominant Fermi-level electron phase shift associated with the localized potential V_1. A similar expression can be obtained for a localized potential outside a metallic surface in which case α and ϵ_c become functions of the hole-surface separation [16].

Proceeding to the surface scattering problem, it can be shown [13] that (2.3) and (3.1) (with $\Sigma_{\tilde{m}} \to \int d\epsilon_*$) yield the exact no-loss intensity

$$P_o(\tau_R) = [1 + (\omega_c \tau_R)^2]^{-\alpha}. \tag{3.2}$$

Equation (3.2) displays in a transparent way all the expected behavior in the limiting cases.

(a) Weak scattering: As $V_1 \to 0$, $\alpha = 0$, and thus $P_o \to 1$ independent of τ_R. Introduction of a zero-strength potential has no effect on the excitations.

(b) Short time: As $\tau_R \to 0$, $P_o \to 1$ independent of α. Turning a potential on and off before the electron gas has time to sample it is equivalent to doing nothing and thus no excitations are generated.

(c) Long time: As $\tau_R \to \infty$, $P_o \to 0$ which is just a statement of the Anderson orthogonality theorem.

In Fig. 1 P_o, given by (3.2), is drawn as a function of $\omega_c \tau_R$, taking α parametrically within a range likely to apply for real-surface scattering situations. Again one notes the intuitively reasonable points that for fixed scattering strength (constant α), the no-loss intensity decreases, and thus the inelastic intensity increases the longer the scattering potential remains on. Similarly, for fixed τ_R, the no-loss line decreases as α and thus the scattering strength increases. With not-unreasonable values of $\alpha \simeq 0.1$ for strongly interacting chemisorbed systems and $\epsilon_c = 10$ eV, the value of τ_R for 50% elastic scattering ($P_o(\tau_R) = 0.5$) is $\tau_R \approx 2 \times 10^{-15}$ sec. For such systems, significant depletion in the no-loss line is to be expected [13].

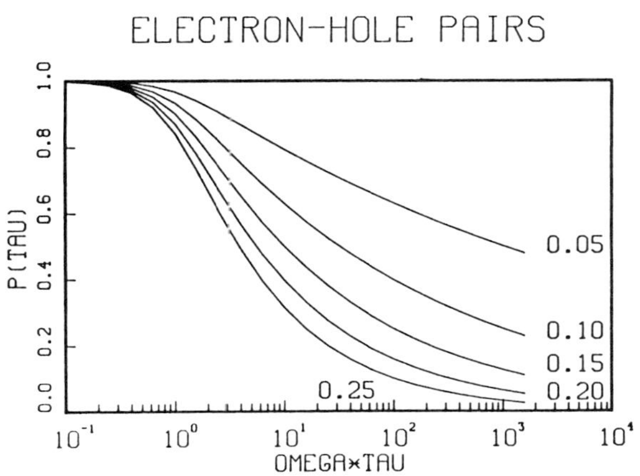

Fig. 1-No loss intensity vs. $\omega_c \tau_R$, from (3.2), with α varied parametrically.

3.2 Bosons/Forced Harmonic Oscillator

The driven harmonic oscillator (HO), indpendent Boson model (IBM) is at the heart of almost any analytic model of a time-dependent process in surface chemical physics [17]. Not only discrete localized intramolecular vibrational modes but also the delocalized continua of elementary excitations (phonons, pairs, plasmons, etc) of a (metallic) solid are usually considered within the harmonic limit. With this in mind, it seems reasonable that the excitation of molecular vibrations or of substrate modes, as a result of a molecule-surface collision involving double curve crossing could be described in terms of a driven oscillator/IBM in which the force $\equiv \lambda(t)$ (i.e. $H_{int}(t)-H_o = q\lambda(t)$ with q the oscillator coordinate). A well known result of the linearly displaced harmonic oscillator model is that the Franck-Condon factors appearing in (2.3) are Poisson distributed as

$$|<0;0|int;\tilde{m}>|^2 = e^{-\beta_o} (\beta_o^{\tilde{m}}/\tilde{m}!) \tag{3.3}$$

where

$$\beta_o = \Delta\epsilon_r/\hbar\omega \tag{3.4}$$

and $\Delta\epsilon_r = \frac{1}{2}\kappa q_o^2$ is the relaxation energy or Franck-Condon shift associated with the harmonic potentials ($\omega = \sqrt{\kappa/\mu}$) displaced with respect to each other by q_o. Since the HO eigenvalues are $\epsilon_{\tilde{m}} = \hbar\omega(\tilde{m} + \frac{1}{2})$, the dynamics problem (2.3) is solved in terms of static properties (3.3) as

$$P_o(\tau_R) = e^{-2\beta_o} |\sum_{\tilde{m}} (\beta_o e^{-i\omega\tau_R})^{\tilde{m}}/\tilde{m}!|^2$$

which nicely yields

$$P_o(\tau_R) = \exp[-2\beta_o(1-\cos\omega\tau_R)]. \tag{3.5}$$

Equation (3.5) gives the probability that a discrete HO, say the intramolecular stretch mode within the molecules, will remain unexcited throughout the double curve-crossing collision. Thus in the zero energy limit, the vibrationally assisted sticking probability is given by (2.5), (3.4), and (3.5). In analogy with Fig. 1, $P_o(\tau_R)$ given by (3.5) is shown in Fig. 2 as a function of $\omega\tau_R$ taking β_o parametrically. (Note: $\beta_o \simeq$ 1-3 for A_2/A_2^- displaced oscillators of C,N,O diatomics.) Unlike P_o due to thecontinuum of electron-hole pairs (3.2) which is a monotonically decreasing function of intermediate state lifetime, P_o from a localized oscillator, (3.5) is periodic. If τ_R is such that $\omega\tau_R$ is an odd (even) integral multiple of π, maximum (minimum) oscillator excitation occurs. Thus the commensurability between the oscillator frequency and the time delay between curve crossings is a crucial factor in controlled selectivity based on the driven oscillator.

Another exact result of the forced HO is that the distribution of excited states is also Poisson in terms of the parameter $\beta = |\lambda(\omega)|^2/2\mu\hbar\omega$ with $\lambda(\omega)$ the Fourier transform of the forcing function. For the present "top-hat" time dependence, it has been shown [6] that the Poisson parameter

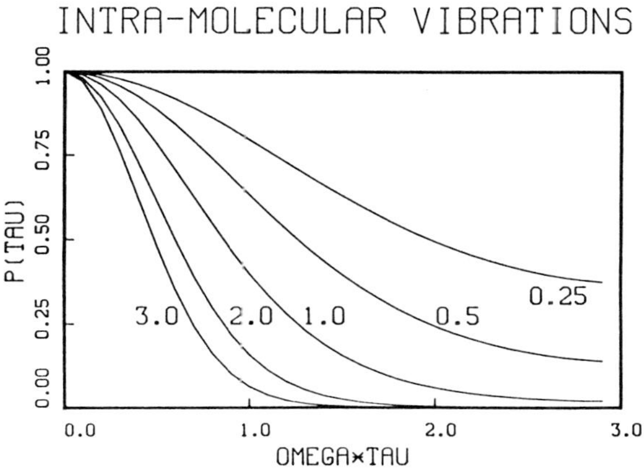

Fig. 2-No-loss intensity vs. $\omega\tau_R$, from (3.5), with βo varied parametrically. Due to periodicity, $P_o(\tau_R)$ is symmetric about $\omega\tau_R = \pi$ (and integer multiples of π).

$$\beta(\tau_R) \equiv 2\beta_o(1-\cos\omega\tau_R)$$

obtained from (3.4) and (3.5) is appropriate. Thus the mean vibrational energy carried by the scattered molecule is $\Delta\epsilon(\tau_R) = \beta(\tau_R)\hbar\omega$, which in principle provides a controllable source of molecules with extreme vibrational population inversion. It has been suggested [3a] that such

a controlled inversion in scattered polyatomic molecules, could provide a feasible alternative to the 2-photon pump-dump proposal for "selectivity in reactivity" [4]. Independent of the ultimate truth, it is a measure of the maturing process in chemical physics at surfaces that such propositions are even being considered.

References

1. R. D. Levine and R. B. Bernstein: Molecular Reaction Dynamics and Chemical Reactivity, (Oxford Univ, N.Y., Oxford, 1987).
2. D. W. Goodman: Ann. Rev. Phys. Chem. 37, 425 (1986).
3. a) J. W. Gadzuk: Chem. Phys. Letters 136, 402 (1987).
   b) J. C. Tulley: J. Elec. Spect. (in press).
4. D. J. Tannor and S. A. Rice: J. Chem. Phys. 83, 5013 (1985).
5. A. W. Kleyn, J. Los, and E. A. Gislason: Phys. Repts. 90, 1 (1982).
6. J. W. Gadzuk: Comments Atom. Mol. Phys. 16, 219 (1985).
7. S. I. Sawada, A. Nitzan, and H. Metiu: Phys. Rev. B 32, 851 (1985).
8. A. Yoshimori and M. Tsukada, eds.: Dynamical Processes and Ordering on Solid Surfaces, Springer Ser. Solid-State Sci. Vol. 59 (Springer, Berlin, Heidelberg 1985); R. B. Gerber: Chem. Rev. 87, 29 (1987).
9. J. W. Gadzuk and S. Holloway: Phys. Scripta 32, 413 (1985); J. Chem. Phys. 84, 3502 (1986).
10. J. W. Duff and D. G. Truhlar: Chem. Phys. 9, 243 (1975).
11. D. Kumamoto and R. Silbey: J. Chem. Phys. 75, 5164 (1981); W. Brenig: Phys. Scripta 35, 329 (1987); J. Harris: Phys. Scripta 36, 156 (1987).

12. W. Brenig: Z. Phys. B 36, 227 (1980); G. R. Brivio and T. B. Grimley: Surf. Sci. 161, L573 (1985); A. Yoshimori and K. Cho: Surf. Sci. 176, 438 (1986).
13. J. W. Gadzuk: Phys. Rev. B 24, 1866 (1981).
14. E. Müller-Hartmann, T. V. Ramakrishman, and G. Toulouse: Phys. Rev. B 3, 1102 (1971); Solid State Commun. 9, 99 (1971).
15. G. D. Mahan: Solid State Phys. 29, 75 (1974).
16. J. W. Gadzuk and H. Metiu: Phys. Rev. B 22, 2603 (1980).
17. S. Waldenstrom and K. Razi Nagvi: Chem. Phys. Letters 85, 581 (1982); K. M. Leung, G. Schön, P. Rudolph, H. Metiu: J. Chem. Phys. 81, 3307 (1984); J. W. Gadzuk and S. Holloway: Prog. in Surf. Sci. (in press).

General Discussion

D. R. HAMANN : Following Peter Toennies' report of surface phonons on GaAs(110), Charlie Duke commented on results of his predicting a very soft mode not present the the spectra shown. Bruce Doak has recently measured the GaAs(110) surface phonons using He–beam scattering in the direction perpendicular to the principal corrugations. In this direction, he sees a second very soft phonon, with an energy of ~ 5 meV at the zone boundary.

J. P. TOENNIES: I find these results very interesting. Because of our more complicated manipulator required to facilitate *in situ* cleavage we were not able to look in a direction perpendicular to the π-bonded chain.

N.D. LANG: I wanted to comment on Ward Plummer's question about the local–density approximation. You know, the essence of the local–density approximation is that the electron sees what appears to be a homogeneous electron gas, and that in particular means it sees an exchange–correlation hole symmetrically disposed around it—roughly spherical. Now if an electron is in the bulk gas, in something that has a surface, it'll see that. If I move the electron way outside into the vacuum, the exchange–correlation hole stays on the surface And then the exchange–correlation hole and the electron are well separated from each other, and in that case this hole just becomes the image deficiency on the surface—it's the image charge. So, one thing you might use to test how things go is to look at the screening of a point charge as it moves from the inside, deep inside, out into the vacuum, and to see when the screening charge begins to be separated from that point charge. When the point charge is far outside the surface, the screening charge will sit on the surface; that'll be the image charge. When the point charge is way inside, the screening charge would be symmetrically disposed around the point charge and that's bulk–like behavior. Now it turns out that in a jellium calculation, even for the position of a point charge a couple of bohr radii outside where the last lattice plane would be, you get a screening charge very symmetrically disposed around the point charge. So that means you're getting rather bulk–like screening behavior much further out than you might ordinarily think you would get it. And if you're looking, for example, at the vibrations of the last lattice plane, that's well inside of that region. So those vibrating atoms are in a region where the screening behavior is very bulk–like, at least as indicated by that calculation.

M. CARDONA: I would like to propose the following qualitative explanation for the pair production by impact. The impact produces a strain with a strong uniaxial component which splits the top of the valence band, raising the upper component above the bottom of the conduction band at unstrained points. Carriers then fall from the "strained" valence bands to the unstrained conduction band through Zener tunneling. The model, which is very qualitative, is susceptible to a number of refinements.

M.J. CARDILLO: Your idea seems quite reasonable and is in large part quite analagous to the concepts employed to describe collisionally-induced curve crossing in molecular excitations. An interesting point in the development of a description of these collisionally induced surface excitations is the extent to which the properties of the bulk solid, such as the bands, can be preserved during the action of the energy transfer.

G. A. SOMORJAI: The speakers have pointed out clearly the excellent agreement obtained in helium scattering and high resolution electron energy loss spectroscopy studies of measuring phonon dispersion curves. I wonder if they could summarize what happens with the phonon dispersion curves (1) in the presence of a large concentration of surface steps (2) in the presence of a monolayer of heteroatoms (a monolayer on Ni for example) and (3) as the adsorbate layers build up, layer by layer say, up to ten monolayers (thin films).

J.P. TOENNIES: Let me answer the questions in the order presented:

(1) In the bulk defects are known to produce localized modes. With the exception of one EELS study [H. Ibach and D. Bruchmann, *Phys. Rev. Lett.* **41**, 958 (1978)] localized modes at surfaces have not been studied experimentally. A number of theoretical studies have however been carried out on the effect of steps on a vicinal surface of type (11m) [e.g., G. Armand and P. Masri, *Surface Sci.* **130**, 89 (1983)]. We are presently preparing an experiment to confirm these predictions.

(2) I know of only one study by Daum [*J. Electron Spectroscopy*, to be published] in which EELS was used to study the modes of a single layer of Cu on Ag(100). The effect of CO chemisorption has been studied by He scattering [A.M. Lahee, J.P. Toennies and Ch. Wöll, *Surf. Sci.* **177**, 371 (1986)]. The disappearance of the Rayleigh mode of the substrate and the increase in intensity of the adsorbate modes could be followed with increasing coverage. At a c(4×2) coverage only a flat dispersionless mode, which was attributed to a frustrated translations, could be observed.

(3) This experiment has indeed already been carried out using He atom scattering from physisorbed Kr and Xe on Ag(111) and Pt(111). In the first

study of this kind [K.D. Gibson and S.J. Sibener, *Phys. Rev. Lett.* **55**, 1514 (1985)] a flat mode with a frequency of about 3meV independent of the wavevector was observed. With increasing number of layers the frequency at the zone boundary drops down, and for 25 layers the usual Rayleigh mode behavior was observed. These experiments have recently been repeated with improved resolution by K. Kern, R. David, R.L. Palmer and G. Comsa [*Phys. Rev. Lett.* **56**, 2823 (1986)].

H. IBACH: We have also completed a study of the phonon dispersion of Ag(111) type mono- and multilayers on Cu(100) [W. Daum, *J. El. Spect. Rel. Phen., in press.*], and also of Fe(100) overlayers on the same surface. It is indeed very interesting to follow the surface phonons in these overlayers as a function of thickness and the signature of surface stress in these layers.

Part VI

Chemical Reactions at Surfaces

Dynamics of Chemical Interactions at Surfaces

G. Ertl

Fritz-Haber-Institut der Max-Planck-Gesellschaft,
Faradayweg 4–6, D-1000 Berlin 33, Germany

1. INTRODUCTION

The atoms in the surface are missing part of their nearest neighbors and exhibit therefore unsaturated valencies. The interaction of particles impinging from the gas phase may hence lead to bond formation (chemisorption) whereby even intramolecular bonds may be dissociated. Adsorbed particles may interact with each other and eventually form new bonds, and finally the surface species may be released into the gas phase. These processes are governed by the multi-dimensional potential surface $V(r_i, R_j)$ (resulting from solution of the Schrödinger equation within Born-Oppenheimer approximation), whereby r_i are the nuclear coordinates of the adparticles and R_j those of the solid. The static properties of the system (equilibrium geometry, bond energies) are determined by the minima of the potential, while the curvature around these minima yield the vibrational properties. The dynamics of chemical interactions are, on the other hand, ruled by passage of the system over the potential surface and the energy exchange processes associated.

These elementary steps will in turn then determine the rates of surface reactions, i.e. the time-evolution of macroscopic systems, which form the basis of processes such as heterogeneous catalysis.

The strategy for experimental as well as theoretical description of the microscopic processes consists in a stepwise reduction of complexity and separation of the variables along the following lines:
i) The use of 'perfect' single crystal surfaces renders the potential periodic with respect to motion parallel to the surface.
ii) In the limit of zero coverage the interactions <u>between</u> adparticles vanish and the interaction between a <u>single</u> particle and the surface can be separated.
iii) With surfaces undergoing no displacements of the atoms of the outermost layers (reconstruction) the coordinates R_j may approximately be considered as being constant (if relaxation of the interlayer spacings is neglected).
iv) Similar arguments hold for the adparticle in non-dissociative chemisorption for which the internuclear distances remain approximately fixed (albeit they may differ from the values of the free molecule in the gas phase).

This contribution will concentrate on selected aspects of the experimental approach to these problems. First the principles governing the interaction dynamics of diatomic molecules with a surface without alteration of the internuclear configurations within both the solid and the adparticles will be discussed in some detail. Next some aspects associated with bond-breaking and their consequences on the rates of surface processes will be outlined, and finally an example of steady-state catalysis will be described which is governed by principles of non-linear dynamics based on the previously discussed elementary processes.

Springer Series in Surface Sciences Vol. 14
Solvay Conference on Surface Science
Editor: F.W. de Wette

2. DYNAMICS OF MOLECULE-SURFACE INTERACTION

The variation of the potential as a function of the distance of the center of gravity of a molecule from the surface is in the simplest case of the type drawn schematically in Fig. 1. The actual shape of this potential will, in addition, be dependent on the orientation of the molecular axis with respect to the surface plane and on the point of impact within the unit cell of the two-dimensional surface lattice. The energetically most favourable of these situations determines the geometry of the adsorption complex, and the depth of this potential well the adsorption energy (in the limit of zero coverage, i.e. without influence of interactions between adparticles). The probability that a particle impinging from the gas phase reaches this state (i.e. is 'trapped') is denoted as (initial) sticking coefficient, s_0. After a certain lifetime on the surface the particle will return back into the gas phase, that is why this is called the trapping/desorption channel of interaction. As an alternative possibility the impinging molecule is reflected back into the gas phase after energy exchange during the collision but without full thermal accommodation at the bottom of the adsorption well (direct-inelastic scattering).

The partitioning between these two channels depends on the shape of the interaction potential, the surface temperature T_s, and the energy content of the incoming particle in its various degrees of freedom. Fig. 2, as an example, shows how s_0 for NO on graphite with fixed incident kinetic energy (80 meV) varies with the surface temperature, while on the other hand a rise of $\langle E_{kin} \rangle$ at constant T_s also causes lowering of the sticking coefficient /1/. Except for very light projectiles (for which quantum effects come into play /2/) theoretical treatment can rather successfully be performed by classical trajectory calculations /3/; the qualitative features can in fact already be modelled by rather simple 'cube'-models /2/. As long as the kinetic energies of the incident particles are in the thermal region, energy exchange with the solid is dominated by (multi) phonon creation/destruction; the excitation of electron-hole pairs becomes only relevant with hyperthermal kinetic energies /4/. For the interacting molecules energy redistribution is not only restricted to the translation motion but comprises also the internal degrees of freedom (rotation and vibration). State-selective experiments are therefore required which can be performed by the use of molecular beams in combination with laser spectroscopic techniques /2,5/.

A set of typical data obtained in this way is reproduced in Fig. 3 /6/. A beam of NO molecules with a narrow kinetic energy distribution around 730 meV and very low rotational energy (15 meV) was scattered at an oxidized Ge

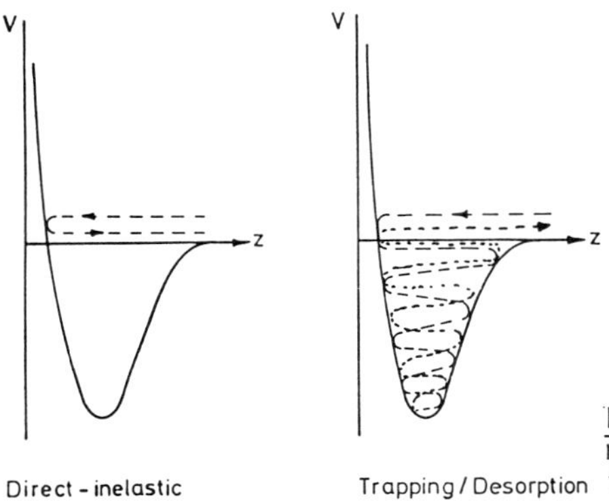

Fig. 1: One-dimensional potential for a molecule approaching a surface and the two channels of interaction

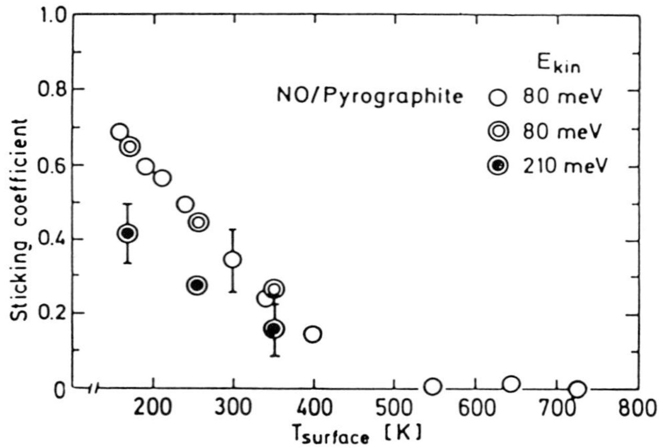

Fig. 2: The sticking coefficient for NO on a graphite surface as a function of surface temperature at initial kinetic energies of 80 and 210 meV, respectively

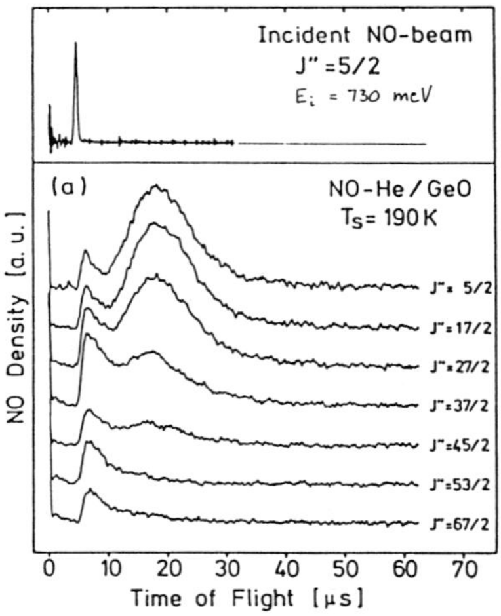

Fig. 3: Time-of-flight spectra of NO molecules in different rotational levels after scattering off an oxygen covered Ge surface. The incident molecules had an average translational energy of 730 meV

surface, and time-of-flight (TOF) spectra of molecules coming off the surface in individual rotational levels J" were recorded by the technique of resonance-enhanced-two photon ionisation. The TOF data are easily converted into velocity distributions, so that the translational energy distributions of the molecules with their respective rotational energies become available. (Practically all molecules were in the vibrational ground state in this experiment, so that population of higher vibrational levels is negligible).

The data of Fig. 3 clearly exhibit for each J" the existence of two groups of molecules with different average translational energies. Detailed analysis revealed that the 'slow' particles (i.e. those with their maximum in the TOF distribution at longer times) originated from the trapping/desorption channel

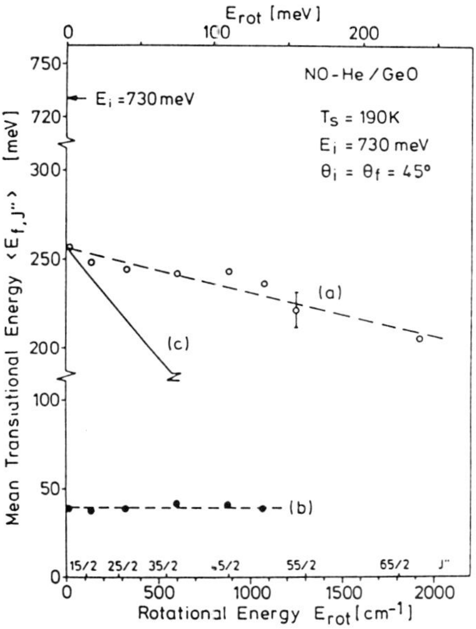

Fig. 4: Analysis of the data of Fig. 3: Mean translational energy versus rotational energy of the NO molecules coming off the Ge surface in the two channels

while the 'fast' molecules (with intensity peaked at short flight times) were undergoing direct-inelastic scattering. Separate analysis for these two channels provides information on the mean translational energy as a function of the internal energy (Fig. 4), as well as on the populations of the various rotational levels. Starting from these results we will discuss in the following the essential features of the dynamics of molecule-surface interactions.

i) Direct-inelastic scattering. Molecules undergoing direct-inelastic scattering exhibit an angular distribution which is peaked near the specular direction but is considerably broadened due to energy and momentum exchange with the solid. As can be seen from Fig. 4, with the present example a high fraction of the primary translational energy goes into the heat bath of the solid, but obviously also a substantial part is transferred into rotation of the molecules which had initially been rotationally cold. If translational energy transfer to the solid and to rotational excitation would be completely decoupled from each other, the variation of $\langle E_{trans} \rangle$ with E_{rot} should follow the full line in Fig. 4 which is obviously not observed. It turns out that those molecules which are rotationally excited in the collision transfer less of their kinetic energy to the solid. Semi-quantitative description of this effect is already achieved by application of a rather simple classical model for non-spherical molecules bouncing off a flat surface. Depending on the angle formed between the molecular axis and the surface upon impact there will be more rotational and less phonon excitation and vice versa. Application of the laws of energy and momentum conservation leads to a simple relation between the energy transferred to the solid, E_p, the rotational energy of the molecule coming off the surface, E_r, and its initial translational energy, E_t /17/: $E_p = 4E_t/[x+(1+E_r/E_p)x]^2$, whereby $x^2 = M/m$ denotes the ratio of the masses of the surface atom and the molecule.

Integration of the TOF spectra yields the relative populations of the various rotational levels. Inspection of Fig. 4 shows that for the direct-inelastic scattering channel even high rotational states are appreciably populated and that there exists obviously no Boltzmann distribution. This 'rotational rainbow' effect has already been investigated extensively /2/. Qualitatively it can again be explained by the same simple model just outlined. If one considers all possible orientations of the molecular axis one may derive the rotational energy distribution of molecules bouncing off the surface which exhibits characteristic 'rainbow' features /8/.

ii) Thermal desorption. Molecules released into the gas phase by thermal desorption exhibit an angular distribution which is symmetric with respect to the surface normal (but not necessarily cosine), i.e. the molecule has completely 'lost' the information about its initial momentum. For the example illustrated by Figs. 3 and 4, the molecules undergoing trapping/desorption exhibit a Maxwell-Boltzmann translation energy distribution which is independent of E_{rot}, and which may be characterised by a translational temperature T_t which turns out (in this case) to be equal to the surface temperature T_s - just as naively expected for molecules which had been in thermal equilibrium with the solid. Similar experiments at higher temperature reveal, however, that now $T_t<T_s$ (e.g. T_t = 450 K at Ts = 550 K). This effect of 'translational cooling in desorption' appears surprising at first glance, but has been reported several times in the literature /2/ and can most conveniently be interpreted by a 'detailed balancing' argument, i.e. if one considers the reverse process of sticking as will be outlined below /9/.

A similar effect is observed with respect to the rotational energy of desorbing molecules. A 'Boltzmann plot' of the rotational states populations vs. E_{rot} yields a straight line whose slope may be identified with a rotational temperature T_{rot}. However, again $T_{rot}=T_s$ only at low surface temperatures while T_{rot} levels off with increasing T_s. Fig. 5 shows data of this type for NO desorbing from Pt(111) /1/ which clearly exhibit this effect of 'rotational cooling in desorption' /3,10/. The incomplete uptake of rotational energy by a molecule leaving the surface may even differ for differing orientations of the rotational axes ('helicopter' vs. 'cartwheel') so that polarisation effects are observed /11/.

The kinetics of thermal desorption in systems of the type of our present discussion can be described by a simple first-order rate expression: $-dn_s/dt=k_d \cdot n_s$, where n_s is the surface concentration of the adsorbed species and k_d the rate constant which is usually written as $k_d=\nu_d \exp(-E_d/RT)$. The reciprocal of k_d equals the mean surface residence time τ which may be determined directly in experiments applying modulated molecular beam relaxation techniques. A plot of log τ vs. 1/T should hence yield a straight line, from which the two kinetic parameters - the preexponential ν_d and the

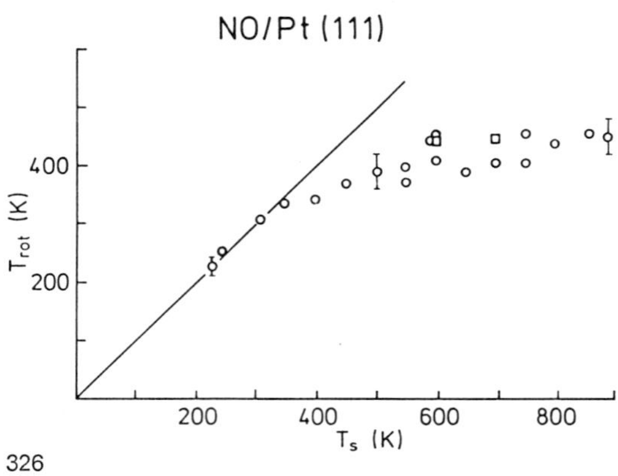

Fig. 5: Rotational temperature of NO molecules thermally desorbing from a Pt(111) surface as a function of surface temperature

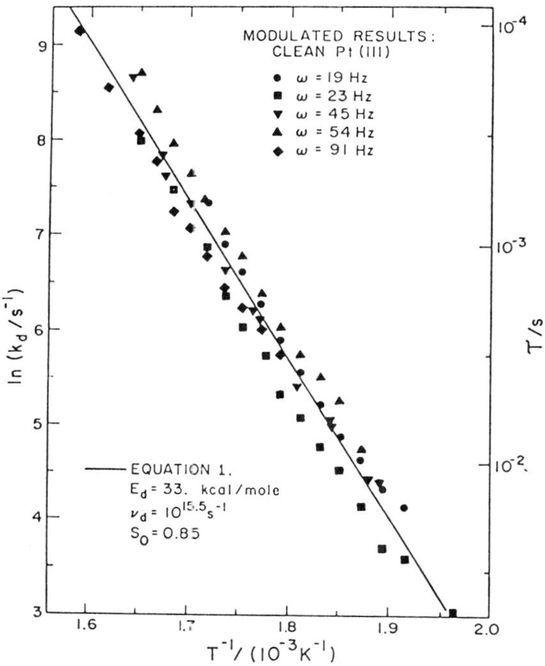

Fig. 6: Arrhenius plot of the rate constant for thermal desorption k_d resp. mean surface lifetime τ in the system NO/Pt(111) as determined by modulated molecular beam experiments

activation energy for desorption E_d = depth of the adsorption well - may be derived. An Arrhenius plot of this type for our system, NO/Pt(111), is reproduced in Fig. 6 /12/. Here one realizes for example that at T_S=540 K, τ is about 10^{-2} sec, which is practically an infinitely long time for reaching thermal equilibrium with the solid. Inspection of Fig. 5 demonstrates, on the other hand, that the molecules desorbing at T_S=540 K exhibit a rotational temperature of only about 400 K, which effect therefore clearly has to be attributed to the dynamics of the gas-solid energy transfer.

This dynamics is contained in the sticking coefficient s, which parameter is not only responsible for the adsorption probability, but also for the reverse process of thermal desorption. The rate constant for desorption as formulated within the framework of transition state theory reads $k_{TST} = (k_B T/h)\cdot e^{\Delta S^{\neq}/R}\cdot e^{-E_d/RT}$, whereby the entropy difference $\Delta S^{\neq}$ between the adsorbed state and the molecule in front of the surface is determined by equilibrium properties, i.e. the respective partition functions. Application of the principle of microscopic reversibility, however, yields that the true rate constant is given by $k_d = s\cdot k_{TST}$ /9/.

This result now tells us that if the sticking coefficient decreases with increasing translational energy of the incident molecules, also fewer molecules with higher velocity will desorb - the average translational energy of desorbing particles will become lower than that corresponding to T_S. Similar arguments also hold for the effect of rotational cooling in desorption: Since the adsorbed molecules are no longer freely rotating on the surface their initial rotational energy has to be transferred to the solid via intermediate conversion into translational-type collision energy, which explains why the sticking coefficient also decreases with increasing rotational energy. Quantitative substantiation of these qualitatively outlined effects can be found in the relevant literature /2,3,10/.

iii) Non-thermal desorption. For the sake of completeness it should be mentioned that desorption may of course also be initiated by non-thermal processes, such as electronic excitations either by electrons or photons /13/. In this case the molecule is lifted to another potential curve and starts to leave the surface from a region which is repulsive with respect to the molecule-surface interaction. In general, also the distribution of energy into the various degrees of freedom of the desorbing molecule will be non-thermal. For example it was recently found that NO molecules desorbing from a NiO surface by UV radiation (λ=193 nm) exhibited a very pronounced spin-orbit selectivity at low rotational quantum numbers /14/. This field is still largely open for further exploration.

3. BREAKING AND FORMATION OF CHEMICAL BONDS

Next we admit also the possiblity for alterations of the internuclear distances within the molecule approaching the surface which may eventually lead to dissociative chemisorption. Fig. 1 has to be extended by at least one additional coordinate. For the simplest case of a diatomic molecule interacting with a surface with a fixed orientation of its axis, the potential will now be of the type shown in Fig. 7. Frequently this potential is again reduced to a one-dimensional version by plotting V(x) at two fixed values of y representing the undisturbed molecule and the fully dissociated state ('Lennard-Jones potential'). Less misleading would, however, be a plot of the variation of the potential along the reaction coordinate . Potentials of the type discussed exhibit usually an activation barrier which the system has to surmount in order to lead to dissociative chemisorption. Following Polanyi's classification for homogeneous reactions in the gas phase /15/ this barrier may be either located in the 'entrance' or the 'exit' channel. In the former case (as drawn in Fig. 7) the activation energy has essentially to be

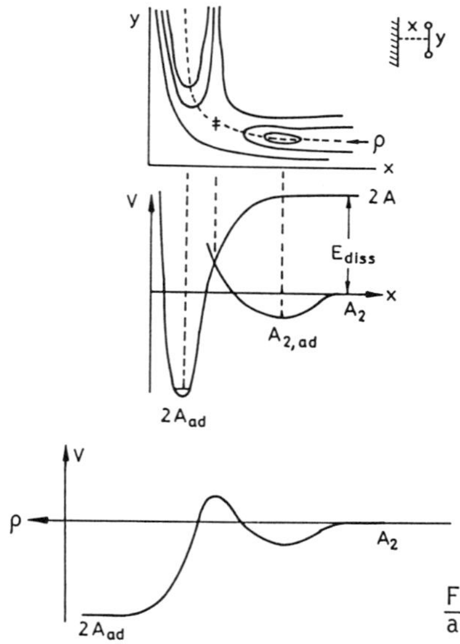

Fig. 7: Schematic potential for dissociative adsorption of a diatomic molecule

provided by motion along the x-coordinate, i.e. by the translation energy of the incoming molecule. Dissociative chemisorption then takes place preferentially by direct collision from the gas phase. If, on the other hand, the activation barrier is in the exit channel, it is surmounted through vibrational activation of the internuclear distance which may take place by uptake of thermal energy from the solid while the particle is in a molecularly adsorbed 'precursor' state.

The actual situation will be far more complex. The interaction potential will, in addition, be dependent on the orientation of the molecular axis as well as on the point of impact within the surface unit cell, and it becomes thus evident that in general the sticking coefficient will not show a sharp thereshold behaviour upon increasing the energy cortent of the molecule-surface complex. Detailed trajectory calculations are in fact required in order to quantitatively explore the efficiency of the excitation of the various degrees of freedom in affecting the sticking coefficient. Nevertheless there exist various experimental examples which show the dominance of one of the discussed channels: As can be seen from Fig. 8 the sticking coefficient for H_2 on Ni(111) increases continuously with increasing translational energy of the impinging molecules, while it is practically

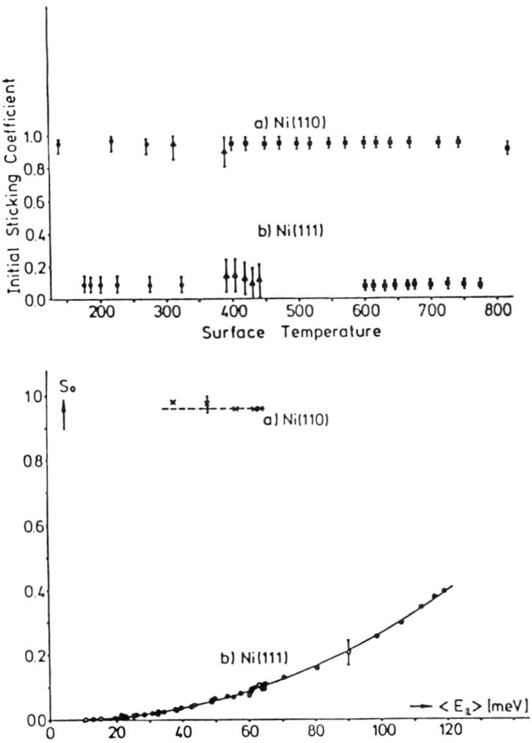

Fig. 8: Initial sticking coefficient for dissociative adsorption of H_2 on Ni(110) and Ni(111) surfaces. *Top*: Variation with surface temperature at fixed initial translation energy of 60 meV. *Bottom*: Variation with initial translational energy at 300 K surface temperature.

independent of the surface temperature. In this case obviously the 'direct' mechanism with dissociation through collision from the gas phase prevails /16/. The dissociative chemisorption of N_2 on Fe(111), on the other hand, is an example in which the molecular 'precursor', $N_{2,ad}$, may even be isolated and characterised in detail /17/. Here the overall sticking coefficient is strongly affected by surface temperature, but also by the kinetic energy of incoming N_2 molecules which affects the trapping probability into the precursor state /17c/.

Since activation energies enter exponentially into rate constants it is quite obvious that even small variations of the activation barrier may have pronounced influence on the sticking coefficient for dissociative adsorption. Such variation may be associated with the surface structure: As can be seen from Fig. 8 the H_2 sticking coefficient for low translational energies is much higher on Ni(110) than on Ni(111). But also the presence of surface defects (steps) or of coadsorbates (promoters) may have marked influence on the kinetics of dissociative adsorption. This is the source of structure sensitivity of surface reactivity. (Note that for non-dissociative adsorption in the absence of an activation barrier the sticking coefficient will be determined by the dynamics of translational energy transfer to the solid which is much less affected by the structure of the surface).

Motion of the system across the potential of Fig. 7 along the reverse path describes associative desorption through thermally activated recombination. Due to the existence of an activation barrier the desorbing molecules may now exhibit excess energy ($>kT_s$) in various degrees of freedom, for which effects a number of experimental results were reported in the literature /2,18/.

If different kinds of species, say A_{ad} and B_{ad} (formed from A_2 and B_2), are present on the surface it may be energetically more favourable to recombine $A_{ad}+B_{ad}$ to a new molecule AB, so that this would be the predominant species leaving the surface.

This is the principle underlying heterogeneous catalysis, which phenomenon will now be further illustrated by means of one of the most simple reactions of this category, namely the oxidation of carbon monoxide at platinum metals. The individual reaction steps may be formulated as follows /19/:

$$
\begin{aligned}
CO + * &\rightleftharpoons CO_{ad} \\
O_2 + 2* &\rightarrow 2O_{ad} \\
O_{ad} + CO_{ad} &\rightarrow CO_2 + 2*
\end{aligned}
$$

* denotes schematically a free adsorption site on the surface which has, however, a different meaning for CO and O_2 adsorption: Dissociative oxygen adsorption requires a larger ensemble of neighboring empty surface atoms, so that this process is inhibited if the CO_{ad} concentration exceeds a critical value. The O_{ad} species, on the other hand, form rather open overlayer structures which exert no noticeable inhibition on the additional uptake of CO. The recombination step $2O_{ad} \rightarrow O_2$ is energetically much less favourable than the process $O_{ad}+CO_{ad} \rightarrow CO_2$, so that practically all O_{ad} species leave the surface in the form of CO_2.

Under steady-state conditions the reaction rate R will be governed by the concentrations of <u>both</u> O_{ad} and CO_{ad}, and these will in turn be dependent on the partial pressures of O_2 and CO in the gas phase and on temperature. As an example, the full line in Fig. 9 shows the variation of R with p_{CO} for a (perfect) Pt(110) surface /20/, if p_{O_2} and T are kept fixed. The shape of this curve can easily be rationalized: At low p_{CO} there will be an excess of oxygen and the CO coverage and hence also R will increase with increasing p_{CO}

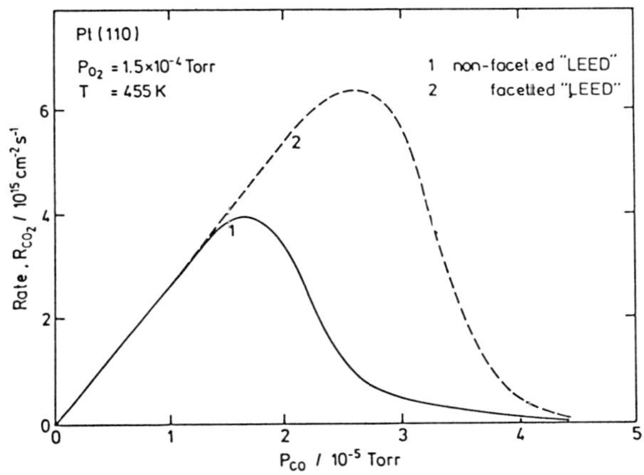

Fig. 9: Catalytic oxidation of CO on Pt(110): Steady-state rate of CO_2 formation as a function of CO partial pressure on a 'perfect' (full line) and a facetted surface (dashed line)

up to a maximum where about equal steady-state concentrations of CO_{ad} and O_{ad} are present. At even higher p_{CO} the further increasing CO coverage inhibits more and more the uptake of oxygen and therefore R decrases continuously in this range. Here each dissociatively adsorbing O_2 molecule will be rapidly transformed into the product, and the reaction rate is determined by the probability for dissociative oxygen chemisorption.

Fig. 9 contains also a second (dashed) curve in which a higher maximum rate is reached and the decrease of R due to inhibition of oxygen adsorption by adsorbed CO takes place only at substantially higher p_{CO}. These data belong to a structurally modified Pt(110) surface which exhibits a higher oxygen sticking coefficient leading to enhanced catalytic activity in the range where oxygen adsorption is rate-determining.

The quoted structural modification was caused by the catalytic reaction itself /20/. Prolonged operation under conditions of high CO coverage caused the formation of facets consisting of arrays of terraces and monoatomic steps. The latter are responsible for the enhanced oxygen sticking coefficient, - the catalytic reaction modifies the surface structure in a way to improve the reactivity, an effect which was already suggested by Langmuir /21/.

This example shows that chemical interactions may not only affect the bonds in the adsorbed particles, but may even cause breaking and formation of new bonds between the atoms of the substrate. More specifically, the first step for facet formation consists in the lifting of the reconstruction of the clean Pt(110) by CO adsorption which process is associated with substantial transport of material.

Reconstruction or lifting of reconstruction of the surface structure by adsorbates is a quite frequently occurring process for which the driving force has to be sought in the overall energetics. Fig. 10 shows the schematic energy diagram for a Pt(100) surface under the influence of CO /22/. The clean surface is reconstructed ('hex') which phase is lower in energy by 10 kcal/mole than the 1x1-phase. The latter exhibits, however, a larger

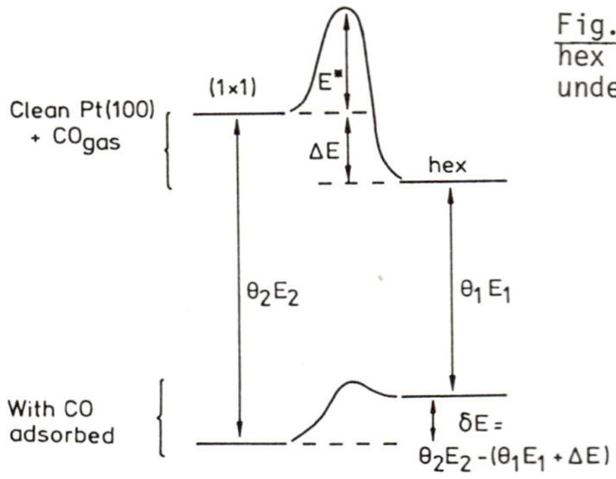

Fig. 10: Schematic energy diagram for the hex and 1x1-phases of the Pt(100) surface under the influence of CO adsorption

adsorption energy for CO and becomes thus stabilized as soon as a critical CO coverage is reached. This hex→1x1 transformation proceeds through nucleation and growth without any noticeable activation energy. If the CO-coverage on the 1x1-phase drops, on the other hand, below another critical value, the 1x1-phase becomes metastable and transforms back into the hex-phase with an activation energy of about 25 kcal/mole. Similarly, the Pt(110) surface may be switched between a reconstructed 1x2-phase at low CO coverage and the 1x1-phase at high CO concentration /23/. For both systems the oxygen sticking coefficient is higher on the non-reconstructed than on the reconstructed surface which will be of decisive importance for the phenomena to be described next.

4. NONLINEAR DYNAMICS

If a catalyst is operated under continuous flow conditions with the partial pressures and temperature kept constant then also the rate of product formation will usually (after some transient period) be constant. On the other hand this is an open system far from equilibrium and hence phenomena of spatial and/or temporal self-organisation may occur as characteristic for the field of non-linear dynamics /24/. The continuous increase of the steady-state rate in combination with progressing facetting of the surface structure as described in the preceding section was already an example belonging into this category.

Under different conditions with the same system (CO oxidation on Pt(110)) the surface structure is not subject to facetting, but nevertheless the reaction rate stays not constant with time. As can be seen from Fig. 11, adjustment of the O_2-pressure from 1.5×10^{-4} Torr to a new constant value of 2.0×10^{-4} Torr (while p_{CO} and T were kept fixed) causes a slow continuous rise of the work fucntion (which here is proportional to the reaction rate), followed by small-amplitude temporal oscillations which finally develop into sustained oscillations with large amplitude /25/.

As another example Fig. 12 shows results obtained with a Pt(100) surface under conditions of self-sustained temporal oscillations /26/. Here data for the variation of the work function (≙ reaction rate), probing the integral behaviour of the whole surface area of 7x4 mm^2, are displayed together with the intensities of various LEED spots monitoring the structural properties of the surface within an area of about 0.1 mm^2. The latter data demonstrate the occurrence of the following local changes within one period of the macroscopic oscillations of the reaction rate: The CO-covered 1x1-phase (≙

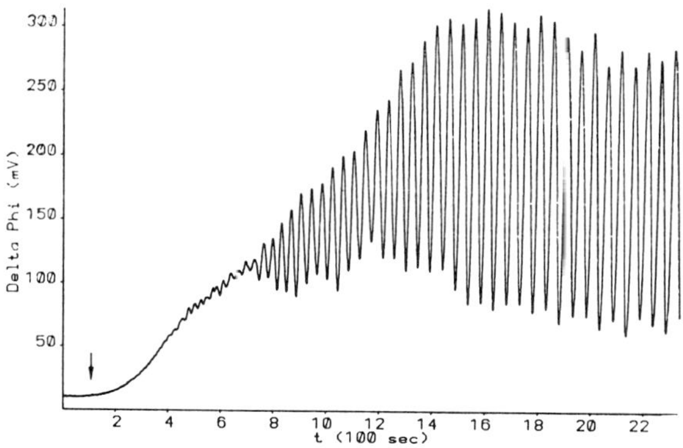

<u>Fig. 11:</u> Work function (reflecting the reaction rate) of a Pt(110) surface during catalytic oxidation of CO under isothermal flow conditions at T=470 K and p_{CO}=2.3x10^{-5} Torr. At the point marked by an arrow p_{O_2} was changed from 1.5 to 2.0x10^{-4} Torr.

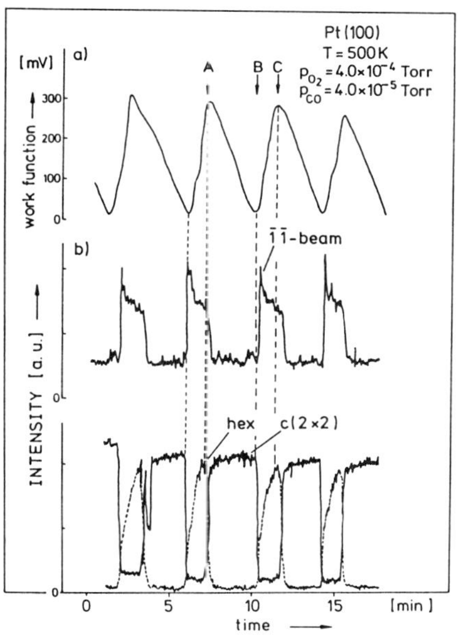

<u>Fig. 12:</u> Kinetic oscillations during the catalytic oxidation of CO on a Pt(100) surface. a) Variation of the work function (≙ reaction rate), b) variations of the intensities of different LEED spots with time

c2x2 spot intensity) remains almost constant over a longer period and then rapidly disappears, while simultaneously a high O-coverage on 1x1 surface (≙ 11 spot intenstiy) is built up. Subsequently the latter decays continuously parallel to the build-up of the hex-phase. When this hex-phase has reached maximum concentration within the area probed by the LEED beam it rapidly disappears and simultaneously the c2x2-1x1 phase is formed again.

These experimental observations can be well reproduced by the numerical solution of a set of four coupled, non-linear differential equations describing the temporal variations of the CO coverages on both the hex and 1x1-phases, of the O-concentration on the 1x1-phase, and the hex$\rightleftharpoons$1x1 transformations driven by critical CO coverages /27/. Alternatively, modelling was successfully achieved by computer simulation /28/. It should be noticed that all the relevant parameters had been derived from independent experimental investigations of the adsorptive properties, whereby the above-mentioned difference of the oxygen sticking coefficient between the hex- and 1x1-phase is of particular relevance. This difference causes the surface to switch between a reactive and an inreactive state, which seems to be the basic requirement also for kinetic oscillations occurring with other systems /29/.

Instead of discussing the theoretical formalism governing the present features in detail, we will merely sketch the qualitative aspects of the underlying mechanism by means of Fig. 13 /30/. If we start with a surface close to completion of the c2x2-1x1 structure (i.e. the conditions have to be adjusted in way that the CO coverage in the absence of O_2 is near Θ=0.5 and the surface is fully transformed into the 1x1-phase), then oxygen adsorption is strongly inhibited and will be restricted to only few sites (X). Once an adsorbed O atom is formed, it will rapidly react with a neighboring CO_{ad}, and the CO_2 formed is immediately released into the gas phase. Thus in an autocatalytic process free adsorption sites are created and the c2x2-phase will be rapidly removed. The bare 1x1-sites exhibit a high sticking coefficient for O_2 and hence the O_{ad}-concentration rises rapidly parallel to the break-down of the CO-c2x2 phase. (The conditions are chosen in a way that the rate of O_2 adsorption exceeds markedly that of CO adsorption on these bare patches). As a consequence the (11) LEED spot intensity rises steeply parallel to the drop of the c2x2-intensity. Due to the continuous reaction between O_{ad} and CO molecules arriving either from the gas phase or by diffusion from neighboring c2x2-domains, the O-coverage will, however, not be high enough to prevent slow (activated) transformation of the 1x1-structure

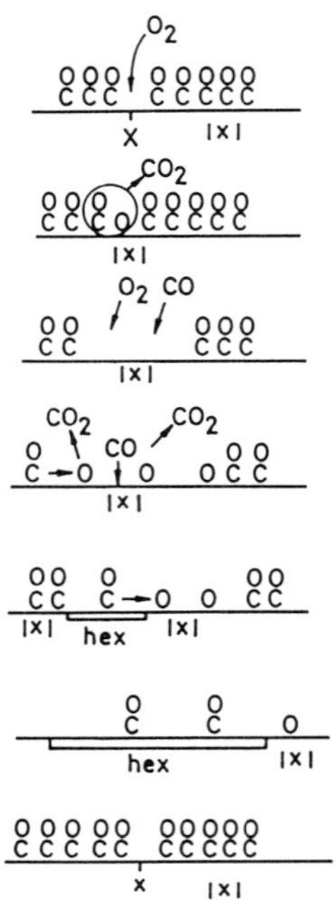

Fig. 13: Schematic sketch of the various steps involved in the structural transformations on Pt(100) associated with the kinetic oscillations

into the hex-phase: The (11)-intensity decreases while simultaneously that of the hex-phase grows. Because of the small O_2 sticking coefficient on the hex-surface there will essentially only CO adsorption take place on this phase. These molecules may be 'trapped' by diffusion to adjacent O-1x1-sites where they react off. As a consequence the hex domains will grow further in size unless loss of CO adsorbed on these domains by the process just mentioned becomes less probable. Then the CO coverage on the hex-patches rises and finally exceeds the critical value for back-transformation into the c2x2-1x1 phase which step completes one cycle.

While in the example represented by Fig. 12 the integral behavior of the reaction rate (≙ work function) changes with time with the same periodicity as the mezoscopic structural properties, the actual shapes of these variations cannot be directly related to each other. This is connected with the problem of spatial synchronization which was investigated experimentally by scanning the electron beam used for LEED across the sample surface area /31/. In this way the spatial distribution of the c2x2-1x1 and hex-domains could be determined every 10 sec. A typical set of data showing the respective intensity distributions is reproduced in Fig. 14 /32/. This shows nicely that the structural transformation propagates wave-like across the whole surface area and thus causes spatial self-organisation which is responsible for the observation of integral effects in the reaction rate. Frequently several trigger waves were observed to emanate at different points on the surface. In these cases (which were more the rule) the macroscopic temporal oscillations became rather irregular, as expected.

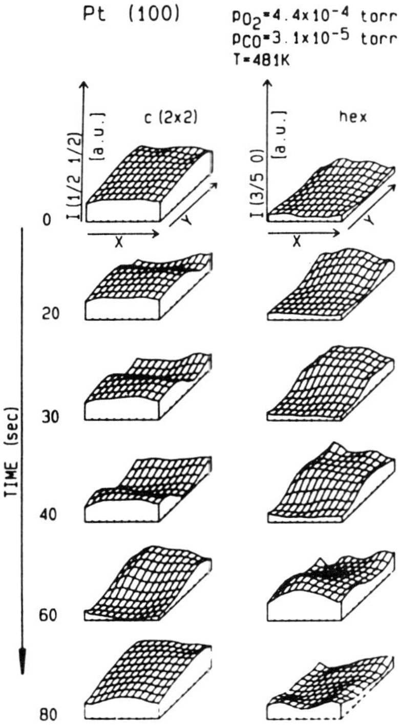

Fig. 14: Scanning LEED data recorded over a 7x4 mm^2 area of a Pt(100) surface at varying intervals during the kinetic oscillations demonstrating the wave-like macroscopic propagation of the structural transformations

This complication is absent with the oscillations observed with the Pt(110) surface. (Here the situation may become complex for another reason, namely if the oscillations occur under conditions which also enable the process of facetting as discussed in the preceding section /20/. These phenomena will be discarded in the present context). With this system the conditions for the occurrence of kinetic oscillations are restricted to very narrow ranges of parameters /33/. At fixed p_{O_2}, T, for example, the CO pressure may typically not vary by more than about 1%. This has an important consequence for the process of spatial self-organization: Variations of the reaction rate associated with temporal oscillations cause modulations of the partial pressures of the reactants of a similar order of magnitude. Hence synchronization is mediated through the gas phase and the whole surface area oscillated in phase without any detectable spatial pattern formation such as in the case of Pt(100).

This extreme sensitivity of the temporal oscillations with Pt(110) on the external parameters is illustrated by Fig. 15 which shows a sequence of time series recorded under conditions in which the CO pressure was varied between 4.2 and 3.9x10^{-5} Torr, while the oxygen pressure and the temperature were kept strictly constant /32/. While at first the amplitudes of the oscillations were very small, slight reduction of p_{CO} causes these to grow (b), which is further followed by period doubling (c) and transition to irregular (chaotic) behavior (d) before the oscillations die off again. The phase portraits constructed from the time series Figs. 15b)-d) as reproduced in Fig. 16 reflect these properties again, namely the transition from a simple limit cycle via bifurcation to a strange attractor. Detailed investigations showed that this sequence is a manifestation fo the Feigenbaum scenario for the transition from regular oscillations to deterministic chaos /34/.

For the sake of completeness we mention that additional effects of non-linear dynamics, namely harmonic resonance, sub- and superharmonic entrainment, as well as quasiperiodic behavior, may be observed with the present system, if under conditions of stable, self-sustained kinetic oscillations it is subjected to low-amplitude periodic modulation of one of the external parameters (e.g. O_2 partial pressure) /33/.

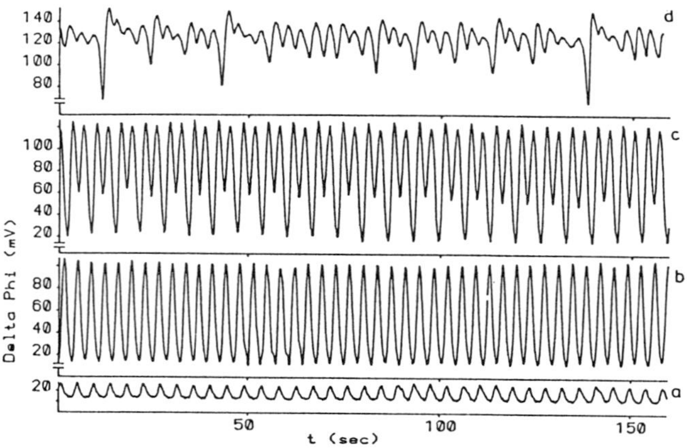

Fig. 15: Kinetic oscillations during the catalytic oxidation of CO on Pt(110) at T=530 K, p_{O_2}=8x10^{-5} Torr. p_{CO} was altered from 4.2x10^{-5} Torr (a) to 3.9x10^{-5} Torr (d)

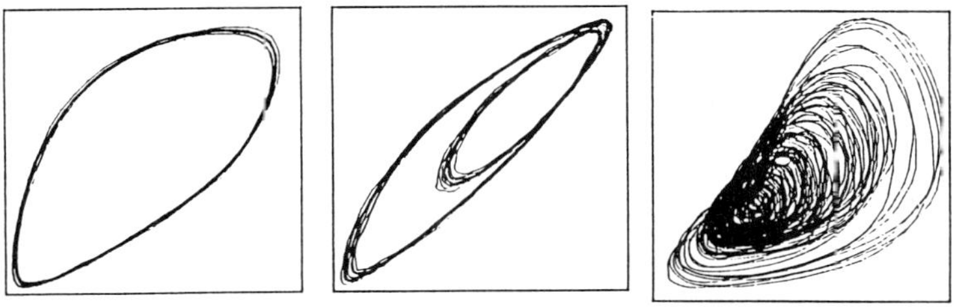

Fig. 16: Phase portraits constructed from the time series of Fig. 15 b)-d) by the delay method, $\Delta\varphi\,(t+\tau)$ vs. $\Delta\varphi(t)$ with τ=0.8 sec.

5. CONCLUSIONS

The aim of this contribution was to illustrate by means of a few selected examples some of the principles which govern the dynamics of chemical interactions at surfaces. The situation appears to be rather complex and far away from the possibility for theoretical treatment based on first principles, although the tools of surface physics, on the other hand, provide rather detailed insights into the microscopic processes. The situation with 'real' systems, as e.g. present in industrial heterogeneous catalysis, is even more complex, mainly for the following reasons:

i) *Surface inhomogeneity.* Large surface area catalysts are inhomogeneous with respect to chemical composition as well as crystallographic structure. Large efforts are therefore made to characterize the actual catalyst surface and to identify the nature of its 'active centers', which knowledge may form the starting point for investigations with model systems along the surface science approach.

ii) *Complex reaction pathways.* In general, different reaction possibilities exist for the adsorbed particles. This required identification of the various surface species whereby molecular complex chemistry may serve as a useful guide.

iii) *The pressure gap.* The surface spectroscopic techniques are usually not applicable under the high pressure conditions of real systems, so that the transferability of data is always a serious question.

Despite this situation which looks rather unfavourable at a first glance, there has recently been encouraging progress in practical catalysis research along the sketched lines, in so far as, for example, the yields of an industrial plant for ammonia synthesis could be predicted with high accuracy on the basis of data obtained with single crystal surfaces under UHV conditions /35/. It appears as if this good agreement is not just accidental, but that the applied procedure may be generalized to calculate rates for non-ideal surfaces /36/. No doubt heterogeneous catalysis is on the transition from 'black art' to a physical science.

REFERENCES

1. J. Segner, H. Robota, W. Vielhaber, G. Ertl, F. Frenkel, J. Häger, W. Krieger, H. Walther: Surf. Sci. 131, 273 (1983)
2. J. A. Barker, D. J. Auerbach: Surf. Sci. Rep. 4, 1 (1985) (Review)
3. C. W. Muhlhausen, L. R. Williams, J. C. Tully: J. Chem. Phys. 83, 2594 (1985)

4. a) C. T. Rettner, F. Fabre, J. Kimman, D. J. Auerbach: Phys. Rev. Lett. 55, 1904 (1985)
b) M. J. Cardillo: this conference
5. a) M. C. Lin, G. Ertl: Ann. Rev. Phys. Chem. 37, 587 (1986)
b) D. S. King, R. R. Cavanagh: In Chemistry and Structure of Interfaces, ed. by R. B. Hall and A. B. Ellis (VCH Weinheim 1986) p. 25
6. A. Mödl, T. Gritsch, F. Budde, T. J. Chuang, G. Ertl: Phys. Rev. Lett. 57, 384 (1986)
7. J. Kimman, C. T. Rettner, F. Fabre, D. J. Auerbach, J. A. Barker, J. C. Tully: Phys. Rev. Lett. 57, 2053 (1986)
8. W. L. Nichols, J. H. Weare: J. Chem. Phys. 63, 379 (1975); 66, 1075 (1977)
9. U. Leuthäusser: Z. Phys. B50, 65 (1983)
10. a) J. M. Bowman, J. L. Gossage: Chem. Phys. Lett. 96, 481 (1983)
b) W. Brenig, H. Kasai, H. Müller: Surf. Sci. 161, 608 (1985)
c) J. C. Polanyi, R. J. Wolf: J. Chem. Phys. 82, 1555 (1985)
11. D. C. Jacobs, K. W. Kolasinski, R. J. Madix, R. N. Zare: J. Chem. Phys. 87, 5038 (1987)
12. C. T. Campbell, G. Ertl, J. Segner: Surf. Sci. 115, 309 (1982); see also J. A. Serri, M. J. Cardillo, G. E. Becker: J. Chem. Phys. 79, 1530 (1983)
13. See e.g. a) Desorption induced by electronic transitions-DIET II, ed. by W. Brenig and D. Menzel (Springer 1985)
b) T. J. Chuang: Surf. Sci. Rep. 3, 1 (1983); Surf. Sci. 178, 763 (1986)
c) A. R. Burns, E. B. Stechel, D. R. Jennison: Phys. Rev. Lett. 58, 250 (1987)
d) K. Domen, T. J. Chuang: Phys. Rev. Lett. 59, 1484 (1987)
14. F. Budde, A. V. Hamza, P. M. Ferm, G. Ertl, D. Weide, P. Andresen, H. J. Freund: Phys. Rev. Lett. 60, 1518 (1988)
15. J. C. Polanyi, J. L. Schreiber: Faraday Disc. 62, 267 (1977)
16. H. J. Robota, W. Vielhaber, M. C. Lin, J. Segner, G. Ertl: Surf. Sci. 155, 101 (1985)
17. a) G. Ertl, S. B. Lee, M. Weiss: Surf. Sci. 144, 515 (1982)
b) M. Grunze, M. Golze, W. Hirschwald, H. J. Freund, H. Pulm, U. Seip, M. C. Tsai, G. Ertl, J. Küppers: Phys. Rev. Lett. 53, 850 (1984)
c) C. T. Rettner, H. Stein: Phys. Rev. Lett. 59, 2768 (1987)
18. G. Comsa, R. David, B. J. Schumacher: Surf. Sci. 95, L 210 (1980); see also M. P. D'Evelyn, R. J. Madix: Surf. Sci. Rep. 3, 413 (1983)
19. a) T. Engel, G. Ertl: Adv. Catalysis 28, 1 (1979)
b) G. Ertl: In Catalysis. Science and Technology, ed. by J. R. Anderson and M. Boudart, Vol. 4 (Springer 1983) p. 209
20. S. Ladas, R. Imbihl, G. Ertl: Surf. Sci. 197, 153 (1988)
21. I. Langmuir: Trans. Faraday Soc. 17, 607; 621 (1921)
22. R. J. Behm, P. A. Thiel, P. R. Norton, G. Ertl: J. Chem. Phys. 78, 7437; 7448 (1983)
23. a) T. E. Jackman, J. A. Davies, D. P. Jackson, W. N. Unertl, P. R. Norton: Surf. Sci. 120, 389 (1982)
b) P. Hofmann, S. R. Bare, D. A. King: Surf. Sci. 117, 245 (1982)
24. I. Prigogine: From Being to Becoming (Freeman, New York 1980)
25. M. Eiswirth, G. Ertl: Surf. Sci. 177, 90 (1986)
26. M. P. Cox, G. Ertl, R. Imbihl, J. Rüstig: Surf. Sci. 134, L 517 (1983)
27. R. Imbihl, M. P. Cox, G. Ertl, H. Müller, W. Brenig: J. Chem. Phys. 83, 1578 (1985)
28. P. Möller, K. Wetzl, M. Eiswirth, G. Ertl: J. Chem. Phys. 86, 3519 (1986)
29 L. F. Razon, R. A. Schmitz: Catal. Rev. Sci. Eng. 28, 89 (1986)
30. G. Ertl: Langmuir 3, 4 (1987)
31. R. Imbihl, M. P. Cox, G. Ertl: J. Chem. Phys. 84, 3519 (1986)
32. M. P. Cox, G. Ertl, R. Imbihl: Phys. Rev. Lett. 54, 1725 (1985)

33. M. Eiswirth, G. Ertl: Phys. Rev. Lett. 60, 1526 (1988)
34. K. Krischer, M. Eiswirth, G. Ertl: Surf. Sci. (in press)
35. P. Stoltze, J. K. Nørskov: Phys. Rev. Lett. 55, 2502 (1985)
36. M. Boudart: Catalysis Lett. 1, 21 (1988)

Discussion

D.R. HAMANN: Burrows and Chabal studied the oscillatory oxidation of CO on Pt under somewhat different conditions, higher pressures and predominantly (111) polycrystaline foils. Using *in-situ* infrared spectroscopy and other methods, they concluded that the build-up and burn-off of chemisorbed C drove the oscillations. This is very different from your structural mechanism, and I wonder whether you dispute their conclusions, or believe that there may be two different mechanisms that produce the same rather unusual phenomenon?

G. ERTL: At higher pressures and less stringent conditions of surface cleanliness other mechanisms like the one quoted by you, or one involving the formation and removal of an oxide layer may become dominant, while such a possibility can definitely be ruled out with our experiments. The common feature of all these mechanisms consists in the possibility of the surface to transform between an active and an inactive state.

M. BOUDART: I wish to comment on the contributions thus far of surface science to heterogeneous catalysis. First, surface science has refined old concepts like reconstruction and activated adsorption. Second, it has quantified the importance or lack of importance of structural factors. Third, it now provides standards to which work done with industrial catalysts can be referred. But since catalysis is first and foremost a kinetic phenomenon, the main potential assistance of surface science to catalysis is through the accumulation of rate constants that ultimately will form a data bank that will help to design new catalytic cycles. Indeed in the past year, it has been shown that the rates of carbon monoxide oxidation on Rh(1) and of ammonia synthesis on Fe(2) both at high pressure on industrial catalysts can be calculated from data obtained at low pressure on single crystals.

(1) S.H. Oh, G.B. Fisher, J.E. Carpenter, and D.W. Goodman, *J. Catal.*, **100**, 360 (1986).
(2) P. Stoltze and J.K. Nørskov, *J. Vac. Sci. Technol.* **A5**, 581 (1987).

J.W. GADZUK: To what extent is interest in the rotational excitation of scattered/desorbed NO dictated by its intrinsic importance in surface chemical dynamics as opposed to merely the compatibility of the NO rotational spectrum with current laser technology?

Secondly, what about intramolecular vibrations? Since the elementary chemical events of bond breaking and formation involve movement of energy into and out of vibrational coordinates, it would seem that the study of vibrational excitation in beam scattering from surfaces should be relevant and important.

G. ERTL:

(1) NO is a prototype molecule whose spectral features are conveniently accessible by current laser tools. Since it is an open–shell molecule, it offers additional interesting properties such as two different spin-orbit manifolds which may be subject to further studies. The results presented here are considered to be representative also for other diatomic molecules.

(2) In direct–inelastic scattering with thermal kinetic energies, respectively non-dissociative trapping/desorption under conditions as discussed in my lecture, vibrational excitation plays a negligible role due to energetic reasons. This is quite different with dissociative adsorption/recombinative desorption processes where vibrational excitation may indeed become important, as for example shown in experiments by Zare *et al.*

P. BORCKMANS: Are the activity waves on Pt(100) characterized by a single or a small band of wavelength? Is it furthermore possible to relate this wavelength to the intrinsic kinetic and transport parameters of the system?

G. ERTL: The waves of structural transformation on Pt(100) were in fact found to frequently start at different points on the surface and to interfere with each other (like with the unstirred Belousov-Zhabotinski reaction) so that the macroscopic wavelengths are not very reproducible. The propagation of a single concentration wave depends on the rate constants of the surface transformation and of the different reaction steps, as well as on the diffusion of adsorbed CO, in a rather complex manner.

M. CARDILLO: If I recall correctly, the CO oxidation oscillates for dispersed Pt catalysts. If true, how does the mechanism you postulate allow for the required spatial communication? If a different mechanism is required, this suggests a concern with separating those processes which drive oscillations from those which follow them.

G. ERTL: Oscillations with supported Pt catalysts are observed at pressures which are several orders of magnitude higher than those applied in our experiments. In addition, these catalysts have a much poorer heat conductivity than our bulk single crystals. As a consequence, the heat released by the reaction causes considerable variations of the temperature and spatial communication is in these cases achieved through these temperature gradients. Our observations refer to isothermal conditions. Once the conditions for oscillations have been established, it may take quite some time until macroscopic effects can be observed (as was shown by a figure), i.e., until different patches on the surface oscillate in phase. The structural transformations are intimately linked to the occurrence of oscillations whereby no distinction between cause and consequence can be made.

I. PRIGOGINE: Is there a dimensionless parameter characterizing the reconstruction of the surface as the result of catalytic action?

G. ERTL: The driving force for the structural transformation of the surface lies in the difference of the CO adsorption energy on the two surface modifications and the local CO coverage which in turn is affected by the catalytic reaction. So this transformation cannot be characterized by a single parameter.

J.P. TOENNIES: I'm wondering if the increase in the H_2/Ni(111) sticking coefficient with kinetic energy might be related to the onset of multiphonon processes? The reason I ask this question is that the curve you showed is very similar to the increase in multiphonon processes which we see in He scattering on similar nickel surfaces.

G. ERTL: Our interpretation is that there exists (in contrast to Ni(110)) an activation barrier for dissociative adsorption which has to be overcome by translation energy of the impinging molecule. This conclusion is supported by the observation that in the reverse process of recombinative desorption the H_2 molecules come off the surface with translational energies higher than corresponding to the surface temperature.

D.A. KING: In your own much earlier work on O_2/CO reacting on Pd surfaces, and in our work on O_2/CO on Pt(111), it was clearly demonstrated that the reaction to form CO_2 occurs at CO island boundaries, but I notice that you treat the adsorbate as randomly distributed in your current work. Is this because the surface temperature in the work you refer to here is considerably higher?

G. ERTL: The steady-state experiments described here were performed at considerably higher temperatures with a higher degree of disorder on the surface. The kinetics were discussed here in terms of random distribution of the adsorbates for the sake of simplicity, but the qualitative features would not be affected by introducing some degree of order.

P. MARCUS: When CO transforms the Pt(100) and Pt(110) one would expect many small fragments of close-packed rows to form, which could have different chemical properties. But perhaps these fragments move to give a simple stepped structure of the kind you showed. Do you observe such behavior?

G. ERTL: The reconstructed ('hex') Pt(100) surface exhibits an atomic density which is about 20% higher than the non–reconstructed 1×1–phase. STM observations demonstrated a layer above the original surface plane during the adsorbate–induced transformation where they cluster together to islands whose size depends on temperature (because of diffusion). Similarly with Pt(110) the observed LEED beam broadening and splitting reflects analogous processes. The facet spots become sharper at higher tempratures, again indicating diffusion limitation.

J.T. YATES, JR.: Your lecture is a beautiful description of studies on single crystal surfaces where site uniformity is encouraged by the experimental method. One of the major factors in technological heterogeneous catalysis is the *active site* and the activity of special sites produced by *promotors* or *poisons*. Could you comment on the role of the methods of surface science in studying the details of the special chemistry present on these sites?

G. ERTL: There exists a large number of investigations in which the effects of well-defined additives on single crystal surfaces as models for promoters (like K) or poisons (like S) with respect to the chemisorption properties of molecules involved in catalytic reactions were studied. These results were able to connect the influence of these additives on the catalytic activity with their physical (i.e., electronic) properties. This field represents a particularly fruitful link between surface physics and chemistry.

G.A. SOMORJAI: Your talk, as well as others (like Sylvia Ceyer's), clearly indicated the importance of molecular excitations (translational, vibration, rotation, or perhaps electronic) in influencing sticking and dissociation probabilities on the surface the catalytic turnover roles and reaction selectivities. Up to now heterogeneous catalysis has been the chemistry of molecules in their ground energy states. Would it not be important to use plasmas or light to pre-excite molecules and then study their altered surface chemistry or catalytic chemistry?

G. ERTL: This will be certainly a very interesting field for future fundamental studies with flat surface onto which the excited molecules may directly impinge. For practical purposes with porous high surface area catalysts the molecules would be largely deexcited before they reach the surface site on which they react.

The Photochemistry of Methyl Halides Adsorbed on Pt(111)

B. Roop, S.A. Costello, Z.-M. Liu, and J.M. White

Department of Chemistry, University of Texas, Austin, TX 78712, USA

We report that photochemistry of methyl halides adsorbed on Pt (111) is observable and that a significant fraction of the photoproducts is, at low temperatures, held at the surface. The evidence favors dissociation of the C-X bond to form CH_3 and X.

1. Introduction

Most investigations of surface photochemistry using modern tools of surface science have been carried out on semiconductor and insulator surfaces [1-5]. There are now a few cases involving metal-adsorbate interfaces, some of which lead to observable photochemistry [6-8] and others which do not [9]. Quenching of fluorescence is well-established [10] but the extent to which other bond breaking processes compete with quenching of electronic and vibrational energy is not well established.

In this paper, we show that ultraviolet light, located in the region of the normal gas or liquid phase absorption spectra of methyl bromide and methyl chloride [11] produces a photoeffect which we interpret as cleavage of the carbon-halogen bond.

2. Experimental

The experiments were carried out in a standard ultrahigh vacuum chamber. The Pt(111) crystal was prepared, cleaned, heated and cooled by standard techniques [12]. Coverages of CH_3Br and CH_3Cl were prepared by controlled dosing and annealing of high purity gases.

The light source, external to the UHV system, was a focussed high pressure 100 watt Hg lamp. After passage through selected filters, the radiation was transmitted through the vacuum wall using high quality quartz windows.

Analysis of the effects of irradiation were made by temperature programmed (5 K/s) desorption (TPD) and Auger electron spectroscopy (AES) after illumination.

3. Results and Discussion

In TPD of both methyl halides, there is no more than 1% <u>thermal</u> decomposition after adsorption at any temperature below 130K. No methane desorbs. On the other hand, after <u>irradiation</u> with the full arc (Fig. 1B), significant amounts of CH_4 were present in TPD for either monolayer or submonolayer amounts of

Springer Series in Surface Sciences Vol. 14
Solvay Conference on Surface Science
Editor: F.W. de Wette

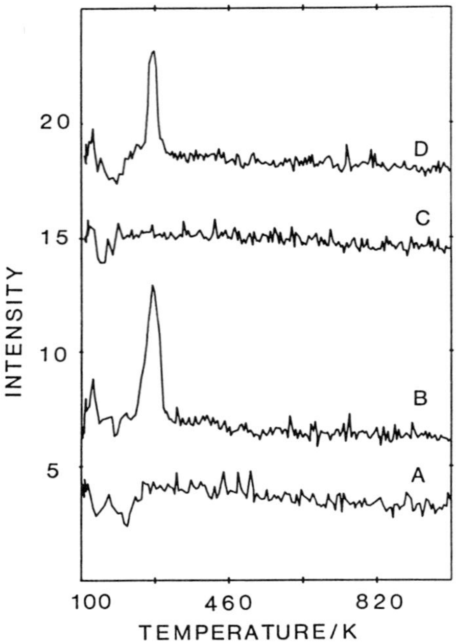

Figure 1. CH_4 (m/e=16) TPD spectra following irradiation of adsorbed CH_3Br at 108 K. The crystal was prepared by dosing a multilayer exposure (2 L) at 108 K and then heating to 130 K to remove the multilayer. (A) no irradiation; (B) irradiation with the full Hg arc; (C) irradiation with 420 nm cut-off filter; and (D) irradiation with 280 nm cut-off filter. During the irradiation period (1700 s), the bulk crystal temperature rose to 115 K. Zeroes are offset for clarity.

the halides. For CH_3Br at 115 K (includes an 8 K temperature rise due to radiative heating) there is a strong CH_4 TPD peak at 285 $\pm$ 5 K. The amount of CH_3Br desorbing was lower after irradiation. A small amount of molecular hydrogen desorbed in two broad peaks at 310 and 430K. In TPD after irradiation, Br desorbed between 750 and 850 K and carbon was detected by AES after TPD. Without irradiation, this C and Br was not present.

For CH_3Br, this methane signal was not present (Fig. 1C) after irradiation through a cut-off filter which removed all radiation below 420 nm. In this experiment, the lamp intensity was increased to deliver the same power (giving the same temperature rise of about 10K) as the full arc. Methane was found when the sample was irradiated through a 280 nm cut-off filter (Fig. 1C). We conclude that photons with energies higher than 2.95 eV are required for CH_3Br photochemistry and that radiative <u>heating</u> alone is insufficient.

For CH_3Cl, there was a methane desorption peak (280K) for the full Hg arc but none when the 280 nm filter was in place.

The appearance of CH_4 at 280 - 285K in TPD supports the photochemical formation of adsorbed methyl groups. The peak position is the same as that observed in the TPD of CH_3I which is known by HREELS to <u>thermally</u> cleave the C-I bond to form adsorbed CH_3 and I [13]. Once formed, these CH_3 groups begin to decompose thermally to provide surface H for the hydrogenation of CH_3.

Small peaks of H_2 desorb and both C and I are detected by AES after TPD.

Based on these results, we suppose that photochemistry is initiated by absorption into CH_3X and that the electronically excited state is repulsive just as it is in the gas phase [11]. After excitation, at least two processes may lead to C-X bond cleavage. Both of these bond breaking pathways can be assisted by the formation of very strong X-M and C-M bonds (M is the metal).

First, motion along the coordinate extending the C-Br bond may be rapid enough to compete effectively with electronic processes involving the metal substrate which would quench this state to the bound ground state. We note that the bond elongation process corresponds to less than one full normal vibration and that a decision to cleave the bond may require no more than a fraction of the normal distance traversed when vibrating in the ground state. If the bond stretches beyond that point, electronic quenching, which should be a very strong function of distance, will no longer be competitive.

Second, the electronically excited CH_3Br could relax to a vibrationally excited ground state species with excess energy in the C-Br vibrational mode. Energy in this mode could well be important in meeting the activation energy requirement for bond dissociation.

Based on molecular CH_3Br TPD before and after illuminating 1ML, we find that 0.7 ML of CH_3Br remains after 1700 s illumination under the conditions of Fig. 1. Of the 0.3 ML lost, more than 0.025 and less than 0.05 ML is accounted for as CH_3 retained by the surface (based on 0.025 ML of desorbed methane determined by TPD peak area and the requirement that, at most, one CH_3 decomposes to provide the H for hydrogenating another CH_3). Based on the delivered power and the wavelength distribution, we estimate an effective cross section for loss of CH_3Br of 3×10^{-21} cm^2 which is of the same order of magnitude (10^{-20} cm^2) as for gas phase photodissociation [11].

Since we have no independent measure of the photon absorption cross section for adsorbed CH_3Br, we crudely estimate the quantum yield using the gas phase molecular absorption coefficient at 254 nm of 1 $M^{-1} cm^{-1}$ [11]. The quantum yield for retained CH_3 then lies between 0.05 and 0.1 as compared to the gas phase photodissociation quantum yield of unity. The remaining 0.25 to 0.28 ML must desorb during irradiation as CH_3Br and/or CH_3.

The rough wavelength dependence for both CH_3Br and CH_3Cl [14] indicates that there is clearly a strong wavelength-dependent absorption but that it is red-shifted compared to the gas phase. The latter is based on the observation of photoproducts after irradiation of adsorbed CH_3Br with a 300 nm cut-off and CH_3Cl with the full arc. In neither case, would yields be detectable, based on the use of gas phase extinction coefficients [11]. Simply red-shifting the gas phase absorption spectra without changing the integrated absorption strength would lead to an increased extinction coefficient and a decreased quantum yield compared to the above calculation.

Our results confirm and complement those of Marsh et al. [7], who concluded that photofragmentation of CH_3Br occurred on Br/Ni(111) irradiated with 193 nm excimer radiation. They detected CH_3 desorbing during irradiation. Because CH_3Br decomposes thermally on Ni but not on Pt(111), the latter is simpler and our postirradiation TPD data clearly show that the metal will retain significant amounts of photodissociated products.

Acknowledgment

This work was supported in part by the National Science Foundation Materials Research Group Program (DMR8418086).

References

[1] R. M. Osgood, Jr.: Ann. Rep. Phys. Chem. **34**, 77 (1983).
[2] E. B. D. Bourdon, J. P. Cowin, I. Harrison, J. C. Polanyi, J. Segner, C. D. Stanners, P. A. Young: J. Phys. Chem. **88**, 6100 (1984).
[3] F. L. Tabares, E. P. Marsh, G. A. Bach, J. P. Cowin: J. Chem. Phys. **86**, 738 (1987).
[4] J. R. Creighton: J. Appl. Phys. **59**, 410 (1986).
[5] N. S. Gluck, Z. Ying, C. E. Bartosch, W. Ho: J. Chem. Phys. **86**, 4957 (1987).
[6] B. Roop, S. A. Costello, C. M. Greenlief, J. M. White: Chem. Phys. Lett. (in press).
[7] E. P. Marsh, F. L. Tabares, M. R. Schneider, J. P. Cowin: J. Vac. Sci. Technol. **A5**, 519 (1987).
[8] F. G. Celii, P. M. Whitmore, K. C. Janda: Chem. Phys. Lett. **138**, 257 (1987).
[9] K. Domen, T. J. Chuang: Phys. Rev. Lett. **59**, 1484 (1987).
[10] R. R. Chance, A. Prock, R. Silbey: J. Chem. Phys. **62**, 2245 (1975).
[11] J. C. Calvert, J. N. Pitts: Photochemistry, (Wiley, N.Y., 1966)
[12] J. R. Creighton, J. M. White: Surface Sci. **136**, 449 (1984).
[13] M. A. Henderson, G. E. Mitchell, J. M. White: Surface Sci. **184**, L325 (1987).
[14] S. A. Costello, B. Roop, Z.-M. Liu, J. M. White: to be published.

Discussion

A. M. BRADSHAW: Are you certain that you are observing a photochemical effect and not excitation via photoelectrons from the substrate? The presence of a very low energy $n - \sigma^*$ transition in these methyl halides suggests that there is a shallow affinity level, which could result in a negative ion state upon electron capture and subsequent decomposition. In the present case the threshold for photoemission could indeed be as low as 420 nm due to the lowering of the work function by the adsorbate. I am reminded of the experiment which Dr. de Lozanne reported with the tunnel microscope where he decomposes an organo-Cd molecule with very low energy electrons. Similarly, the primary process in laser writing on semiconductor surfaces has been shown by the Kompa group in Garching to be a photoelectron–induced decomposition of the organometallic.

J.M. WHITE: For the case of ketene we have shown that wavelengths larger than those required to overcome the work function are still capable of causing a photoeffect. Thus, we are able to eliminate photoejected electrons. Further, nanoamps of electron current from a filament are not effective in causing the formation of products. We have no data which would rule out optical excitation directly from the metal to the σ^* orbital of CH_3X.

R. GOMER: It is interesting that no C_2H_6 is observed, suggesting that diffusion of CH_3 is so activated that dissociation into $H + CH_{n<3}$ predominates over diffusion even over the small distances required for CH_3 recombination to C_2H_6.

J.M. WHITE: You are quite correct. C-H bond cleavage always predominates on Pt(111) and the mobility of H(a) below 280K is high. Thus, H(a) can easily find a CH_3(a) with which to react and form CH_4. CH_3(a) mobility is quite low between 100 and 280K.

J.G. DASH: Have you examined the possible dependence of quantum yield on the direction of polarization of the light? If the molecules are oriented, and the molecular cleavage is a direct photon process, the yield should be strongly dependent on the relative polarization–surface normal direction.

J.M. WHITE: You are quite correct that the angular dependence should be studied, but we have not yet done so.

J. T. YATES, JR.: Is the red shift observed for methyl halide surface photolysis on Pt(111) comparable to red shifts observed for these molecules in condensed phase or in complexes?

J.M. WHITE: Red shifts for these molecules in typical solvents are small. The optical spectra of their complexes with metals are, to my knowledge, unknown.

S.J. TEICHNER: For the decomposition of methyl halides into methane some extra hydrogen is required. This means that a more severe decomposition of the methyl halide occurs giving hydrogen and carbon (which stays on the surface).

J.M. WHITE: Our temperature programmed desorption of CH_3I and photochemical decomposition of CH_3Br and CH_3Cl all indicate that there is some C-H bond cleavage in $CH_3(a)$ between 200 and 280K to provide mobile surface hydrogen which reacts with $CH_3(a)$ to form CH_4.

G. ERTL: The relatively large cross sections reported appear to be inconsistent with the Menzel–Gomer model for electronic excitation of a molecule adsorbed on a metal surface, whereafter there is a high probability for filling the hole in the ground state by electron transition from the metal. This could, however, mean in the present case that indeed a negative ion resonance is intermediately formed from where decomposition takes place. This state would be indistinguishable from that proposed by Bradshaw as being the result of capture of secondary electrons by the ground–state adsorbate.

J.M. WHITE: I agree that negative ion resonances are a viable candidate for the mechanism of the photoeffect. More detailed wavelength–dependent studies and electron energy loss experiments are underway to help us resolve some of the mechanistic issues.

C.B. DUKE: Following up prior questions, I might note that the large red shift as well as the large cross section for the photodissociation of adsorbed $BrCH_3$ favors a mechanism based on the capture of photoelectrons to form a negative adsorbed ion which subsequently dissociates as opposed to the formation of an excited state of a neutral adsorbate which subsequently dissociates. Which occurs, of course, is a quantitative issue depending upon both the electronic energetics of the excited neutral versus the negative ion and the relative dissociation cross sections of the two species. One might expect that both factors favor a negative–ion pathway for a strongly red–shifted dissociation with large cross section. What is the status of the quantitative analysis of the photoinduced negative–ion dissociation pathway?

J.M. WHITE: The results we have cannot yet distinguish the negative–ion resonance path from the dissociation of an excited neutral. We intend to systematically vary the work function of the adsorbate substrate system, keeping the excitation energy constant. Perhaps this will help resolve which of these paths dominates.

J.P. TOENNIES: Concerning the electron capture cross section, it's well known that these capture cross sections for SF_6 are at zero energy. They have a maximum there, so your electron beam experiment doesn't really simulate the correct experimental conditions that you do have in photoemission. Because of the large trapping cross section at low energies SF_6 is used in transformers to suppress discharges.

M. CARDILLO: The experiments of Cowin's group at Santa Barbara do a reasonably good job at accounting for the energetics of photodissociation of CH_3Br adsorbed on surfaces. Is there a cross section available from their work? If so, that will help decide whether the negative–ion resonance from photoelectrons is a significant contribution to the chemistry you observe for CH_3Br.

J.M. WHITE: There is not a *total* cross–section available from Cowin's work.

Hydrogen Spillover on Oxide Surfaces

S.J. Teichner

University Claude Bernard (Lyon I), 43, Bd. du 11 Novembre 1918, F-69622 Villeurbanne Cedex, France

The migration of a sorbed species, like hydrogen, from one solid phase, like a metal, where it is easily adsorbed and dissociated, onto another solid phase, like an oxide, in contact with the first, where it does not directly adsorb, has been defined as spillover.

Spillover may produce various changes in the non-metallic phase onto which it occurs. First, the spiltover species, like hydrogen, is transported across the surface as a two-dimensional gas. It may exchange with similar surface species : OH + D(sp) $\rightarrow$ OD + H(sp). In a stronger interaction the spiltover species may react with the oxide surface : H(sp) + M-O-M' $\rightarrow$ M-O-H + M', which leads to the creation of surface defects and/or active sites. Further, the bulk of the oxide may be transformed into a different structure. In this way an oxide can be reduced into metal at a temperature which is lowered by the presence of spiltover hydrogen, or it may be transformed into hydrogen bronze (MoO_3 or WO_3). In each of these cases the oxide phase cannot be considered as inert. It is not serving to promote the inherent activity of the metal. The oxide phase is participating directly in the transport, exchange and reaction with the spiltover species. In many cases, the oxide phase is able to become catalytically active on its own and thereby to participate directly in catalytic reactions.

1. Enhanced adsorption

Litterature data are mainly centered around hydrogen adsorption on group VIII, VI and I-metals supported on Al_2O_3, SiO_2, zeolites or carbons. If the metal alone is sorbing hydrogen which does not migrate on the oxide support the ratio H/M, where H stands for the number of H atomic species adsorbed per total number of supported metallic atoms M, rarely exceeds 2, for the best dispersions of the metal. Now ratios as high as H/Pt $\approx$ 200 can only be understood if the oxide support is also acceptor of hydrogen [see for the review ref. (1) and (2)] . Simultaneously, the rate of adsorption increases for spillover enhanced adsorption (3). The extent of this increase may depend on the contact between the metallic component and the second phase. Indeed the metal can be either supported on this second phase or it can be simply mechanically admixed to this phase. For metal-on carbon systems, as active carbon adsorbs hydrogen by itself (at T > 300°C) the effect of spillover is to increase the rate of uptake but not the extent. With the inorganic oxides spillover most often results in an increase in the amount of adsorption. The rate determining step in the overall process does not seem to be the surface diffusion (4) but the spillover itself i.e. the passage from the metal to the support (2).

Some differences among the litterature data concerning the nature of factors influencing the spillover can be noticed (2) : the range of the

Springer Series in Surface Sciences Vol. 14
Solvay Conference on Surface Science
Editor: F.W. de Wette

temperatures of hydrogen adsorption (below room temperature or significantly above) ; the possible necessity of a cocatalyst such as water or other proton acceptors (alcohols) ; the amount and the percentage dispersion of the source of hydrogen spillover ; the nature of the contact between the source and the acceptor ; the specific areas of the coupled source-acceptor ; the possibility of migration through the gas phase and not exclusively on the surface ; the role of the partial pressures of H_2 ; the strength of the bond between spiltover hydrogen and acceptor sites ; the chemical nature of the acceptor and source and the nature of spiltover hydrogen.

The first step in the modification of the surface of the acceptor by spillover is the isotopic exchange. This reaction, when involving hydrogen and deuterium, is easily monitored by IR spectroscopy, chromatography and mass spectroscopy. The exchange between D_2 in the gas phase and OH's on alumina cannot take place in the absence of the metal (Pt)(5). Diffusion over a distance from the source of spillover (Pt spot) on a surface of alumina or silica pellet was clearly shown by IR and FTIR techniques (2).

2. Weak and strong interactions with oxides

The lowering of the temperature of reduction by H_2 of numerous metal oxides by addition of a transition metal is well documented and is attributed to spillover [see for the review ref. (1) and (2)]. Among all the metal dopants added, Pt and Pd are seen to be the most efficient. For H_2 reduction without spillover (without the metal activator), the well-known sigmoïdal kinetic curves are often found, whereas for spillover induced reductions no evidence for an induction period was found. The apparent activation energies of the spillover induced reductions are similar to those of the "normal" reductions (without the activator). The effect of the activator (Pt) was found to increase the available reactive hydrogen and/or to increase the rate of the nucleation (pre-exponential factors). Thus, this "catalysis" of reduction in the presence of the activator increases the availability of H but does not "classically" decrease the activation energy.

The formation at fairly low temperatures of tungsten or molybdenum oxide hydrogen bronzes, induced by spillover from Pt, has been thoroughly studied [see for review ref (1) and (2)]. A new, hexagonal variety of $H_{0.4}WO_3$ can be obtained by H spillover from Pt whereas the non spillover hydrogen bronzes of WO_3 belong to the monoclinic variety. The rate determining step in the formation of these bronzes is the penetration of the spiltover species below the surface of the initial corresponding oxide (6).

In the strong metal-support interaction (SMSI)(chiefly on metal/TiO_2 systems), hydrogen spillover can be considered a necessary condition for creation of the SMSI state. The reduction of the oxide (Ti_4O_7) occurs near 500°C in the presence of a metal able to chemisorb and dissociate molecular hydrogen. The function of Pt is to provide a source of H atoms which spillover to titania and are responsible for its conversion to Ti_4O_7. The SMSI state tends to suppress the chemisorption of H_2. Pt particles, with or without trapped electrons (from reduced TiO_2) are always able to dissociate H_2 (H_2-D_2 exchange is not suppressed) but the particles of Pt enriched in electrons (SMSI state) do not bond strongly with hydrogen (7). In other words the SMSI state is observed with a support of high electronic conductivity. Even alumina in the case of Pt/Al_2O_3 can be reduced by H_2 at 500°C or even below (8) and a Pt-Al alloy is formed. Another explanation of the SMSI state of TiO_2 involves the migration of titanium suboxide onto Pt particles (9).

3. Catalytic activity induced on refractory oxides

Initially the spiltover hydrogen was considered only as a reactant for various hydrogenations. In particular, it has been presumed only to increase the amount of available hydrogen due to spillover from a supported metal, admixed or deposited on the support. Recent experiments show however that the spiltover hydrogen is not very reactive and that the true agent of hydrogenation is molecular hydrogen activated on the spillover treated surface of the oxide (10). A partial reduction of the host oxide by hydrogen spillover can induce or modify the catalytic activity of the host material. To yield unambiguous results, not screened by an eventual catalytic activity of the metal activator, a reactor was imagined in which the activator (Pt/Al_2O_3 catalyst) was first in contact with the acceptor (Al_2O_3 or SiO_2) and exposed at ~ 400°C to H_2. After this hydrogen spillover activation the catalyst may be withdrawn and isolated from the acceptor (fig. 1).

The interaction between the hydrocarbon to be hydrogenated and hydrogen, molecular and spiltover, is therefore observed in the absence of the metal, with only the acceptor oxide present in the reactor.

Silica may be converted by hydrogen spillover activation into a very unusual hydrogenation catalyst, active at 150°C for ethylene hydrogenation and not poisoned by O_2, H_2O or NH_3. The reaction is initially slow (induction period) due to the presence of spiltover hydrogen. This is no longer the case for the second dose of ethylene or for hydrogenation performed after the evacuation and desorption of the spiltover hydrogen. This behaviour shows that the spiltover hydrogen is strongly adsorbed on silica and is less reactive than molecular hydrogen. This last reactant is adsorbed at saturation at sites distinct from those adsorbing ethylene (order zero with respect to each reactant for a stoichiometric feed). The conversion of acetylene into ethylene and ethane is sequential and a maximum in the conversion to ethylene is found (12). Although 1,4-cyclohexadiene does not affect the hydrogenation of ethylene into ethane, it completely inhibits the conversion of acetylene into ethylene. The 1,3-cyclohexadiene has no effect on either reaction. This behaviour

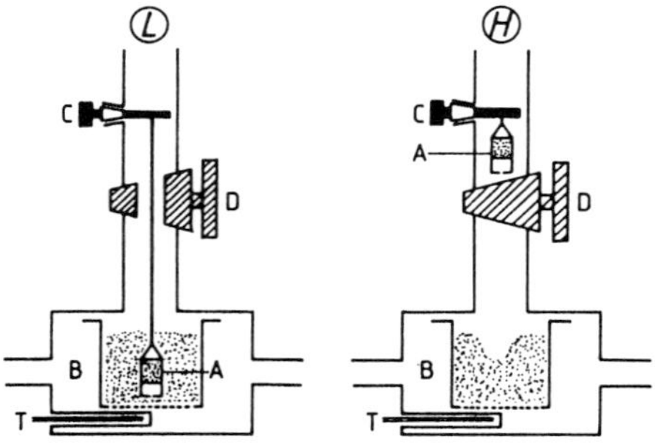

Fig. 1. Scheme of the reactor with an "elevator". Position L : catalyst (Pt/Al_2O_3 : 20 mg) in a porous pyrex bucket in contact with the acceptor oxide B (SiO_2 or Al_2O_3 : 1 g). Position H : the bucket A is withdrawn from the reactor by a lifting device C and the stopcock D isolates the catalyst (11).

shows that silica activated by hydrogen spillover acquires at least two types of hydrogenating sites, those poisoned by 1,4-cyclohexadiene in the conversion of acetylene into ethylene and those unpoisoned by this cycloolefin in the conversion of ethylene into ethane.

With benzene on activated silica the spiltover hydrogen can be simply added to this reactant (forming cyclohexene and cyclohexane) in a non-catalytic reaction which entirely exhaust this spiltover hydrogen. The catalytic reaction in the presence of molecular hydrogen at 170°C leads to ethane with the intermediate production of acetylene. In the presence of He the activated silica converts benzene only into acetylene. This cracking of benzene is poisoned by ammonia and inhibited by oxygen pretreatment (no effect on hydrogenation of ethylene). Therefore a third type of sites (for cracking) seems to be created on silica activated by hydrogen spillover and may be of an acidic nature (13). Silica activated by hydrogen atoms from a plasma, instead of spiltover hydrogen species from Pt/Al_2O_3, behaves exactly in the same way in the reactions with ethylene or benzene (14). 1,3-cyclohexadiene, like previously benzene, is hydrogenolyzed on activated silica into ethane, whereas 1,4-cyclohexadiene is only cracked to acetylene because, as pointed out before, it is a poison for the hydrogenation of acetylene.

The spillover activated alumina behaves in the same way as silica in the hydrogenation of ethylene at 150°C, whereas in the reactions with benzene and cyclohexadienes only cyclohexane is formed. Contrary to silica no cracking was observed in the presence of He. Ammonia and oxygen at 430°C did not affect the activity. The acid sites which were developed by the activation of silica do not seem to be induced on alumina (12). The activation by hydrogen spillover is, therefore, specific to the nature of the oxide to be activated. This conclusion is reinforced by a still different behaviour in the previous reactions and poisonings of magnesia activated by hydrogen spillover (15).

It appears therefore that some new and unforeseen catalytic properties may be induced on refractory oxides by hydrogen spillover. If the activator, like Pt/Al_2O_3, is not separated after the spillover activation from the activated oxide, these new properties could be masked by the catalytic activity of Pt/Al_2O_3.

4. Retention of catalytic activity

The spiltover hydrogen may be continuously supplied during the reaction because its source (like Pt/Al_2O_3) has not been removed from the oxide. This is the most studied example of the synergy in the catalytic reactions, observed when a supported metal (like Pt/Al_2O_3) is diluted with pure support (i.e. Al_2O_3)(16). The spiltover hydrogen may create, like previously, new catalytic centers on the support ; it may regenerate old and new centers by removing coke or carbonaceous material ("remote control" concept). The spillover of hydrogen from Pt towards carbonaceous material or even solid carbon, to form methane, is well documented [for review see ref. (2)]. Spillover is therefore a primary step in the elimination and/or control of coke for metal containing catalyst systems. The spillover of oxygen from a supported metal has also been implicated in the removal of coke and gasification of solid carbon phases. Many reactions involve redox couples in which reduction may be promoted by spillover from another phase as well as by direct adsorption from the fluid phase. Spiltover hydrogen may also be involved in promoting specific mixed metal or defected phases on which catalytic activity is found.

Finally, spillover may be involved in the removal of poisons to the catalytic activity that may form or adsorb on the surface. As an exemple, sulfur has been shown to be removed from the surface as H_2S via the spillover of hydrogen (17). All these processes contribute to the retention of activity.

5. Mechanims of spillover

The dissociative sorption of hydrogen onto a metal is still the object of controversy concerning its nature : homolytic or heterolytic. This can result in four possible species being created and subsequently transported on the sorbing surface. The atomic species can be a positive or negative ion, a radical or a bonded species. Various experimental arguments in favour of each of these species are found in the litterature [For review see ref. (2)]. It is therefore probable that hydrogen spillover does not occur via a single entity and that more than a single form of hydrogen can exist as a spiltover species under similar conditions. However many authors agree in the estimates of the concentration of spiltover hydrogen on the surface which, for many oxides, is of the order of 10^{12} cm^{-2} whereas for different carbons concentrations of 10^{13} to 10^{16} cm^{-2} were reported (2). For most systems the spillover from metal to support is the rate controling step whereas the surface diffusion onto the oxide surface is easy, as demonstrated by OH → OD exchange (2). At least for silica surface the primary mode of transport involves associated hydroxyls ("bucket brigade") whereas the isolated hydroxyls exchange with the spiltover deuterium in a secondary process (random surface diffusion). This hypothesis is supported by the observation that the rate of transport is decreased if the sample is pretreated by degassing at temperatures above 480°C where the associated hydroxyls are beginning to be removed (18).

In conclusion, the interphase transport of an activated species onto a surface (and sometimes into the bulk), where it is unable to be formed without the activator, can induce a variety of changes on, and reactions with, the surface. All the reactions of atomic hydrogen are found to be induced by spillover : exchange, bronze formation, reduction, demethoxylation, induction of catalytic activity. The effects are dependent on the nature of the receiving surface (and the temperature of the system). In the case of a catalytic reaction a distinction should be made between the effects of the spiltover species on the receiving solid phase and their effects on other adsorbed reactants.

1. P.A. Sermon and G.C. Bond, Catal. Rev., 8, 211 (1973).
2. W.C. Conner, G.M. Pajonk and S.J. Teichner, Adv. in Catal., 34, 1 (1986).
3. K. Fujimoto and S. Toyoshi, Proc. 7th Intern. Congr. Catal., 235 (1981).
4. R. Kramer and M. Andre, J. Catal., 58, 287 (1979).
5. W.J. Ambs and M.M. Mitchell Jr., J. Catal., 82, 226 (1983).
6. R.E. Taylor, M.M. Silva-Crawford and B.C. Gerstein, J. Catal., 62, 401 (1980).
7. B.H. Chen and J.M. White, J. Phys. Chem., 86, 3534 (1982).
8. K. Kunimori and T. Uchijima : in Spillover of Adsorbed Species, eds. G.M. Pajonk, S.J. Teichner and J.E. Germain, Vol. 17 (Elsevier, Amsterdam 1983) p. 197.
9. D.E. Resasco and G.L. Haller, J. Catal., 82, 279 (1983).
10. S.J. Teichner, G.M. Pajonk and M. Lacroix : in Surface Properties and Catalysis by non Metals, eds. J.P. Bonnelle et al Vol. C105 (Reidel Publ. Dordrecht, Netherlands 1983) p.457.

11. G.E.E. Gardes, G.M. Pajonk and S.J. Teichner, Comptes-rendus Acad. Sci., 227C, 191 (1973).
12. M. Lacroix, G.M. Pajonk and S.J. Teichner, Bull. Soc. Chim. France, 94 (1981).
13. M. Lacroix, G.M. Pajonk and S.J. Teichner, Bull. Soc. Chim. France, 265 (1981).
14. J.P. Nogier, J. Bonardet and J. Fraissard : in Spillover of Adsorbed Species, eds. G.M. Pajonk, S.J. Teichner and J.E. Germain, Vol. 17 (Elsevier, Amsterdam 1983) p. 233.
15. M. Lacroix, G.M. Pajonk and S.J. Teichner, React. Kinet. Catal. Lett., 12, 369 (1979).
16. B.K. Hodnett and B. Delmon : in Catalytic Hydrogenation, ed. L. Cerveny, Vol. 27 (Elsevier, Amsterdam 1986) p. 53.
17. T. Fleisch and R. Abermann, J. Catal., 50, 268 (1977).
18. J.F. Cevallos-Candau and W.C. Conner, J. Catal., 106, 378 (1987).

Discussion

K. TAMARU: Could you give us any estimated values of the rates of spillover for hydrogen and deuterium and also the differences in the temperature dependence (or activation energies) of the the two rates?

S.J. TEICHNER: The spillover of hydrogen or deuterium over silica or alumina surface is an easy process (diffusion coefficient in the range of $10^{-5} cm^2 sec^{-1}$; activation energy of the order of a few kcal/mole). The difficult step is the migration of hydrogen, dissociatively adsorbed on platinum, onto the silica or alumina surface, on which platinum is deposited or in contact. The nature of this contact between platinum particles and the surface of the oxide determines this difficulty. As an example of two opposite situations, platinum may be mechanically admixed to the oxide (bad contact), or it can be chemically deposited onto the oxide surface (good contact).

J.M. WHITE: Could you comment on the interpretation of the Maier experiments at Berkeley in which oxide overlayers on Pt foils showed high C–H bond cleavage even though no Pt was present at the surface? Is this result consistent with your suggestion of an active oxide formed by H–spillover?

S.J. TEICHNER: We have shown a few years ago [see ref. (2) of the paper] that the C-H bond may be cleaved by silica, previously activated by hydrogen spillover and in the absence of platinum and of the spiltover hydrogen. The reactivity of this spiltover hydrogen is rather low compared with the reactivity of molecular hydrogen activated by spiltover activated silica. The real catalyst is silica containing catalytic sites developed by hydrogen spillover activation. In a few cases, the spiltover hydrogen may induce a chain reaction with the hydrocarbons (like n-heptane), always in the absence of Pt, which finally gives various products of the C-H bond cleavage [ref. (2)].

M. BOUDART: Hydrogen spillover is an intriguing, old, and real phenomenon that is begging for a surface science approach for its elucidation. It is elusive because of the low number density of sites created by H spillover, $10^{12} cm^{-2}$. It is also a very slow phenomenon so that it does not compete with the catalytic reactions proceeding much faster on the metal itself. But it may contribute to slow phenomena, such as the removal of precursors of carbon residues on the support surface.

S.J. TEICHNER: The comparison between relative rates of catalytic processes occuring on a metal surface and on the surface of an oxide activated by hydrogen spillover is given in [ref. (2)] of the paper. For a typical catalyst containing less than 1% of Pt deposited on an oxide, like alumina or silica, the rate of the reaction which proceeds on the oxide surface may be very significant. The selectivity profile may be dramatically changed for a reaction which proceeds on a metal surface and the reaction with the same reactants on a spillover activated surface. This was, in particular, demonstrated in the case of the hydrogenolysis of n-heptane on hydrogen spillover activated erionite (A. El Tanany, G.M. Pajonk, K.H. Steinberg and S.J. Teichner, *Appl. Catal.*, in press).

G.A. SOMORJAI: As you suggest, a transition metal is needed to dissociate dihydrogen to atomic hydrogen which then spills over to the oxide that cannot dissociate dihydrogen on its own. After the oxide is "activated" this way, can it now dissociate dihydrogen without the transition metal as judged perhaps by its ability to carry out $H_2 - D_2$ exchange?

S.J. TEICHNER: One has to differentiate between the exchange $H_2 + D_2 = 2HD$ and the exchange $OH + D_{spill.} \rightarrow OD + H_{spill.}$. The first one requires the dissociation of hydrogen (which is easy on platinum) but which does not occur on silica, even activated by hydrogen spillover. It does not seem, indeed, that this activated silica is able to reform the spiltover type hydrogen from molecular hydrogen, in the absence of platinum. But in the case of alumina activated by hydrogen spillover, this spiltover type hydrogen can be reformed from molecular hydrogen (in the absence of Pt) [see ref. (2) of the paper].

The second type of exchange requires the platinum "gate" in contact with the oxide (like silica). Through this "gate" the spiltover (and dissociated) hydrogen or deuterium migrates onto the surface of the oxide containing hydroxyls and easily exchanges with these OH's (or OD's), independently of the fact that the oxide was or was not previously activated by hydrogen spillover.

The Sticking of Ne on Ru(001): Quantum vs. Classical Behavior

H. Schlichting [1], *D. Menzel* [1], *T. Brunner* [1], *W. Brenig* [1], *and J.C. Tully* [2]

[1] Physik-Department, Technische Universität München, D-8046 Garching, Fed. Rep. of Germany

[2] AT&T Bell Laboratories, Surface Physics Research, 600 Mountain Avenue, Murray Hill, NJ 07974, USA

The adsorption, desorption and scattering of atoms and molecules from solid surfaces have been extensively and successfully simulated by numerical solution of classical mechanical equations of motion [1]. Classical calculations can be carried out in full dimensionality, including large numbers of surface atoms, non-zero coverage, steps, defects and impurities, in contrast to quantum theories which in practice are limited to simple models of reduced dimensionality. Nevertheless, the correct dynamics is quantal, and it is important to ascertain the limitations of classical mechanics. In some situations the limitations are obvious; e.g., diffraction which is observed in scattering of light species (H_2 or He) from periodic surfaces, or tunneling of H atoms during dissociative adsorption. For more massive species which do not exhibit such striking quantum effects, however, it is quite difficult to determine the importance of quantum effects. The problem arises from two sources. First, it is difficult to make a definitive comparison between exact quantum and classical solutions for a realistic problem. While simple models that are exactly solvable by quantum mechanics may exhibit substantial deviations from classical mechanics, it can be argued that classical mechanics improves with complexity. That is, average quantities are frequently well described by classical mechanics, and averaging over many impact sites, thermal fluctuations, molecular orientations, etc., may wash out most quantum details. Second, it is difficult to make a definitive comparison with experiment. Calculated results can depend so sensitively on uncertain interaction parameters that agreement or lack of agreement between a classical calculation and experiment is not a definitive test.

We have carried out experiments on the sticking of Ne atoms on a flat Ru(001) surface at low temperature [2]. The results of these experiments cannot be adequately described by classical mechanics, irrespective of the interaction parameters assumed.

Details of the experiments and the associated quantum and classical calculations are reported elsewhere [2]. Briefly, a thermal beam of Ne atoms was directed at a clean, cold (7K) Ru(001) surface for a fixed time, and the accumulated coverage then determined by thermal desorption [3]. Zero-coverage absolute sticking probabilities were obtained by this procedure for gas temperatures, T_g, between 30 and 600 K. Results are shown in Fig. 1.

Our analysis of these results rests on one assumption: for a fixed surface temperature the sticking probability depends only on the energy, E_z, associated with normal motion of the Ne atom. This is generally assumed to be true for flat surfaces such as Ru(001), but has not been adequately demonstrated. Indeed, classical calculations [4] suggest that at incident angles greater than 45° with respect to the surface normal, scaling with E_z can break down significantly. However, the experimental configuration employed for this

Springer Series in Surface Sciences Vol. 14
Solvay Conference on Surface Science
Editor: F.W. de Wette

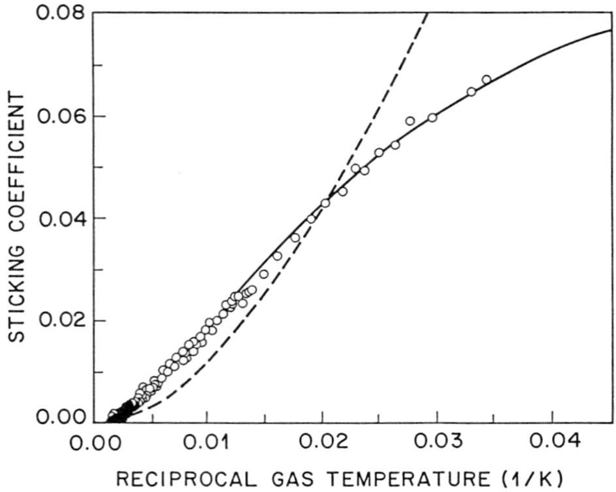

Fig. 1 Absolute sticking coefficients (circles) measured for Ne on Ru(001) vs. $1/T_g$, for T_s=7K. Solid and dashed curves are fits to data as described in text.

experiment restricts the incident angle to less than 45°. Furthermore, modest deviations from E_z scaling will not effect the final conclusion.

With this caveat, we proceed. The sticking probability $S(T_g)$ is then given by

$$S(T_g) = \int_0^\infty dE_z \; P_s(E_z)N(E_z), \tag{1}$$

where $P_s(E_z)$ is the sticking probability of a monoenergetic beam of Ne at E_z, and $N(E_z)$ is the Maxwell-Boltzmann flux distribution of E_z. For this experiment [2],

$$N(E_z) = 2.44k_B^{-1}T_g^{-1}[\exp(-E_z/k_B T_g)-\exp(-1.69E_z/k_B T_g)] \tag{2}$$

where k_B is Boltzmann's constant.

We parametrize $P_s(E_z)$ by a Gaussian centered at E_z=0:

$$P_s(E_z) \simeq P_o \exp[-(E_z/E_c)^2] \tag{3}$$

This simple Gaussian form provides an acceptable, albeit not quantitative, fit to a wide range of classical mechanical calculations of rare gas sticking probabilities on flat surfaces [4]. More importantly, the thermally averaged sticking probability, $S(T_g)$ of Eq. (1), is insensitive to the detailed shape of $P_s(E_z)$ within the experimental temperature range. An exponential or step-function form fits the data almost as well, with the same final conclusions. Varying the Gaussian parameters P_o and E_c to best reproduce the experimental data produces the solid curve of Fig. 1, with P_o=0.11 and E_c=65K. Constraining P_o to be unity, as required by classical mechanics so that $P(E_z)\rightarrow 1$ as $E_z \rightarrow 0$, produces the dashed curve of Fig. 1, with E_c=13K. The accuracy of the measured data is sufficient to conclude unambiguously that the dashed curve is inconsistent with experiment.

The monoenergetic sticking probabilities derived from experiment by this simple procedure are shown in Fig. 2. The dashed curve is the best "classical" fit; i.e., the best fit

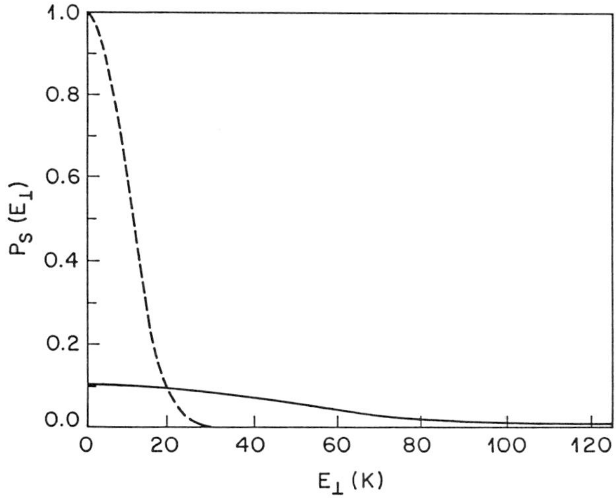

Fig. 2 Monoenergetic sticking probabilities, $P_s(E_z)$. Solid curve: $P_s(E_z)=0.11 \exp[-(E_z/65K)^2]$. Dashed curve: $P_s(E_z)=1.0 \exp[-(E_z/13K)^2]$.

constrained such that $P_s(E_z)=1$ as $E_z \to 0$. This curve is in serious disagreement with the best fit to experiment (solid curve). No classical calculation based on any reasonable choice of parameters could significantly improve on the dashed curve. This result is further supported by extensive classical, quasi-classical and quantum calculations using realistic interaction parameters, as reported elsewhere [2].

Classical mechanics is deficient in two important respects [5,6]. First, every classical event exchanges some infinitessimal energy, in contrast to quantum scattering which exhibits a finite magnitude elastic zero-phonon line. This is responsible for $P_s(E_z)$ attaining a non-unit value as $E_z \to 0$ for quantum mechanics. The best fit value $P_o=0.11$ demonstrates that zero-phonon scattering is a dominant factor in this system. Second, $P_s(E_z)$ remains significant to higher values of E_z in quantum mechanics due to high frequency phonon excitations that do not occur in classical mechanics. This is analogous to what are termed "classically forbidden" processes in gas phase translational-to-vibrational energy exchange [7]. Finally, we note that a second mechanism for reducing $P_s(E_z)$ below unity as $E_z \to 0$, reflection of the atomic wave packet from the attractive well, is of negligible significance for the range of energies probed in this experiment [8].

1. J. C. Tully: In Many–Body Phenomena at Surfaces, ed. by D. Langreth and H. Suhl (Academic Press, N.Y., 1984) p.377
2. H. Schlichting, D. Menzel, T. Brunner, W. Brenig, J. C. Tully: Submitted to Phys. Rev.
3. P. Feulner, D. Menzel: J. Vac. Sci. Technol. 17, 662 (1980)
4. J. C. Tully: Unpublished
5. W. Brenig: Z. Phys. B36, 81 (1979)

6. J. Böheim, W. Brenig: Z. Phys. B41, 243 (1981)

7. W. H. Miller: Adv. Chem. Phys. 25, 69 (1974)

8. J. Böheim, W. Brenig, J. Stutzki: Z. Phys. B48, 43 (1982)

Discussion

W. KOHN: At the end of your talk you mentioned briefly the issue of the zero energy limit of the sticking coefficient, P_s (E→0), in particular the possibility that it may vanish. Could you say a little more about the status of this question?

J. TULLY: I have not studied this question myself. From discussion with M. Stiles, W. Brenig and J. Harris, I think the situation is the following. At low energies a quantum particle can reflect off the front of the attractive potential, preventing sticking, if the potential is short-ranged. Early calculations using Morse potentials exaggerated this effect because Morse potentials are too short–ranged. Calculations using the correct inverse z^3 potential show that quantum reflection does not become significant except at very low incident energies, much less than $1K$ for realistic systems.

J.W. GADZUK: Actual experiments may soon be carried out to investigate the issue raised in John Tully's talk concerning quantum reflection in the zero kinetic energy limit, from the long range attractive tail of the atom-surface potential, using a trapped Na atom with ~ mK energies and thus several 100 Å thermal wavelengths (W. Phillips, NBS).

J.B. PENDRY: There is a series of classic experiments on the quantum zero energy sticking coefficient by D.O. Edwards on the ^{4}He/liquid ^{4}He system at very low temperatures. He shows that the sticking coefficient does indeed tend to zero. This agrees with theory by Echenique and myself.

Atomic and Electronic Structure of Compound Semiconductor Interfaces

C.B. Duke

Xerox Webster Research Center, 800 Phillips Road, 0114/38D, Webster, NY 14580, USA

The cleavage faces of zincblende and wurtzite structure compound semiconductors constitute the best understood of all semiconductor interfaces. Structure determinations by low-energy electron diffraction and ion-scattering of the zincblende (110) cleavage surfaces reveal not only greatly distorted surface geometries relative to the bulk, but also the unanticipated result that III-V and II-VI compounds exhibit comparable surface structures in contrast to their quite different molecular geometries. This observation motivated an extensive theoretical effort to predict these structures, with the result that an electronically driven atomic relaxation phenomenon provides a common quantitative description of all the experimental data. Extension of this theory to wurtzite $(10\bar{1}0)$ and $(11\bar{2}0)$ cleavage faces reveals an analogous mechanism for relaxations at these surfaces, as well as the existence for each surface of a unique atomic geometry, determined by the atomic connectivity of the truncated bulk surface, which depends on the specific material only via a linear scaling with the bulk lattice constant. Further extensions of the theory and structure analyses to interface geometries, specifically Sb on GaAs(110) and InP(110), reveal the existence of a new type of epitaxically mediated pi bond of (1x1) overlayers of Sb on GaAs and InP. Therefore the determination and prediction of surface atomic geometries have disclosed unanticipated types of chemical bonding which exhibit neither bulk nor molecular analogs.

1. EPITAXICALLY-CONSTRAINED SURFACE BOND REHYBRIDIZATION

Unlike most metals, whose surfaces exhibit only small oscillatory relaxations of the interlayer spacings associated with surface smoothing /1/, the vacuum surfaces of semiconductors exhibit large atomic rearrangements caused by the rehybridization of the surface chemical bonding to accommodate the extra charge which would have existed in localized directional "dangling" bonds at the surface of a truncated bulk solid /2/. Perhaps the most significant experimental indication of the surface (as opposed to material) specificity of this phenomenon is the fact that the (110) cleavage faces of zincblende structure compound semiconductors exhibit the same atomic rearrangements for III-V (AlP, AlAs, AlSb, GaP, GaAs, GaSb, InP, InAs, InSb) as for II-VI (ZnS, ZnSe, ZnTe, CdTe) materials /2/, i.e., the only distinction between the various surface structures is the experimentally measured /3/ and theoretically predicted /4,5/ linear scaling of the magnitude of the surface structural parameters with the bulk lattice constant, a_0. The small-molecule chemistries of III-V and II-VI compounds differ dramatically, however. For III-V molecules the anions prefer threefold distorted p^3 configurations (e.g., the As in AsH_3) and the cations planar sp^2 configurations (e.g., the Ga in $Ga(CH_3)_3$). Small molecules of the anion species alone form tetrahedrons (e.g., P_4). The II-VI atomic species, in contrast, form two-fold-coordinated chain-like structures in either closed (e.g., S_6, S_8, monoclinic Se (Se_8)) or open (trigonal Se) geometries. Thus, if one envisages the surface chemical bonding of the cleavage faces of zincblende structure compound semiconductors

Springer Series in Surface Sciences Vol. 14
Solvay Conference on Surface Science
Editor: F.W. de Wette

as intermediate between small molecule bonding and the bulk sp^3 tetrahedrally-coordinated bonding, the occurrence of common surface geometries for III-V and II-VI materials is truly remarkable.

In order to describe the surface rehybridization phenomena a quantum chemical model which incorporates both the bulk sp^3 bonding and an accurate description of the surface is required. For the (110) surfaces of zincblende structure compounds two such models have been reported in the literature: an extended tight-binding model which has been applied to the III-V compounds /4,5/ as well as to ZnTe /6/, ZnSe /6,7/ and ZnS /8/, and a first principles pseudopotential model which has been applied to GaAs /9/ and ZnSe /10/. These calculations reveal that the surface relaxation is driven by the lowering in energy of a surface state from near the middle of the band gap to the top of the valence band. This state changes its character as the relaxation proceeds, going from a dangling-bond character to surface-bond character in which added charge is deposited in the surface anion-cation bonds. The major characteristics of this surface state are determined by the tetrahedrally-coordinated bulk atomic geometry as manifested in the atomic connectivity at the surface. The details of the atomic potentials modify it only slightly; hence, the close relationship between the surface structures of III-V compounds (e.g., GaAs) and their isoelectronic II-VI counterparts (e.g., ZnSe). Therefore the surface chemical bonding is driven by the substrate geometry (as reflected in the surface atomic connectivity) more than by the small-molecule coordination chemistry. Stated alternatively, the atomic geometry of the substrate "template" which the surface bonding must match epitaxically dominates the character of this bonding and gives rise to a new type of "surface" chemical bond which is unknown in either small molecules or bulk solids.

An important step in establishing the origin of surface rehybridization for tetrahedrally-coordinated compound semiconductors was the extension of the analyses to the cleavage faces of the wurtzite as well as zincblende structure materials /7,8,11,12/. These two allotropes differ only in the packing sequence normal to the c (wurtzite) or (111) (zincblende) axis. Thus, for ideal c/a ratios, the nearest-neighbor coordinations are the same, but the next-nearest-neighbors are not. Wurtzite-structure crystals exhibit two cleavage faces. The $(10\bar{1}0)$ surface consists of isolated dimers each atom of which exhibits one surface and two substrate bonds. The $(11\bar{2}0)$ surface consists of zig-zag anion-cation chains, much like zincblende (110), each atom in which possesses two surface and one substrate bond. Unlike zincblende (110), the wurtzite $(11\bar{2}0)$ surface unit cell contains four inequivalent atoms, so the chains pucker, the anion relaxing out of the surface and the cations into the substrate, upon surface rehybridization. The $(10\bar{1}0)$ surfaces rehybridize via a nearly-bond-length-conserving rotation of the surface dimers, anion upwards and cation downwards. Thus, both surfaces exhibit distinctly different relaxed structures from the tilted-surface-chain structure of zincblende (110). Yet the mechanism of surface relaxation is the same in all three cases: a near-mid-gap dangling-bond surface state converts into a back-bonding state thereby lowering its energy, and moving into the top of the valence band. The electronic energy gained by the electrons occupying this surface state overcomes the energy increase due to small bond length and angle distortions incurred in the relaxation, converting the anions to a distorted p^3 conformation and the cations to a nearly planar sp^2 conformation. Moreover, the model predictions of these surface structures also exhibit face universality: I.e., the geometry of each cleavage face depends on the specific material only via a linear scaling of the independent surface structure variables on the bulk lattice constant /13/. Hence each crystallographic surface exhibits a "universal" (i.e., material independent) structure.

The polar surfaces of tetrahedrally-coordinated compounds also exhibit surface-rehybridization-driven reconstructions. These depend, however, in a complex way on surface stochiometry and process history /2,14/. Specific models have been advanced for GaAs(111)-(2x2) /15-17/, ZnSe(111)-(2x2) /16/, GaAs(111)-(2x2) /17,18/, GaAs(100)-(2x4) /19/, and InP($\bar{1}\bar{1}\bar{1}$)-(1x1) /20/. A local distorted p^3 conformation of the anions and sp^2 conformation of the cations are characteristic of these structures also. As in the case of the cleavage surfaces, these local conformations are exhibited by II-VI materials (for which they are unexpected) as well as III-V materials (for which they are characteristic of small-molecule geometries).

2. EPITAXICALLY-CONSTRAINED ADSORBATE BONDING

In addition to novel surface chemical bonding on clean surfaces, geometrical constraints imposed by epitaxical growth also induce new types of adsorbate-substrate bonds /21/. One of the most extensively studied semiconductor adsorbate systems is the p(1x1) Sb saturated monolayer structure on the (110) surfaces of GaAs /21-31/ and InP /22,32-34/. The Sb occupy surface sites in zig-zag chains roughly where the top layer anions and cations would have been. The Sb species are larger than Ga and As, however, so that in order to accommodate the large Sb-Sb bond length (2.8A), the angle between neighboring Sb species closes to 91° from its value of approximately 109° in GaAs. In this structure, three of the Sb electrons participate in bonds and two generate a sp hybrid lone pair charge density which is analogous to the charge density for bonding to the substrate species but pointing out of the surface. Thus, the valences of both surface Sb species are fully saturated at monolayer coverage so that the main driving force for the relaxation of the GaAs(110) substrate is eliminated by the Sb-substrate bonding. Indeed, both the Sb chains and the GaAs(110) substrate are nearly unrelaxed. The Sb bonded to the substrate As is lowered by 0.1A relative to that bonded to the substrate Ga and the uppermost As is relaxed downward by a corresponding 0.1A /23/. These are small distortions relative to the relaxations on a clean GaAs(110) surface, however, and reveal the profound influence of a saturated surface valence on the corresponding atomic geometry.

An unusual feature of the Sb-substrate bonding is the p^2 local bonding conformation within the Sb chain (i.e., approximately 90° Sb-Sb-Sb bond angles) and distorted sp^3 local Sb conformation (approximately 105° bond angles) relative to the substrate. Since there are not enough independent p orbitals to support simultaneously both types of local valence, the detailed nature of the bonding is unlike that in either small molecules or the bulk tetrahedrally-coordinated solids. A detailed study /21/ revealed that the Sb-substrate bond is a hybrid between pi orbitals of a p^2 bonded Sb chain and sp^3 dangling bond orbitals of an undistorted GaAs substrate. The occurrence of this bond is driven by the combination of the large size of the Sb relative to the GaAs substrate, the saturated valence of the Sb monolayer, and the occurrence of epitaxical ordering of the monolayer relative to the substrate. Thus, the pi-sp^3 hybrid bond is a unique type of surface bonding associated with the constraint of epitaxical monolayer growth on the compound semiconductor substrate /21,35/. Direct confirmation of this type of bonding has been given during the past few years by virtue of direct observations of the surface-state bands characteristic of the bonding charge density via angle resolved photoemission spectroscopy for p(1x1) Sb monolayers on GaAs /27-29, 31/, GaP /29/, and InP /32/. These states are the analogs for the Sb saturated monolayer systems of the back bonding rehybridized surface states which drive the surface rehybridization on the clean cleavage faces of tetrahedrally-coordinated compound semiconductors /4,7,8,21/.

3. SYNOPSIS

The study of the atomic geometries and electronic structures of the surfaces of compound semiconductors has led to four major conceptual innovations. First, it is now recognized that, at least for the cleavage faces of these materials, each crystallographically inequivalent face exhibits a unique "universal" geometry which varies from one material to another only via a linear scaling of the independent structural variables with the bulk lattice constant /3,13/. Second, the mechanism of the surface reconstruction has been shown to be universal for all the cleavage surfaces of both III-V and II-VI compounds /7,8,13/. This observation, in turn, led to the recognition of a new type of chemical bonding, which exhibits neither bulk nor molecular analogs, for II-VI compounds /7,8,11-13/. Fourth and finally, extension of analogous models and concepts to the III-V(110)-p(1x1)-Sb(1ML) system led to the discovery of another new type of chemical bonding in that case as well /21,35/. Thus, the study of semiconductor interface atomic and electronic structures have broadened signficantly our repetoire of types of surface and interface chemical bonding.

4. REFERENCES

1. S.P. Chen, A.F. Voter and D.J. Srolovitz: Phys. Rev. Lett. 57, 1308 (1986).
2. C.B. Duke: Surface Properties of Electronic Materials, ed. by D.A. King and D.P. Woodruff (Elsevier Science Publishers, Amsterdam, 1987), chpt. 3.
3. C.B. Duke: J. Vac. Sci. Technol. B 1, 732 (1983).
4. C. Mailhiot, C.B. Duke and D.J. Chadi: Surf. Sci. 149, 366 (1985).
5. C.B. Duke, C. Mailhiot, A. Paton, D.J. Chadi and A. Kahn: J. Vac. Sci. Technol. B 3, 1087 (1985).
6. D.J. Chadi: Phys. Rev. B 9, 2074 (1979).
7. Y.R. Wang, C.B. Duke and C. Mailhiot: Surf. Sci. Lett. 188, L708 (1987).
8. Y.R. Wang and C.B. Duke: Phys. Rev. B 36, 2763 (1987).
9. A.C. Ferraz and G.P. Srivastava: Surf. Sci. 182, 161 (1987).
10. A.C. Ferraz and G.P. Srivastava: J. Phys. C: Solid State Phys. 19, 5987 (1986).
11. Y.R. Wang and C.B. Duke: Surf. Sci. 192, 209 (1987).
12. Y.R. Wang, C.B. Duke, A. Paton, K. Stiles and A. Kahn: Phys. Rev. B in press.
13. C.B. Duke and Y.R. Wang: J. Vac. Sci. Technol. A 6 (1988) in press.
14. A. Kahn: Surf. Sci. Repts. 3, 193 (1983).
15. D.J. Chadi: Phys. Rev. Lett. 52, 1911 (1984).
16. D.J. Chadi: J. Vac. Sci. Technol. A 4, 944 (1986).
17. E. Kaxiras, Y. Bar-Yam, J.D. Joannopoulos and K.C. Pandey: Phys. Rev. B 33, 4406 (1986).

18. D.J. Chadi: Phys. Rev. Lett. 57, 102 (1986).

19. D.J. Chadi: J. Vac. Sci. Technol. A 5, 834 (1987).

20. X.-Y. Hou, G-S. Dong, X-m Ding and X. Wang: Surf. Sci. 183, 123 (1987).

21. C. Mailhiot, C.B. Duke and D.J. Chadi: Phys. Rev. Lett. 53, 2114 (1984); Phys. Rev. B 31, 2213 (1985).

22. J. Carelli and A. Kahn: Surf. Sci. 116, 380 (1982).

23. C.B. Duke, A. Paton, W.K. Ford, A. Kahn and J. Carelli: Phys. Rev. B 26, 803 (1982).

24. C.M. Bertoni, C. Calendra, F. Manghi and E. Molinari: Phys. Rev. B 27, 1251 (1983).

25. K. Li and A. Kahn: J. Vac. Sci. Technol. A 4, 958 (1986).

26. W. Pletschen, N. Esser, H. Munder, D. Zahn, J. Geurts and W. Richter: Surf. Sci. 178, 140 (1986).

27. M. Mattern-Klossen, R. Strümpler and H. Lüth: Phys. Rev. B 33, 2259 (1986).

28. A. Tulke, M. Mattern-Klossen and H. Lüth: Solid State Commun. 59, 303 (1986).

29. A. Tulke and H. Lüth: Surf. Sci. 178, 131 (1986).

30. R. Strümpler and H. Lüth: Surf.Sci. 182, 545 (1987).

31. P. Martensson, G.V. Hansson, M. Lähdeniemi, K.O. Magnusson, S. Wiklund and J.M. Nicholls: Phys. Rev. B 33, 7399 (1986).

32. C. Maani, A.C. McKinley and R.H. Williams: J. Phys. C: Solid State Phys. 18, 4975 (1985).

33. C.B. Duke, C. Mailhiot, A. Paton, K. Li, C. Bonapace and A. Kahn: Surf. Sci. 163, 391 (1985).

34. D. Zahn, N. Esser, W. Pletschen, J. Geurts and W. Richter: Surf. Sci. 168, 823 (1986).

35. C. Mailhiot, C.B. Duke and D.J. Chadi: J. Vac. Sci. Technol. A 3, 915 (1985).

Discussion

D.R. HAMANN: Contour plots of surface state charge densities on relaxed GaAs (110) and other III-V's indicate that the S_1 surface state has the character of a lone pair localized above the anion, and not of "back bond" states. S_2 and S_3 appear to be back bond states in these calculations. I believe these results are at variance with your qualitative interpretation. In addition, spectroscopic STM studies bear out the calculated results on the spatial localization of S_1.

C.B. DUKE: The surface–state eigenfunctions associated with the figures which I showed are fully compatible with literature calculations for GaAs(110) and other III–V compounds as described in detail on pp. 2225-6 of C. Mailhiot, C.B. Duke and D.J. Chadi, *Phys. Rev. B* **31**, 2213 (1985). The point emphasized in my presentation was the change in the character of these eigenfunctions upon surface relaxation. The state to which you refer as S_1 changes from almost pure top-layer arsenic "p" orbitals, which are oriented along the "broken-bond" directions for the unrelaxed ("truncated-bulk") surface into a more complex linear combination of top-layer As "p" orbitals which exhibits components along the surface As-Ga bond directions for the relaxed surface. Moreover, these latter orbital components are hybridized with some Ga p-orbital character at certain regions of the surface Brillouin zone. The states to which you refer as S_2 and S_3 also differ from their unrelaxed-surface counterparts in analogous fashions but involving different detailed charge redistributions. There is, therefore, no incompatibility between my interpretation of the relaxation-induced changes in the surface charge density and the literature on the surface-state eigenvalues and eigenfunctions on the (110) surfaces of zincblende structure III–V compounds.

I am mystified by your reference to scanning tunneling microscopy (STM) studies. The only quantitative analysis of STM spectroscopy for GaAs(110) of which I am aware (i.e., R.M. Feenstra and J.A. Stroscio, *J. Vac. Sci. Technol.* **B5**, 923 (1987)) does not invoke surface states at all. Since our calculations reproduce the bulk energy band structure of GaAs, they are manifestly compatible with the Feenstra-Stroscio interpretation of their STM results.

M. CARDONA:

(1) There are no data known to me for the I–VII compounds (CuCℓ,... AgI). It would be quite interesting to check your universality ideas for them. I actually would expect universality to break down for these I–VII compounds because of the strong p-d hybridization.

(2) There is another parameter which, surprisingly, is the same for the III–V as for the II–VI compounds: the *dynamical transverse charge* e_T^*. This is particularly surprising since the ionicities are quite different. Since the motion related to the reconstruction is actually a (surface) optical phonon, I wonder whether there is any connection between this reconstruction and the charge e_T^*.

C.B. DUKE:

(1) The universality of the surface structures is a direct consequence of the commonality of the qualitative character of the chemical bonding in the materials considered, just as you imply. In cases for which this bonding can change for materials of a given structure, such universality is not anticipated. A case which we have studied explicitly is the (110) cleavage face of rutile-structure oxides, i.e., GeO_2, SnO_2, PbO_2, and TiO_2. The Ti "d" electrons

do change the detailed results for TiO_2 from those obtained in the $GeO_2 - SnO_2 - PbO_2$ series.

(2) The only attempt of which I am aware to correlate semiconductor surface structures with effective charges is that reported by R.V. Kasowski, M.-H. Tsai and J.D. Dow, *J. Vac. Sci. Technol.* **B5**, 953 (1987) for Zincblende (110) surfaces. These authors considered the longitudinal effective charge which yields a structure-ionicity relationship. The proposed relationship fails to fit the known structural data (see, e.g., Fig. 2 in Kasowski *et al., loc. cit.*) as have prior proposals of structure-ionicity relationships (see, e.g., C.B. Duke, A. Paton, and A. Kahn, *J. Vac. Sci. Technol.* **A1**, 672 (1983)). A correlation of the Zincblende (110) structures with the transverse effective charge, e_T^*, would be successful if and only if this charge scales linearly with the bulk lattice constant. I do not know whether or not such is the case.

G.A. SOMORJAI: LEED and other surface science studies indicate that considerable non-stoichiometry exists at alkali halide surfaces. You indicate that the stoichiometry is maintained at cleavage surfaces of III–V and II–VI compounds. Could you comment on the possibility of non-stoichiometry at these surfaces?

C.B. DUKE: In my presentation I considered only cleavage surfaces. If these actually are prepared by vacuum cleavage, stochiometry is presumed to occur and is "verified" by Auger electron spectroscopy. Indeed, such surfaces often are used to calibrate measured Auger intensities. In those cases for which cleavage surfaces are prepared by ion-bombardment and anneal cycles, non-stochiometry often occurs. It must be guarded against in LEED studies of surface geometry by careful attention to the specification of these cycles. Polar surfaces exhibit much more complex behavior, yielding multiple surface stochiometries and structures, depending upon the preparation conditons. A recent review of surface preparation and structure may be found in A. Kahn, *Surf. Sci. Repts.* **3**, 193 (1983).

General Discussion

H. IBACH : J.E. Müller has recently proposed a possibility for hydrogen molecules to become self-trapped in a precursor state. The suggested mechanism involves conversion of zero point energy (which is reduced near the surface because at the flatter H–H potential curve near the surface) into translational energy along the surface, part of which would be lost due to enhanced phonon scattering. This would self-trap the molecule in a molecular state of high mobility. Is this a reasonable proposition and does one have any direct or indirect evidence for it?

G. ERTL: Those H_2 molecules which are not dissociatively adsorbed on Ni surfaces exhibit highly elastic behavior (specular reflection and diffraction) which could not be reconciled with the suggested mechanism of intermediate population of a precursor state.

J. TULLY: One comment that's not conclusive about that is that I think the experimental evidence is that D_2 traps with a higher probability than H_2, whereas this effect would be larger for H_2 because of the bigger difference in zero point energy. But of course the reason D_2 traps more favorably is because of the subsequent energy transfer when it hits the surface.

H. IBACH: One has to be a bit careful in arguing with sticking because the final sticking is always a dissociation process. What I'm talking about is sticking in a precursor state which might lead to dissociation or to desorption back into the gas phase.

D.A. KING: I realize I'm wandering a bit off beam here, but we've looked at hydrogen and deuterium sticking on tungsten (100) and looked very, very carefully for an isotope effect there, and within an accuracy of .01 for a sticking probability of .6, we can see no isotope effect in that system at all.

G.A. SOMORJAI: Professor Rieder from Berlin reported on the strong diffraction of Neon from smooth metal surfaces as compared to helium or other inert gases. This observation seems surprising. Is there an explanation from other surface science studies?

M. CARDILLO: There was a rather careful study by the Genoa group analyzing inelastic scattering by Ne and He from LiF as a function of surface temperature, i.e., looking at the coupling to phonons as the basis for the Debye-Waller factor. One of their conclusions was that phonon excitation due to Ne scattering peaked at Q=0 for each zone, thus giving the appearance of increased diffraction in the angular distributions in the absence of energy loss analysis.

J.B. PENDRY: I think you probably expect neon to show stronger diffraction effects, because if you look at the experiment of putting an electron in liquid helium or liquid neon you find that—I think they both have negative affinity, but the helium has far, far stronger negative affinity for electrons than does neon. And therefore, the turning point at which the interaction between the atom and the solid becomes repulsive is closer–in for neon than it is for helium; helium really hates electrons in solids; and the further in you get, the stronger the corrugations of the density distribution.

Part VII

Solid-Solid Interfaces and Superlattices

Solid-Solid Interfaces and Superlattices

A.C. Gossard

University of California, Santa Barbara, CA 93106, USA

1. INTRODUCTION

In this paper, we shall discuss recent developments in interface growth techniques and present data on the structure of semiconductor interfaces and the related optical and electronic transport properties of the interfaces. We shall also describe progress toward one- and zero-dimensional quantum confinement structures.

Recent developments in the means for the growth and preparation of layered solids and thin films have permitted access to an increasing range of solid-solid interfaces and multiple layered structures. Progress has been especially rapid in semiconductors, where epitaxial growth techniques permit growth of high quality single crystal films that have a low density of structural defects, high chemical purity, and thickness control approaching the atomic monolayer level.

Concomitant advances in transmission electron microscopy technique and electron and x-ray diffraction methods give fine scale structural information on these new materials. Furthermore, the fine structural control and the high quality of the materials allow new ways of confining and manipulating electrons and light within the structures that are leading to new optical and electrical transport properties and in turn to more critical means of examining the interfaces.

While the semiconductor systems in which the most perfect interfaces and layered structures can be synthesized are the lattice-matched valence-matched compounds, the phenomena which occur at interfaces between lattice-mismatched and valence-mismatched materials are interesting both from the fundamental and applied points of view, and much recent progress has been made with these systems as well. Interfaces between semiconductors and metals or insulators are also receiving increased attention, but the levels of control accessible in these systems are not yet as high as in the semiconductor-semiconductor case.

Although the processes of thin film crystal growth provide a natural means of formation of layered structures and interfaces, there is thus

Springer Series in Surface Sciences Vol. 14
Solvay Conference on Surface Science
Editor: F.W. de Wette

far no comparable technique for the formation of fine structures in the lateral dimensions. Generation of structures with artificial carrier confinement in two and three dimensions is thus more difficult. But progress on this interesting problem is nonetheless beginning to be made, and the additional novel phenomena expected in these structures may soon be accessible.

2. GROWTH AND STRUCTURE OF INTERFACES

2.1 Growth Techniques

The concept of artificial synthesis of multilayered superlattice structures is not particularly new. In the 1930's, for example, Jesse Dumond and his students at California Institute of Technology prepared multilayered copper/gold structures by intermittent copper and gold evaporation, with the goal of preparation of artificial x-ray gratings [1]. They employed various substrate materials and obtained crystallographically oriented multilayers which gave superlattice x-ray diffraction but which interdiffused substantially at room temperature.

The concepts of current evaporation techniques, such as molecular beam epitaxy, are very similar to the earlier work. But the technology is greatly enhanced, with features like stainless steel ultrahigh vacuum chambers, heated, temperature-controlled single-crystal substrate materials, cryoshielded high temperature evaporation sources, computer-controlled molecular beam shutters, evaporant flux monitors, and electron diffraction surface analysis within growth chambers. Many of these features have derived, in fact, from surface science tools and understanding. A principal driving force for the development of molecular beam epitaxy was the willingness to devote surface analysis equipment to the evaporation of materials and study of surfaces and interfaces [2].

Coincident with the advances in the evaporation techniques was the development of vapor phase epitaxial crystal growth techniques which also can grow thin, smooth layers under close control and with high purity. A principal advance in vapor phase epitaxy was the development of metal-organic chemical vapor deposition which enabled growth of mixed-composition materials with high purity [3].

The distinction between the evaporation techniques and the chemical vapor deposition techniques has blurred, however, with introduction of vapor deposition processes which operate at less than atmospheric pressure on the one hand, and molecular beam epitaxy processes which operate with molecular gas sources, on the other hand [4].

2.2 Surface Migration and Interface Structure

A common feature of all the epitaxial growth techniques which can produce smooth surfaces is the presence of lateral surface mobility of the atoms which are forming the materials. Without surface mobility, deposited atoms would produce progressively rougher surfaces, whereas with mobility, the atoms can migrate to lower spots on the surface, where there are energetically more favorable bonding sites and where the atoms can form a progressively smoother surface. The effects can be seen very clearly in transmission electron cross-sectional images of the smoothing of interfaces of superlattices grown on initially rough surfaces [5].

The smoothing process can also be detected in the reflection electron diffraction from the surface of a growing crystal, wherein a progressive lengthening of electron diffraction streaks occurs during growth if the surface smooths [6]. The smoothing can be quite temperature and growth condition dependent and is a valuable tool in monitoring crystal growth. The behavior is especially remarkable under conditions of marginal surface mobility, in which case the roughness of the surface oscillates after commencement of growth with a periodicity equal to the deposition time for a single atomic monolayer [7]. This provides a useful means for calibration of crystal growth rates. It apparently results from the nucleation of growth in islands and the subsequent filling of the space between the islands to form completed monolayers. The maximum roughness occurs at half monolayer coverage. The oscillations become damped as phase memory of the starting point is lost because of nonuniformities in evaporant beam flux intensities or because of imperfect filling of the space between islands.

Although reflection electron diffraction can sensitively measure differences in surface roughness and periodicities in deposited surface coverage, it is less useful for absolute determinations of surface structure and still less useful for determinations of interface structure. Cross-sectional transmission electron microscopy of interfaces, on the other hand, can provide images of interface planes as well as images and diffraction from superlattices. Atomic plane imaging has shown abrupt interfaces at the atomic level, and transmission electron diffraction from cross sections of alternate monolayer superlattice depositions have shown superlattice spots characteristic of alternate monolayer structural periodicity [8]. X-ray diffraction from superlattices also has shown superlattice diffraction lines from alternate single monolayer depositions [9]. But quantitative diffraction intensity measurements are necessary in order to infer the absolute degree of order and have been done in only a limited number of cases, all of which to date have shown substantial disorder. The atomic plane imaging, although suggestive of atomic ordering, is not quantitative, and averages over a number of atomic positions within each imaged row of atoms.

Consequently, all of the direct methods of probing structure of interfaces are limited in accuracy. The electrical and optical properties of interfaces provide an alternate means of probing interface structure though. As we shall discuss in the following sections, these properties are sensitively dependent on interface geometries. The electrical and optical properties are thus useful probes of structure as well as being of interest in their own right for some of their unusual behaviors.

2.3 Surface Orientation Effects

The most extensive studies to date have been done on lattice-matched, valence-matched interfaces grown on cubic (100) crystal faces. But there are many interesting phenomena which are observed in interfaces grown in other configurations. For example, while (Al,Ga)As alloy layers grown on (100) surfaces are generally observed to be random alloys, those grown on (110) faces show a strong tendency to segregate into their GaAs and AlAs constituents [10]. This requires an extensive amount of atomic motion and suggests that (110) interfaces between GaAs and AlAs have low energy. Spontaneous formation of alternate monolayer ordered material has been observed under both MBE and MOCVD growth on (110) and (100) surfaces, in fact [11]. Since alternate monolayer order observed spontaneously on the (110) surface also implies a concurrent (100) alternate monolayer structure, these ordering phenomena may be related to the ability to produce alternate monolayer order by alternate deposition on (100) surfaces.

2.4 Polar/Non-polar Interfaces

An interface of considerable technological interest is the interface between elemental and compound semiconductors, or more specifically silicon and gallium arsenide. Such interfaces may allow integration of silicon architectures with microwave or optical GaAs structures. A persistent problem in such interfaces has been the occurrence of antiphase domains within the compound semiconductor grown on the elemental semiconductor [12]. The antiphase domains consist of adjacent regions of GaAs in which the sublattice occupations of Ga and As are reversed (Fig 1). This leads to high energy interfaces between the domains in which there are many Ga-Ga and As-As bonds. These degrade surface smoothness and electrical neutrality. An effective means of suppression of antiphase domains has been the use of off-axis substrate crystalline orientation, thermal annealing, and Group V saturation in order to stabilize double-height steps on the surface [12].

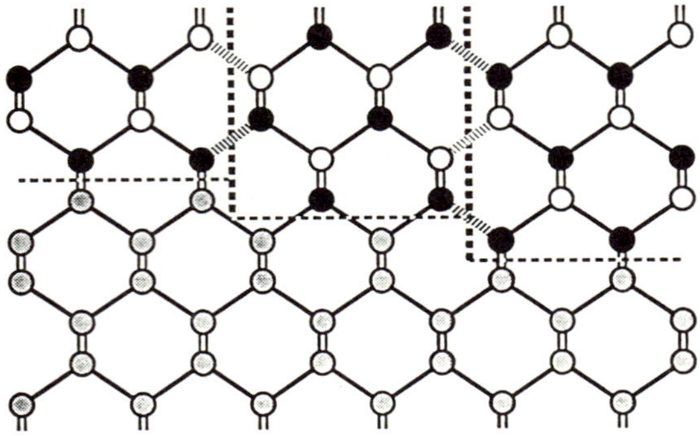

Fig. 1. Antiphase domains of polar on nonpolar crystal

2.5 Lattice-mismatched Interfaces

Strained layer interfaces allow the attractive possibility of further band edge engineering by means of stress, but entail the added problem of the formation of misfit dislocations in epitaxial layers of greater than critical thicknesses [13]. The misfit dislocation problem may be one of the most severe limiting factors in the preparation of GaAs on silicon substrates, where a substantial lattice mismatch between the constituents exists.

3. OPTICAL PROBES OF INTERFACES

3.1 Quantum Wells

By virtue of the dissimilar electronic structures of materials forming heterostructure interfaces, discontinuities in energy levels occur at interfaces. Charge carriers at the interfaces experience interface potential steps which may confine them on one side of the interface. Quantization of the energy levels of confined carriers occurs, and this quantization is sensitively dependent on the dimensions of the confining potentials, which in turn are determined by the dimensions of the confining layers. The quantization of the energy levels may be probed by optical transitions between them, which thus provide another probe of interface geometry. In the simplest case of a high square well confining potential, where the confined energy levels are nearly $E_n = h^2n^2/8m^*L^2$ for a particle of mass m^* in a well of width L (Fig. 2), a nonuniformity in the layer width will lead to nonuniformities in the energy levels and inhomogeneous broadening of optical transitions among the square well states.

This broadening is routinely observed and is a dominant source of linewidth in quantum well optical spectra. It is dependent on growth

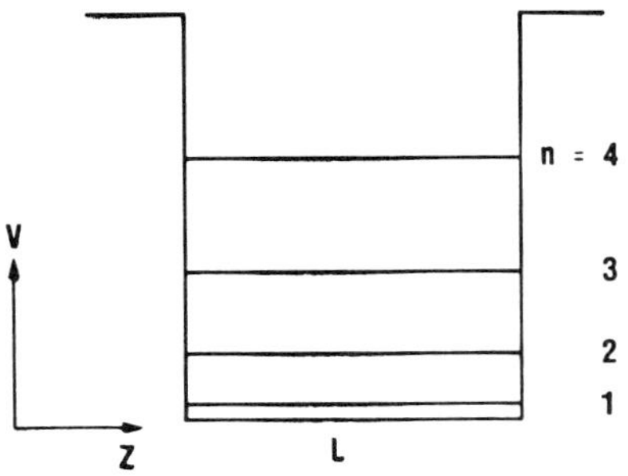

Fig. 2. Electron energy levels in square potential well of width L.

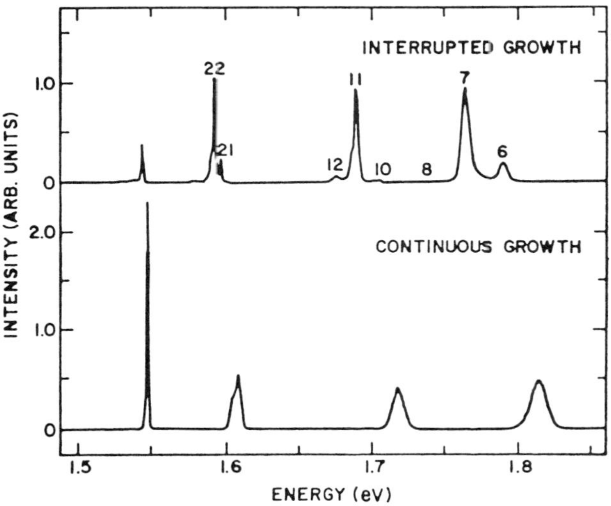

Fig. 3. Photoluminescence emission intensity from four single quantum wells grown by molecular beam epitaxy with and without growth interruption (from Ref. 14)

conditions such as substrate temperature, and is especially sensitive to growth interruptions at quantum well interfaces. This is illustrated in Figure 3, which shows luminescence spectra observed from GaAs quantum wells grown with interrupted and continuous growth [14]. In the case of the interrupted growth of a sample containing a GaAs buffer layer and single quantum wells comprising approximately 22, 11, and 7 monolayers of GaAs, discrete luminescence lines corresponding in position to integral numbers of monolayers are observed. The growth interruption is produced by shuttering off the gallium and aluminum molecular beams while allowing the arsenic molecular beam to reflect from the crystal surface. The lines are interpreted as exciton recombination luminescence from regions of uniform thickness with diameters larger than exciton diameters. Satellite peaks arise from regions of the quantum well layers that are one monolayer thicker or thinner than the regions giving the central peaks. The spectra provide evidence of nearly monolayer smoothness of the quantum well interfaces. For the quantum wells formed by continuous growth with no interruptions, single broadened luminescence lines are observed at positions which do not necessarily correspond to integral numbers of monolayers. This is interpreted as excitonic recombination luminescence from quantum wells whose thickness fluctuations are on a scale shorter than the exciton diameter [15]. Each exciton spans a number of thickness fluctuations and has an energy corresponding to an average thickness within its wavefunction. These two configurations are illustrated in Figure 4. These data give strong evidence of monolayer smoothing of at least one quantum well interface during growth interruptions. It does not prove perfect order, however, and in no case to date has a growth procedure resulted in a single sharply quantized monolayer-smooth emission line.

As mentioned above, surface and interface smoothness are strongly dependent on growth conditions and particularly on substrate temperature, which provides the thermal atomic surface mobility needed to produce smooth growth. At the same time, though, elevated substrate temperatures limit surface and interface smoothness because of thermal surface roughening and thermal atomic interdiffusion at interfaces. Growth interruption allows time for surface migration to occur and permits growth at a lower temperature, where these effects are reduced.

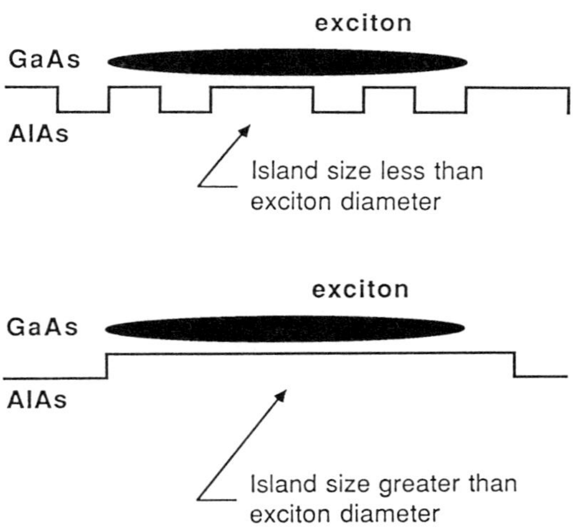

Fig. 4. Excitons at GaAs/AlAs interfaces with island size less than and greater than exciton diameter

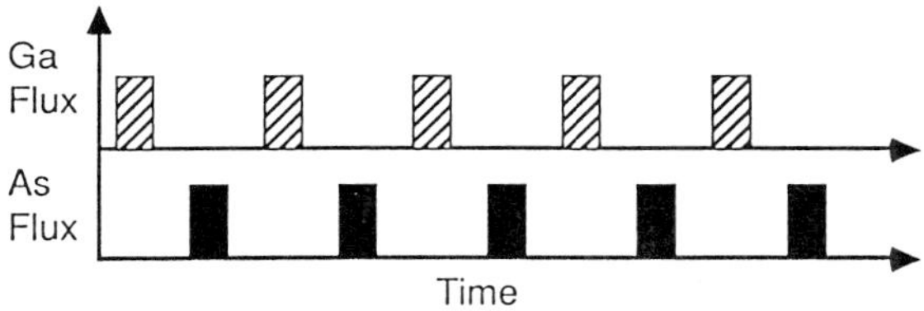

Fig. 5. Flux programs in migration-enhanced atomic later epitaxy

A promising means for still further reduction of growth temperatures and increase in control of epitaxial growth has recently been introduced. It is known variously as migration enhanced epitaxy [16] and atomic layer epitaxy [17]. It consists of alternate deposition of single atomic monolayers of the atomic constituents of a compound material and differs from previous interrupted growth techniques in that it alternately interrupts each source beam forming a compound semiconductor (Fig. 5). It has grown smooth, optically efficient GaAs and AlAs layers at substrate temperatures as low as 200 C and 300 C respectively, which are more than 100 degrees cooler than possible with previous growth techniques. The surface mobility of group III atoms on the epitaxial surface is enhanced by interruption of group V flux, and this is the apparent factor that permits growth at lower temperatures.

3.2 Graded Interfaces

Although the largest amount of work on quantum well structures has involved growth of structures with abrupt interfaces and efforts to achieve abrupt interfaces, there are many interesting phenomena that are associated with structures with graded interfaces. For example, grading of an interface can produce potential gradients that act on carriers at interfaces. This has produced artificial rectifying structures [18] and lasers with enhanced performance produced by graded index separate confinement heterostructures [19]. Shaping of quantum wells to parabolic profiles leads to quantum well energy levels with uniform energy spacings and increases the number of allowed quantum well optical transitions (Fig. 6) [20].

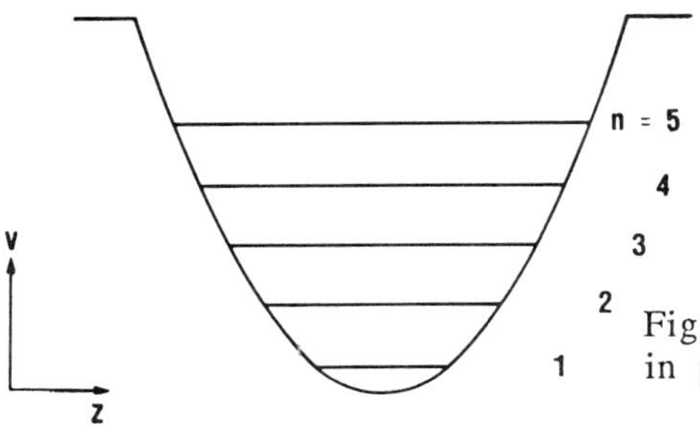

Fig. 6. Uniformly spaced energy levels in parabolically shaped quantum well.

This allowed absolute determination of energy level spacings that in turn led to the first accurate determination of the band offsets in the GaAs-(Al,Ga)As heterostructure system [20]. A further interesting possibility with graded potential well structures is the production of potentials equivalent to the potentials that would be produced by fixed distributions of background charges. It thus should become possible to mimic the potential of a positive background charge density as assumed in models of jellium, without the necessity of the presence of strongly scattering ionic charge backgrounds which occur in conventional doped semiconductor structures.

Several means exist for the preparation of graded interfaces. In one approach, a shutter can be gradually moved across the orifice of a molecular beam source, which typically consists of an open-mouthed crucible. In a second approach, the temperature of a molecular beam source can be ramped up and down to produce desired changes in beam flux intensities. A third approach employs a superlattice of graded composition to approximate the desired profile. The superlattice structure used to approximate a 500 Angstrom parabolic quantum well in such an approach is shown in Figure 7 [20]. Electrons in this graded superlattice structure tunnel through the various layers and feel an average composition profile which is parabolic in shape. The quantized energy level spacings calculated for this graded superlattice differ by only a few percent from those calculated for the smooth parabolic potential well. Still closer approximations to the smooth parabola may be achieved simply by use of shorter superlattice periods. The graded superlattice approach has the attractive feature that it is suitable for flexible implementation with computer control of shutters.

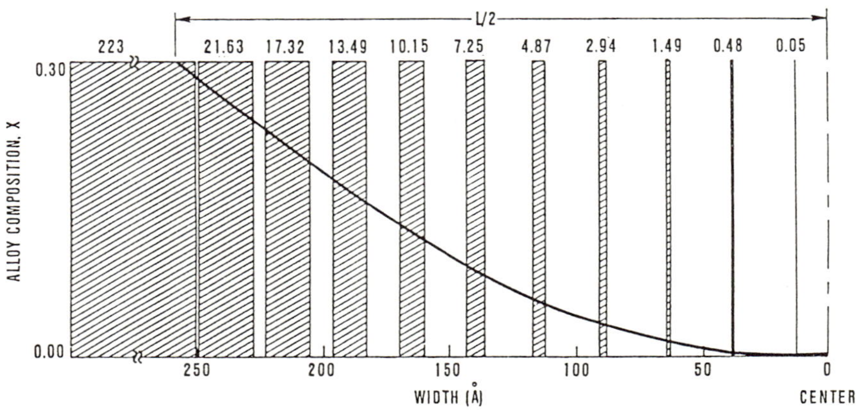

Fig. 7. Superlattice with graded composition for synthesizing parabolic potential profile

3.3 Electro-optic Effects

Electric fields at interfaces may also be introduced by the direct means of external application of a potential across the interfaces. This may be accomplished by inclusion of the interfaces in an undoped or depleted region between n and p semiconducting regions under a reverse bias. Electrons and holes in a quantum well may be attracted to opposite faces of the well under the bias, reducing their overlap and luminescence recombination probability as well as reducing the energy gap for creation of an electron-hole pair (Fig. 8) [21]. Referred to as the quantum confined Stark effect, this is the quantum well equivalent of the Franz-Keldysh effect. For large wells it reduces to the Franz-Keldysh effect, but for smaller wells is substantially larger and is attractive for applications to optical modulator and optical logic functions. At a staggered bandgap interface, as occurs between InAs and GaSb or between layers of (Al,Ga)As with high aluminum content, both conduction and valence band edges are higher on one side of the interface. This leads to the interesting and exactly opposite situation under electric field that 1) electrons and holes reside on opposite sides of the interface, 2) they are drawn toward each other by an electric field, 3) their luminescent recombination is increased by an electric field, and 4) their bandgap energy is increased by the electric field [22].

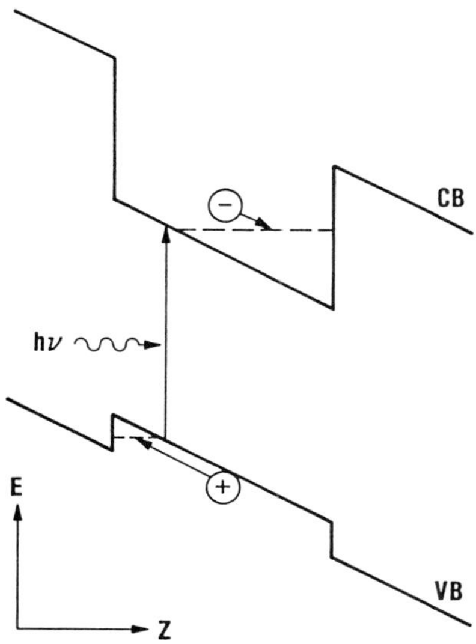

Fig. 8. Quantum well in electric field. CB and VB are conduction band and valence band edges.

4. ELECTRON TRANSPORT AT INTERFACES

The energy band offsets at interfaces play a number of important roles in electron transport at interfaces. They can act as barriers, serving to block electron transport between layers. Or, in the case of thin layers, they can allow controlled amounts of tunnelling transport between layers. In the case of asymmetric barriers, they can provide asymmetric conductivity and rectification. The energy band offsets also play important roles in the transport of electrons within layers along interfaces.

4.1 Two-dimensional Electronic Conduction:

By confining the perpendicular motion of electrons to quantized states at interfaces, the dimensionality of free motion of carriers parallel to heterostructure interfaces is reduced. This is of interest for the study of conductivity in low dimensions and for application to devices that guide and control electron flow. Electrons can be introduced to the interface region without the necessity of doping within the current-carrying channel layer where the electrons flow. This may be accomplished by the use of gates which invert or accumulate carriers to the interface layer or by modulation doping in which dopant atoms in barrier layers supply carriers which transfer to the current carrying channel regions on the low potential sides of the interfaces (Fig. 9) [23]. In either case, much higher electron mobilities may be obtained than would be found in uniformly doped materials of comparable carrier density.

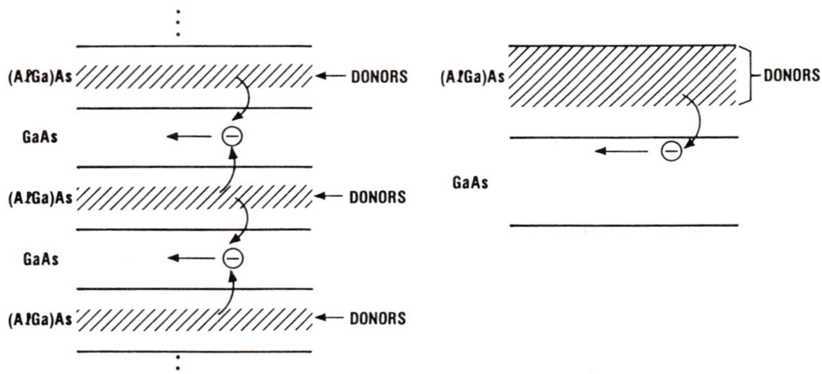

Fig. 9. Modulation doped superlattice and modulation doped heterostructure single interface

4.2 Modulation Doping and Atomic-plane Doping

Modulation doping at heterostructure interfaces was first introduced in the GaAs/(Al,Ga)As system and is extensively used in this system for high speed heterostructure transistors. A diagram of the energy band edges, dopant energy levels and conduction electron energy levels at a modulation doped single interface is shown in Figure 10. The mobilities obtained by the modulation doping technique have steadily increased

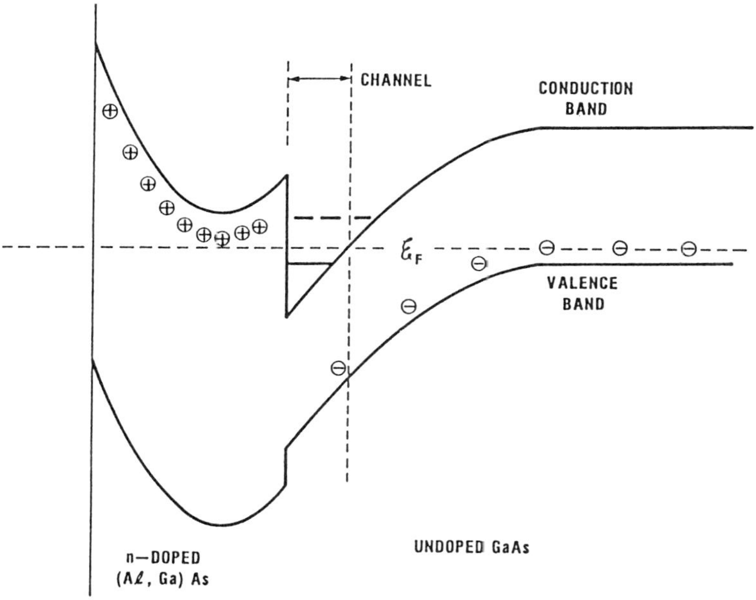

Fig. 10. Band edges and energy levels at modulation doped heterostructure single interface

with improvements in epitaxial growth techniques and design of the heterostructures. The highest electron mobility (5.0×10^6 cm^2/Volt sec at T = 0.3 K) yet observed in a semiconductor has been produced using this technique [24]. The mobility and carrier concentration vs. temperature for this sample are shown in Figure 11. along with a sketch of the layers comprising its active region. The channel region lies at the top of a one micrometer layer of undoped GaAs grown by molecular beam epitaxy on a superlattice buffer on an undoped GaAs substrate. The channel is defined by the interface with a 75 nm undoped (Al,Ga)As setback layer, above which lie fourteen sheets of silicon donor atoms embedded 5 nm apart in 65 nm of undoped (Al,Ga)As (Fig. 12).

The dopant sheets were formed by an atomic plane doping technique in which aluminum and gallium beams were shuttered off to stop growth while the silicon dopant beam and arsenic beam impinged on the surface. Each dopant sheet contained a concentration of silicon atoms which would produce 1×10^{11} electrons/cm^2 in GaAs. The goal of the interrupted-growth sheet doping technique was to enhance the fraction of silicon atoms incorporated onto gallium sites, where they act as donors. By contrast, if the silicon atoms occupy arsenic sites, they become acceptors, compensate the donors and act as ionized scattering

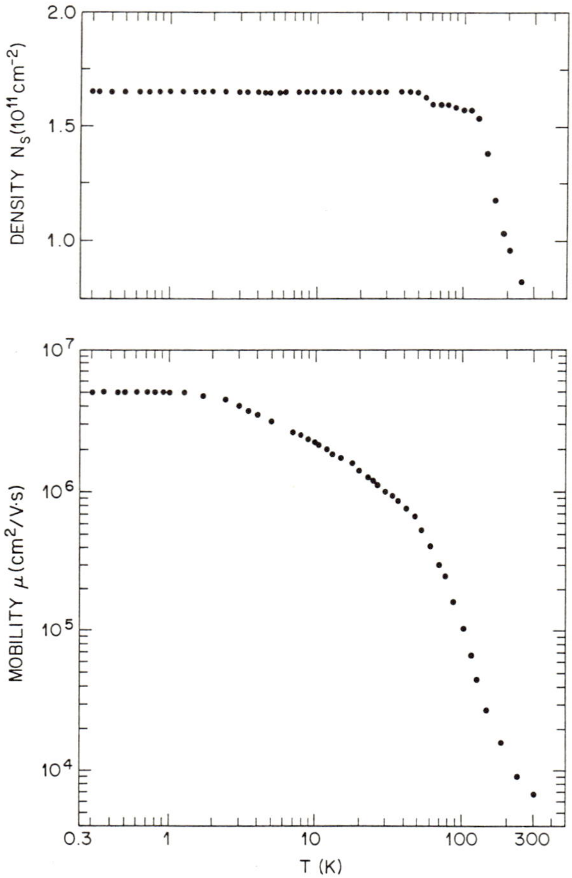

Fig. 11. Carrier density and mobility vs. temperature under illumination for modulation doped heterostructure single interface depicted in Fig. 12

centers for the conduction electrons in the channel. The atomic plane sheet doping technique has also been used in previous work to increase the maximum conduction electron density which can be achieved with silicon doping of GaAs [25].

4.3 Electron Scattering

There are several sources of scattering of two dimensional electrons at heterostructure interfaces [26, 27]. The principle source of scattering which limited the low temperature electron mobility in the above sample was the ionized impurity scattering from the dopant planes,

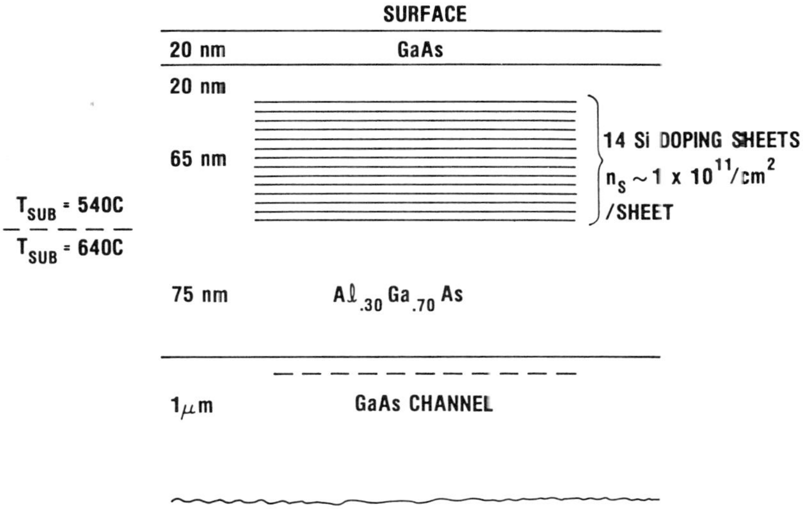

Fig. 12. Channel, barrier, and atomic plane sheet doping sequence of modulation doped single interface referred to in Fig. 11

which can nearly account for the observed mobilities. All of the dopant sheets were ionized, either by depletion to form the conducting channel, by depletion to the surface, which lay 40 nm above the top doping sheet, or by depletion to deep states within the (Al,Ga)As material. Measurements were made under illumination, in order to empty photosensitive traps.

Other sources of scattering of the electrons are phonons, background impurities in the channel, interface charges, and alloy impurity scattering. The phonon scattering is temperature-dependent, being the principal mechanism at room temperature, but being completely frozen out at low temperature, as illustrated by the fact that the mobility is entirely temperature independent below T = 1K. This is a result of the fact that all phonons with wave vector sufficient to scatter an electron across the Fermi surface are exponentially frozen out. The material is in the residual resistance regime, which had not previously been observed in a semiconductor. Background impurity charge concentrations in the sample must be less than 10^{14} cm^{-3} and interface charge density at the channel interface must be less than 10^{8} cm^{-2} in order to allow the observed mobility values.

4.4 New Quantum Hall Effect Phenomena

The high mobilities obtainable with the modulation doped heterostructures reflect long scattering times and long electron mean free paths in the structures. These allow detailed study of magnetotransport phenomena and the observation of new effects of the two dimensional electron gas in magnetic fields. Modulation-doped GaAs structures show strong quantum Hall effect phenomena and are currently slated to become the new quantum standard of electrical resistance [28], replacing the present wire-coil resistance standards. They also led to the discovery of the fractional quantum Hall effect, which occurs in the extreme quantum limit in which electrons in the lowest Landau and spin levels condense into ordered states having reduced resistance and quantized Hall resistance at fractional fillings of the Landau levels [29].

The high mobilities are of special importance in observations of the fractional quantum Hall states because the condensation energies and energy gaps for the fractional states are smaller than for the integral quantum Hall effect states. With lower mobility, extrinsic collision broadening of the Landau levels reduces the gaps, rendering ordering effects difficult or impossible to observe and making measurement of intrinsic energy gaps of the fractional quantum Hall effect difficult.

The ultrahigh mobility samples described above have allowed several interesting new effects to be observed in the fractional regime. In one study, it has allowed the first quantitative test of theories of the gap energy in the fractional quantum Hall effect [30]. The comparison requires the spatial extent of the two dimensional electron gas to be known, which was determined from infrared intersubband observations of the two dimensional gas energy levels. It confirms the basic correctness of the result that the energy gap is $0.1\ e^2/\varepsilon l_0$ for the ideal two-dimensional system (where ε is the dielectric constant and l_0 is the magnetic length).

In a second study with the ultra high mobility samples, an even-denominator fractional quantum Hall effect quantum number was confirmed for the first time [31]. It was observed at quantum number 5/2, corresponding to filling of 2 1/2 Landau levels, and may suggest the involvement of spin-reversed electrons in the ordered states. A theory of the state does not yet exist.

5. LATERALLY CONFINED STRUCTURES

With sufficiently fine in-plane dimensions, the motion of carriers in quantum wells will become quantized for motion in the lateral as well as the perpendicular directions. Confinement in one dimension within the layer plane will produce a quantum well wire, and in two dimensions within the layer plane will produce a quantum box. Both optical and electrical properties will be strongly modified by the new confinement.

Formation of microstructures with lateral confinement is only beginning to be developed, however, and progress in their fabrication has been slow. This is because fine control of lateral dimensions is more difficult than control of film thickness, and no clear choice has emerged for a preferred method to form sharp lateral interfaces.

A number of approaches, most of which start from layered structures, have been used or suggested to form such structures. A direct approach is use of lithography to remove selected areas of a layered structure, leaving behind posts or ridges containing quantum wires or boxes [32]. But clear quantization of energy levels has not yet been observed with these techniques. A technique which does not require lithography is epitaxial overgrowth on an etched or cleaved edge of a previously grown quantum well [33]. This technique has also not yet produced observable quantized states.

Another proposed means of quantum wire formation employs alternate submonolayer depositions in the step growth mode of epitaxial crystal growth on an off-axis substrate [34]. If the off-axis crystal surface contains a regular array of atomic terraces, the submonolayer coverages will partially cover the steps. The remainder of the step width could then be covered with a second material. With subsequent repetition of the process, arrays or superlattices of quantum wires could be produced (Fig. 13). Preferential nucleation at step edges has been observed with this technique and lateral superlattice formation has been reported [35], but quantum wires have not yet been achieved.

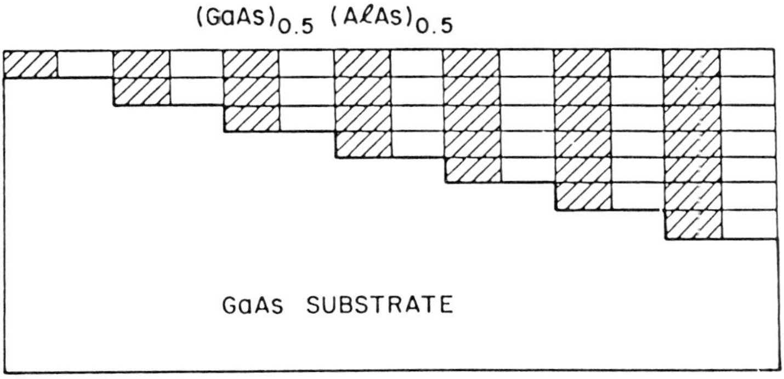

Fig. 13. Quantum wire structure proposed by growth of sequential half monolayer depositions on off-axis substrate

A recently developed technique for low-dimensional structure fabrication employs enhanced interdiffusion of layers in selected areas defined by electron beam lithography and ion bombardment [36]. This technique has given the most pronounced quantum wire and box spectral effects yet observed. Upward luminescence shifts and discrete luminescence lines attributed to lateral quantization have been seen with decreasing wire and box size. Aluminum/gallium interdiffusion in ion bombarded regions upon thermal annealing produces the laterally confining regions. The processes produce a graded aluminum interface profile and thus a graded lateral potential profile. Quantization in the lateral direction is produced by the graded profile and yields a nearly uniform energy ladder of quantized states, similar in some respects to the spectra from the parabolic-interface quantum wells described previously. Both quantum wires and quantum boxes have been formed with this technique.

ACKNOWLEDGEMENT

I wish to acknowledge the collaboration of my many colleagues at AT&T Bell Laboratories and the University of California, Santa Barbara who have made my parts of the research described here possible.

REFERENCES

1. Jesse DuMond and J. Paul Youtz: J. Appl. Phys. 11, 357 (1940)
2. J. R. Arthur: J. Appl. Phys. 39, 4032 (1968)
3. M. Razeghi and J. P. Duchemin: In Springer Series in Solid-State Sciences 53, 100 (Springer, Berlin, Heidelberg 1984)
4. M. B. Panish: J. Crystal Growth 81, 249 (1987)
5. P. M. Petroff: J. Vac. Sci. Technol., 14, 973 (1977)
6. A. Y. Cho and J. R. Arthur: Prog. Solid State Chem. 10, 157 (1975)
7. P. J. Dobson, B. A. Joyce, J. H. Neave and J. Zhang: J. Crystal Growth 81, 1 (1987)
8. A. C. Gossard, P. M. Petroff, W. Wiegmann, R. Dingle and A. Savage: Appl. Phys. Lett. 29, 323 (1976)
9. R. M. Fleming, D. B. McWhan, A. C. Gossard, W. Wiegmann and R. A. Logan: J. Appl. Phys. 51, 357 (1980)
10. P. M. Petroff, A. Y. Cho, F. K. Reinhart, A. C. Gossard and W. Wiegmann: Phys. Rev. Letters 48, 170 (1982)
11. T. S. Kuan, T. F. Keuch, W. I. Wang and E. L. Wilkie: Phys. Rev. Letters 54, 201 (1985)
12. H. Kroemer: J. Crystal Growth 81, 193 (1987)
13. J. C. Bean: Science 230, 127 (1985)
14. C. W. Tu, R. C. Miller, B. A. Wilson, P. M. Petroff, T. D. Harris, R. F. Kopf, S. K. Sputz and M. G. Lamont: J. Crystal Growth 81, 159 (1987)

15. M. Tanaka and H. Sakaki: J. Crystal Growth 81, 153 (1987)
16. Yoshiji Horikoshi, Minoru Kawashima and Hiroshi Yamaguchi: Appl. Phys. Letters 50, 1686 (1987)
17. A. Salokatve, J. Varrio, J. Lammasniemi, H. Asonen and M. Pessa: Appl. Phys. Letters 51, 1340 (1987)
18. A. C. Gossard, W. Brown, C. L. Allyn and W. Wiegmann: J. Vac. Sci. Technol: 20, 694 (1982)
19. W. T. Tsang: Appl. Phys. Letters 39, 134 (1981)
20. R. C. Miller, A. C. Gossard, D. A. Kleinman and O. Munteanu: Phys. Rev. B 29 7085 (1984)
21. D. A. B. Miller, D. S. Chemla, T. C. Damen, A. C. Gossard, W. Wiegmann, T. H. Wood and C. A. Burrus: Phys. Rev B 32, 1043 (1985)
22. E. J. Caine, S. Subbanna, H. Kroemer, J. L. Merz, and A. Y. Cho: Appl. Phys. Letters 45, 1123 (1984)
23. A. C. Gossard and A. Pinczuk: In Synthetic Modulated Structures, ed. by Leroy L. Chang and B. C. Giessen (Academic Press, Orlando 1985) p. 215
24. J. H. English, A. C. Gossard, H. L. Stormer and K. W. Baldwin: Appl. Phys. Letters 50, 1826 (1987)
25. Klaus Ploog: J. Crystal Growth 81, 304 (1987)
26. P. J. Price: Surface Science 113, 199 (1982)
27. W. Walukiewicz, H. E. Ruda, J. Lagowski and H. C. Gatos: Physical Review B 30, 4571 (1984)
28. B. N. Taylor, J. of Research of the National Bureau of Standards: 92, 55 (1987)
29. D. C. Tsui, H. L. Stormer, A. C. Gossard: Phys. Rev. B 48, 1559 (1982)
30. R. L. Willett, J. P. Eisenstein, H. L. Stormer, D. C. Tsui, A. C. Gossard and J. H. English: Phys. Rev. Letters
31. R. L. Willett, H. L. Stormer, D. C. Tsui, A. C. Gossard and J. H. English: submitted for publication
32. M. A. Reed, R. T. Bate, K. Bradshaw, W. M. Duncan, W. R. Frensley, J. W. Lee and H. D. Shih: J. Vac. Sci. Technol. B 4, 358 (1986)
33. Hiroyuki Sakaki: Jap. Jour. Appl. Phys. 19, L735 (1980)
34. P. M. Petroff, A. C. Gossard and W. Wiegmann: Appl. Phys. Letters 45, 620 (1984)
35. Takashi Fukui and Hisao Saito: Appl. Phys. Letters 50, 824 (1987)
36. J. Cibert, P. M. Petroff, G. Dolan, S. J. Pearton, A. C. Gossard and J. H. English: Appl. Phys Lett. 49, 1275 (1986)

Discussion

M. CARDONA: You did not have time to comment about phonons so I would like to say a few words. In optical spectroscopes for bulk materials one can only see phonons with $k \simeq 0$, a rather severe restriction. For the superlattices one can see k$\simeq$0 plus a vector of the reciprocal lattice of the *superlattice*, i.e., a much larger number of phonons. One can use this fact to obtain the *bulk* dispersion relations with optical spectroscopes (e.g., Raman), which are much simpler and have better resolution than neutron scattering. This feature can be thought of as related to the partial two-dimensionality of these systems. In a similar way we have seen that surface–phonon dispersion relations can be measured with spectroscopes such as HAS or EELS, much simpler than neutron scattering. For superlattices, one measures with Raman scattering *folded* acoustic phonons, *confined* optic phonons, and interface phonons, partly localized around the interfaces and propagating along them and also repeating along the superlattice axis so as to fulfill Bloch's theorem.

W. KOHN: Has anybody seen the quantization of phonons in the wells?

M. CARDONA: Yes.

W. KOHN: Optically, or...?

M. CARDONA: Yes, optically. Some beautiful work has been done on this.

H.C. SIEGMANN: I am also interested to hear about your idea to make a really highly polarized electron source. I think I want to say this: There are many solids in which you have fully polarized electrons—many magnetic materials. And you also can get them out and measure their polarization. However, if you want to make an electron source, you have to do it in great quantities. And I think this is the problem with these quantum wells, that you have a relatively thin solid from which you have to extract electrons, and I think this is the reason why you never get high intensities. Now, please explain to me what I did not understand here. I just did not understand how you can penetrate the surface, make negative electron affinity, and get these electrons out quantitatively—or, in the huge quantities which you actually need to compete with the present gallium–arsenide source.

A.C. GOSSARD: In the experiments at SLAC, as far as the efficiencies were concerned, the number of electrons that were obtained were larger, comparable to what you get in ordinary gallium arsenide. You have several thousand angstroms of penetration depth. If you figure in your superlattice you can maybe penetrate through twenty barriers; there you already have 2000 Å, so you can quite readily imagine absorbing half of your light. I think the problem, though, is to get the electrons out before they spin depolarize; so, the things that we've considered either have been to sacrifice half of the emitted electrons by letting them just go into the material and disappear, or another approach we mean to try has been to build a reflector which is maybe 2000Å into the material, and then hope that your spin depolarization time is longer than what it takes to travel from 2000Å. That is the crucial question.

H.C. SIEGMANN: Can we repeat, what are the prospects to make a detector for spin depolarization by looking at cathodo–luminescence? This would actually interest me much more, to be quite honest. That would be more useful. Because the gallium arsenide source is quite good. Is there a prospect that you can do luminescence and observe the recombination of a polarized electron with this polarized hole?

A.C. GOSSARD: We talked about that at lunch one day this week, in fact; and what I don't understand is how you get the electron in at a sufficiently low energy, and with its spin still preserved.

M.G. LAGALLY: I have three questions.

1. You indicated that in reflection electron diffraction of MBE GaAs layers, at best short streaks along hemispheres are observed. This indicates that a certain degree of roughness remains in the surface. Based on what we have observed by comparing RHEED patterns from cleaved GaAs(110) with such patterns, we would estimate a step density of one per several hundred Å or so. Is this sort of step density significant in electrical properties? Can one, or has one, done better in the growth?
2. RHEED intensity oscillations indicate layer-by-layer growth and are caused by the oscillating statistical roughness of the surface, as you indicate. However, their existence also implies that the atomic mobility is low. Would it not be better, from the point of view of electrical properties, to grow at temperatures where there are no oscillations? Has anyone compared carrier mobilities of films grown at temperatures where oscillations are present with ones grown at temperatures where there were no oscillations?
3. Can one make quantum wires by using a vicinal surface as a substrate?

A.C. GOSSARD:

1. The density of surface steps that you mention could be decreased by further development of the technique of polishing very flat substrate surfaces and accurate orientation of cut substrate crystal surfaces with respect to crystal axes, together with low temperature growth by atomic layer epitaxy or interupted growth at interfaces. Anisotropies in electron scattering at interfaces with a denisity of steps of one per several hundred Angstroms would probably be observable in very high mobility modulation doped heterostructures.
2. I agree that growth carried out at temperatures where RHEED oscillations are maximum is generally at a temperture that is too low for optimal optical and electronic properties. Luminescence efficiencies in GaAs–based heterostructures are generally higher at higher growth temperatures. Thermal roughening will be larger at higher temperatures, however.

3. Arrays of quantum wires can be formed by sequential alternate submonolayer depositions on vicinal surfaces, as proposed originaly by Pierre Petroff and as observed very recently by Gaines, Petroff and Kroemer and by Fukui.

K. TAKAYANAGI :

Streaks in RHEED pattern means *no indication of atomically flat surface*; we see surface flatness by reflection electron microscopy. Si(111)7×7, Si(001)2×1 with atomically flat terraces separated by atomic-high steps show spots (on circles of Ewald sphere) (*Jpn. J. Appl. Phys.*, 1987). MBE prepared GaAs(001) surfaces which give REM images indicating "wavy" surface geometry show a streaky RHEED pattern (unpublished).

A.C. GOSSARD: I agree with your interpretation.

G. DASH: Some strained layer epitaxial films are surprisingly metastable. Bean and colleagues, for instance, have observed that Ge-Si films on Si can be grown without dislocations, although intrinsic misfits may be as great as a few percent, to thicknesses much greater than expected. Has this unusual metastability been explained?

A.C. GOSSARD: More recent data from Kohama, Fukuda and Seki (*Applied Physics Letters* **52**, 380 (1988)) suggest that the region of metastability is much smaller than originally reported by Bean. More sensitive transmission electron microscope and EBIC studies have shown that dislocations are indeed present at smaller thicknesses than observable with the earlier x–ray diffraction and Rutherford backscattering measurements of Bean and of Kasper.

Electronic Structures of Strained Layer Superlattices $(Si)_m/(Si_{1-x}Ge_x)_m(100)$ with m=1 and m=2n (n=1–10)

Xie Xide, Shen Dingli, and Zhang Kaiming

The Institute of Modern Physics, Fudan University,
Shanghai, People's Republic of China

The empirical tight-binding method is used to calculate the electronic structures of type-II strained layer superlattices $(Si)_m/(Si_{1-x}Ge_x)_m(100)$ with m=1 and m=2n (n=1-10). Band folding is found to play an important role for the Si-like characteristics of strained Si/Ge, particularly for the quasi-direct gap of $(Si)_{2n}/(Ge)_{2n}$ when n>5. The electronic property of the strained monolayer $(Si)_1/(SiGe)_1$ is an intermediate case between that of the bulk zincblende semiconductor and the ordinary semiconductor superlattices.

1. Introduction

The rapid development of semiconductor technology has resulted in great progress in low-dimensional materials. The techniques of MBE (molecular beam epitaxy) and MOCVD (metalorganic chemical vapour deposition) could provide people with semiconductor superlattice having "arbitrarily" tailored gaps [1]. Among those commensurately grown strained layer superlattices (SLS), Si/SiGe has gained much interest for its practical importance in device applications [2]. Despite the large lattice mismatch between Si and Ge (~4.2%), defect-free commensurate interfaces of Si/Ge have been grown through elastic deformation. G. ABSTREITER et al. [3] have reported the two dimensional electron gas induced by built-in stress in selectively doped $Si/Si_{0.5}Ge_{0.5}$ superlattice. Recently, T. P. PEARSALL et al. [4] have observed new optical transitions in $(Si)_4/(Ge)_4$. It is desirable to have high-mobility, direct-gap superlattice synthesized from indirect host matrices such as Si and Ge [5].

Since it has already been proved to be possible to grow ultrathin SLS Si/Ge, it might be likely that strained monolayer superlattices could be available also in the future. $(Si)_1/(SiGe)_1$ might be quite popular because it could retain certain features of both superlattices and bulk zincblende semiconductors. In this paper we shall give some of our results of theoretical studies on the electronic structures of SLS $(Si)_m/(Si_{1-x}Ge_x)_m$ for both m=1 and m=2n(n=1-10).

Springer Series in Surface Sciences Vol. 14
Solvay Conference on Surface Science
Editor: F.W. de Wette

For simplicity, the uniaxial stress along [100] is neglected in the calculation and only hydrostatic strain is considered. The crystal will take the lattice constant of bulk Si (Ge) if the upper layer is grown on the Si (Ge) substrate with strain. Under the mode of "free standing", the lattice constant parallel to the interface is given by [6-8]

$$a'' = a_{Si} [1 + f\, G_{alloy}\, h_{alloy}/(G_{Si}\, h_{Si} + G_{alloy}\, h_{alloy})] \qquad (1)$$

where h is the layer thickness, G the shear moduli, f the lattice misfit between alloy and Si.

In the present tight binding calculations, the Slater-Koster parameters were adopted from Y. LI et al. [9] with interactions taken into account as far as to the next nearest neighbours. The interaction between Si and Ge is approximated by the average of that of both Si and Ge. The non-linear VCA used for the nearest neighbour parameter due to stress is as follows [10]

$$a''^2 E^{alloy}(1/2,1/2,1/2) = (1-x)\, a_{Si}^2 E^{Si}(1/2,1/2,1/2) + x\, a_{Ge}^2 E^{Ge}(1/2,1/2,1/2) \qquad (2)$$

Valence band discontinuity ΔE_v between Si and Ge should be added to Si diagonal elements for certain fixed Ge parameters. From the recent synchrotron radiation data ΔE_v=0.16ev [11-12]. Other authors suggested from both the experiment [3] and theoretical calculation [13] that SLS Si/Ge should be type -II. We have used the value from Ref.13. According to our calculation, energy gaps for the strained bulk Si and Ge in three different strain cases mentioned above change dramatically with the existence of strain. Gaps of strained Si and Ge are quite close under each strain distribution (within 0.1ev).

2.1 Results for $(Si)_{2n}/(Si_{1-x}Ge_x)_{2n}$ (n=1-10)

Figure 1(a) gives the curve of E_g vs. n for $(Si)_{2n}/(Ge)_{2n}$ under free standing mode, it can be seen that for ΔE_v=0.51ev E_g drops from 0.96ev (n=1) to 0.60ev (n=10). It is interesting to note when n=5, 6 and 8, the gap of SLS becomes direct. This fact could be understood in terms of the band folding. Table I shows the position of the bottom of the conduction band (BCB) along [100] and also that of the bulk Si after folding in this direction. It can be seen that the folded BCB of bulk Si is rather close to Γ when n=5, 10. The calculation shows that though top of the valence band remains fixed at Γ, superlattice BCB is determined mainly by that of the folded Si. The role of Ge is to change slightly the position of the superlattice BCB. For cases of n=5, 6 and 8, it can be seen from Table I that BCB is at Γ and the BCB of SLS will not exceed $0.1\times2\pi/a$ for n>5, then the superlattice becomes quasi-direct.

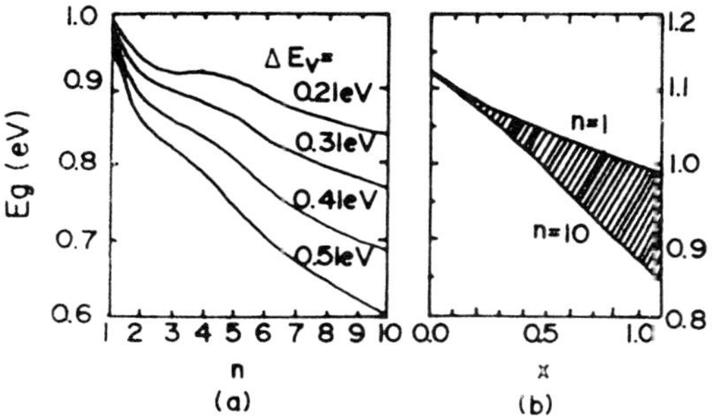

Fig.1 Gap variation with n=1-10 in $(Si)_{2n}/(Si_{1-x}Ge_x)_{2n}$ (100) under free standing mode (a) $x=1, \Delta E_V=0.21$, 0.31, 0.41 and 0.51ev (b) $0<x<1, \Delta E_V=0.21$ev

Table 1. Conduction band minimum of $(Si)_{2n}/(Ge)_{2n}$ (100) and that of bulk Si after folding along [100] direction. (Unit: $2\pi/a$)

n	1	2	3	4	5	6	7	8	9	10
$(Si)_{2n}/(Si)_{2n}$	0.21	0.21	0.13	0.04	0.008	0.04	0.07	0.04	0.01	0.008
$(Si)_{2n}/(Ge)_{2n}$	0.21	0.25	0.17	0.13	0.00	0.00	0.07	0.00	0.06	0.05

If ΔE_V is varied from 0.21ev to 0.51ev, the curve in Fig.1(a) will move downwards, implying a stronger binding of the carrier. Although in some cases BCB would change slightly with ΔE_V, the quasi-direct feature of SLS for n>5 remains unchanged. Fig.1(b) is the calculation of gap curve for $(Si)_{2n}/(Si_{1-x}Ge_x)_{2n}$(100) ($\Delta E_V$=0.21ev). The hatched area represents the region of gap variation.

Electroreflectance studies [4] have shown that new transitions exist at 0.76, 1.25 and 2.31ev for $(Si)_4/(Ge)_4$ when $a''=a_{Si}$. According to the present calculation (ΔE_V=0.74ev [13]), direct transition at Γ and X point is 1.09 and 2.01ev respectively and BCB occurs at the X point. For $(Si)_6/(Ge)_6$ ($a''=a_{Si}$) the value for BCB is $0.17x2\pi/a_{Si}$ which is very close to the $0.3\pi/a_{Si}$ given by self-consistent pseudopotential result [14]. In the energy band we also find some interface states which is to be confirmed by future experiment.

2.2 Results for $(Si)_1/(Si_{1-x}Ge_x)_1$

Figure 2(a) shows the gap curve of strained monolayer superlattice when Si is grown under the tetrahedrally tensile stress to match the Ge substrate. The

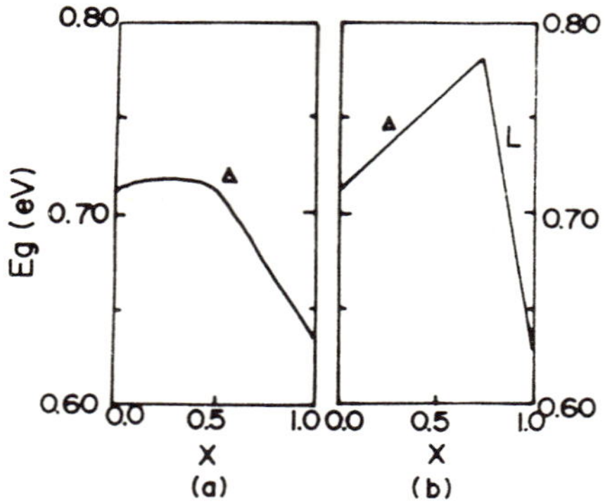

Fig.2 Gap variation with composition x (a) for $(Si)_1/(Si_{1-x}Ge_x)_1$ (100) when Si is strained to cubic Ge (ΔE_V= 0.31 ev) (b) for the alloy $Si_{1-x}Ge_x$ in the same strain

corresponding curve for alloy under the same strain is plotted in Fig.2(b) which shows the well-known shift of the conduction band minimum from Δ to L($\Delta$is the [100] axis and L is (1/2,1/2,1/2)$2\pi/a$ point, all belongs to zincblende Brillouin zone) when $x>0.8$. For the monolayer superlattice, the conduction band minimum is at (0.8,0,0)$2\pi/a$ for $x\leq0.5$ and shifts towards the X point when $x>0.5$. At $x=0.3$, a maximum of gap is reached. Such a maximum could be attributed to similar phenomenon in the gap curve of alloy which is subject to the same strain. It is worth mentioning that this kind of variation seems to be quite unusual for superlattices.

When the alloy $Si_{1-x}Ge_x$ is added to cubic Si with a compressive stress, the gap curve of superlattice is given in Fig. 3 (solid line denotes ΔE_V=0.84ev, dashed line denotesΔE_V=0.74ev). Again one can see the non-monotonical variation. After using the coherent potential approximation (CPA), it is found that for n=1 the whole curve shifts downwards with the maximum point changing from x=0.5 to x=0.3. For the purpose of comparison the gap curve of $(Si)_n/(Si_{1-x}Ge_x)_n$ (100) is plotted for the case of n=2 and n=4, assuming also that the superlattice is grown on the Si (100) substrate. It is found that the gap variation is greatly influenced by the quantum size effect and the maximum of the gap was almost nonexisting for the n=4 case. The gap decreases considerably with the increase of n for a fixed composition.

The energy band of the free standing mode $(Si)_1/(Ge)_1$ is shown in Fig. 4. The strong coupling between adjacent layers results in large dispersion in the energy bands. The Si characteristic is revealed mainly through the conduction band minimum. Band splitting around -9 - -10ev along the XU direction shows the characteristics of the T_d symmetry (not the diamond symmetry of the matrix). Therefore special attention should be paid to the 3-dimensional feature of this material.

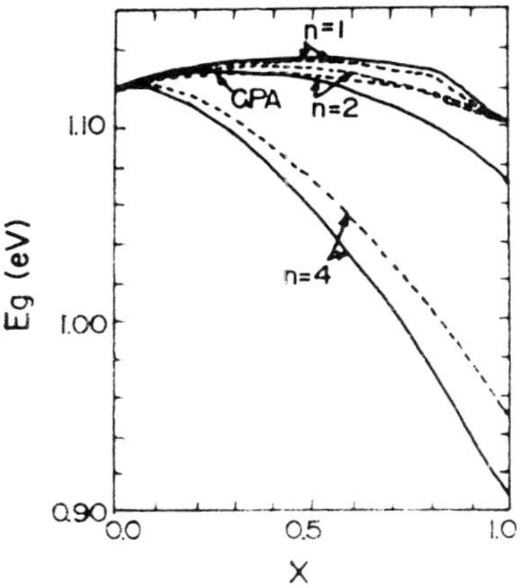

Fig.3 Gap variation of $(Si)_1/(Si_{1-x}Ge_x)_1$ (100) with composition x when alloy is strained to cubic Si. Solid line denotes ΔE_V = 0.84 ev (dotted line is the CPA result). Dashed line denotes ΔE_V= 0.74 ev

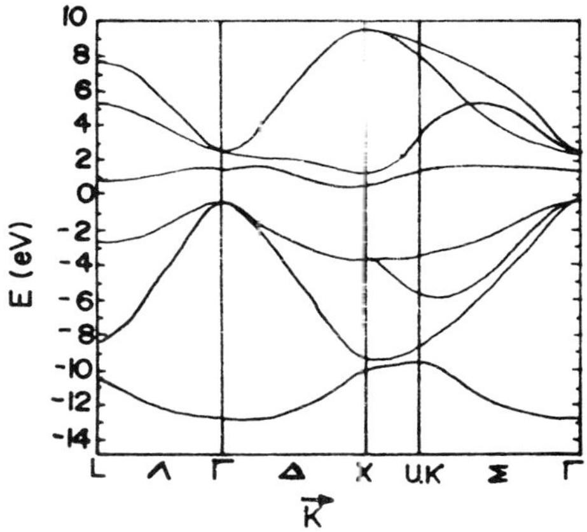

Fig.4 Energy band of $(Si)_1/(Ge)_1$ (100) for the free standing mode. ΔE_V =0.61ev

3. Discussion

In spite of the approximation a''=a$^{\perp}$ used here, the results obtained agree reasonably well with available experimental data and results of other calculation,expeci ally in predicting the position of conduction band minimum.

We would like to thank Dr. R. M. Martin for his stimulating discussion. We are also obliged to F. Lu and X. Zhang for their valuable suggestions and assistances.

1. L. Esaki: IEEE Journal of Quantum Electronics. QE-22, 1611 (1986)
2. G. C. Osbourn: IEEE Journal of Quantum Electronics. QE-22, 1677 (1986); R. People: ibid., 1696 (1986)
3. G. Abstreiter, H. Brugger, T. Wolf, H. Jorke, H. J. Herzog: Phys. Rev. Lett. 54, 2441 (1985); Surf. Sci. 174, 640 (1986)
4. T. P. Pearsall, J. Bevk, L. C. Feldman, J. M. Bonar, J. P. Mannaerts, A. Ourmazd: Phys. Rev. Lett. 58, 729 (1987); Bull. Am. Phys. Soc. 32, 773 (1987)
5. U. Gnutzmann, K. Clausecker: Appl. Phys. 3, 9 (1974)
6. J. W. Matthews, A. E. Blakeslee: J. Vac. Sci. Technol. 14, 989 (1977)
7. G. C. Osbourn: J. Appl. Phys. 53, 1586 (1982)

8. A. Ishida, M. Aoki, H. Fujiyasu: J. Appl. Phys. 58, 797 (1985)
9. Y. Li, P. J. Lin-Chung: Phys. Rev. B27, 3465 (1983)
10. K. E. Newman, J. D. Dow: Phys. Rev. B30, 1929 (1984)
11. R. People, J. C. Bean, D. V. Lang, A. W. Sergent, H. L. Stormer, K. W. Wecht, R. T. Lynch, K. Baldwin: Appl. Phys. Lett. 45, 1231 (1984)
12. G. Margaritondo, N. G. Stoffel, A. D. Katnani, F. Patella: Solid State Commun. 36, 215 (1980)
13. C. G. Van de Walle, R. M. Martin: Phys. Rev. B34, 5621 (1986); J. Vac. Sci. Technol. B3, 1256 (1985); J. Vac. Sci. Technol. B4, 1055 (1986)
14. S. Froyen, D. M. Wood, A. Zunger: Bull. Am. Phys. Soc. 32, 906 (1987)

Discussion

M. CARDONA: In the unstrained superlattices, you assume the type II case. I seem to recall that most calculations yield the valence band of Ge above that of Si in this case, i.e., type I. Is this correct? What is the basis for your assuming type II?

XIE XIDE: Early experiments indicated that Si/Ge showed type-I features. However, recent experiments (G. Abstreiter, H. Brugger, T. Wolf, H. Jorke and H.J. Herzog, *Phys. Rev. B* **34**, 2441 (1985); *Surface Science* **174**, 640 (1986)) and theoretical calculations (C.G. Van de Walle and R.M. Martin, *Phys. Rev. B* **34**, 5621 (1986) and others) show that Si/Ge should be type-II.

First-Principles Calculations of Quasiparticle Energies at Surfaces and Interfaces: Semiconductor Surface-State Spectra and Band Offsets

S.G. Louie

Department of Physics, University of California, Berkeley, CA 94770, USA, and
Materials and Chemical Sciences Division, Lawrence Berkeley Laboratory, Berkeley, CA 94720, USA

A first-principles theory for calculating surface-state energies and semiconductor band offsets is described. Within a quasiparticle interpretation of excitation spectra, the approach provides well-founded energies which can be compared directly with spectroscopic measurements. Results for the As-capped Si(111) and Ge(111) surfaces and for the GaAs-AlAs(001) heterojunction are discussed and compared with experiment.

1. INTRODUCTION

A detailed description of the electronic excitation spectrum is an essential ingredient in understanding the physical and chemical properties of surfaces and interfaces. Local density functional (LDA) calculations have been quite successful in determining the structure of many surface systems.[1,2] However, since the density functional formalism is a ground-state theory, these calculations do not provide direct information on the excitation spectra. Indeed, as in the case of bulk solids, the common practice of comparing LDA eigenvalues to spectroscopic data has often led to large discrepancies in the surface-state energies. The proper interpretation of the excited-state spectra requires the concept of quasiparticles, the particle-like excitations in an interacting many-electron system.[3,4]

In this paper, we show that a recently developed quasiparticle theory [4] which has been successful in calculating the optical and photoemission spectra of bulk crystals [4-6] extends to surfaces and interfaces. The approach is based on a first-principles evaluation of the electron self-energy operator to first order in the dynamically screened Coulomb interaction and the dressed electron Green's function. The results for the two prototypical systems discussed here illustrate that the theory is accurate and of predictive power. For the As-capped Si(111) and Ge(111) surfaces,[7] the calculated energies are in excellent agreement with data from angle-resolved photoemission experiments and show a substantially larger gap between the empty and occupied surface states in comparison with LDA calculations. The predicted features of the empty surface states have been verified by a recent scanning tunneling microscopy study.[8] Our results for the band offsets of the GaAs-AlAs(001) interface [9] are also in very good agreement with recent experimental values. We find that there is a substantial many-body correction to the value of the valence band offset calculated using the LDA.

Springer Series in Surface Sciences Vol. 14
Solvay Conference on Surface Science
Editor: F.W. de Wette

2. THEORY OF QUASIPARTICLE ENERGIES

In the Green's function approach, the quasiparticle properties are determined by solving [3]

$$(T + V_{ext} + V_H)\,\psi_{n\vec{k}}(\vec{r}) + \int d\vec{r}'\,\Sigma(\vec{r},\vec{r}';E_{n\vec{k}})\psi_{n\vec{k}}(\vec{r}') = E_{n\vec{k}}\psi_{n\vec{k}}(\vec{r}) \quad . \tag{1}$$

where T is the kinetic energy operator, V_{ext} is the external potential due to the ions, and V_H is the Hartree potential. The exchange and correlation contributions are included in the self-energy operator Σ. Σ, in general, is nonlocal, energy-dependent, and nonHermitian with the imaginary part giving the lifetime of the quasiparticles. Our approach [4] is based on the GW approximation in which a first order expansion for the self-energy operator is taken:

$$\Sigma(\vec{r},\vec{r}';E) = i \int \frac{d\omega}{2\pi}\, e^{-i\delta\omega} G(\vec{r},\vec{r}';E - \omega)W(\vec{r},\vec{r}',\omega) \quad , \tag{2}$$

where δ is a positive infinitesimal. The major components of the theory are the fully interacting Green's function for which we use a quasiparticle approximation,

$$G(\vec{r},\vec{r}';E) = \sum_{n\vec{k}} \frac{\psi_{n\vec{k}}(\vec{r})\psi^{*}_{n\vec{k}}(\vec{r}')}{E - E_{n\vec{k}} - i\delta_{n\vec{k}}} \quad , \tag{3}$$

and the dynamically screened Coulomb interaction,

$$W(\vec{r},\vec{r}';\omega) = \Omega^{-1} \int d\vec{r}''\,\varepsilon^{-1}(\vec{r},\vec{r}'';\omega)V_C(\vec{r}'' - \vec{r}) \quad , \tag{4}$$

where ε is the time-ordered dielectric matrix whose off-diagonal elements in Fourier space describe the local fields (variations in the screening through the unit cell) and V_C is the bare Coulomb interaction.

As can be seen from the structure of Eqs. (1)-(4), the quasiparticle energies together with Σ and G must be obtained in a self-consistent fashion. In the calculations, the electron Green's function is constructed initially using the LDA Kohn-Sham eigenfunctions and eigenvalues and is subsequently updated with the quasiparticle spectrum from Eq. (1). The static dielectric matrix $\varepsilon(\vec{r},\vec{r}',\omega=0)$ is obtained as a ground-state quantity from the LDA calculation and extended to finite frequencies using a generalized plasmon pole model together with exact sum rules.[4]

The present approach has been applied successfully to a variety of crystals including semiconductors, ionic insulators, and alkali metals.[4-6] Highly accurate band gaps, optical transition energies, and photoemission spectra have been obtained. It is found that, for semiconductors and insulators, the use of the crystalline Green's function and inclusion of both local fields (the full dielectric matrix) and dynamical screening effects are important factors for quantitative results.[4] For the alkali metals, the quasiparticle bandwidths are shown to be also sensitive to the correct treatment of exchange-correlation effects in the dielectric screening.[6]

3. SEMICONDUCTOR SURFACE-STATE ENERGIES

A survey of the literature shows that, for semiconductor surfaces, even in cases where the structure is well-determined, there are rather severe discrepancies between the LDA surface-state energies and experimental values. In general, three areas of systematic disagreement exist: (1) the LDA surface-state band dispersion is qualitatively reasonable but is misplaced relative to the valence band maximum; (2) the calculated dispersion of the surface-state band is too small in some cases; and (3) the gap between the empty and occupied surface-state energies is often dramatically underestimated. This last feature is analogous to the "band gap" problem in bulk semiconductors and insulators.

The systems we examined to address the problem of predicting surface-state spectra are the As-chemisorbed Si(111) and Ge(111) surfaces. At saturation coverage, As is found to substitute for the outermost layer atoms. The surface then becomes chemically inactive and is stable against reconstruction showing a 1x1 periodicity. These are ideal prototype systems for the many-body calculation because of their geometric simplicity and the availability of detailed experimental data.[10] Moreover, they are of intrinsic importance as initial stage of GaAs growth on Si and Ge.

The calculations are carried out using a supercell geometry with a 12-layer thick slab. The surface geometry and the corresponding ground-state charge density are obtained from a LDA calculation using <u>ab initio</u> pseudopotentials. After the structure is determined, the quasiparticle energies for both the bulk states and surfaces states are calculated as described in Section 2. For the Si(111) and Ge(111) surfaces, equilibrium geometry is achieved with an outward displacement of the As atoms by 0.16 Å and 0.17 Å respectively. This structural relaxation is in excellent agreement with the recent experimental result of 0.17 Å from x-ray standing wave measurements on the Si(111):As surface.[11]

The calculated results are depicted in Figs. 1-3. In Fig. 1, the quasiparticle surface-state bands (solid lines) together with the LDA surface-state bands (dashed lines) are plotted against the projected quasiparticle band structure of Si. As expected, the fully occupied surface band corresponds to the dangling bond (lone pair) states of the As acatoms. These lone pair states have been carefully studied using angle-resolved photoemission measurements.[10] The theory also shows an empty surface state band in the gap. These empty states are split off the continuum states and, near $\overline{\Gamma}$, have atomic character at the surface similar to the L_{1c} conduction bands. Very similar results are obtained for the Ge(111):As system.

In comparison with the LDA results, the occupied quasiparticle surface-state band is lower in energy and has a broader dispersion. Both are needed

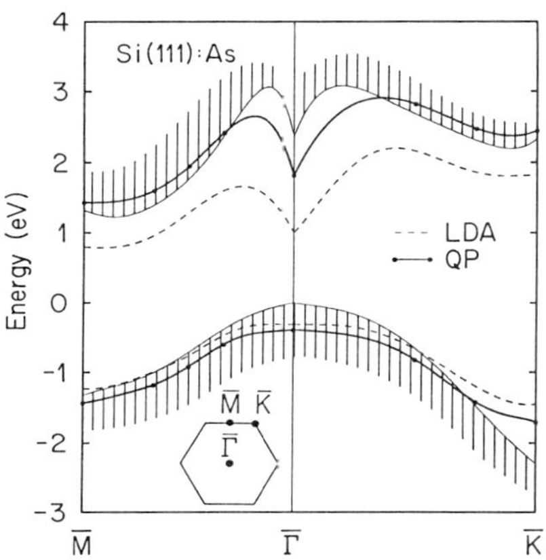

Fig. 1. Quasiparticle surface-state energies compared to LDA surface-state energies for the Si(111):As surface. Also shown is the quasiparticle bulk projected band structure.

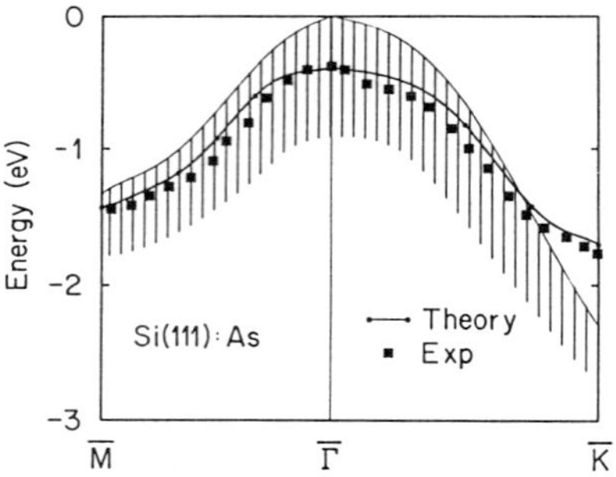

Fig. 2. Calculated filled quasiparticle surface-states energy for Si(111):As compared to data from photoemission (Ref. 10).

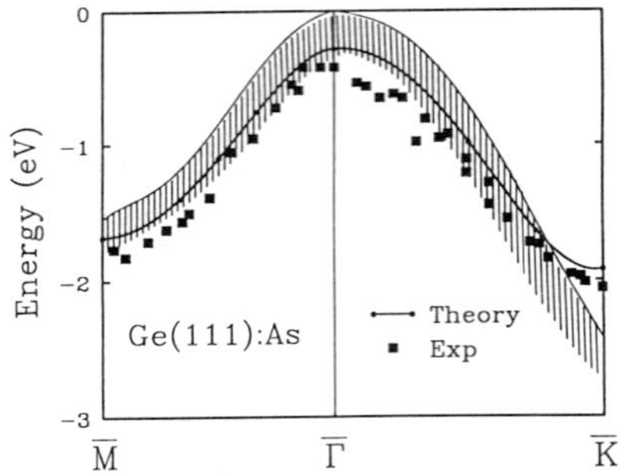

Fig. 3. Same as Fig. 3 except for Ge(111):As.

to bring theory to agreement with experiment. For Ge(111):As, the quasiparticle calculation also removes an unusual dispersion in the LDA results near $\overline{\Gamma}$.[7] Figures 2 and 3 compare the calculated lone pair surface-state bands with angle-resolved photoemission data. For both systems, the agreement is excellent in both the placement and the width of the band and is well within the estimated errors of ±0.1 eV with experiment and theory.

As seen in Fig. 1, the effect of the many-body correction on the empty surface states is quite dramatic. These states are substantially shifted upwards, opening up the gap between the empty and filled surface states by nearly 1 eV at some k-points. The position of the empty surface states is a prediction of the present theory. They should be experimentally accessible using angle-resolved inverse photoemission measurements, surface optical transition measurements, or scanning tunneling spectroscopy. The scanning tunneling experiment has recently been done for the Si(111):As surface.[8] Not only is the 1x1 surface structure observed directly, the observed surface-state gap of 1.9 - 2.3 eV in the normalized differential conductivity agrees very well with the predicted value of 2.2 eV.

The many-body results here, thus, address the many difficulties associated with the LDA surface-state bands. In particular, the gap between the empty and filled surface states is substantially opened in comparison with the LDA gap. Analysis of the calculated results, however, shows that the size of the correction depends on the detailed character of the surface states.[7] Hence, the shortcomings of the LDA surface-state spectrum cannot be corrected by a simple rigid shift.

4. SEMICONDUCTOR BAND OFFSETS

The present approach has also been applied to compute the band offsets of the GaAs-AlAs(001) interface.[9] The band discontinuities at a semiconductor interface are simply the differences in the quasiparticle energy E_{qp} across the junction for the band edge states. Since previous bulk and surface calculations showed that the quasiparticle wavefunctions are virtually identical to the LDA wavefunctions, we may write E_{qp} near an interface as

$$E^{qp}_{n\vec{k}} = \varepsilon^{LDA}_{n\vec{k}} + \Sigma_{n\vec{k}} - V^{n\vec{k}}_{xc} \tag{5}$$

where $\Sigma_{n\vec{k}}$ and $V^{n\vec{k}}_{xc}$ are respectively the expectation value of the self-energy operator and the LDA exchange-correlation potential for a given state. Then, the valence band offset ΔE_v becomes [12]

$$\Delta E_v = E^{LDA}_v + \delta_{vbm} \tag{6}$$

where ΔE^{LDA}_v is the LDA calculated valence band offset and δ_{vbm} is a many-body correction given by

$$\delta_{vbm} = \left(\Sigma - V_{xc}\right)^{vbm}_{GaAs} - \left(\Sigma - V_{xc}\right)^{vbm}_{AlAs}. \tag{7}$$

Since E_{qp} in Eq. (5) should be evaluated at a distance away from the interface in determining the band offset and both Σ and V_{xc} are short range interactions, $(\Sigma - V_{xc})^{vbm}$ can be replaced by bulk values. To calculate the LDA band offset, we perform a fully converged 12-layer superlattice calculation using an approach similar to that of Van de Walle and Martin.[13]

For the GaAs-AlAs(001) interface, we find that the many-body correction is quite significant; that is, δ_{vbm} = 0.12 eV. It is more than 29% of the LDA result for the valence band offset which is ΔE^{LDA}_v = 0.41 eV. Combining the two gives a calculated value for the valence band offset of ΔE_v = 0.53 eV. This result is in excellent agreement with the most recent experimental values of 0.53 - 0.56 eV.[14] The many-body correction can be understood in terms of a more localized valence band wavefunction for AlAs which leads to a more negative self-energy for AlAs as compared to GaAs and, thus, a positive δ_{vbm}. We expect the many-body correction to play an even more important role for cases where the junctions are made of materials with lesser chemical similarities than GaAs and AlAs.

5. SUMMARY AND CONCLUSIONS

A self-energy approach for calculating quasiparticle surface-state energies and band offsets from first principles is described. This development allows, for the first time, <u>ab initio</u> calculation of electronic excitation energies at surfaces and interfaces which can be directly compared with spectroscopic measurements. Results for the Si(111):As and Ge(111):As surfaces and the GaAs-AlAs(001) heterojunction show that the theory is accurate and of predictive power.

ACKNOWLEDGEMENTS

This work was supported by National Science Foundation Grant No. DMR8319024 and by the Director, Office of Energy Research, Office of Basic Energy Sciences, Materials Science Division of the U.S. Department of Energy under Contract No. DE-AC03-76SF00098. I would like to acknowledge valuable discussion and collaboration with M. L. Cohen, M. S. Hybertsen, D. Tomanek, and S. B. Zhang in the work discussed here.

REFERENCES

1. D. R. Hamann, this proceedings.
2. S. G. Louie in Electronic Structure, Dynamics and Quantum Structural Properties of Condensed Matter, eds. J. Devreese and P. Van Camp, Plenum, New York (1985), p. 335.
3. L. Hedin, Phys. Rev. **139**, A796 (1965); L. Hedin and S. Lundqvist, Solid State Phys. **23**, 1 (1969).
4. M. S. Hybertsen and S. G. Louie, Phys. Rev. Lett. **55**, 1418 (1985); Phys. Rev. B **34**, 5390 (1986).
5. R. W. Godby, M. Schluter, and L. J. Sham, Phys. Rev. B **35**, 4170 (1987); Phys. Rev. B (to be published).
6. J. E. Northrup, M. S. Hybertsen, and S. G. Louie, Phys. Rev. Lett. **59**, 819 (1987).
7. M. S. Hybertsen and S. G. Louie, Phys. Rev. Lett. **58**, 1551 (1987); Phys. Rev. B (to be published).
8. R. S. Becker, B. S. Swartzentruber, J. S. Vickers, M. S. Hybertsen, and S. G. Louie, Phys. Rev. Lett. (in press).
9. S. B. Zhang, D. Tomanek, S. G. Louie, M. L. Cohen, and M. S. Hybertsen, to be published.
10. R. D. Bringans, R. I. G. Uhrberg, R. Z. Bachrach, and J. E. Northrup, Phys. Rev. Lett. **55**, 533 (1985); R. I. G. Uhrberg, R. D. Bringans, M. A. Olmstead, R. Z. Bachrach, and J. E. Northrup, Phys. Rev. B **35**, 3945 (1987).
11. J. R. Patel, J. A. Golovchenko, and P. E. Freeland, Phys. Rev. B, to be published.
12. Equation (6) is based on the assumption that the electrostatic potential (the Hartree term) at the interface is well-reproduced in the LDA. This should be a very good approximation since LDA is known to give excellent charge densities and also surface energies and work functions.
13. C. G. Van de Walle and R. M. Martin, J. Vac. Sci. Technol. B **3**, 1256 (1985).
14. P. Dawson, K. J. Moore, and C. T. Foxon in Quantum Well and Superlattice Physics, Proceedings of SPIE, **792**, eds. G. H. Dohler and J. N. Schulman (SPIE, Washington, 1987), p. 208, and D. J. Wolford, private communications.

Discussion

L. KLEINMAN:

(1) I just want to mention that the recent experimental valence band offsets quoted in the next paper (including one by Gossard) are about 0.1eV less than those you quote.

(2) I think that if the quantum wells are thick, the valence band offset can be considered a ground state problem since the energy to remove an electron from the top of either the GaAs or AlAs valence band is the difference between the N and N-1 electron total energies. The valence band offset also involves a double layer potential which density functional theory should give correctly because it gives charge densities correctly. Also, because the GaAs and AlAs wave functions are so similar, it is surprizing that you find their self energies are so different.

S.G. LOUIE: Let me first answer the second question: I do not believe the valence band discontinuity is a ground–state property. If you look at a superlattice and ask "Where is the top of the valence band?" then you could say that from the density functional Kohn–Sham type of equation you could determine that state. But, for the band discontinuity across a junction, you have to do an experiment which is a transport experiment and the measured quantity is an excited state property. To answer your first question, I have quoted the latest numbers from Bell Labs which is Dawson's number and also from IBM which is Don Wolford's number, and both groups are giving numbers around the value of .55 eV.

W. KOHN: Certainly, there's no question that, logically speaking, there are such many–body corrections to the LDA quasi–particle, energies; their magnitude will vary from case to case.

L. KLEINMAN: Yes, but Kohn–Sham is excellent if you're comparing charge densities that are almost identical and gallium arsenide and aluminium arsenide are so identical to one another that you would expect Kohn–Sham to work very well.

W. KOHN: Well, I think that's more or less what Mr. Louie also said, that it's a relatively small correction.

S.G. LOUIE: Yes, it's a small correction if you look at the absolute value of the self–energy. Let me show you a table, here, if I may.

L. KLEINMAN: It's a fairly large correction...

S.G. LOUIE: But it's a large correction for the quanity that you're looking at. For example, if you look at the top of the valence band and examine the expectation value of the LDA exchange–correlation operator, because it only depends on the charge density and the gallium arsenide and aluminium arsenide charge densities are so similar, you get exactly the same number. Whereas, Σ being a nonlocal operator—which is the correct operator to consider for quasi–particle energies, gives you a difference in energy and that's where the *correction* is coming from.

P. MARCUS: You mentioned that you first made an LDA calculation to determine the relaxation, but you didn't tell us the size of that relaxation and whether it made a significant difference for the surface state. I should also like to suggest that it would be interesting to do the calculation for the Ge(111) 1×1-H surface, for which we have recently determined the relaxation from LEED analysis. This is also a perfect semiconductor surface and perhaps more perfect than the As covered surface, since the H does not disrupt the Ge lattice they way As does.

S.G. LOUIE: This is the calculation which shows the relaxation of the arsenic layer from the ideal position. We find that it relaxes outward by .16Å;

there's a recent, very accurate, x-ray standing wave measurement by Patel and Golovchenko, and they find a result of .17Å for the As layer relaxation..

D.R. HAMANN: I'd like to comment on L. Kleinman's comment. The valence band maximum is a ground state property and is given exactly by exact density functional theory. Local density functional theory is an approximation, and the GW self-energy, which presumably is more exact, should give corrections to LDA.

C.B. DUKE: This question concerns the use of basically the perfect r-r' screening in the kernel for the Coulomb interaction in Dyson's equation. Specifically, a lot of folks have spent a lot of time worrying about the fact that you really need a proper non-local dielectric in the vicinity of an interface, and of course they're all fairly schematic calculations, but I think there is widespread agreement that within a couple of atomic layers of the interface you cannot use the nice r-r' form, you've got to go to a more complex form of the dielectric function and I wondered, did you examine this? If so, why is this so small in your calculation when there are all those folks out there who make a living doing the corrections?

S.G. LOUIE: In the quasiparticle energy calculations, we use the full dielectric matrix $\epsilon(\underline{r},\underline{r}';\omega)$. That is, ϵ is not just a function of $|\underline{r}-\underline{r}'|$. The full matrix takes into account fully the local-field effects and the changes in dielectric screening at a surface. It turns out that, for the band offset problem, one only needs the ϵ for the individual semiconductors. But, for the surface case, we need to calculate the surface dielectric matrix using a supercell technique.

Comparison of Interfacial Charge Densities, Valence Band Offsets, and Formation Enthalpies of $(GaAs)_n(AlAs)_n(001)$, (110), and (111)

D.M. Bylander and L. Kleinman

Department of Physics, The University of Texas, Austin, TX 78712, USA

We became intrigued with the thermodynamic stability of GaAs-AℓAs superlattices when Kuan[1] reported that long range $(GaAs)_1(A\ell As)_1(100)$ (110) ordering occurred during continuous codeposition of Aℓ,Ga and As without any imposition of special growth sequences. (Note that the (001) and (110) monolayer superlattices are identical, differing only by a 90° rotation). He suggested that this superlattice might be the thermodynamically stable state, but since the ordering occurred much more strongly and over a wider range of deposition parameters with a (110) rather than a (001) substrate, we thought this ordering was more likely a result of growth kinetics. In this talk we will present formation enthalpies of $(GaAs)_n(A\ell As)_n(001)$, (110) and (111) with n=1 and 3 calculated with the scalar relativistic norm conserving pseudopotential method.

A second matter intrigued us and that was the question of valence band offsets. Because the zero of Coulomb potential in an infinite crystal is completely arbitrary, one cannot determine the offsets from a knowledge of the energy bands of the semiconductors making up the superlattice. Until 1984 the experimentalists [2] told us the valence band offset was 0.19 eV and many theorists had vastly oversimplified models that were in fair agreement. Recently however the experimental value[3-5] has changed to between 0.43 and 0.47 eV. Tersoff[6] and Cardona and Christiansen[7] have suggested that the offsets could be obtained by lining up effective midgap points (EMP). Although they defined the EMP in different and *ad hoc* ways, fairly good agreement with experiment is obtained for a large number of superlattices. Since the EMP are bulk properties, it follows that the band offsets should be face independent, which seemed to be the case both experimentally[8] and theoretically[9]. What we did not understand at first (and which we shall presently explain) was how the offsets could be identical if, as we presumed, there occurred vastly different dipole double layers at each of the interfaces.

To obtain the accuracy of a few meV per 12 atom unit cell that we need, one has to work extremely hard. For example, rather than computing matrix elements by performing three-dimensional integrals over the large unit cell, we fit the potential at a large number of points with a mixed set of

Springer Series in Surface Sciences Vol. 14
Solvay Conference on Surface Science
Editor: F.W. de Wette

functions and calculate the matrix elements analytically. We generate a set of random points in the zinc blende unit cell. By symmetry, each random point represents 24 points in the cell. In the (110) superlattice each random point represents only four points so each zinc blende point is replaced by six points which represent the 24 points. Then each zinc blende cell must be repeated six times to fill the superlattice cell. Because of its lower symmetry we fit the (110) superlattice at more points (23,988) using more independent symmetrized combinations of fitting functions (1708) than for the (100) or (111) cases. To check our fitting accuracy, we generated two different sets of random points for the three layer (100) superlattice obtaining a discrepancy of 17 meV in cohesive energy. But because we generate our superlattice random points in many-to-one correspondence with the zinc blende, when we subtracted the GaAs and AℓAs cohesive energies from that of the superlattice to obtain the formation enthalpy, the discrepancy became only 1 meV. Similar cancellation of errors occurs throughout the calculations[10-13].

In Table I we display the formation enthalpies we have calculated. The first thing we note is that all the monolayer superlattices are unstable to disproportionation. Although we were the first to obtain this result several other workers have done (001) and are in agreement[14-17]. (Some of these get larger instability than we because their GaAs and/or AℓAs equilibrium lattice constants are poor, resulting in a strained superlattice.) We have very preliminary results[18,19] which indicate these superlattices are stable against random alloy formation up to 700°C. The 0.2 meV discrepancy between the monolayer (001) and (110) superlattices (which differ only by a 90° rotation) is due to the different BZ **k**-point samples and is another example of our great accuracy. We note that (001) and (110) are the easiest and hardest directions in which to grow thick superlattices, which corresponds to our calculated n=3 formation enthalpies.

We plot contours of $\Delta\rho$, the charge density in the superlattice minus the charge density of the bulk constituent semiconductors. Figure 1 displays $\Delta\rho$ in the (011) plane of $(GaAs)_3(A\ell As)_3(001)$. This plane which cuts the interface at 45° was chosen because it contains the covalent bonds. Note that although the AℓAs bond gains charge and the GaAs loses it, there is a net flow of charge from the AℓAs side to the GaAs. In Fig. 2 we show

Table I. Formation enthalpy in meV per unit cell (i.e. per pair of interfaces) for n=1 and n=3 $(GaAs)_n(A\ell As)_n(001), (110)$ and (111) superlattices.

	(001)	(110)	(111)
n=1	15.1	14.9	11.4
n=3	1.7	27.8	15.4

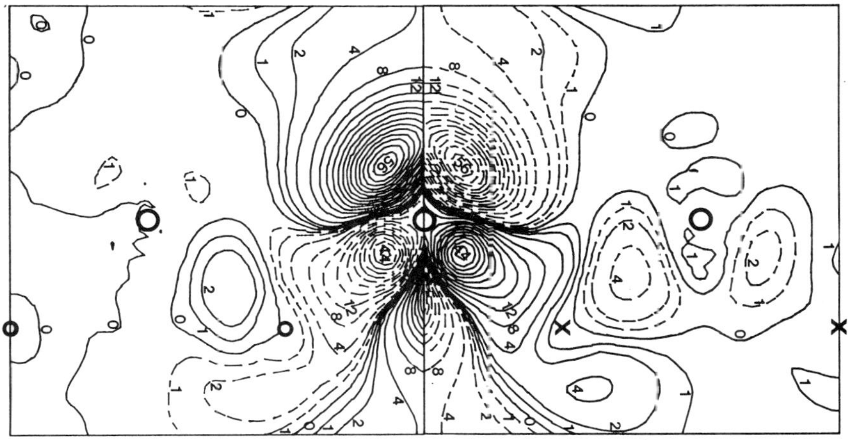

Fig. 1. Contours of constant $\Delta\rho$ in the (011) plane where $\Delta\rho$ is the $(GaAs)_3(A\ell As)_3(001)$ charge density minus that of bulk GaAs on the left and $A\ell As$ on the right side of the figure. Large circles are As atoms, small circles are Ga and X's are $A\ell$. Solid and dashed lines are positive and negative contours in units of $2.5 \times 10^{-5} e^-/\text{bohr}^3$.

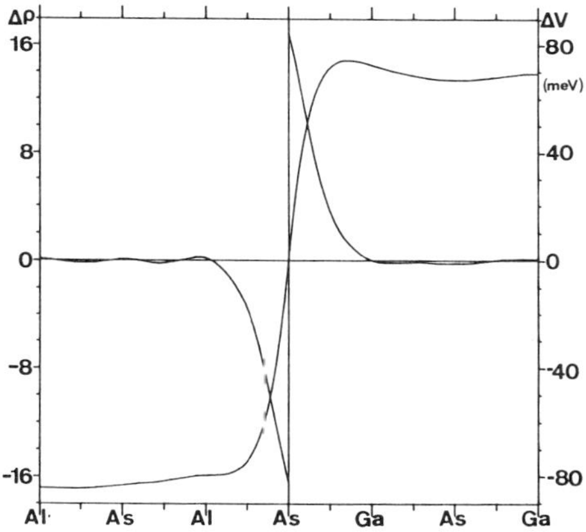

Fig. 2. Plot of the planar average of $\Delta\rho$ in units of $2.5 \times 10^{-5} e^-/\text{bohr}^3$ and the double layer potential it engenders in meV for $(GaAs)_3(A\ell As)_3$ (001).

the planar average of $\Delta\rho$ and the potential ΔV it engenders. See Ref. 12 for similar (110) figures. The difference between ΔV in the central planes of the two wells is Δ_{dip} which is given in Table II along with the valence band offsets. These are obtained by calculating the difference between the average potential of either the central plane or the central cell in the two quantum wells and then calculating the spin-orbit split Γ_8 levels in the two semiconductors with respect to these average potentials. ΔE_v is the average of the results obtained using the two different average potentials and δ is the difference. When the wells are sufficiently thick that the central unit cell is bulk-like, δ will vanish. We note that δ is consistent with the (111) well being thickest and the (110) thinnest.

Table II. Interfacial double layers Δ_{dip} and valence band offsets $\Delta E_v \pm \delta/2$.

	(001)	(110)	(111)
Δ_{dip}	154	314	254
ΔE_v	446	447	444
δ	-0.9	5.0	0.1

In spite of the wide variation in Δ_{dip}, we see that ΔE_v is even more face independent than previously obtained[9]. If one takes the bulk electronic charge density (plus positive point charges) with its arbitrary zero of potential and slices it into slabs, the slabs have well defined Coulomb average potentials. In Fig. 3 are juxtaposed the planar-averaged Coulomb potentials of such GaAs and AℓAs (001) slabs with their averages shown and the Γ_8 levels calculated with respect to the average Coulomb potentials. These juxtaposed slab charges differ from the superlattice charge density by $\Delta\rho$ and their Coulomb potentials differ by ΔV. Thus $\Delta \mathrm{E_v} =$ $\Gamma_8^{(0)}(\text{GaAs}) - \Gamma_8^{(0)}(\text{A}\ell\text{As}) + \Delta_{\text{dip}} = 12.8484 - 12.5661 + 0.1538 = 0.4461$ eV. These slab Coulomb potentials are strongly face dependent. One sees that what must be face independent for $\Delta \mathrm{E_v}$ to be face independent is $\Delta_{\text{dip}} + \Delta_{\text{Coul}}$, where Δ_{Coul} is the difference between the GaAs and AℓAs slab-averaged Coulomb potentials.

Because the two faces of (111) slabs are not equivalent, there is a potential difference of 63.514 volts across the AℓAs slab and 58.779 volts across the GaAs. Because $\mathrm{V} = \mathrm{V_o z}$ is a solution of Laplace's equation, it can always be added to the potential obtained from the slab charge. Thus in Fig. 4 we choose $\mathrm{V_o}$ to cancel the sum of the potentials across the two slabs, leaving a sawtooth potential of height 2.368 volts corresponding to

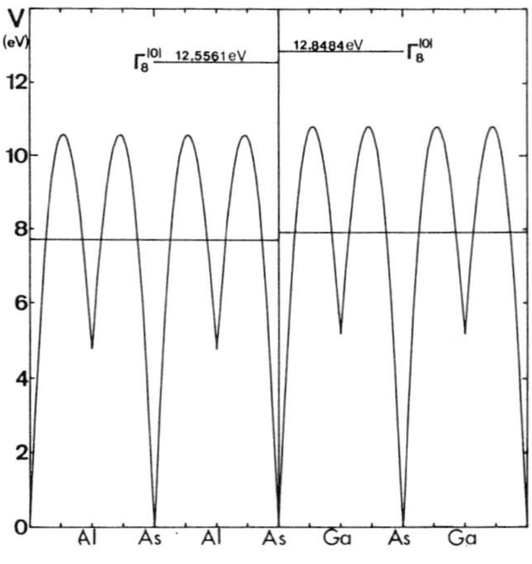

Fig. 3. (001) planar average of Coulomb potential arising from point ions and bulk pseudocharge densities of GaAs and AℓAs. The long horizontal lines represent the average of these potentials (7.7119 eV for AℓAs and 7.9139 eV for GaAs). The shorter lines represent the tops of the valence bands when the zero of Coulomb potential is chosen according to this prescription.

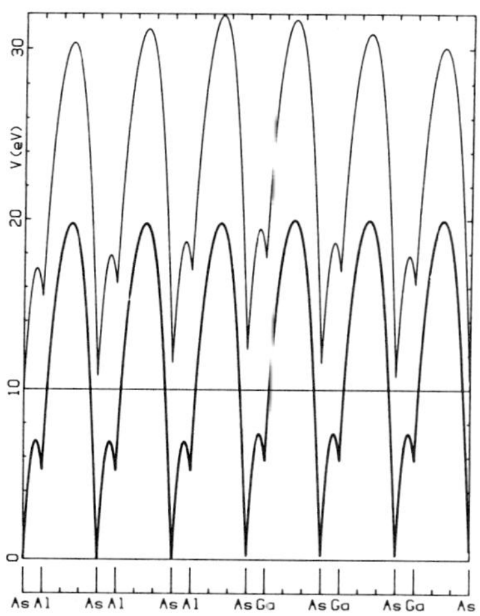

Fig. 4. The planar average of the Coulomb potential of adjacent three layer (111) slabs of bulk GaAs and AℓAs charge densities. The lower curve is the sum of the upper curve plus ΔV of Fig. 5

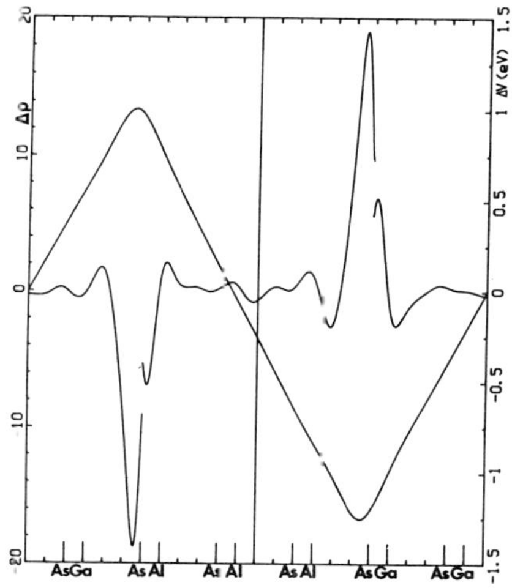

Fig. 5. Plot of the planar average of $\Delta\rho$ in units of 2.5×10^{-5}e/bohr^3 and ΔV, the potential it engenders in eV for $(\mathrm{GaAs})_3(\mathrm{A\ell As})_3$ (111).

electric fields of $\pm 2.414 \times 10^9$V/m in the two slabs. Figure 5 displays $\Delta\rho$ and ΔV for the (111) superlattice. The discontinuity in $\Delta\rho$ arises from subtracting the AℓAs charge density from the AℓAs well and the GaAs charge density from its well. ΔV has been zeroed at the center of the GaAs well and when it is added to the slab potential in Fig. 4, the superlattice planar averaged Coulomb potential shown at the bottom of Fig. 4 is obtained.

We see there are large interfacial charge densities which set up long range screening fields to cancel the long range fields obtained by a juxtaposition of constituent semiconductor charge densities. Because the interfacial charges extend into the central cells of the three-layer superlattice, we cannot tell if this cancellation is exact. The $\pm 2.414 \times 10^9$V/m becomes 1.50×10^7 and -2.08×10^7V/m across the central cells. However, the interfacial charge obtained by integrating from cell center to cell center is $\pm 4.287 \times 10^{-2}$Coul/m^2 which results in $E_{sc} = \mp 2.421 \times 10^9$V/m which slightly overscreens, giving $E = \mp 7 \times 10^6$V/m. Thus we can say that in spite of the general rule that whatever is not forbidden by symmetry exists, large interfacial charge densities exist in (111) superlattices which very nearly, if not exactly, screen out the long range electric fields. We suspect this screening also occurs for piezoelectric fields in strained crystals.

This work has been supported by the Robert A. Welch Foundation and the National Science Foundation under Grant No. DMR-8412408.

References

1. T.S. Kuan, Bull. Am. Phys. Soc. **31**, 558 (1986).
2. H. Kromer, *Proceedings of the NATO Advanced Study Institute on Molecular Beam Epitaxy and Heterostructures*, Erice, Sicily, 1983, L. Chang and K. Ploog ed. (Martinus Nijhoff, Netherlands, 1985).
3. R.S. Bauer and G. Margaritondo, Physics Today **40**, 27 (1987).
4. W.I. Wang and F. Stern, J. Vac. Sci. Technol. B**3**, 1280 (1985).
5. J. Menendez, A. Pinczuk, D.J. Wender, A.C.Gossard, and J.E. English, Phys. Rev. B**33**, 8863 (1986).
6. J. Tersoff, Phys. Rev. B**30**, 4874 (1984); Surf. Sci. **168**, 275 (1986).
7. M. Cardona and N.E. Christiensen, Phys. Rev. B**35**, 6182 (1987).
8. W.I. Wang, T.S. Kuan, E.E. Mendez, and L. Esaki, Phys. Rev. B**31**, 6980 (1985).
9. C.G. Van de Walle and R.M. Martin, J. Vac. Sci. Technol. B**4**, 1055 (1986).
10. D.M. Bylander and L. Kleinman, Phys. Rev. B**34**, 5280 (1986).
11. D.M. Bylander and L. Kleinman, Phys. Rev. B**34**, 5280 (1986).
12. D.M. Bylander and L. Kleinman, Phys. Rev. Lett. **59**, 1091 (1987) and Erratum, to be published.
13. D.M. Bylander and L. Kleinman, to be published.
14. B.I. Min, S. Massidda and A.J. Freeman, to be published.
15. A. Oshiyama and M. Sato, Phys. Rev. B**36**, 6156 (1987).
16. S. Ciraci and I.P. Batra, Phys. Rev. B**36**, 1225 (1987).
17. D.M. Wood, S.-H. Wei, and Z. Zunger, Phys. Rev. Lett. **58**, 1123 (1987).
18. L. Kleinman, Phys. Rev. B**35**, 3854 (1987).
19. Y.-T. Shen, D.M. Bylander and L. Kleinman, to be published.

Discussion

S.G. LOUIE: I would like to comment on your statement regarding the screening of ionic movement for q→0 phonons. Linear screening *still* holds for this case. However, one must take into account the full dielectric matrix. The acoustic sum rule guarantees that the perturbation will be screened out and that vanishingly small restoring force is obtained for q→0 acoustical phonon displacements.

L. KLEINMAN: We agree. What I said is that linear screening with the full dielectric matrix holds for superlattices if you think of them as virtual crystals with a long wavelength perturbation. This I think can explain both the face independence of the valence band offset and the lack of strong internal electric fields in the (111) superlattice. Surfaces can be calculated as vacuum-semiconductor superlattices. In that case the perturbation is strong, linear screening is not valid, and one should obtain face dependent work functions.

M. CARDONA: To this point. A longitudinal phonon produces a potential wave proportional to the deformation potential. And there are different deformation potentials for the conduction band and for the valence band. Thus you have to consider an average one. This is the best way of viewing the problem. You have to consider an average potential for the valence band and an average potential for the conduction band. But what you have to screen with the dielectric constant, is neither one of the two because holes and electrons are completely equivalent. What you have to screen is the average of the two average deformation potentials of the valence and the conduction bands. If you screen that, dividing by the electric constant, you get the correct screened electron–phonon interaction. Christensen and I have shown this in *Phys. Rev. B* **35**, 6182 (1987).

L. KLEINMAN: I don't understand that.

M. CARDONA: You have a phonon, and the phonon produces a wave according to the deformation potential *that* has to be screened. The question is, which wave do you screen? Each electron has a different wave produced by the phonon. What do you screen? What you have to screen, by the dielectric response, is actually the average wave produced in the valence band and the conduction band, because that is what produces the dielectric response: excitations between valence and conduction band. It is hard to put it in words, but it is easy to understand when you look at the formalism given in the above reference.

L. KLEINMAN: Well I agree with you that the dielectric function depends on excitations between the valence band and conduction band, but you're talking about scattering of an electron or a hole when you talk about deformation potentials, and so that electron or hole has a particular deformation potential which in fact you want to calculate self–consistently so it's got screening built in it.

M. CARDONA: Yes, if you do it, you get exactly what I am saying. The result is, when you do the calculation for a semiconductor like germanium and silicon numerically but for no particular reasons, that the top of the valence band has almost zero deformation potential. But I repeat, this is a quantitative result, there is no particular reason for it. The bottom of the conduction band has a fairly large screened deformation potential for optical phonons.

L. KLEINMAN: I still don't understand what you're saying.

P. MARCUS: Could you tell us again the calculated band offset values and how they compared with the values found by Louie?

L. KLEINMAN: His LDA value is .41eV, if I remember correctly. His self energy correction increases it by .12eV, which I think is too large because of the similarity of GaAs and AlAs. Our LDA value is .445eV.

Surface Segregation at Metal/III-V Compound Semiconductor Interfaces

Zhangda Lin, F. Xu, and J.H. Weaver*

Department of Chemical Engineering and Materials Science,
University of Minnesota, Minneapolis, MN 55455, USA
*Permanent address: Institute of Physics, Chinese Academy of Science, Beijing, People's Republic of China

1. Introduction

This paper focuses on the interesting observation that semiconductor atoms segregate to the free surface for many evolving metal/semiconductor interfaces. Their presence contributes to a range of intriguing properties of interfaces and their study offers insight into complicated interface behavior. The purpose of this paper is to develop a simple model for surface segregation for evolving metal/semiconductor interfaces. The starting point for this modeling is recent theories developed to qualitatively interpret grain-boundary segregation and surface segregation on alloy systems. As we will show, our approach will yield excellent agreement when compared with the wealth of experimental information available in the literature, including our own experimental results and those of others. We focus on interfaces grown under the cleanest of conditions. We do not summarize the large device literature because devices are grown under less controlled conditions and the presence of impurities will alter the atom distribution.

2. Summary of Experimental Results

Direct measurements of the spatial distribution of the various elements in an evolving interface are difficult. In some cases, such information has been obtained from sputter depth profiling or from polar-angle-dependent photoemission intensity profiling. A third technique involves the measurement of the core-level intensity as the kinetic energy of the photoelectron is changed, and hence the probe depth is varied. Indirect studies include measurements of the core level emission intensity of a particular element as a function of nominal metal coverage. Careful inspection of attenuation curves based on intensity variations at high metal coverages gives insight into the change in atom concentration on the free surface and in the near-surface region.

The Ti/GaAs(100) system is an example of a reactive metal overlayer on GaAs(100). Recent sputter-profile studies of the intensity variation of the As 3d, Ga 3d, and Ti 3p emission measured after the growth of 100 Å of Ti on GaAs(100) showed that a significant amount of As accumulates on the free surface and only a very small quantity of As is dissolved in the Ti matrix[1]. This interesting observation points to the possibility of anion segregation to the surface region of a reactive metal, a phenomenon which has only recently been noted. Detailed depth profiling for this Ti/GaAs system conducted for various metal coverages indicates that for sufficiently thick metal overlayers (>20 Å), As keeps "floating" on the free surface, strongly suggesting a driving force which lowers the free energy of the system.

Springer Series in Surface Sciences Vol. 14
Solvay Conference on Surface Science
Editor: F.W. de Wette

In studies of a chemically very different system[2], Au/GaAs(100), polar-angle-dependent x-ray photoemission was used to investigate the elemental distribution in the overlayer. These results showed a high concentration of both Ga and As on the free surface and in the near-surface region. A series of recent investigations of metal overlayer formation on GaAs(110) and InP(110) based on core level studies using high resolution synchrotron radiation photoemission[3,4] showed that the anion or cation intensities generally attenuate much more slowly than they should if they were covered up by the growing metal overlayer. Further, the slopes for these attenuation curves at high coverages differ from metal to metal and from substrate to substrate.

It is now well known that chemical reactions induce surface morphology changes and intermixing. In most cases these chemical reactions dominate in the low-metal-coverage range at room temperature but, as the amount of deposited metal increases, the reaction becomes diffusion limited and can be neglected. If the intermixed region is covered up in a layer-by-layer fashion by the metal overlayer, then the attenuation of the substrate emission should have the mean free path of the photoelectrons as a characteristic length scale. If the experimental results show slower attenuation at high coverages, there is reason to consider metal-cluster formation, continuous out-diffusion of semiconductor atoms into the overlayer, or semiconductor-atom segregation to the free surface.

With the exception of a few metals, such as Al, Ag, and In deposited on III-V semiconductor substrates, there is no strong evidence for metal-cluster formation at high coverages (this is especially true for the reactive metals where the photoemission intensity of the substrate signal is rapidly attenuated, even if the total is not). On the other hand, the low solubilities of semiconductor atoms in the metal matrix strongly argues that continuous out-diffusion with high concentration in the overlayer is very unlikely. We can therefore regard the slow attenuation of emission from substrate elements as indicating a concentration of these atoms on and near the free surface, i.e. surface segregation. We can also use the intensity attenuation curves with multiple photon energies (surface sensitive and bulk sensitive synchrotron radiation photoemission) to distinguish surface segregation from out-diffusion[5].

From the literature[3,4,6,7] we can see that for the same metal overlayer (Fe, Cr, Ti, for example), the attenuation curves for In have lower slopes than those of Ga, implying that In has a greater tendency for surface segregation than Ga. Similarly, comparing metal-InP, metal-GaAs, and metal-InSb (Refs. 3 and 4) systems for the same metal overlayer, Sb has the strongest surface segregation, As has medium surface segregation, and P has very weak, or negligible, surface segregation. In addition, we have examined the experimental results from a wide variety of studies of the type described above for metal overlayers grown sequentially on GaAs and InP under ultrahigh-vacuum conditions. These results are the basis for comparison with the theoretical model developed in the next section.

3. Theoretical Model and Discussion

The phenomenon of surface segregation in metal/semiconductor systems has been recognized for several years. However, because the problem is so complex, no theoretical model has been proposed to describe the segregation process in these evolving interfaces. Recently, we surveyed the experimental results for a large number of interfaces and developed a simple, qualitative model to explain, from an energetic point of view, why semiconductor atoms segregate to the free surface of some overlayers but not others[8].

It is now generally accepted that the deposited metal atoms induce substrate disruption and release the semiconductor species into the overlayer in most cases. For the reactive metals, there are reactions involving the metal and the anion. The resulting reaction products can be quite complicated and more than a single phase can be formed. Likewise, the cations released by substrate disruption can mix in the growing metal overlayers where they can form solid solutions. Therefore, the interface is an intermixed region where metal compounds, solid solution, released cations and anions which escaped from compound formation, and metal atoms may coexist. If the metal compounds are energetically stable against dissociation, the participating atoms will be trapped[9], thus ruling out surface segregation for the participating atoms. The possibility of escape then depends on the chemical reactivity between the metal and the substrate species, the energetics and the stoichiometry of compound formation, and the degree of interface heterogeneity. The escaping atoms and the other atoms, similar to the elements in an alloy, may segregate to the surface, depending on the energy considerations. Therefore, based on the recent theories for surface segregation on alloy systems we developed a simple model for evolving metal/semiconductor interfaces.

Basically, the treatment of segregation has considered the preferential decoration of an AB alloy surface by atoms of species A. One reason that this occurs is that the A atoms have lower surface energy than do B atoms. Another involves the relative sizes of A and B, thus the strain energy associated with dissolving larger A atoms in the B matrix. When there exists a C species which covers the surface of the AB alloy, then A atoms have a greater tendency to move to the surface to react with C if A atoms are more reactive with species C. From a thermodynamic point of view, surface segregation in alloys is due to lowering the surface free energy by atom exchange between surface and bulk sites to gain the bonding and the strain energies and to the chemical activity of the different species, if there is any. Since the photoemission experiments that we will discuss were all conducted in ultrahigh vacuum, we can assume that there are no chemically active C atoms on the surface to induce preferential segregation so that the third factor of chemical reactivity can be neglected.

According to the theory for alloys, the free energy of segregation can be written[10]

$$\Delta E = \gamma_A - \gamma_B + 2\Omega[Z_l(x_A^b - x_A^s) + Z_v(x_A^b - \frac{1}{2})] \quad , \tag{1}$$

where γ_A and γ_B are the molar surface energies of pure A and B elements, Z_l and Z_v are the atomic coordination number in one layer and that to one of the adjacent layers, $x_A{}^b$ and $x_A{}^s$ are bulk and surface concentration of A atoms in AB alloy, respectively, and Ω is a function depending on the bulk concentration of A and B in AB alloy, the enthalpy of mixing and the coordination number. Seah evaluated γ_A and γ_B in terms of the melting points of the segregant and matrix atoms. In our treatment of the parameters of Eq. (1), we correlated the surface segregation energy with the cohesive energies rather than with the melting points of the species involved. This seems more reasonable since the main difference between solid and liquid states involves the transition from long-range to short-range atomic order without atomic bond breaking. Thus, it induces only a small change in bond strength. Instead, the cohesive energy involves bond breaking and conversion to the atomic orbital configuration from a solid-state configuration and should represent a better approach in modeling the surface-segregation mechanism.

To calculate the strain energy released when solute A atoms are allowed to segregate from the bulk to the surface, Burton and Machlin[11] wrote the difference of strain energy of a solute atom A at the surface and in the bulk as

$$E_{str} = \frac{24\pi k G a_B a_A (a_B - a_A)^2}{3ka_A + 4Ga_B} , \tag{2}$$

where k is the solute bulk modulus, G is the solvent shear modulus, and a_B and a_A are the appropriate radii for solvent and solute atoms in their pure states. From Eq. (2) we can see that the strain energy is always released when atom A is replaced from bulk to surface, in spite of whether a_A is bigger than a_B, because ΔE depends on the square of the difference in atomic radii. However, Tsai *et al.*[12] pointed out that current atomic calculations indicate that there is no appreciable relaxation of atomic positions about a vacancy because of the hard-core repulsion of nearest-neighbor interactions. Thus, the strain-energy term includes only the surface free energy for solute atoms larger than the solvent atoms. For smaller solute atoms, the strain term can be ignored.

The total surface free energy of segregation can now be written

$$\Delta E = (\gamma_A - \gamma_B) + V\Omega - W(a_A - a_B)^2 , \tag{3}$$

where the coefficients V and W must be determined by comparison with experiment. Seah[10] showed that V is nearly unity and the second term is typically only 4% of the first term for diluted alloy systems. The magnitude of W depends on a_A and a_B through the atomic size dependence of the moduli k and G; W will be equal to zero for $a_A < a_B$. Because of complications in evaluating W, we do not try to determine which term in Eq. (3) is more important, cohesive energy or strain energy, for a given system. Instead, we include the predictions from bonding energy and strain energy separately in Tables II and III. However, as discussed above, the strain energy plays a role only when a_A is bigger than a_B, and in many cases it is the bonding energy (cohesive energy) that plays a key role in determining surface segregation. With Eq. (3) we can predict whether surface segregation is energetically favorable for any metal III-V-compound-semiconductor system.

In Table I we summarize the melting temperatures T^{melt}, the cohesive energies E_{coh}, and the atom sizes a_A, of elements of importance for the metal III-V-semiconductor systems under examination here. T^{melt} and E_{coh} were taken from Ref. 13. a_A is calculated in angstroms.

According to Eq. (3), we use data taken from Table I to predict the existence of surface segregation shown in Tables II and III for metal-GaAs and metal-InP systems, respectively. The first column shows the prediction with account taken of only the cohesive energy; the second considers only the melting temperature, and the third for strain energy. The fourth column gives experimental results where Y means the existence of surface segregation, N means no surface segregation, and A means that we are not sure (abstain). From Table I we see that the cohesive energy of In is lower than that of Ga and the atomic size of In is larger than that of Ga. From our model, considering either the gain in cohesive energy or the gain in strain energy, we should expect that In has a greater tendency to surface segregate than Ga for a

TABLE I: Melting temperatures, cohesive energies, and atom sizes of elements of importance for the metal/III-V semiconductor systems under examination here. Some of the data were taken from Ref. 13. Atom sizes were obtained by calculations.

Element	Melting Temp. (°C)	Cohesive Energy (kJ/mole)	Atomic Size (Å)	Element	Melting Temp. (°C)	Cohesive Energy (kJ/mole)	Atomic Size (Å)
Ag	961.9	284.9	2.57	Ni	1455	430.1	2.22
Al	660	329.7	2.55	P	590	332.2	2.67
As	817	302.5	2.79	Pd	1554	376.6	2.45
Au	1064	368.2	2.57	Pt	1772	565.7	4.36
Ce	799	423	3.27	Sb	630.5	264.4	3.12
Co	1495	428.4	2.22	Sc	1541	377.8	2.92
Cr	1857	398	2.29	Si	1410	446	2.72
Cu	1063	337.6	2.28	Sm	1077	206.7	3.21
Fe	1535	415.5	2.28	Ta	2996	782	2.63
Ga	29.8	273.6	2.70	Ti	1660	469.9	2.61
In	156.6	243	2.97	Tl	303.5	182	3.06
Ir	2410	670	2.42	V	1890	514.2	2.42
Mn	1244	283.3	2.33	Yb	819	152.1	3.46

given metal overlayer. Indeed, when we compare the results for overlayers of Ti, Cr, and Fe on GaAs and InP, we can clearly see this trend. The In intensities decrease much more slowly than do those of Ga, and In has a greater tendency to segregate to the surface.

We have already discussed the characteristics of chemical trapping at metal/semiconductor interfaces. These effects are not taken into account in our simple treatment. The absence of P surface segregation for Co, Cr, Fe overlayers on InP(110) is plausibly connected with phosphide compound formation and trapping near the buried interface. However, the observed As segregation in metal/GaAs interfaces, despite metal-As compound formation, indicates that surface segregation can still be predicted by our simple model once these reactive atoms have escaped from chemical trapping.

The chemical reactivity of P is greater than that of As and that of As is greater than that for Sb, as can be seen from their electronegativities and the heats of metal-anion compound formation. The greater the reactivity of the anion, the less intense will be surface segregation. This implies that the lower chemical reactivity of Sb would induce more intense surface segregation compared to As and P. Furthermore, Sb atoms are the largest among the pnictides and P atoms are the smallest (excluding Bi). From the point of view of strain energy, we should also expect that Sb would segregate most readily. This would predict that the tendency to surface segregate would diminish from Sb to As to P for the same metal overlayer. The results from Refs. 4,14-16 confirm this since most P atoms are trapped near the interface for metal/InP systems; we have observed no P surface segregation for Fe, Cr, Co, and Ti overlayers on InP. For metal/GaAs systems, we see As surface segregation for the same Fe, Cr, Co, and Ti overlayers, although it is weak. In metal/InSb systems we found strong Sb surface segregation for Cr, Co[17] and other metals.

From Tables II-III we can see the agreement between our predictions and the experimental results is quite good, with deviation observed for only a few systems. Our considerations are based on the thermodynamic equilibrium state, although we have noted above that these systems need not reach equilibrium and interface formation is limited by diffusion. It is worth

TABLE II - Metal/GaAs: Comparison between predictions and experimental results of Ga and As surface segregation for metal/GaAs semiconductor systems. Y denotes that segregation is predicted, N denotes that there is no segregation, and A denotes that we cannot predict (abstain).

Metal over-layer	Ga seg. pred. by coh. energy	Ga seg. pred. by melt. temp.	Ga seg pred. by strain energy	Exp.	As seg. pred. by coh. energy	As seg. pred. by melt. temp.	As seg pred. by strain energy	Exp.	Ref.
Al	Y	Y	Y	Y	Y	N	Y		18,19
Au	Y	Y	Y	Y	Y	Y	Y	Y	2,20,21
Ce	Y	Y	A	Y	Y	N	A	Y	20,22
Co	Y	Y	Y	N	Y	Y	Y	Y	23
Cr	Y	Y	Y	Y	Y	Y	Y	Y	24
Cu	Y	Y	Y	Y	Y	Y	Y	Y	25
Fe	Y	Y	Y	N	Y	Y	Y	Y	26
In	N	Y	A		Y	N	A		
Ir	Y	Y	Y		Y	Y	Y	Y	27
Pd	Y	Y	Y	Y	Y	Y	Y	Y	28
Sm	N	Y	A	N	N	Y	A	N	29
Ta	Y	Y	Y		Y	Y	Y	Y	27
Ti	Y	Y	Y	Y	Y	Y	Y	Y	30,31
V	Y	Y	Y	Y	Y	Y	Y	Y	32
Yb	N	Y	A	N	N	A	A	N	33

noting that the segregation process in these evolving metal/semiconductor interfaces is different from that in ordinary alloys. The kinetic process is associated with the diffusion of the segregant. In turn, this is related to the amount of substrate disruption, metal-anion compound formation, and the distribution of the reaction products. Clearly, these interfacial regions are more complicated than AB alloy surfaces and the details differ from system to system. It is therefore not surprising that our simple model fails for a few systems. Indeed, it is remarkable that it works so well in so many systems! That success implies that the cohesive energy and the strain energy are key factors which control the process of surface segregation, in addition to chemical trapping effects for compound formation. This demonstration is equivalent to demonstrating the essence of surface segregation for metal/semiconductor interfaces.

We have recently applied our simple model to metal/Si, metal/Ge, and other interface systems and obtained the same successful agreement with experiments[3].

TABLE III - Metal/InP: Comparison between predictions and experimental results of In and P surface segregation for metal/InP semiconductor systems, analogous to those of Table II.

Metal over-layer	In seg. pred. by coh. energy	In seg. pred. by melt. temp.	In seg. pred. by strain energy	Exp.	P seg. pred by coh. energy	P seg. pred. by melt. temp.	P seg. pred. by strain energy	Exp.	Ref.
Ag	Y	Y	Y		N	Y	Y	Y	34,35
Al	Y	Y	Y	Y	A	Y	Y		34,36-39
Au	Y	Y	Y	Y	Y	Y	Y	Y	4,38,40,41
Ce	Y	Y	A	Y	Y	Y	A	N	14
Co	Y	Y	Y	Y	Y	Y	Y	N	15
Cr	Y	Y	Y	Y	Y	Y	Y	N	4,16,42
Cu	Y	Y	Y	Y	Y	Y	Y	Y	39,41,43
Fe	Y	Y	Y	Y	Y	Y	Y	N	4
Ga	Y	N	Y	Y	N	N	A		44
Mn	Y	Y	Y	Y	N	Y	Y		42
Ni	Y	Y	Y	Y	Y	Y	Y	Y	34,41,45
Pd	Y	Y	Y	Y	Y	Y	Y	Y	34,46
Pt	Y	Y	A		Y	Y	A		
Sm	N	Y	A		N	Y	A		
Ti	Y	Y	Y	Y	Y	Y	Y	N	4
Tl	N	Y	A	N	N	N	A		36

4. Acknowledgments

The authors thank C. Aldao, I.M. Vitomirov, Y. Shapira, and J.J. Joyce for helpful and useful discussions. This work was supported by the Office of Naval Research under ONR N00014-87-K-0029 and N00014-86-K-0427.

REFERENCES

1. F. Xu, D.M. Hill, Zhangda Lin, and J.H. Weaver, Phys. Rev. B (submitted) Ti/GaAs(100); F. Xu, D.M. Hill, Zhangda Lin, Steven G. Anderson, Yoram Shapira, and J.H. Weaver, Phys. Rev. B (submitted) Ti/GaAs.
2. F. Xu, Yoram Shapira, D.M. Hill, and J.H. Weaver, Phys. Rev. B **35**, 7417 (1987).
3. J.H. Weaver, Zhangda Lin, and F. Xu, Chapter 10 in *Surface Segregation and Related Phenomena*, edited by P.A. Dowben and A. Miller (CRC Press - in press).
4. C.M. Aldao, I.M. Vitomirov, F. Xu, and J.H. Weaver, Phys. Rev. B (submitted).
5. F. Xu, C.M. Aldao, I.M. Vitomirov, and J.H. Weaver, Appl. Phys. Lett. (submitted) Au/Si.
6. A. Franciosi, J.H. Weaver, D.G. O'Neill, F.A. Schmidt, O. Bisi, and C. Calandra, Phys. Rev. B **28**, 7000 (1983).
7. J.J. Joyce, M. del Giudice, M.W. Ruckman, F. Boscherini, and J.H. Weaver, J. Vac. Sci. Technol. A **5**, 2019 (1987).
8. Zhangda Lin, F. Xu, and J.H. Weaver, Phys. Rev. B **36**, 5777 (1987).

9. L.J. Brillson, C.F. Brucker, N.G. Stoffel, A.D. Katnani, and G. Margaritondo, Phys. Rev. Lett. **46**, 838 (1981).
10. M.P. Seah, J. Catal. **57**, 450 (1979).
11. J.J. Burton and E.S. Machlin, Phys. Rev. Lett. **37**, 1433 (1976).
12. N.H. Tsai, G.M. Pound, and F.F. Abraham, J. Catal. **50**, 299 (1977).
13. *Handbook of Chemistry and Physics*, 66th edition, edited by R.C. Weast (CRC, Boca Raton, Florida, 1985-1986).
14. J.J. Joyce, C. Aldao, I. Vitomirov, and J.H. Weaver (unpublished) Ce/InP.
15. F. Xu, C. Aldao, Zhangda Lin, I.M. Vitomirov, and J.H. Weaver, Phys. Rev. B **36**, 3495 (1987) Co/InP.
16. J. Nogami, T. Kendelewicz, I. Lindau, and W.E. Spicer, Phys. Rev. B **34**, 669 (1986).
17. F. Boscherini, Yoram Shapira, C. Capasso, C.M. Aldao, and J.H. Weaver, J. Vac. Sci. Technol. B **5**, 1003 (1987).
18. N.G. Stoffel, M. Turowski, and G. Margaritondo, Phys. Rev. B **30**, 3294 (1984).
19. N.G. Stoffel, M.K. Kelly, and G. Margaritondo, Phys. Rev. B **27**, 6561 (1983); R. Ludeke and G. Landgren, J. Vac. Sci. Technol. **19**, 667 (1981).
20. M. Grioni, J.J. Joyce, and J.H. Weaver, J. Vac. Sci. Technol. A **4**, 965 (1986).
21. P.W. Chye, I. Lindau, P. Pianetta, C.M. Garner, C.Y. Su, and W.E. Spicer, Phys. Rev. B **18**, 5545 (1978); W.G. Petro, I.A. Babalola, T. Kendelewicz, I. Lindau, and W.E. Spicer, J. Vac. Sci. Technol. A **1**, 1181 (1983); W.G. Petro, T. Kendelewicz, I. Lindau, and W.E. Spicer, Phys. Rev. B **34**, 7089 (1986).
22. J.H. Weaver, M. Grioni, J.J. Joyce, and M. del Giudice, Phys. Rev. B **31**, 5290 (1985).
23. F. Xu, J.J. Joyce, M.W. Ruckman, H.-W. Chen, F. Boscherini, D.M. Hill, S.A. Chambers, and J.H. Weaver, Phys. Rev. B **35**, 2375 (1987).
24. J.H. Weaver, M. Grioni, and J.J. Joyce, Phys. Rev. B **31**, 5348 (1985).
25. J.J. Joyce and J.H. Weaver, Mater. Res. Soc. Symp. Proc. **54**, 359 (1986).
26. S.A. Chambers, F. Xu, H.W. Chen, I.M. Vitomirov, S.B. Anderson, and J.H. Weaver, Phys. Rev. B **34**, 6605 (1986); M.W. Ruckman, J.J. Joyce, and J.H. Weaver *ibid.* **33**, 7029 (1986).
27. J.R. Waldrop, S.P. Kowalczyk, and R.W. Grant, J. Vac. Sci. Technol. **21**, 607 (1982).
28. A. Oustry, M. Caumout, and A. Escaut, Thin Solid Films **79**, 251 (1981).
29. M. Grioni, J.J. Joyce, and J.H. Weaver, Phys. Rev. B **32**, 962 (1985).
30. R. Ludeke and G. Landgren, Phys. Rev. B **33**, 5526 (1986).
31. M.W. Ruckman, M. del Giudice, and J.H. Weaver, Phys. Rev. B **33**, 2191 (1986).
32. M. Grioni, J.J. Joyce, and J.H. Weaver, J. Vac. Sci. Technol. A **3**, 918 (1985).
33. J. Nogami, M.D. Williams, T. Kendelewicz, I. Lindau, and W.E. Spicer, J. Vac. Sci. Technol. A **4**, 808 (1986).
34. T. Kendelewicz, W.G. Petro, I. Lindau, and W.E. Spicer, J. Vac. Sci. Technol. B **2**, 453 (1984).
35. W.G. Petro, T. Kendelewicz, I.A. Babalola, I. Lindau, and W.E. Spicer, Mater. Res. Soc. Symp. Proc. **25**, 329 (1984).
36. T. Kendelewicz, M.D. Williams, W.G. Petro, I. Lindau, and W.E. Spicer, Phys. Rev. B **31**, 6503 (1985).
37. T. Kendelewicz, W.G. Petro, I.A. Babalola, J.A. Silberman, I. Lindau, and W.E. Spicer, Phys. Rev. B **30**, 5800 (1984).
38. Y. Shapira, L.J. Brillson, A.D. Katnani, and G. Margaritondo, Phys. Rev. B **30**, 4856 (1984).
39. F. Houzaz, J.M. Moison, M. Bensoussan, J. Vac. Sci. Technol. B **3**, 756 (1985).

40. I.A. Babalola, W.G. Petro, T. Kendelewicz, I. Lindau, and W.E. Spicer, J. Vac. Sci. Technol. A **1**, 762 (1983); T. Kendelewicz, W.G. Petro, I. Lindau, and W.E. Spicer, J. Vac. Sci. Technol. B **2**, 453 (1984); In their earlier paper they argued that the experimental results indicate the formation of a metallic-like In-Au alloy with In on or near the surface. In the later paper they pointed out that they did not observe any evidence of metallic In segregation. We believe that, as for other Au - III-V-semiconductor systems, In segregates but does not appear as pure metallic-like In. Insufficient numbers of In atoms should be present to form metallic-like In-In bonds.
41. L.J. Brillson, C.F. Brucker, A.D. Katnani, N.G. Stoffel, and G. Margaritondo, J. Vac. Sci. Technol. **19**, 661 (1981).
42. R.S. List, T. Kendelewicz, M.D. Williams, I. Lindau, and W.E. Spicer, J. Vac. Sci. Technol. A **3**, 1002 (1985).
43. T. Kendelewicz, W.G. Petro, I.A. Babalola, J.A. Silberman, I. Lindau, and W.E. Spicer, Phys. Rev. B **27**, 3366 (1983); T. Kendelewicz, G. Rossi, W.G. Petro, I.A. Babalola, I. Lindau, and W.E. Spicer, J. Vac. Sci. Technol. B **1**, 564 (1983).
44. R.H. Williams, A. McKinley, C.J. Hughes, T.P. Humphreys, and C. Maani, J. Vac. Sci. Technol. B **2**, 561 (1984).
45. T. Kendelewicz, M.D. Williams, W.G. Petro, I. Lindau, and W.E. Spicer, Phys. Rev. B **32**, 3758 (1985).
46. T. Kendelewicz, W.G. Petro, I. Lindau, and W.E. Spicer, Phys. Rev. B **28**, 3618 (1983); J. Vac. Sci. Technol. A **2**, 542 (1984); T. Kendelewicz, R.S. List, K.A. Bertness, M.D. Williams, I. Lindau, and W.E. Spicer, J. Vac. Sci. Technol. B **4**, 959 (1986).

Discussion

H. IBACH: Are you sure that you have observed the surface segregation with equilibrated systems and that in cases where you did not observe segregation the segregation was not kinetically inhibited?

ZHANGDA LIN: The surface segregation in evolving interface systems is a special kind of surface segregation, where interfaces were formed at room temperature after very low rate deposition of metal ($\sim 1 Å$/min). Due to the intensive migration of the atoms on the developing interface region, the distribution of atoms in this region – although it was not in a thermal-equilibrium state as expected – should be very close to the equilibrium state where segregation of specific species occur, if its segregation can lower the free energy of the system.

D.R. HAMANN: It seems possible that some of your data could be explained by cluster growth of the metal overlayer rather than by surface segregation. Can you distinguish between these two cases?

ZHANGDA LIN: With the exception of a few metals, such as Al, Ag, and In deposited on III-V semiconductor substrates, there is no strong evidence for metal-cluster formation at high coverages. This is especially true for the reactive metals where the photoemission intensity of the substrate signal is rapidly attenuated, implying no metal-cluster formation, even if the total is not. Therefore, from the comparison of attenuation curves of both total signal and signal of substrate only, one can distinguish whether the cluster growth takes place. Of course, a careful analysis of the details for each of these systems is always needed before we can definitely attribute the observed low attenuation slope to surface segregation.

General Discussion

M. CARDONA: (to S. Louie and L. Kleinman.)

The *observable* band offset is, in my opinion, not a ground state property for the N-electron system. As measured with photoelectron spectroscopy, for instance, the valence band offset may be viewed as the difference in ground state energies of N-1 electron systems. Similarly, the conduction band offset is the difference in ground state energies of N+1 electron systems. The density functionals, however, are not the same, strictly speaking, for N-1, N, and N+1 particles. However, because of the similarity of the wave function at the top of the valence band and the rest of the valence band wave functions, the density functionals for N-1 and N particles are rather close while a big discontinuity exists between N and N+1. This is, in my opinion, the reason why the correction found by S. Louie is small (~0.1eV) for the valence offset but much larger (~1eV) for the conduction bands. I would like to emphasize the fact that the correction is small for the valence bands. Most calculations performed on the LDF basis give results which agree within scales with experiments and Steve has shown us here why. He claims that the correction will be much larger for pairs of more different semiconductors. I think he should calculate the ZnSe – Ge pair in order to prove or disprove that statement.

S.G. LOUIE: I agree with Dr. Cardona that the valence band offset is *not* a ground state quantity. I also would like to comment that the 0.12eV correction that we obtained for the GaAs - AlAs interface is significant because GaAs and AlAs are such similar materials. I expect that the corrections to the LDA band offsets will be larger for cases where the junction is made of materials which are chemically less similar than GaAs and AlAs.

L. KLEINMAN: I don't agree too much with either of the previous two comments, except that I think that .1 eV is fairly large, when you consider how similar the two wave functions are; I mean, when you compare valence band with conduction band, you're comparing bonding functions with anti–bonding functions, which are tremendously different. And I would expect that the difference between GaAs and AlAs is about 1% of the difference between bonding and anti–bonding. So, that's really my only objection. Also, I would like to ask Steve, did you use the *full* dielectric matrix in the superlattice or are there some approximations that you were forced to make because of the complexity of your calculation, and, if so, what sort of errors do you think they might have introduced?

S.G. LOUIE: We make use of the fact that the quasi-particle wave functions are extremely close to the LDA wave functions. We know this from our previous bulk calculations and surface calculations. If you take that as a good assumption and make use of the fact that the charge densities are given—highly accurately—by LDA, then you could decouple the problem. One needs only to calculate the many-body correction by doing the bulk calculations. So the calculations of δ in my formula are in fact obtainable from bulk calculations of GaAs and AlAs.

M. CARDONA: We are now in better agreement I think that .1 eV is of course a significant number, and it's nice to get it. Whether it can be very large—I mean I think that your comment on the LiF dimer is irrelevant. I think that we are talking here about systems that you can make experimentally, which means that the materials are rather similar anyhow. I think it would be nice if you could calculate a system like ZiSeGe that you can make, but the materials are rather different, but they have the same lattice constant, and see what the correction is there; and if you find .5 eV, then I will really take my hat off and say 'Wonderful!'. But if you find .1 eV also, then I say, 'Well, you know. It's nice to get it, but I mean it's not that earth-shaking.'

Part VIII

Surface Studies with Synchrotron Radiation

Dynamical Response of a Metal Surface

E.W. Plummer, In-Whan Lyo, K.-J. Song, and K.-D. Tsuei

Department of Physics, University of Pennsylvania,
Philadelphia, PA 19104, USA

The dynamical response of the electrons at or near a surface is intimately related to many of the chemical and physical properties of a surface. We have used various electron and photon spectroscopies to probe the dynamical response of the electrons in nearly free electron (NFE) metals. The NFE metals can be treated theoretically using interacting electron gas theories, so the experimental data can be used to evaluate various theoretical approximations. Existing theories were first tested in the bulk where the charge density gradient existing at or near the surface is not important. Experimental data for the quasiparticle band structure of Na (obtained from angle resolved photoemission) compared to theoretical calculations clearly indicate that the response of an electron gas must be treated beyond RPA type approximations. At the surface the wavelength and penetration depth of the longitudinal density waves created in response to an external field have been measured as a function of frequency using Second Harmonic Generation. The characteristics of these longitudinal fields do not agree with existing RPA type calculations. Finally we have measured the surface plasmon dispersion using inelastic electron scattering, and compared the measurements to developing theories.

I. INTRODUCTION

The dynamical response of the electrons at a metal surface to a time varying external probe is intimately related to many physical and chemical properties of the surface such as Van der Waals forces, sticking coefficients, energy transfer, or lifetimes of excited states, optical reflectance and absorption, nonlinear surface optical response, surface plasmon dispersion and the surface photoelectric effect[1-13]. Since simple metals can be represented by the electron gas model (Jellium) there have been extensive theoretical investigations of the dynamical response of an electron gas near its surface[1-13], yet almost nothing is known experimentally about the dynamic screening at a crystal surface[1]. The two exceptions to this statement are the measurement of the local field enhancement of the surface photoelectric effect[14] and the observation of standing plasmon waves in thin films[15,16]. The ability to prepare clean surfaces of the simple metals coupled with the more sophisticated experimental techniques available today enables the experimentalist to begin to probe the microscopic properties of the surface dynamical screening.

Springer Series in Surface Sciences Vol. 14
Solvay Conference on Surface Science
Editor: F.W. de Wette

There is a fundamental difference between the screening at a surface and in the bulk since at a surface a long wavelength transverse external electromagnetic field can induce a short-wavelength longitudinal response inside of the metal[1]. The easiest way to illustrate this phenomena is to start at the static limit in the response[17,18], that is to apply a static field and vary the field slowly enough that the electrons respond adiabatically to the time varying external field[19]. In the classical limit this slowly oscillating field would induce a "Surface Charge", but in the quantum mechanical regime the screening occurs by excitation of the electron-hole pairs or plasmons[1]. Figure 1 shows the normalized charge density induced by a static field at the surface of a Jellium metal of the density of Al, calculated by Lang and Kohn using the Local Density Approximation (LDA)[17,18]. The large peak at the surface is the analog of the classical "Surface Charge", while the damped oscillations going into the solid are the Friedel oscillation. These oscillations are a direct result of the limit on the momentum transfer q that is available in the electron-hole pair excitation spectra at very low excitation energies. This cutoff is at $2k_F$, creating damped oscillations with a wavelength $\lambda = \pi/k_F = \lambda_F/2$. In the low frequency or static limit ($\omega << \omega_p$) the induced charge follows adiabatically the applied field so that an external field of frequency ω induces a longitudinal field inside of the metal with a wave length approximately one half of the wave length of a Fermi energy electron. As the external frequency increases the response of the electron gas begins to deviate from the static limit, and in the limit where the driving frequency is above the energy of the bulk plasmon, longitudinal waves can propagate into the bulk.

Figure 2 shows a calculation by Liebsch for the induced density fluctuations in a metal whose electron density is approximately the same as that of Rb, as a function of the external frequency[5,19]. Table I lists the relevant

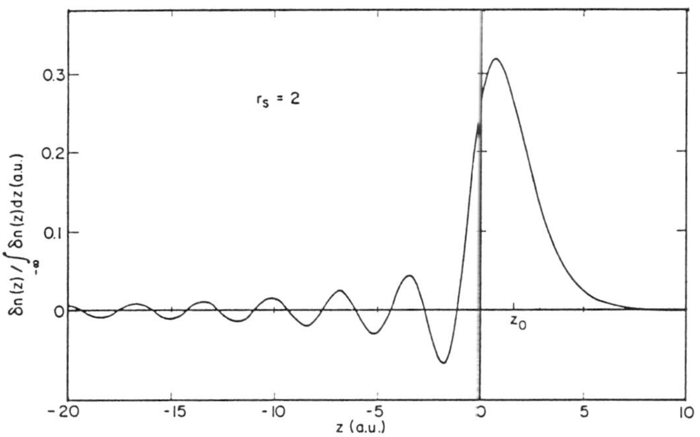

Fig 1. Screening charge in the surface region of the uniform background model induced by a weak uniform normal electric field. After Ref.[17,18]

properties of the simple metals. This figure clearly illustrates that for frequencies $\omega/\omega_P \leq 0.5$ there is a short wavelength response denoted λ_1 in Figure 2. This is basically the static limit response at $2k_F$. As the external frequency reaches approximately 0.7 ω_P a new and longer wavelength λ_2 is seen in the response of the electron gas. The basic physics underlying the frequency dependence of the response is easy to illustrate using Figure 3, where the excitation spectra of a bulk metal with the density of Rb is plotted as a function of the momentum transfer q. Since the presence of the surface acts like a source (or sink) of momentum an incident wave of frequency ω can excite excitations

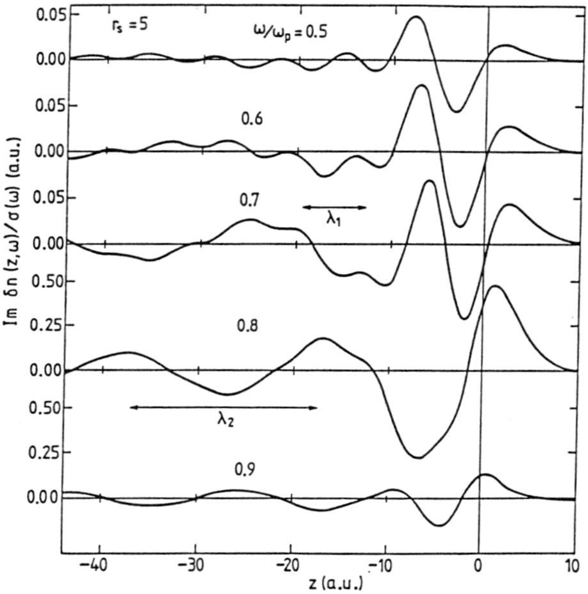

Fig. 2. Imaginary part of the normalized electron density induced by an external field of frequency ω. The positive ionic background is to the left of the zero. The two wavelength λ_1 and λ_2 of the Friedel oscillations are described in the text[5,19].

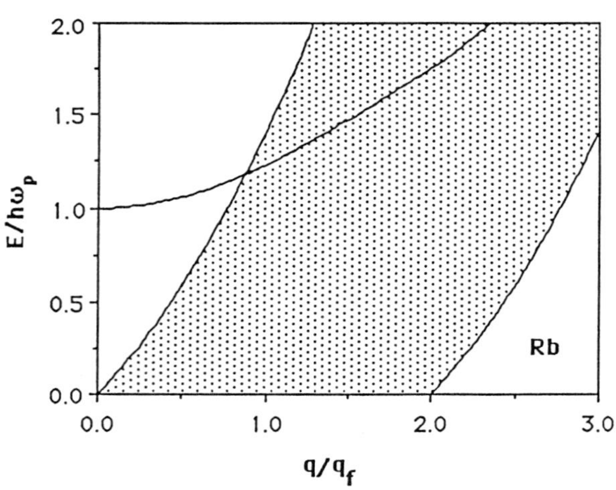

Fig 3. Schematic drawing of the electron-hole pair and plasmon excitation in (q, ω) regime. The shaded region is where the electron-hole pairs are allowed.

Table I: Properties of Simple Metals
Calculated from free electron theory

Metal	r_s/a_o	k_F ($Å^{-1}$)	$\lambda_F/2$ (Å)	E_F (eV)	$\hbar\omega_P$ (eV)
Cs	5.62	0.65	9.67	1.59	3.53
Rb	5.20	0.70	8.98	1.85	3.97
K	4.86	0.75	8.38	2.12	4.39
Na	3.93	0.92	6.83	3.24	6.03
Li	3.25	1.12	5.61	4.7	8.03
Mg	2.66	1.36	4.62	7.1	10.8
Al	2.07	1.75	3.59	11.7	15.8
Be	1.87	1.94	3.23	14.2	19.5

in the near surface region with any allowed value of q(for a given ω) shown in fig. 3. In the low energy limit, $\omega \ll \omega_p$, the cutoff in the electron-hole pair excitation spectra occurs at $q_1 = k_F + \sqrt{k_F^2 + 2m\omega/\hbar}$ which goes to the static $2k_F$ oscillations as $\omega \to 0$. In principle the induced longitudinal fields could oscillate with a wavelength λ equal to any value of $2\pi/q$ allowed by the electron-hole pair continuum shown in Fig. 3, but the oscillations will be dominated by the values of q that are the limits of the integral over the response of the electron hole pair continuum[21]. Therefore if the driving frequency is less than the plasmon frequency there are only two primary wavelengths λ_1 and λ_2 associated with the high and low cutoff in the electron-hole continuum shown in Fig. 3. We will refer to a short wavelength λ_1 as the static like response and a longer wavelength λ_2 ($q_2 = \sqrt{k_F^2 + 2m\omega/\hbar} - k_F$) in the longitudinal field as the dynamic response.

The objective of this paper is to summarize our present ability to measure directly the microscopic response of an electron gas to a time varying external probe and to use the presently available experimental data to evaluated our theoretical understanding of this surface phenomenon. Section II discusses our present ability to understand the excitations in the bulk, by discussing the quasiparticle spectrum of Na measured with angle resolved photoemission. The conclusion of this section will be that the measurements can be reconciled with theory only if the theory is done beyond the usual Random Phase Approximation (RPA) [22]. Section III describes recent measurements of the wavelength and penetration depth of the longitudinal induced fields using Second Harmonic Generation (SHG) [20]. Again the data will point out that RPA type theories are inadequate in predicting the transition from the static to the dynamic response discussed above. Finally in section IV, recent measurements and calculations of the surface plasmon dispersion for simple metals will be compared.

II. BULK QUASIPARTICLE SPECTRUM

Angle resolved photoemission combined with the tunability of synchrotron radiation sources has produced a technique capable of (in principle) measuring the band structure of any conducting crystalline material[23,24]. At the same time, theory has developed to the stage where the band structure of almost any material can be calculated in a self-consistent fashion[25]. Yet there is growing evidence that systematic differences exist between calculated band structures and measurements even for the simple metals[26,27]. Jensen[27] measured the band width of Na to be 2.5 eV wide compared to a theoretically calculated width of 3.2 eV[28], i.e. 22% narrower than calculated[28]. This observation was even more disturbing when compared to the many body correction to the single particle band structure calculated by Hedin[29] which gave only a 10% narrowing. In general these findings on the simple metals have raised several serious questions. (1) Is there some problem associated with the angle resolved measurements that is not fully appreciated[30]? (2) Do we understand the meaning of the band structure calculated in the LDA scheme[31]? (3) Is there a fundamental problem with the many body corrections to the single particle band structure [29]? Surely if this problem in the bulk cannot be reconciled there is no need to discuss the correlation of experiment and theory for surface excitations. The problems are much simpler in the bulk where at least for the simple metals the electron gas should be homogeneous in contrast to the surface region.

The angle resolved measurements for Na[27] have generated many theoretical explanations ranging from surface interference effects[30] to many body corrections to the single particle band structure[32-34]. Before discussing either the data or the theoretical calculations it is essential to carefully define the procedure for comparing the experimental angle resolved photoemission data with theory so that it will be possible to understand the origin of any disagreement. First a single particle band structure $E_0(k)$ is calculated for the metal of interest. Then the many body corrections are calculated via the self energy operator $\Sigma(E,k)$ [29]. The measured quasiparticle band structure $E(k)$ is then defined by,

$$E(k) = E_0(k) + \mathrm{Re}\{\Sigma(E,k)\}. \qquad (1)$$

The self-energy operator is complex and the real part gives the many body distortion of the single particle band structure, which is the quasiparticle band structure.

If the crystal potential is strong then there are effects both on E_0 and Σ due to the ion cores and it is not legitimate to calculate the self energy within a Jellium model[34]. On the other hand if a LDA scheme is used to calculate E_o then there is not a clear prescription for calculating Σ, since some part of the exchange-correlation potential is already included[18]. Since our purpose is to evaluate the response of an electron gas, a simple metal was used where $E_o(k)$ is given by a free electron band and Σ can be evaluated using only the properties of an interacting electron gas. Na is such a material, where the LDA calculations give an occupied band identical to the free electron band. We should note here that it is possible that this comparison does not prove that Na is really a free electron metal. The crystal potential terms in the LDA theory could accidentally cancel out with the local exchange-correlation potential giving a free electron band for the wrong reason. But for the case of Na this is not true since Davenport's Hartree calculation for Na gives a free electron band[35].

The normal procedure for converting angle resolved photoemission data into dispersion curves for the occupied bands requires a detailed knowledge of the final bands, usual obtained by ansatz or from theory[23]. Neither of these alternatives is acceptable if the measurements are to be used as a test of present theoretical models. The band structure of Na has been measured using an extremum searching procedure first outlined by Wohlecke et al[36]. This procedure relies on the fact that $k_{//}$ is conserved as the electron is emitted from the solid. If a series of spectra are recorded as a function of photon energy keeping $k_{//}$ fixed then the extremum point in the measured binding energy is the extremum point in the band structure for the rod in three dimensional k space with the preset value of $k_{//}$. Figure 4 shows a set of energy distributions from Na taken at a value of $k_{//}$=0.68 Å^{-1} as a function of photon energy[37]. The inset shows a plot of the binding energy vs photon energy for these spectra, showing a maximum binding energy of 1.2 eV at 18 eV photon energy. This gives one point on the quasiparticle dispersion curve, E(0.68 Å^{-1})=1.2 eV. The value of $k_{//}$ is changed and a new set of spectra are collected to obtain another data point on the quasiparticle dispersion curve.

Figure 5 presents the measured quasiparticle band structure of Na measured with the extremum searching procedure described above. The solid curve is either the free electron band structure or the band structure calculated with LDA[28] or Hartree theory[35]. The measured quasiparticle band

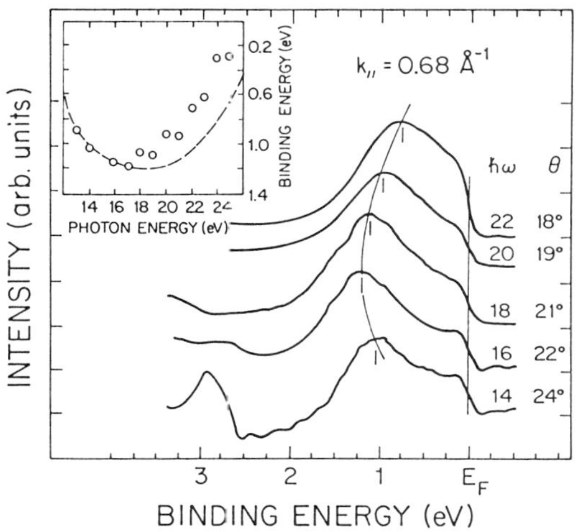

Fig. 4 Photoemission spectra from Na(110) as a function of photon energy for a fixed value of $k_{//}$ =o.68 Å^{-1} [37]. Dashed lines in the inset represent the free-electron gas for Na.

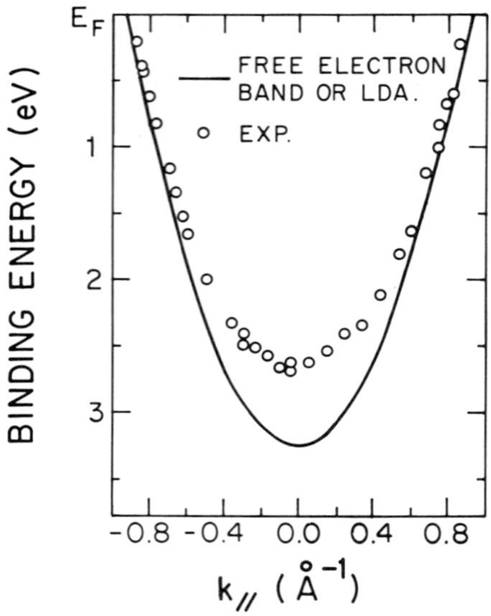

Fig. 5 Measured quasiparticle band structure of Na compared to the free electron band structure[37].

structure shown in fig. 5 is parabolic but with a narrower band width than predicted theoretically. The band width in this measurement is 18% narrower than the calculated width which is equivalent to an effective mass of 1.21.

The information in Fig. 5 can be used to determine an experimentally derived self-energy. For a free electron metal, the difference between the experimental data (E(k)) and the free electron band (E_0(k)) is the real part of the self-energy

$$\mathrm{Re}\{\Delta\Sigma(k)\} = E(k) - E_0(k), \qquad (2)$$

with $\Delta\Sigma(k) = \Sigma(k) - \Sigma(k_F)$.

This experimental self-energy is compared in Fig. 6 to the various self-energy calculations for Na. The curve denoted RPA is from Hedin[29], while all of the other curves (a->c) are calculations beyond RPA. Curve (a) is the paramagnon model calculation of Zhu and Overhauser[32], where spin fluctuations are included in the correlation energy. While the usual excitation via the plasmon mode is spin-symmetric, the new channel via the paramagnon mode is antisymmetric. This new channel generates a rapidly varying k dependent correlation energy which accounts for the oddly shaped self-energy shown in curve (a). It is clear that our data is not consistent with the band shape predicted by this model.

Curve (b) of Fig. 6 is the k dependent correction to the Na band structure calculated by Northrup, Hybertsen and Louie[34]. The self consistent ground state charge density calculated in LDA was used to derive the dynamic dielectric response function, which was then used to construct the self-energy operator in a self-consistent scheme. Here, only the first order term in the expansion of the self-energy operator in terms of the screened Coulomb potential is taken to

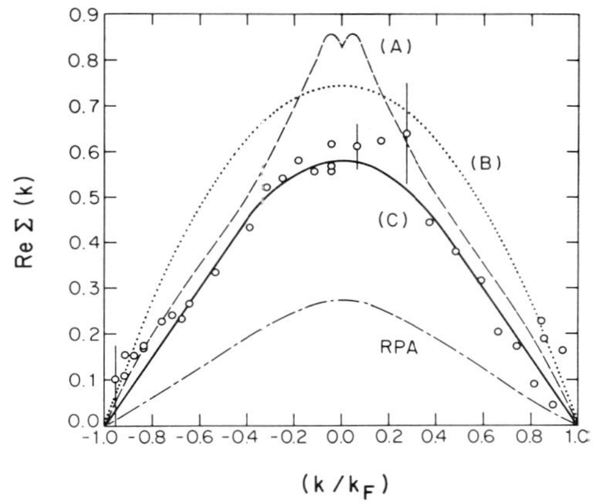

Fig. 6 A comparison of the experimental and theoretical self-energy correction $\Delta Re\{\Sigma(k)\}$ for Na [37].

represent the entire interaction. This step is the so called GW approximation. This calculation showed the importance of including exchange-correlation (beyond RPA) in the dielectric screening, but also indicated that it was important even in the case of Na to include correctly the inhomogeneous charge distribution in the crystal. Curve (b) in Fig. 6 shows that the self-energy correction calculated by Northrup, Hybertsen and Louie[34] is in good agreement with the data. This calculation uses the LDA approximation for the static dielectric function. Therefore, there are still unresolved questions, concerning the importance of the functional dependence of the dielectric function on Σ and the limitations of an electron gas calculation.

In an attempt to elucidate the essential physics in the self-energy correction to the band structure, we performed self-consistent numerical calculations of the exchange-correlation energy for an electron gas with the density of Na, using different models for the dielectric function $\varepsilon(q)$[38]. In our calculation the electron gas excitation spectrum was represented by Overhauser's plasmon pole model[39], where both the electron hole pair continuum and the plasmon excitation are replaced by an effective plasmon branch given by $\omega(q)$. Figure 7 shows the plasmon and electron-hole pair continuum replaced by an effective plasmon branch (dashed line). However our calculation differs from Overhauser's model in not including the vertex correction term[32], which in principle is the same order in the expansion as the exchange-correlation hole. This term that we have ignored represents the electron-plasmon coupling. The model presented here without the vertex correction is equivalent to the GW approximation[29], facilitating the comparison of our calculation to the results of Northrup, Hybertsen and Louie[34] and Hedin[29]. The calculational procedure is a self consistent loop where the exchange energy is calculated explicitly in terms of the Lindhard function and the correlation energy from the plasmon pole model. The coupling term in the plasmon excitation contains the dielectric response $\varepsilon(q)$, which can be written as,

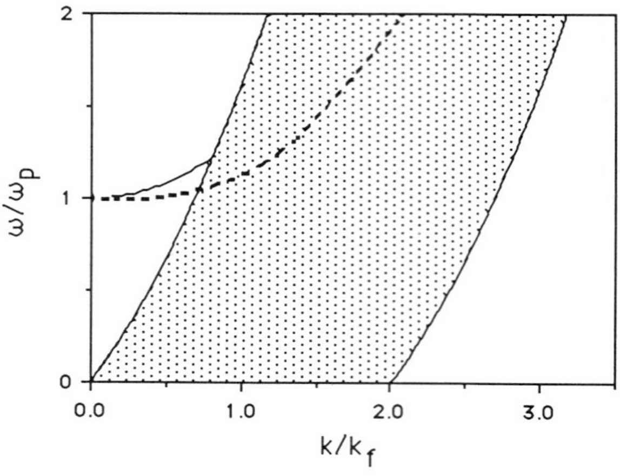

Fig. 7 Schematic drawing of the Overhauser plasmon pole representation [40] of the excitation spectrum of a metal like Na. The dashed line is the effective plasmon pole.

$$\varepsilon(q) = 1 - \frac{\upsilon(q)\chi_o(q)}{1 + G(q)\upsilon(q)\chi_o(q)} \tag{3}$$

where $\upsilon(q)$ is the Coulomb potential and $\chi_o(q)$ the Lindhard polarizability of the non-interacting electron gas. The function G(q) is called the local field correction term[38] and contains the effects of the modified polarization of electron gas due to exchange-correlation hole. In the RPA, G=0 and $\varepsilon(q)$ is given by the Lindhard dielectric function. As a test of this calculational scheme the RPA self-energy correction was calculated, and it was nearly identical to the Hedin's results shown in Fig. 6.

The self-energy correction to the band structure of Na was calculated using a variety of representations of the dielectric response function[38]. In general any $\varepsilon(q)$ that gave a good representation of $G(q)/q^2$ in the small q regime (i.e. $\approx$ 0.28 for Na) produced a self-energy correction in excellent agreement with the data. Curve (c) in Fig. 6 is calculated using LDA type $\varepsilon(q)$ where $G(q)/q^2$ is constant and taken from the interpolated Ceperley-Alder's[38] Monte Carlo data, and displays excellent agreement with the measured data. Using a dielectric function with a more accurate representation of the large q region, such as the Ichimaru-Utsumi's or Vashishta-Singwi's[38], makes little change in the self-energy correction curve shown in curve (c).

The importance of this simple calculation is two fold. (1) It is important to include the effects of the exchange-correlation hole in the the self-energy beyond the RPA type approximation. (2) The correction to RPA is a local correction (independent of q).

III. SURFACE LONGITUDINAL WAVES

In a recent set of experiments we have shown that Second Harmonic Generation (SHG) can be used to directly measure the wavelength and penetration depth of the longitudinal charge density waves created at a metal surface by an external electromagnetic field of frequency ω [20]. Nonlinear optical processes are becoming more widely used as probes of surfaces. In general the complexity of the nonlinear optical process compared to a linear process is compensated by the increased surface sensitivity. Second Harmonic Generation has surely established itself as a tool capable of yielding valuable information about the surface chemistry[40-43] as well as the geometry at the surface[44,45]. It is quite easy to understand the origin of the surface sensitivity on SHG. In a medium with inversion symmetry, like the simple metals, SHG is forbidden in the bulk within the electric dipole approximation. The surface on the other hand is non-centrosymmetric (centroasymmetric) allowing for a dipole allowed SH signal, creating an inherent surface sensitivity for SHG.

The details of the measurement are present in ref. 20 and we will discuss the results and interpretation in this section. Fig. 8 shows the SHG from a Ag(110) surface as a function of the thickness of a Rb ($r_s = 5.2\ a_o$) for two different incident wavelengths. There is an increase of about 1600 in the SH signal with respect to the Ag surface for one monolayer coverage of Rb for an incident infrared wavelength of 1064 nm. Both frequencies show a damped oscillatory signal with a wavelength of 12 Å for 532 nm and 22 Å for 1064 nm. Fig. 9 shows the Fourier transform of the infrared data after a smoothly varying background has been remove. The dominant wave length is seen at 22Å with a minor contribution at about 32Å. These oscillations are a result of the longitudinal density waves set up by the external driving field.

The measurements shown in Figures 8 and 9 were made with the incident light 60° from the normal and p polarized so that there is an appreciably strong component of the light vector normal to the surface. If s polarized light is used there is almost no measurable change in the SH signal as the Rb is deposited, proving that the oscillations are not due to a rough surface morphology. The film thickness was determined by measuring the rate of deposited Rb in a mass spectrometer and calibrating this in terms of film thickness by using the work function change, thermal desorption spectra and photoyield intensity [20].

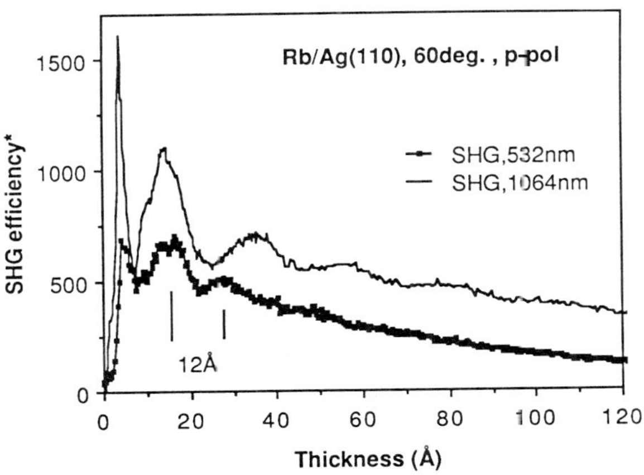

Fig. 8. SH signal as a function of Rb film thickness for two different wavelengths. The SH signal is normalized to the Ag(110) signal at 1064 nm. Efficiency is the SH energy flux divided by the incident energy flux squared[20].

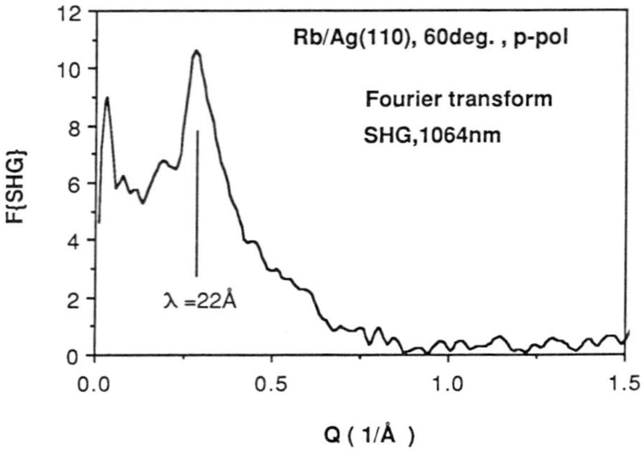

Fig. 9. Fourier transform of the 1064 nm data shown in Fig. 8.

Other experiments have shown that the strength of the short wave length oscillations is very sensitive to having very smooth films. For example if Rb is deposited onto a freshly sputtered Ag surface before it is annealed there are no oscillation in the SH signal vs. coverage[20].

The explanation of the short wavelength oscillations seen in Figures 8 and 9 is easily understood by using the schematic diagram in Fig. 10 of the fields involved in this problem. An incident field of frequency ω is denoted by $E_1(\omega)$. When this transverse wave interacts with the surface a transmitted transverse wave $E'_T(\omega)$ propagates into the film while $E_R(\omega)$ is reflected. At the same time a damped longitudinal field field $E'_L(\omega)$ is created that propagates almostperpendicular to the surface. In the simplest model of SHG the two interfaces are the only place where a second harmonic signal can be generated.

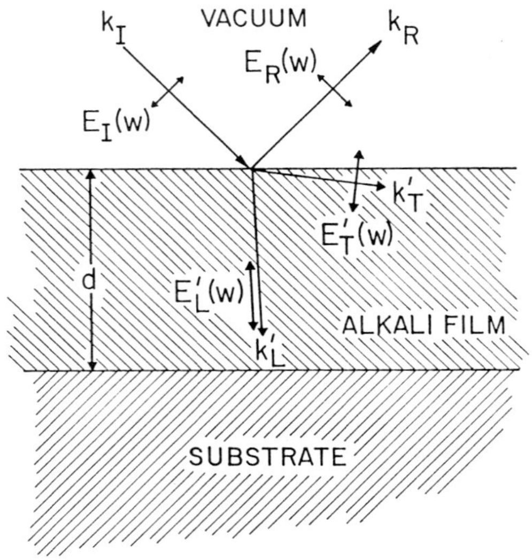

Fig. 10. Schematic drawing of the electromagnetic fields present in the thin film used to obtain the data shown in Fig. 8.

If we denote the second order susceptibility at the vacuum-Rb interface by $\chi_1^{(2)}$ and the susceptibility of the Rb-Ag interface by $\chi_2^{(2)}$, then the second order signal I(2ω) can be described by the following equation.

$$I(2\omega) \propto | E_1^2(\omega)\chi_1^{(2)} + E'^2_T(\omega)\chi_2^{(2)} + E'_L(\omega)E'_T(\omega)\chi_2^{(2)} + E'^2_L(\omega)\chi_2^{(2)} |^2 \qquad (4)$$

We have ignored the effects of the longitudinal wave created at interface 2 on the signal at interface number 1 in this equation, but it would not make a fundamental difference in the interpretation that follows. The last term in equation 4 is assumed to be weak and is dropped and the third term is expanded giving the following relationship

$$I(2\omega) \propto | E_1^2(\omega)\chi_1^{(2)} + E'^2_T(\omega)\chi_2^{(2)} |^2 + 2\,\mathrm{Re}\{ [E_1^2(\omega)\chi_1^{(2)} + E'^2_T(\omega)\chi_2^{(2)}]^* E'_L(\omega)E'_T(\omega)\chi_2^{(2)} \}. \qquad (5)$$

Since the wavelengths of the transverse waves are so long compared to the variations in film thickness being described here none of the terms in equation 5 change rapidly with film thickness except the longitudinal field $E'_L(\omega)$. The gradual variation in the transmitted transverse field $E'_T(\omega)$ as a function of film thickness due to the absorption of the incident light produces the slow variations in the SH signal seen in Fig. 8, and can be explained quantitatively using Fresnel equations for the fields [20]. Eqn. 5 shows that the oscillations measured in the SH signal as a function of film thickness are linearly related to $E'_L(\omega)$ so the wavelength of the oscillations in the data measures the wavelength of the longitudinal field. Also the attenuation of the oscillations is a measure of the penetration depth of this field [20].

Figure 11 compares the measured wavelength of the longitudinal waves at the two frequencies used to the allowed wavelength for the electron-hole pair continuum previously shown in Fig. 3. The data clearly shows that the electron gas responds to the external field by creating an internal longitudinal field with the longest (shortest q) wavelength possible, consistent with the allowed electron-hole pair continuum in the bulk. This wavelength is denoted by λ_2 in the calculations by Liebsch shown in Fig. 2 [5,20]. This long wavelength limit is what we referred to in the introduction as the dynamic limit of the response, since in the static regime the response is at λ_1 (near $2k_F$).

When the experimental data shown in Fig. 11 are compared to any of the present theoretical calculation[1,5,7,8] a fundamental discrepancy is observed. All of the calculations would predict that the response of the metal would stay in the static limit λ_2 for frequencies $\omega/\omega_P < 0.6$ while our data clearly shows that the system is in the dynamic limit (λ_1) for frequencies as low as $\omega/\omega_P = 0.3$. The dynamic oscillation are larger at this frequency than they were at $\omega/\omega_P = 0.6$

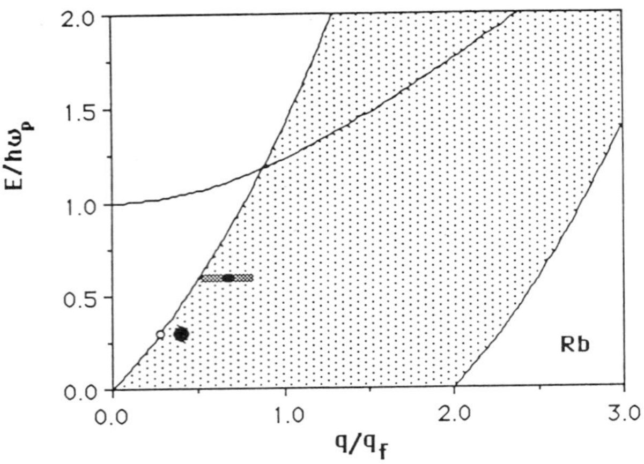

Fig. 11. Drawing of the excitation spectra for a free electron metal with the density of Rb. The individual points represent the data taken from Fig. 8 and 9 [20].

where according to theory they should be at a maximum (see Fig. 2). It is very likely that the origin of this discrepancy between theory and experiment is the same as presented to explain the quasiparticle spectrum in the previous section. If a better representation of the exchange-correlation hole is put into the theory then the system will probably have a dominant response at finite frequencies that is of a longer wavelength than the short wavelength present in the static limit. This means that the present RPA type theories are restricted to a much narrower range of frequencies around the static limit than has been previously expected.

IV. SURFACE PLASMON DISPERSION

Plasmons and surface plasmons are fundamental excitations of a terminated electron gas [1,3,5,9,10,12,46,47], yet there has been very little experimental data published for the surface plasmon dispersion on simple metals. Recent theoretical calculations have shown that the dispersion of the surface plasmon for $q_{//}$ values around $q_{//}=0$ can be related in a self-consistent fashion to the properties of the longitudinal waves discussed in section III as well as the frequency dependence of the photoyield [1,5]. We have used angle resolved inelastic electron scattering to measure the dispersion of the surface plasmon for thick films of Alkali metals as well as the development of the surface plasmon as a function of film thickness[48].

The top of Fig. 12 shows a schematic of the scattering geometry in this experiment and the bottom curve shows a typical loss spectrum, taken for a thick film of Na evaporated onto Al(111)[48]. The momentum transfer parallel to the surface $q_{//}$ is defined by the measurement geometry and the energy loss $\hbar\omega$ by the following equation.

$$\hbar q_{//} = \sqrt{2m}\,(\sqrt{E}\,\sin\theta_i - \sqrt{E-\hbar\omega}\,\sin\theta_e) \qquad (6)$$

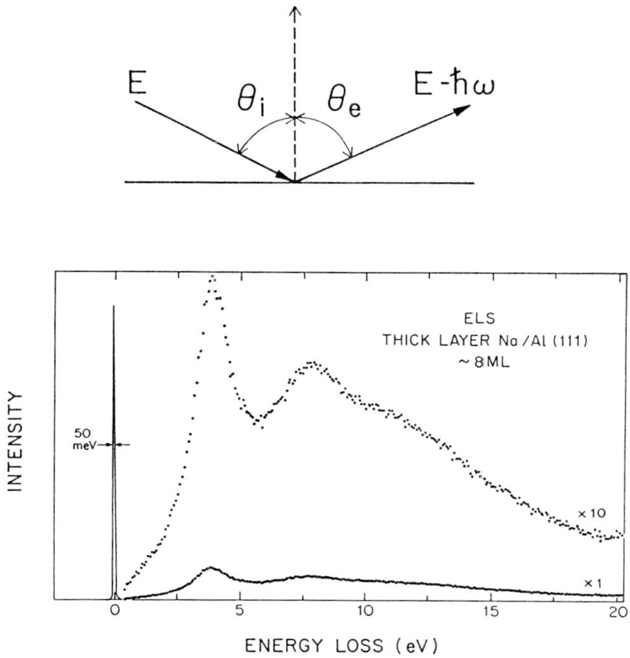

Fig. 12. Illustration of inelastic electron scattering experiment. Top figure shows schematically the scattering geometry. Bottom shows characteristic loss spectrum from a thick Na film[49].

Therefore by changing the angle between incident beam θ_i and the analyzer θ_e the surface plasmon as a function of $q_{//}$ can be measured. The losses at the bottom of Fig. 12 are from the surface plasmon of Na at 3.8 eV and from the excitation of two surface plasmons at 7.6 eV.

Figure 13 plots the loss spectra from a two layer thick Na film on Al(111) as a function of the incident angle θ_i for a fixed collection angle of $\theta_e = 58.5^{\circ}$. The vertical lines show the maximum in the loss peak, obtained after a smooth background has been subtracted. It is quite easy to see that the surface plasmon is dispersing as the value of $q_{//}$ is changed. Using eqn. 6 the energy loss and momentum can be determined for each curve. Fig. 14 shows a plot of the dispersion of the Na surface plasmon for two film thickness, a two layer film and a thick film. In both cases the dispersion is negative as expected theoretically[1,3,5,47], but the dispersion for the thin film is significantly larger than for the thick film.

Recent theoretical calculations by Liebsch[5] have shown that the dispersion of the surface plasmon energy (at $q_{//}=0$) in simple metals can be related to the dynamic screening at a metal surface through the following equation.

$$\omega_S(q_{//}) = \omega_S \left\{ 1 - \frac{q_{//}\, d(\omega_S)}{2} + O(q_{//}^2) \right\} \qquad (7)$$

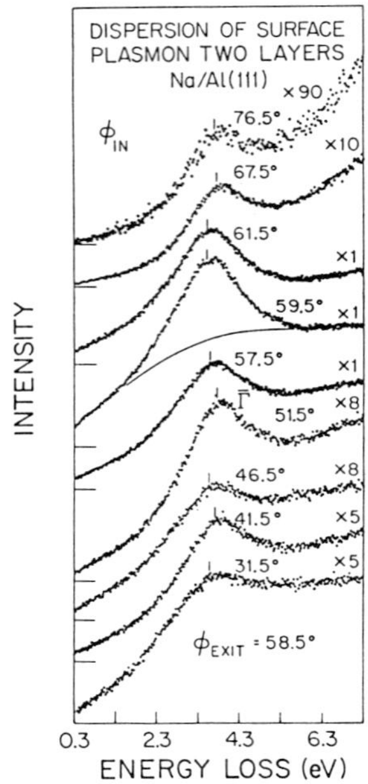

Fig. 13. Loss spectra as a function of incident angle for a two layer film of Na on Al(111). The incident energy was 30 eV[49].

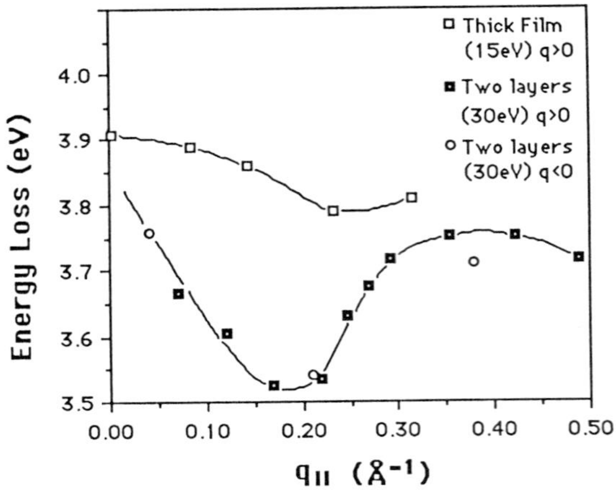

Fig. 14. Measured dispersion of the surface plasmon for a thick Na film and a two layer Na film, both grown on Al(111)[49].

Where ω_S is the surface plasmon energy at $q_{//}=0$, i.e. $(1/\sqrt{2})$ of the bulk plasmon energy, and $d(\omega_S)$ is the first moment of the screening charge induced by the dynamic external field. d is measured from the edge of the positive jellium background and is marked at z_0 ($d(0)=z_0$) in Fig. 1 for the static limit of $\omega_S=0$. The analytical result of Liebsch presented by Eqn. 7 is in agreement with Feibelman's numerical calculation[47] when both theoretical approaches use RPA. The comparison to be made with theory concerns the magnitude of the negative dispersion and its dependence upon the electron density of the substrate.

Figure 15 displays recent calculations by Liebsch of the plasmon dispersion at $q_{//}=0$ as a function of r_S. The results of two different calculation schemes are shown. The open circles are from an RPA calculation and the squares are from a time dependent LDA calculation[5]. In both cases the normalized dispersion increases as the electron density decreases, with the LDA type calculation giving a much larger dispersion than predicted by the RPA theory[5]. The triangles show the data for Na indicating that both theories have appreciably overestimated the magnitude of the dispersion. For example, the RPA calculcuation for Na gives a normalized dispersion of 0.4Å compared to

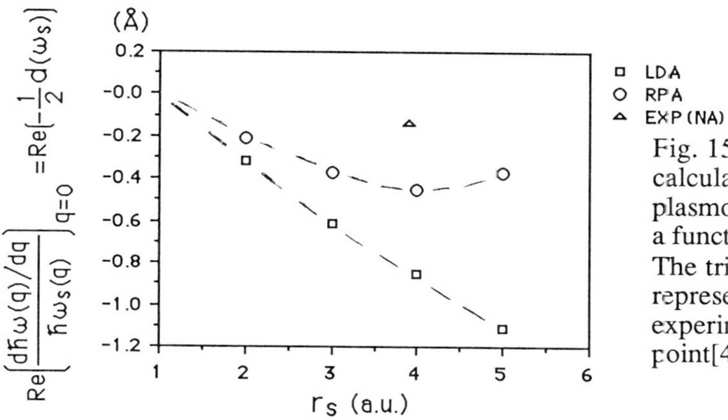

Fig. 15. Theoretical calculations of surface plasmon dispersion as a function of r_s[50]. The triangle represents an experimental data point[49].

the experimental value of 0.14Å. The same discrepancy exists for the dispersion of the two layer film shown in Fig. 14. The experimental data yields a normalized dispersion of 0.78Å while the theoretical prediction of Equiluz and Campbell[9] was 1.59Å. Our data for Al indicates very little dispersion, if any, while earlier results indicated a slight positive dispersion for evaporated films[49], or a slightly negative dispersion for Al(100) obtained from fitting inelastic LEED data[50].

It is likely that this disagreement between theory and experiment originates from the same basic problem discussed in sections II and III. The RPA type theories all seem to overemphasize the coupling to the high q components of the electron-hole pair continuum, where the experiments indicate that the low q region dominates. If this is also true for the calculations of the surface plasmon dispersion it is possible to explain why the theory predicts too large of an effect. It would be illustrative to project out the strength of the coupling of the surface plasmon to different portions of the electron-hole pair continuum as a function of the value of $q_{//}$.

V. CONCLUSION

Hopefully this paper has demonstrated two facts. (1) It is now possible to measure the microscopic properties of the dynamical response at a metal surface for NFE metals. (2) That these new results show a systematic deficiency of RPA based theories. Data on a much wider variety of electron densities needs to be obtained so that the systematics of the screening as a function of r_s can be determined and compared to theory. On the theoretical front it is important to begin to develop theoretical approaches beyond RPA and eventually to incorporate the crystal potential into these theories. The calculation of Northrup, Hybertsen and Louie[34] for the quasiparticle spectrum of Na and Li indicate that theories of this nature can be developed and do in fact agree with experimental data.

Acknowledgements:

This work was supported by the National Science Foundation through grand DMR-8610491 and the synchrotron beam line is maintained by the Penn MRL (DMR-8519059). EWP would like to thank the Guggenheim Foundation for their support during the last year.

REFERENCES

1. P. J. Feibelman, Prog. in Surf. Sci. **12**, 287 (1982)
2. K. L. Kliewer, Surf. Sci., **101**, 57 (1980)
3. F. Forstmann and R. Gerhardts, Festkorper Probleme/Advances in Solid State Physics **XXII**, 291 (1982)
4. G. Mukhopadhyay and S. Lundqvist, Physica Scripta, **17**, 69 (1978)
5. A. Liebsch, Phys. Rev. B, to be published.
6. M. Corvi and W. L. Schaich, Phys. Rev. B**33**, 3688 (1986)
7. B. N. J. Persson and N. D. Lang, Phys. Rev. B**26**, 5409 (1982)
8. C. Schwartz and W. L. Schaich, J. Phys. C**17**, 537 (1984); K. Kempa and W. L. Schaich, Phys. Rev. B**32**, 8375(1985)
9. A. G. Equiluz and D. A. Campbell, Phys. Rev. B**31**, 7572 (1985)
10. D. M. Newns, Phys. Lett. **38A**, 341 (1972)
11. P. Gies and R. R. Gerhardts, Phys. Rev. B**33**, 982 (1986)
12. J. E. Inglesfield and E. Wikborg, J. Phys. F. **5**, 1706 (1975)
13. P. Apell, A. Ljungbert and S. Lundqvist, Physica Scripta, **30**, 367 (1984)
14. H. Levinson, E. W. Plummer and P. J. Feibelman, Phys. Rev. Lett. **43**, 952 (1979)
15. I. Lindau and P. O. Nilsson, Phys. Lett. **31A**, 352 (1970)
16. M. Anderegg, B. Feuerbacher, and B. Fitton, Phys. Rev. Lett. **27**, 1565 (1971)
17. N. Lang and W. Kohn, Phys. Rev. B **1**, 4555 (1970): B **7**, 3541 (1973)
18. *Theory of the Inhomogeneous Electron Gas*, edited by S. Lundqvist and N. H. March (Plenum, NY, 1983)
19. M. Weber and A. Liebsch, Phys. Rev. B**35**, 7411 (1987)
20. J. Song, D. Heskett, E. W. Plummer, H.-L. Dai and A. Liebsch, to be published.
21. P. J. Feibelman, Phys. Rev. B**12**, 1319 (1975)
22. D. Pines and P. Nozieres,*The Theory of Quantum Liquid*, Vol.1 (W. A. Benjamin, 1966)
23. E. W. Plummer and W. Eberhardt, Adv. in Chemical Physics, **49**, 533 (1987)
24. F. J. Himpsel, Adv. Phys. **32**, 1 (1983)
25. V. L. Moruzzi, J. F. Janak and. R. Williams, *Calculated Electronic Properties of Metals* (Pergamon, NY 1978)
26. E. W. Plummer, Physica Scripta, T**17**, 186 (1987)
27. E. Jensen and E. W. Plummer, Phys. Rev. Lett. **55**, 1912 (1985)
28. W. Y. Ching and J. Callaway, Phys. Rev. B**11**, 1324, (1975)
29. L. Hedin, Phys. Rev. **139**, A796 (1965)

30. K. W.-K. Shung and G. D. Mahan, Phys. Rev. Lett. **57**, 1076 (1986)
31. J. P Perdew and M. R. Norman, Phys. Rev. B**26**, 5445 (1982).
32. X. Zhu and A. W. Overhauser, Phys. Rev. B**33**, 925 (1986)
33. T. K. Ng and K. S. Singwi, Phys. Rev. B**33**, 7743 (1986)
34. J. E. Northrup, M. S. Hybertsen and S. G. Louie, Phys. Rev. Lett. **59**,819 (1987)
35. J. Davenport, private communication. The Hartree theory puts in the crystal potential in a self consistent manner but there is no exchange or correlation term.
36. W. Wohlecke, A. Baalmann and M. Neumann, Solid State Comm. **49**, 217 (1984)
37. I.-W. Lyo and E. W. Plummer, to be published.
38. For a recent review see K. S. Singwi and M. P. Tosi, in Solid State Physics, **36**, edited by H. Ehrenreich, F. Seitz, and D. Turnbull (Academic, NY, 1981); S. Ichimaru, Rev. Mod. Phys. **54**, 1017 (1982)
39. A. W. Overhauser, Phys. Rev. B**3**, 1888 (1971)
40. Y. R. Shen, The Principles of Nonlinear Optics (Wiley, 1984).
41. Y. R. Shen, J. Vac. Sci. and Tech. B**3**, 1464 (1985)
42. H. W. K. Tom, C. M. Mate, X. D. Zhu, J. E. Crowell, T. F. Heinz, G. Somorjai, and Y. R. Shen, Phys. Rev. Lett. **52**, 348 (1984)
43. H. W. K. Tom, C. M. Mate, X. D. Zhu, J. E. Crowell, Y. R. Shen and G. Somorjai, Surf. Sci. **172**, 466 (1986)
44. T. F. Heinz, M. M. T. Loy, and W. A. Thompson, Phys. Rev. Lett. **54**, 63 (1985)
45. D. Heskett, K.-J. Song, A. Burns, E. W. Plummer and H.-L. Dai, J. Chem. Phys. **85**, 7490 (1986)
46. H. Raether, "*Excitation of Plasmons and Interband Transitions by Electrons*", Vol 88, Springer Tracts in Modern Physics (Springer-Verlag, 1980)
47. P. J. Feibelman, Phys. Rev. B**9**, 5077 (1974)
48. K.-D. Tsuei, D. Heskett, A. P. Baddorf and E. W. Plummer, To be published.
49. K. J. Krane and H. Raether, Phys. Rev. Lett. **37**, 1355 (1976)
50. C. B. Duke, L. Pietronero, J. O. Porteus and J. F. Wendelken, Phys. Rev. B**12**, 4059 (1975); they found a negative dispersion of 0.19±0.10 Å normalized to $\hbar\omega_s(q=0)$ for Al(100).

Discussion

D.R. HAMANN:

1. The "local field" correction to the dielectric response in the sense you're using it is, in fact, omitted by Louie and Hybertsen. I believe that the reason their self energy gives good results for the band structure of the semiconductors is that they get nearly the correct $\bar{q} = 0$ dielectric constant

without the correction. This may be due to the larger r_s of the semiconductors or the charge-inhomogeneity local fields they do include. Their results are completely independent of the LDA.

2. For the second harmonic generation experiments, have you considered the quantum size effect? This box quantization of the electrons between the front and back surfaces of the alkali film could complicate the comparison with theories for the dielectric response of semi-infinite metals.

E.W. PLUMMER:

1. In the original calculations by Louie for the band gaps in semiconductors the dielectric response beyond RPA was not included, yet in the later calculation for metals this was the most important term. I would be very interested in seeing the semiconductor calculations done over with the dielectric response done in the same way as it was for the nearly free electron metals, to see what effect it has.

2. We have investigated the effect on SHG due to the quantized levels in a quantum well of variable thickness. In these model calculations we find that the resonant excitation energy from the n to n+1 levels becomes too low in energy for a film of approximately five layers to be compatible with our observations. The question is what effect the substrate will have on the energy position and level width of the states in this quantum well.

J.B. PENDRY: You associate the periodicity in second harmonic generation as a function of film thickness with the leading edge of the RPA continuum of excitations. Since the RPA itself does not confirm this, would it not be more reasonable to say there was a new excitation present, possibly a spin density wave?

E.W. PLUMMER: My personal view is that the RPA does not go from the static to the dynamic limit as rapidly as it should. But it surely is possible that there is another excitation like a spin density wave.

J.B. PENDRY: These results for the thin film second harmonic generation, where you see an apparent excitation, if I understood you correctly, of the wavelength 22Å—I mean, I'm sure you have a lot more to tell us about that than you had the time to tell us. Would you like to speculate a little bit about that, perhaps with respect to spin–wave type ideas, or...?

E.W. PLUMMER: I don't think you need anything *fancy* There is an external field of frequency w driving the electron gas. The response of the gas is a surface charge oscillating at w. The metal can't screen this oscillating charge, so there are Friedel oscillations. The only question is, what's the wavelength? And, my contention is that RPA fails in certain frequency regions. I don't thing you need anything fancy.

J.B. PENDRY: But if you have a well-defined wavelength, you then have to think of an excitation. The RPA tells you what the response of that continuum...

E.W. PLUMMER: No! Just look at Feibelman's calculations; you integrate across the whole electron-hole pair continuum and the limits of the integral give you a dominant wavelength. I think that's all that's happening here. Feibelman points out in his paper that you can just plot out what wavelengths you should see by the limits of your integral. I don't think it's anything fancy. I'd be glad if it *was*, but I don't think so.

J.B. PENDRY: Then let me ask a more specific question. If it's as you say, the limits of the integral, then the wavelength disperses linearly with frequency, is that true?

E.W. PLUMMER: Yes.

J.B. PENDRY: Okay. Thank you.

E.W. PLUMMER: Be careful, we have two points, which defines a line. This line goes to $\gamma = 0$ at $\omega = 0$.

H. IBACH: The dispersion of surface plasmons is much smaller than the width of the plasmon loss. This constitutes a problem in comparing theory and experiment unless the theory also predicts the width or even the shape of the loss peak.

E.W. PLUMMER: There is no inherent problem experimentally in defining the peak position to an accuracy much better than the peak width, if you know what the line shape should be. We do this routinely with the photoemission data. The real question concerns the theoretical predictions for the line width, since in all of these systems the dispersion will be related to the line width. At present the theoretical calculations underestimate the line width and overestimate the dispersion, which is your point.

Y. PETROFF:

1. If I remember correctly, the agreement between theory (Northrup *et al.*, P.R.L. **87**), and your experiment on Na was better than the one that you showed.

2. My impression is that those "simple" metals as Na, K, or even Al seem sometimes to be much more complicated to understand than expected. Could you comment, in particular, on the strong structure seen in ARPES near the Fermi level, which has been a "hot" topic recently?

E.W. PLUMMER: The anomalous peak at the Fermi energy for photoemission from Na reported by Jensen (ref. 27) still exists in the data. The higher resolution data recently obtained by Lyo (ref. 37) shows that the peak is even narrower and more intense than we had originally thought. There have been many theoretical explanations for this peak, but at present it is my opinion that none of them will quantitatively explain the data.

I believe that it is possible that this anomalous structure is due to a surface property of Na. For example, since a charge density wave is close to being the ground state configuration in the bulk, the presence of the surface could create a pinned charge density wave that penetrates deeper than we see. The Friedel oscillation of the electron gas and the oscillatory interplanar relaxations at the surface could be what drives a surface charge density wave. It is also possible that the surface has a different structure than the bulk. Most of these alkali metals want to go to a hcp structure at low temperature, usually induced by strain. The surface could furnish the necessary strain on the lattice causing the surface (110) plane to slip parallel to the surface with respect to the next (110) plane. It is an interesting material and there may be altogether new physics here.

P. MARCUS: What message do you have about the validity of LDA for obtaining correct band structures?

E.W. PLUMMER: It is my present understanding from comparison of theory (LDA) and experiment that there is always a disagreement between theory and experiment and based upon the recent calculation of the self–energy correction terms, the discrepancy is expected. Therefore, I will say it again. If a theoretical calculation agrees in detail with an angle resolved photoemission measurement, then there is something wrong with the theory.

It is clear that the calculated d–bands in the noble metals are higher than found experimentally. In contrast the s–band in the simple metals is deeper theoretically than experimentally.

L. KLEINMAN: The only thing I'd argue with you on is, I think LDA does give the correct work function, that's one measurement I think it gives correctly.

E.W. PLUMMER: I don't really understand why LDA gets the work function correct, since you don't get the long range part of the potential current.

A.M. BRADSHAW: Why is that? Is that because the work fnction is not an excited state property of the system? You're just taking an electron off the top?

E.W. PLUMMER: You measure it with an excited state.

L. KLEINMAN: The image potential goes to zero when you take the electron to infinity, so it really doesn't enter into the work function, it enters into the shape of the potential, that the electron sees as it goes out, but it doesn't enter into the work function.

J.P. TOENNIES: I would like to ask you to tell us how you determined the thicknesses of your Na layers.

E.W. PLUMMER: We use several ways to measure the film thickness and the quality of the film. First, the time dependence of the source is measured by always evaporating into a mass spectrometer behind the sample. The film thickness is monitored as it is being deposited by measuring the work function and the photoyield (two photon process of the infrared). The films are thermally desorbed into a mass spectrometer so that we have the TDS of each thickness. Finally, the SHG for the thick films as a function of film thickness is calculated using Fresnel type equations with the film thickness as a parameter. All of these procedures give the film thickness presented in this paper.

As important as film thickness is the smoothness of the film. We monitor the specularly reflected fundamental light intensity and have used low energy inelastic electron scattering to monitor the reflectivity as a function of preparation conditions.

A.M. BRADSHAW: We all use these alkali metal dosers but with varying degrees of success. This is also true in my laboratory, but we've had the most difficulties with Na—in getting clean Na films. I don't know whether you've experienced this as well?

E.W. PLUMMER: We can get clean Na by long and careful outcassing. It's always to your advantage to use the alkous ones that sublime lower temperatures, because less dirt comes off. Besides Li, Na is the highest–temperature alkali.

J.B. PENDRY: Given that you can prepare perfect rubidium films, can you observe Fabry-Perôt resonances in the hole states created by photoemission as suggested by Loly and Pendry?

E.W. PLUMMER: We have been looking for what you predicted but at the present time I cannot report any positive observations to you.

Advances in Surface Crystallography by Multishell SEXAFS

*D.J. Holmes, D.R. Batchelor, and D.A. King**

The Donnan Laboratories, University of Liverpool, Liverpool L69 3BX, UK
*Present address: Department of Physical Chemistry,
Lensfield Road, Cambridge, UK

Multishell analyses of SEXAFS data involves complete calculations of the EXAFS spectrum, with major input variables the shell spacings r_j and shell coordination numbers N_j, and iterative comparison with experimental data until a good fit is obtained. Three tests of the reliability of r_j and N_j obtained from such analyses are described: a reproducibility test, involving the analysis of 14 different data sets from the same structure (Cl on Ag{110} at θ=0.75); a test involving the analysis of simulated data obtained from known input parameters; and a statistical significance test. The tests show that as many as four reproducible bond lengths may be obtained, with sufficient accuracy to distinguish between shells separated by as little as 0.3 Å. Nearest neighbour coordination N_1 is accurate to within ± 0.2, but a higher degree of inaccuracy is obtained for N_j when j ≥ 2. Some results of successful structural analyses utilising this powerful technique are presented.

1. SEXAFS: the Technique

Although the technique is experimentally more demanding than LEED, and experiments can only be conducted at high energy synchrotron radiation sources, surface EXAFS has a number of advantages over LEED as a surface structural tool.[1-4] It is dominated by single scattering, and single-electron, short-range order theories suffice to extract accurate data. It can be applied to both well ordered and, under certain conditions, highly disordered systems. Here we examine the accuracy of the bond lengths and coordination numbers which can be extracted from SEXAFS data using a multishell data analysis technique. Further details are to be made available elsewhere.[5,6]

Based on a harmonic potential approximation, the magnitude of the X-ray absorption fine structure above a K adsorption edge is:[7]

$$\chi (k) = - \Sigma \frac{N_j^{eff} F_j(k)}{k r_j^2} \sin (2k r_j + 2\delta_j + \psi) \exp (-2\langle u_j^2\rangle k^2 \exp (-2Vr_j/k) \quad (1)$$

where the summation is over the number of shells which contribute to the EXAFS, k is the photoelectron wave-vector, δ_j and ψ are phase shifts of absorber and backscatterer atoms, respectively, and $F_j(k)$ is the backscattering amplitude; N_j^{eff} is the effective coordination number, taking account of the incident A vector; and r_j the interatomic bond length for the jth shell. V is the imaginary part of the inner potential of the sample.

Two methods of data analysis are currently employed. The first relies on the determination of the Fourier transform F(r) from the background-subtracted $\chi(k)$ data, reading off bond distances from peaks in the F(r)

Springer Series in Surface Sciences Vol. 14
Solvay Conference on Surface Science
Editor: F.W. de Wette

spectrum for each shell, appropriately corrected for phase shifts. The second involves calculation of $\chi(k)$ using a suitable number of shells, and iterative comparison with the experimental $\chi(k)$ spectrum. In practice, the comparison is made with a spectrum which has had high Fourier component noise removed by taking a Fourier transform of the data, placing a suitable window function over the transform, and back-transforming to produce $\chi(k)$. In this paper, we briefly review the results of accuracy tests on the second of these techniques,[5,6] and some of the surface structural data obtained.[3,10,12-15]

2. Phase shift transferability

Phase shifts were obtained by initial calculation (using a muffin-tin formalism) and then empirically adjusting these calculated phase shifts until good fits were obtained with low temperature EXAFS data from compounds of known crystal structure, with the known r_j and N_j structural parameters. The validity of this approach[8] was tested by utilising phase shifts obtained from one set of bulk compounds of known crystallinity to calculate the r_j and N_j parameters for other bulk compounds of known crystallinity.[9] For example, chlorine absorber and backscatterer phase shifts obtained by fitting to NaCl, KCl, RbCl, CsCl, $CdCl_2$ and $LaCl_3$ were transferred to analyse a AgCl EXAFS spectrum; first-neighbour (Cl - Ag) spacings were reproduced to within ± 0.018Å in each case, and second neighbour spacings to within ± 0.015Å except for CsCl (-0.032Å) $CdCl_2$ (-0.06Å) $CoCl_2$ (+0.083Å) and $LaCl_3$ (-0.087Å). In general, we conclude that absolute errors arising from phase shift effects are ≤0.02Å in first neighbour spacings.

3. Reproducibility test: Cl on Ag{110}

Recently we completed a structural study of the incommensurate structure formed by Cl on Ag{110} at 0.75 monolayers[10] using SEXAFS on the Cl K-edge step at 2,820 eV. To avoid anharmonicity effects,[3] data were taken at 110 K. Fourteen independent SEXAFS data sets were obtained, 7 with the incident A vector along the <100> azimuth and 7 with the A vector along <110>. Each data set was independently subjected to a multishell analysis, including four shells out to 4.5Å. A typical best-fit comparison between the experimental and calculated $\chi(k)$ is shown in figure 1. In table 1 we show the interatomic spacings up to 4 shells for each

Table 1. Comparison between experimental and calculated bond distances for vacancy domain wall structure models: Cl on Ag{110}.[10]

Run/model	A	R_1(Ag)	R_2(Cl)	R_3(Ag)	R_4(Cl)
1	⟨100⟩	2.57	3.18	3.49	3.84
2		2.57	3.18	···	3.93
3		2.56	3.16	3.62	4.03
4		2.54	3.18	···	···
5		2.60	3.18	3.54	4.11
6		2.55	3.11	3.51	4.27
7		2.58	3.18	···	4.12
Mean	⟨100⟩	2.57 ± 0.03	3.17 ± 0.04	3.54 ± 0.06	4.05 ± 0.20
8	⟨110⟩	2.56	3.20	3.44	4.15
9		2.56	3.12	···	···
10		2.62	3.16	3.58	···
11		2.58	3.20	···	3.92
12		2.53	3.01	3.44	4.15
13		2.55	3.14	3.44	4.02
14		2.52	3.19	3.40	4.24
Mean	⟨110⟩	2.56 ± 0.04	3.14 ± 0.04	3.46 ± 0.06	4.09 ± 0.17

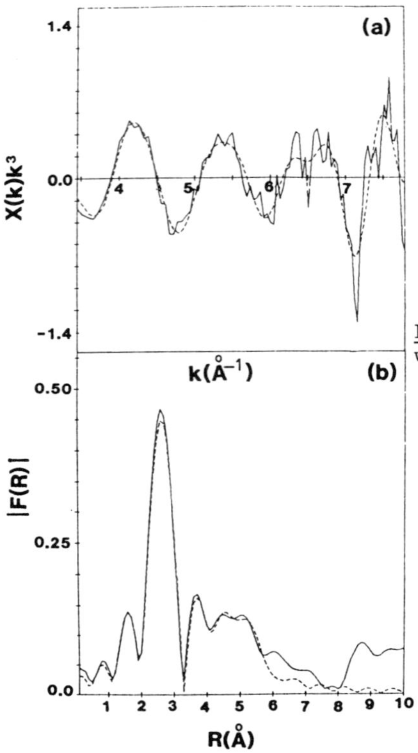

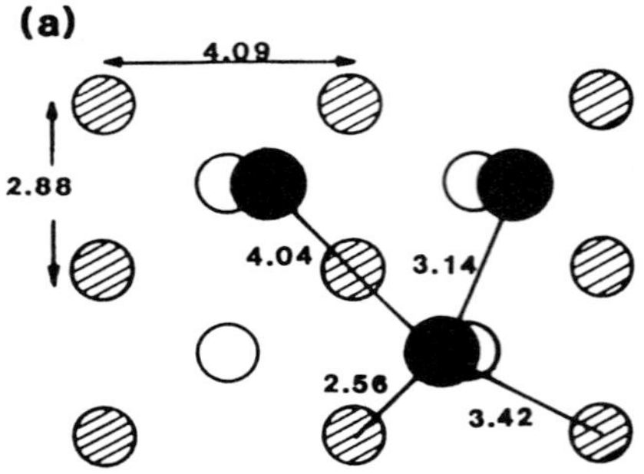

Figure 2(a) Plan view of the vacancy domain wall structure for Cl/Ag{110} at θ=0.75

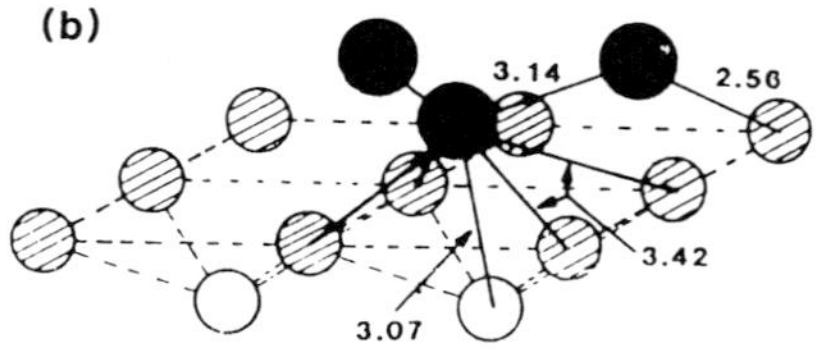

Figure 2(b) Isometric projection of the structure in 2(a) with experimental bond distances.

Figure 1. SEXAFS data (dashed line) and best-fit calculated[10] spectrum (full line) for Cl/Ag{110}. Fourier transforms are also shown.

independent analysis. The reproducibility of these spacings led us to conclude that all four spacings were significant. By triangulation, we were therefore able to arrive at the structure depicted in figure 2.

4. Analysis of Simulated Data[6]

Assuming that the first shell Cl-Ag spacing for the structure shown in figure 2 was correctly obtained as 2.56Å, the second (Cl), third (Ag) and fourth (Cl) shell distances calculated for this model are 3.14Å, 3.41Å and 4.05Å, respectively. Effective coordination numbers were also calculated for this model, pertaining to the $\underline{A}$ vector in both <100> and <110> directions. With the Debye-Waller factors A_j (=2 $\langle u_j^2 \rangle$ in equation (1)) set to 0.02Å^2 for all Ag shells and 0.05Å^2 for all Cl shells, these parameters were used to calculate simulated, noise-free SEXAFS spectra from equation (1).

The simulated spectra were then analysed by the multishell analysis techniques. Initial values of r_j were varied, by as much as 0.5Å either side of the input distances, so as not to bias the end result of the analyses, as was the order in which the contribution from each shell was added to the calculation. Thirteen independent analyses of these simulated SEXAFS data were performed, six with $\underline{A}$ in <110> and seven with $\underline{A}$ in <100>. The mean values of r_j and N_j obtained are shown in Tables 2 and 3. Only the first four shell spacings were found to reproduce the input spacings to within reasonable error limits. Comparison with the input spacings indicates that first shell spacings are regenerated to within ± 0.01Å; the second and third shells, although only 0.3Å apart, are reproduced to within ± 0.05Å, and the fourth shell again to about ± 0.05Å. Coordination numbers are reproduced to within ± 0.2 for the first shell, but agreement for higher shells shows poor reliability.

A further set of seven simulated spectra were generated, each with different random noise at a level typical of that in our experiments. The results of these analyses were parallel to those described above. Random noise to this level did not affect the analyses.

The results demonstrate the validity of the multishell procedure for extracting bond lengths for up to 4 shells. On the other hand, reliability in shell coordination numbers beyond the first shell was relatively poor, due to a correlation in the amplitude function between the Debye-Waller term A_j and N_j (equation (1)).

Table 2: mean values obtained from seven independent analyses of simulated noise-free SEXAFS data ; A vector lying along <100> direction.

	$r_j/\text{Å}$	N_j^{eff}	$A_j/\text{Å}^2$
j=1(Ag)	2.55±0.01	2.1±0.2	0.022±0.003
2(Cl)	3.11±0.04	2.1±0.6	0.038±0.009
3(Ag)	3.44±0.06	1.8±0.6	0.024±0.015
4(Cl)	4.05±0.08	4.1±1.6	0.031±0.018

Table 3: mean values obtained from six independent analyses of simulated noise-free SEXAFS data ; *A* vector lying along <110> direction.

	$r_j/\text{Å}$	N_j^{eff}	$A_j/\text{Å}^2$
j=1(Ag)	2.56±0.01	2.1±0.2	0.021±0.001
2(Cl)	3.22±0.09	2.2±2.1	0.036±0.018
3(Ag)	3.42±0.03	4.1±1.8	0.022±0.014
4(Cl)	4.03±0.02	3.1±1.1	0.030±0.013

5. Significance tests on data analysis

In order to decide how many coordination shells should be included in the analysis of the Cl/Ag(110) data[10] described in section 3 above, we adopted a statistical test suggested by Joyner et al.[11] This test examines the statistical significance of adding the (n + 1)th shell by calculating the probability of obtaining a reduction in the fitting index as large as that actually observed when the added shell is not physically meaningful. The data referred to above for Cl on Ag{110} were subjected[6] to this test; the first three shells were found to be significant to within a 5% confidence limit, and the fourth to within a 10% confidence limit. This analysis strongly supports those given in sections 3 and 4 above.

6. Structural results for Cl and Cs on Ag planes

Using multishell SEXAFS we have analysed surface structures for Cl on Ag{111}[12] {100},[13] and {110}[6] and for Cs on Ag{111}.[14,15] On Ag{111}, Cl atoms are found to occupy the 3-fold hollow site at $\theta = 0.33$ and 0.67; the LEED pattern at the latter coverage is $(\sqrt{3}x\sqrt{3})R30^o$, and is shown to correspond to a graphitic or honeycomb overlayer structure. On Ag{100}, our analysis supports a LEED intensity analysis for this system,[16] showing a simple overlayer model for the θ=0.5 structure, with Cl atoms occupying 4-fold hollow sites. On Ag{110}, as we have seen above (figure 2), Cl atoms occupy distorted out-of-plane 3-fold coordination sites. The Cl-Ag bond length is found to be shorter than that in bulk AgCl (2.77Å) in each case; and independent of coverage. For Ag{100} and Ag{111}, the bond length is 2.70Å; while for Ag{110} it is 2.56Å. We conclude that adsorbed Cl is in each case more covalent than in AgCl; and Cl/Ag{110} is the least ionically bonded (Cl bonding radius 1.11Å).

For Cs on Ag we find[14,15] that the 3-fold hollow site is occupied at all submonolayer coverages; at half monolayer coverage, the Cs-Ag bond length is 3.20Å, but this increases to 3.50Å at the monolayer coverage. We associate this large bond length change with an ionic to covalent bonding transition as the coverage is increased.

Acknowledgements

We are grateful to the SERC for an equipment grant, the provision of beamtime at the SRS, Daresbury, and for a studentship (D.J.H.) and a postdoctoral fellowship (D.R.B.).

References

1. P.A. Lee, P.H. Citrin, P. Eisenberger and B.M. Kincaid, Rev. Mod. Phys. 53 (1981) 769.
2. J. Stöhr, in 'X-ray Absorption: Principles, Applications and Techniques of EXAFS, SEXAFS and XANES', ed. Prins and Konigsberger, Wiley, New York, in press.
3. G. M. Lamble and D.A. King, Phil. Trans. R. Soc. London A318 (1986) 203.
4. D. Norman, J. Phys. C19 (1986) 3273.
5. G.M. Lamble and D.A. King, to be published.
6. D.J. Holmes, D.R. Batchelor and D.A. King, Surface Sci., submitted.
7. P.A. Lee and J.B. Pendry, Phys. Rev. B11 (1975) 2795.
8. P.H. Citrin, P. Eisenberger and B.M. Kincaid, Phys. Rev. Lett. 36 (1976) 1346.
9. G. M. Lamble and D.A. King, to be published.
10. D.J. Holmes, N. Panagiotides, R. Dus, D. Norman, G.M. Lambe, C.J. Barnes, F. della Valle and D.A. King, J. Vac. Sci. Technol. A5 (1987) 703.
11. R.W. Joyner, K.J. Martin and P. Meehan, J. Phys. (in press).
12. G.M. Lamble, R.S. Brooks, S. Ferrer, D.A. King and D. Norman, Phys. Rev. B34 (1986) 2875.
13. G.M. Lamble, R.S. Brooks, J.C. Campuzano and D.A. King, Phys. Rev. B36 (1987) 1796.
14. G.M. Lamble, R.S. Brooks, D.A. King and D. Norman, Europhys. Lett., submitted.

15. G.M. Lamble and D.A. King, in 'Lectures on Surface Science', ed. G.R. Castro and M Cardona, Springer-Verlag, Berlin, 1987, p.250.
16. E. Zanazzi, F. Jona, D.W. Jepsen and P.M. Marcus, Phys. Rev. B14 (1976) 432.

Discussion

H. IBACH: Surface phonons are highly correlated motions, the eigenvectors are frequently such that the nearest neighbor distances change very little while the second-nearest-neighbor distances do. Such surface phonons have low frequencies and therefore these vibrations contribute most to the total spectrum at finite T. I would like to suggest that, rather than assigning a particular mean-square displacement to individual atoms, one should change the calculation such that one uses different mean-square displacements for the different bonds. This may remove some of the problems with the determination of the coordination numbers in different shells.

D.A. KING: Your point is well taken. However, in our multishell analyses of experimental data we allow the Debye-Waller factors A_j to vary, in addition to the structural parameters r_j and N_j, in the iterative process leading to a best fit. In this process, A_j and N_j are strongly coupled to each other through the term

$$N_j^{eff} \exp(-A_j k^2)$$

which dominates the amplitude of the fine structure for each shell in the expression for $X(k)$ (equation (1) of our paper). Hence we make the point that without *a priori* knowledge of the Debye–Waller term, the reliability of coordination numbers is poor for $j \geq 2$. In our analysis of simulated data, we artificially assumed that the Debye–Waller factor was the same for each shell of identical atoms, only for the purposes of calculating simulated data. The results of the analysis (Tables 2 and 3) show that we (incorrectly) obtain differing values for the two Ag and two Cl shells.

F.W. DE WETTE: A related question is, does the anisotropy of the mean square displacements have anything to do with it?

D.A. KING: Again, you're talking about this correlation between the coordination number and how accurately we can get it out. My feeling is that, if you're going to do an analysis on amplitudes, based on amplitudes, you're in difficulty. There's experimental difficulty in getting reproducibility in the amplitudes as well. So before we start worrying about these effects we've got bigger problems. In other words, you can pick up peaks in your spectra and shoulders on the spectra more accurately than you can determine relative amplitudes. I think that's another problem that we have.

J.G. DASH: Have you considered the effect of static disorder on the apparent coordination numbers?

D.A. KING: We *assume*—you're quite right, we *assume* a perfect system, and in our analysis we can't distinguish between static disorder and dynamic disorder.

J.G. DASH: Except perhaps through the Q–vector or temperature dependence.

D.A. KING: Except through the temperature dependence...

J.G. DASH: Yes, the temperature dependence would do.

D.A. KING: We can't look at temperature because it gets so anharmonic. That's our problem.

E.W. PLUMMER: I didn't catch whether you said it, but have you investigated with your model what the unpolarized part of the light is going to do to you in these shells? You could have 25% unpolarized light—circularly polarized, elliptically polarized... Does it do anything to this determination of the shell structure?

D.A. KING: The polarization has a certain percentage to it, is what you're saying. And we're working at 2,800 volts with this particular experiment where it's something like 95% polarized.

E.W. PLUMMER: Are you sure?

D.A. KING: Well it has been measured.

E.W. PLUMMER: These machines can mess up the polarization.

D.A. KING: Certainly, it would be very important in terms of the final structural analysis that we do. For example, for Cl on Ag(110), if the Cl were simply aligned along the rows, we should have seen a *much* stronger polarization effect than we saw; and if the polarization was only 50% our analysis would be wrong.

G. ERTL: The observation of a shorter bond length for Cl - Ag(110) parallels the results obtained by LEED analysis of the H - Ni(110) and H - Ni(111) systems, where the same three–fold sites are occupied with a difference in bond lengths of the same order of magnitude reported by you. This might indicate the operation of a general principle.

J.B. PENDRY: You note a correlation of chlorine/silver bond length with surface. Could this simply be a correlation with *coverage*? The coverage correlates with the charge state and hence with the radius of chlorine.

D.A. KING: We find that Cl on Ag{111} and {110} the bond length is independent of coverage, at least for coverages above 1/3 monolayers to within the reproducibility of our measurements, which is $\pm$ 0.01Å. This result is surprising, since there is clear evidence of strong repulsion between Cl adatoms in nearest neighour sites. However, in our studies of Cs on Ag{111} we have found a strong dependence of bond length with coverage, which does correlate with a change in charge state (from Cs^+ to Cs^0) as the coverage increases, in accordance with your expectation.

M.G. LAGALLY: With regard to John Pendry's question (just prior) about a relationship between bond lengths and surface charge, and your response that you observe no coverage dependence: do you not need to know more about the overlayer microstructure? It could be that you simply have more patches of the same ordered structure as the coverage increases and thus the local environment of each atom remains the same.

D.A. KING: Of course in the case of island formation no change in bond length with coverage should be expected. However, in the two cases under discussion here, Cl on Ag{111} and on Ag{110}, we have independent evidence from both LEED and, more particularly, from SEXAFS that island formation does not occur. For example, in the case of Cl on Ag{111}, there is no LEED pattern associated with 1/3 monolayer coverage, and a $(\sqrt{3} \times \sqrt{3})\ R30°$ pattern develops at 2/3 monolayer. The SEXAFS analysis at these two coverages (reference 3 of our paper) shows that at $\theta = 0.67$ there is a Cl-Cl spacing of 2.93Å corresponding to the close–packed Cl-Cl distance in the honeycomb structure at this coverage, whereas this shell is completely absent at 1/3 monolayer coverage. At both coverages, however, the SEXAFS data reproduce a Cl-Cl spacing of 4.83Å, corresponding to the *next* nearest neighbor in the honeycomb structure. Thus, the *n.n.* interaction is repulsive, and at low coverages these sites are avoided.

Chemical Effects in Surface-Enhanced Raman Scattering

Xudong Jiang, K.G. Lloyd, and A. Campion

Department of Chemistry, The University of Texas, Austin, TX78712, USA

We have investigated chemical contributions to the total enhancement in surface-enhanced Raman scattering under conditions where the electromagnetic contribution is small. On stepped and kinked silver surfaces with surface atoms of high coordination number neither chemisorption of pyridine nor enhancement is observed. On a stepped surface with moderate coordination, chemisorption is observed but there is still no enhancement of the Raman scattering. Finally, pyridine chemisorbs strongly on essentially isolated silver adatoms on a rhodium substrate and enhancements of a factor of fifty or so are observed. We have obtained the electronic spectra of the latter two silver-pyridine systems using high resolution electron energy loss spectroscopy and find no correlation between the strength of the charge transfer bands observed and the enhancement factor.

1. Introduction

Despite more than a decade of research by many groups throughout the world, a complete, quantitative understanding of the mechanism of surface-enhanced Raman scattering has not been achieved. Several recent reviews summarize the state of our understanding of the phenomenon [1-4]. There are two classes of mechanisms thought to be responsible for the six orders of magnitude enhancement which is typically found for molecules adsorbed on roughened surfaces of the coinage metals and a few others. The dominant source of enhancement is electromagnetic in origin; roughness allows the excitation by light of either propagating (grating surfaces) or localized (small bumps) surface plasmons. The roughness features act to amplify both the exciting laser fields and the scattered Raman fields, contributing perhaps four out of the six orders of magnitude.

In addition to the electromagnetic mechanism, it is widely believed that a second source of enhancement exists which is confined to the molecules directly chemically adsorbed on the surface. Various theories have been proposed for this source of enhancement but perhaps the most plausible is the formation of what is essentially a surface complex with an electronic transition that is resonant with the exciting laser [5-8]. This would lead to ordinary resonance Raman scattering, and indeed metal-molecule charge transfer resonances have been observed in electron scattering experiments [9]. Factors of a hundred or so have been estimated to arise from this mechanism.

In an effort to study systematically the chemical mechanism, while minimizing the magnitude of electromagnetic enhancement, we have conducted a series of

Springer Series in Surface Sciences Vol. 14
Solvay Conference on Surface Science
Editor: F.W. de Wette

experiments on the silver-pyridine system using flat, smooth single crystal surfaces, on which the excitation of surface plasmons by light is forbidden. We have created surfaces with well-characterized atomic scale surface roughness and investigated the chemisorption, Raman scattered intensity and, most recently, the electronic absorption spectra of adsorbed pyridine. We find enhanced cross sections as much as a factor of fifty higher than the free molecule for pyridine chemisorbed on isolated silver adatoms but find no correlation between the intensity of the charge transfer bands observed and the enhancement factor in contrast to the expectations of the model described above.

2. Experimental

Experiments were conducted in standard ultrahigh vacuum systems, equipped with the usual surface diagnostics. The identity of stepped and kinked surfaces was confirmed by low-energy electron diffraction. Sample cleanliness and the coverage of silver on rhodium were determined by Auger electron spectroscopy. The electronic spectra were measured using a Leybold-Hereaus ELS-22 spectrometer operating at 40 meV resolution, using 8-10 eV primary electrons and specular collection. The Raman spectra were taken using 0.5-1 W argon ion radiation (usually 5145 Å) with the scattered light analyzed by either a Spex Triplemate or an ISA HR-320, each of which is equipped with an EG&G PARC 1420 intensified diode array.

3. Results and Discussion

The first studies established that there is no chemisorption of pyridine nor enhancement on the low-index faces of silver [10]. Following the hypothesis that sites of lower coordination number than those found on the low index planes might offer the possibility of chemisorption, we prepared the Ag (987), Ag (521) and Ag (540) surfaces [11]. The first two surfaces each had the same local coordination of seven atoms but the (987) surface had wider terraces. In neither case was chemisorption or enhancement observed. On the (540) surface we observed clear evidence of chemisorption but no enhancement. Taking this idea to its limit, we prepared essentially isolated silver adatoms by evaporating silver onto a Rh (100) substrate[12]. For this system, both strong chemisorption and an enhancement factor of between 15 and 65 was observed. This was the first unambiguous evidence for a purely chemical enhancement.

We now had two clear cases with which to test the importance of charge transfer excitations in the enhancement mechanism; chemisorption with enhancement and chemisorption without enhancement. The electronic absorption spectra of these systems were taken using high resolution electron energy loss spectroscopy with surprising results. For the pyridine/silver/rhodium system in which strong enhancement was observed, the charge transfer band was weak and, in fact, indistinguishable from that observed when pyridine is adsorbed on bare rhodium. On the other hand, for pyridine adsorbed on Ag (540) in which no enhancement was observed, an extremely intense charge transfer excitation was observed centered at 1.8 eV. This is exactly the opposite behavior expected if the charge-transfer/resonance Raman mechanism is operating and further experiments need to be conducted to understand fully the nature of the chemical mechanism. One possibility is that the charge transfer resonance observed for this system is simply too broad to contribute much in the way of enhancement (FWHM = ~1.8 eV) since the enhancement factor scales roughly as the square of the ratio of the laser frequency to the width of the intermediate electronic state [13]. Other systems are now being studied to explore the generality of this tentative conclusion.

Acknowledgment

Grateful acknowledgment is made to the Donors of The Petroleum Research Fund for partial support of this work as well as to the Robert A. Welch Foundation.

References

1. A. Campion: Comments Solid State Phys. **11**, 107 (1984).
2. H. Metiu: Progr. Surf. Sci. **17**, 153 (1984).
3. A. Otto: In Light Scattering in Solids, Vol. IV, ed. by M. Cardona and G. Guntherodt (Springer, Berlin, 1984) p. 289.
4. Surface Enhanced Raman Scattering, ed. by R.K. Chang and T.E. Furtak (Plenum, New York, 1982).
5. H. Ueba: In Surface Enhanced Raman Scattering, ed. by R.K. Chang and T.E. Furtak (Plenum, New York, 1982) p. 173.
6. B.N.J. Persson: Chem. Phys. Lett. **82**, 561 (1981).
7. K. Arya, R. Zehyer: Phys. Rev. B**24**, 1852 (1981).
8. F.J. Adrian: J.Chem. Phys. **77**, 5302 (1982).
9. J.E. Demuth, P.N. Sanda: Phys. Rev. Lett. **47**, 57 (1981).
10. D.R. Mullins, A. Campion: Chem. Phys. Lett. **110**, 565 (1984).
11. A. Campion, D.R. Mullins: Surf. Sci. **158**, 263 (1985).
12. X. Jiang, A. Campion: Chem. Phys. Lett. **140**, 95 (1987).
13. S.S. Jha, J.C. Tsang: Phys. Rev. B**28**, 4216 (1983).

Discussion

M. CARDONA: Wouldn't you expect, in the case of the steps at least to see the electromagnetic enhancement (lightning rod effect)?

M. BOUDART: Can you speculate about the discrepancy between the positive SERS reported by NRL and IBM on the (521) and (987) planes of Ag and your negative results on these same planes?

Magneto-Optic Kerr Effect Studies of Two-Dimensional Magnetism

J. Araya-Pochet, C.A. Ballentine, T.-Y. Hsieh, and J.L. Erskine

Department of Physics, The University of Texas, Austin, TX 78712, USA

The Magneto-optic Kerr effect (MOKE) is used to investigate novel magnetic properties of epitaxial Fe films grown on Ag(100). Variable parameters of the study include temperature, film thickness, strength of the applied magnetic field (parallel to the surface) and photon energy. Two layer films (n=2) exhibit a linear dependence of magnetic moment and coersive force with temperature, and an (extrapolated) Curie temperature of 670°C, approximately 35% lower than the Curie temperature of bulk Fe (1043°C). Kerr effect signals scale linearly with film thickness for $3 \leq n \leq 10$ atomic layers when the film is magnetized parallel to the surface, but for n<3, the Kerr effect signal departs significantly from linear behavior, and appears to vanish for n=1.

1. Introduction

Magnetic properties of ultra-thin magnetic films are expected to depart significantly from the properties of bulk magnetic materials. As the thickness of a magnetic film approaches one monolayer, the magnetic coupling between atoms is forced to assume a genuine two-dimensional character. Two-dimensional structures have lower coordination (number of nearest neighbor atoms) than three-dimensional structures, and since the two-dimensional epitaxial layers are stabilized by a substrate, the substrate can introduce additional important effects. For example, an epitaxial film is forced to adopt the lattice constant of the substrate. These constraints can produce dramatic effects on the magnetic properties of ultra-thin epitaxial films.

Several investigators have begun to explore novel magnetic properties of ultra-thin epitaxial layers based on first principles calculations. FU, et. al. [1] have predicted enhanced surface atom moments and layer dependent interplanar coupling (ferromagnetic in some cases, antiferromagnetic in others). Surface spin anisotropy characterized by a preferred spin direction perpendicular to the surface for a single layer film (n=1), and parallel to the surface for thicker films has been predicted by GAY and RICHTER [2] based on ab-initio calculations that include spin-orbit effects. Magnetic phase transformations of two-dimensional systems are also predicted to be

Springer Series in Surface Sciences Vol. 14
Solvay Conference on Surface Science
Editor: F.W. de Wette

qualitatively different from corresponding transitions in bulk (3-dimensional) materials. Curie temperatures, the temperature dependence of magnetization, and critical exponents characterizing the transformation are expected to be different from corresponding bulk material parameters. Recent experimental studies of ultra-thin magnetic films have been reviewed by GRADMANN, et. al. [3].

2. Experiments

We are exploring magnetic phenomena associated with ultra-thin magnetic films using the magneto-optic Kerr effect (MOKE). [4] Spin-orbit coupling and a net spin-polarization produces an interaction between electron spins and the spatial component of electron wave functions in a magnetic material which is manifested by a magnetization (M) dependence of the properties of reflected polarized light (MOKE). The Kerr effect is proportional to M of the sample, and is also subject to geometrical factors related to the direction of M in relation to the polarization vector of light. The sensitivity of the technique is sufficient to probe monolayer thick films.

Figure 1 displays MOKE derived hysteresis curves for a 2.5 layer thick epitaxial Fe film grown on a Ag(100) substrate by molecular beam epitaxy (MBE). Film thickness was determined by monitoring Fe and Ag Auger line intensities during film growth, and by calibration procedures based on quartz microbalances located near the source and near the Ag substrate. The Ag(100) substrate was maintained at room temperature during the epitaxy. Low-energy electron diffraction (LEED) observations before and after the film deposition assured a well-ordered template and served to check the quality of the epitaxial film. The height of the hysteresis loop is proportional

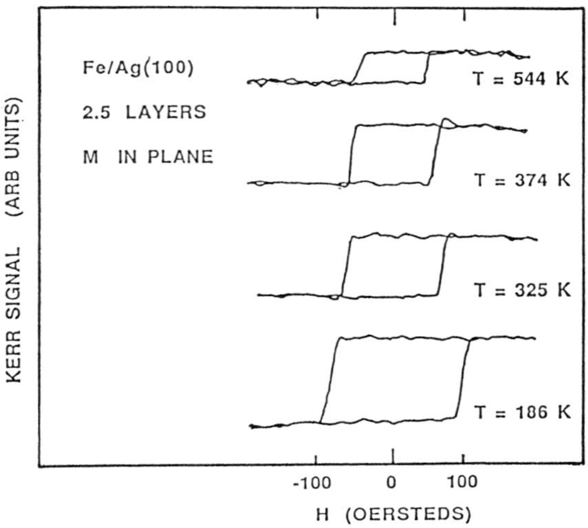

Figure 1 Hysteresis loop determined from MOKE

to the magnetization of the sample, and the width of the loop characterizes the coersive force of the film.

Figure 2 displays the temperature dependence of the magnetization and of the coersive force as a function of temperature. The temperature range is limited by the capability of our liquid nitrogen cooled sample holder (105°K) and the highest temperature that does not produce significant interdiffusion of the film into the substrate (500°K). The apparent linear temperature dependence is one indication of the two-dimensional nature of the magnetic system. Note that the extrapolated Curie temperature is approximately 600°K, far below the Curie temperature for bulk Fe (1043°K).

Figure 3 displays the layer dependence of the magnetization based on the Kerr effect signals. For film thickness much smaller than the wavelength of the light used (λ=6328) (assuming equal and uniform magnetism in each layer versus thickness), the Kerr signals should be proportional to thickness for two reasons: 1) no interference enhancements are expected until the optical path length approaches a half wavelength in the film, (~500Å) and, 2) the attenuation length (also ~500Å) is very large compared to the film thickness. An explanation of the departure of Kerr effect data from a straight line lies in the surface spin anisotropy predicted by GAY and RICHTER, [2] and possibly in the layer dependencies of magnetic moments predicted by FU, et. al. [1].

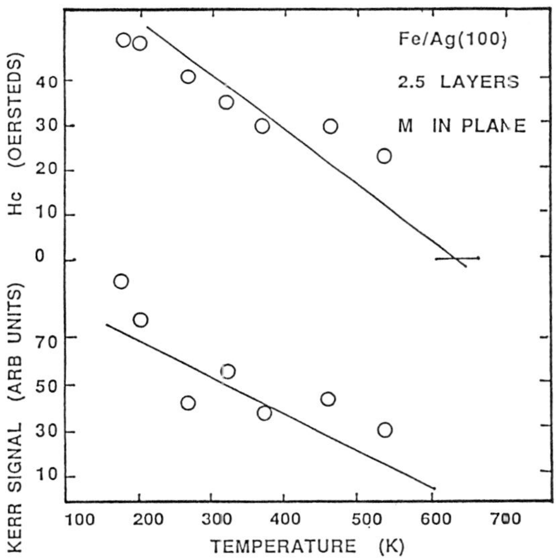

Figure 2 Temperature dependence of the Kerr effect signal and the coersive force of a 2.5 layer Fe film

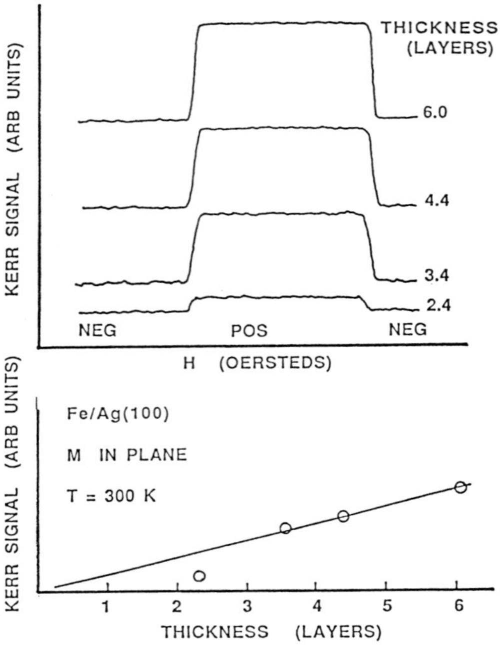

Figure 3 Thickness dependence of Kerr signal

3. Discussion

The linear temperature dependence of M and the reduced Curie temperature apparent in Fig. 2 represent interesting departures from bulk magnetic properties, but the behavior is not unexpected. Similar effects have been observed and reported [3] in other thin film systems that simulate two-dimensionality. The departure of Kerr effect signals from linear thickness dependence for $n<3$ monolayers is more novel. The linear dependence for $n\geq3$ can be explained by assuming that all epitaxial layers for $n\geq3$ are magnetized in the plane and contribute the same magnetic moment per atom to the total magnetization. For $n=2.5$, apparently either the magnetic moment per atom in one of the layers departs significantly from the value for $n\geq3$, (i.e., layer dependent magnetic moments as predicted by FU, et. al.), or surface spin anisotropy causes one of the layers to polarize perpendicular to the direction of macroscopic magnetization probed by the longitudinal Kerr effect configuration used in this experiment. The second possibility, based on predictions of GAY and RICHTER, is most likely the case because layer dependent changes in magnetic moments are typically small, of the order of 10% or less.

4. Conclusions

The magneto-optic Kerr effect offers excellent opportunities for probing novel magnetic properties of ultra-thin epitaxial magnetic films. The technique offers sufficient sensitivity to accurately study monolayer films. Adding the capability to switch between polar (M⊥ surface) and longitudinal ($M_{//}$ surface) configurations will permit analysis of spin anisotropy effects as functions of applied field, temperature, and thickness. Since the probe depth of MOKE is defined by optical penetration rather than electron escape depth, the macroscopic magnetic properties of films can be studied as film thickness is varied. Thin film sandwiches and superlattice structures can also be studied. We have also demonstrated that variable wavelength MOKE spectroscopy can be carried out using an arc lamp and monochromator at the same sensitivity as we have achieved with a laser source. From our experiments, it is clear that the Kerr effect provides a relatively simple method for accurate temperature dependent studies of magnetic properties of ultra-thin films, and of two-dimensional magnetic phase transformations. Systems such as Gd on W(110) which do not exhibit interdiffusion at the Curie point will provide excellent opportunities for measuring critical exponents of two-dimensional systems and testing phase transition models in magnetic systems.

Acknowledgment: This work was supported by JSEP/AFOSR-F49620-86-C-0045 and by NSF-DMR-86-03304.

References

1. C.L. Fu, A.J. Freeman and T. Oguchi: Phys. Rev. Lett. 54, 2700 (1985).
2. J.G. Gay and R. Richter: Phys. Rev. Lett. 56, 2728 (1986).
3. U. Gradmann, Appl. Phys. 3, 161 (1974); U. Gradmann, R. Bergholz, and E. Bergter, Thin Solid Films 126, 107 (1985).
4. A.V. Sokolov: Optical Properties of Metals, (American Elsevier, New York 1967).

Discussion

H. IBACH: A mean field model for a semi-infinite material predicts a critical exponent of the surface magnetism to be 1. The same model applied to a thin magnetic film yields 1/3 as for the bulk. Is this difference an artifact of the mean field model or is one permitted to conclude that the critical exponent for a thin film is fundamentally different from the critical exponent for a surface?

J.L. ERSKINE : I believe that Binder and Hohenberg have shown that mean field treatments are inconsistent with more general (and more accurate?) scaling theories. Predictions based on series expansions and Monte Carlo calculations are consistent with scaling theory, but are also model dependent, that is, the results depend on parameters in the model such as the surface or bulk exchange constants. Therefore, it appears at present that one cannot support the notion of a universal critical exponent for a two-dimensional magnetic film.

H.C. SIEGMANN: The Kerr-effect has not totally been neglected. It forms the basis of the magneto-optic recording industry. The Kerr effect is not proportional to the magnetization. It is proportional to the intensity times magnetization. If one includes ellipticity, the relationship can be very complicated. In fact, nonmagnetic layers evaporated on top of the magnetic layer are used to enhance Kerr rotation.

L. KLEINMAN: Rau, using electron pickup experiments, has been finding for many years now that the surface of paramagnetic vanadium is ferromagnetic. Freeman's calculations say that it is not. Why has no experimentalist checked Rau's results with some other technique?

H. SIEGMANN: At ETH-Zürich, Dr. Felix Meier and my students have been investigating whether Vd is magnetic when evaporated epitaxially on various substrates, including I think Ag, which was used by Rau. No magnetism was found down to 20K with spin polarized photoemission.

Surface vs. Bulk Hysteresis Loops

H.C. Siegmann

IBM Research, Almaden Research Center, 650 Harry Road, San Jose, CA 95120, USA, and Labor für Festkörperphysik, E.T.H.-Hönggerberg, CH-8093 Zürich, Switzerland

Very recently, it has been established that a large number of common magnetic materials exhibit hysteresis loops that depend on the probing depth of the magnetic measurement done at the surface. This indicates that important magnetic properties like the magnetic anisotropy K and the exchange energy JS_1S_2 where S_1 and S_2 are the spins of 2 neighboring atoms along a path perpendicular to the surface, may vary dramatically within a few lattice constants from the surface. The argument can be understood as follows: With bulk values for K and the exchange stiffness $A = JS_1S_2/a$ where a is the distance of the atoms, the magnetization M will change in direction over distances of the order of $\sqrt{A/K}$ which is the width of a magnetic domain wall. It typically amounts to 10-100 nm in materials like Fe or magnetite Fe_3O_4. Hence, if surface hysteresis loops are taken with techniques that differ in probing depth by 10 nm or less, we expect on the basis of the bulk constants that the hysteresis loops must be identical. However, M. Aeschlimann et al.[1] showed that a hysteresis loop taken on Fe_3O_4 using the magneto-optic Kerr-effect with a probing depth of ~20nm shows a remanent magnetization as expected, whereas the magnetization curve taken with spin polarized photoelectrons with a probing depth of ~ 1nm exhibits no remanence at all. Even more dramatic differences between Kerr-effect and spin polarized photoemission occur with the magneto-optic recording materials FeTb and GdCo.[2,3]. Furthermore, Allenspach and coworkers[4] reported that the shape of the hysteresis loops taken on Fe(100) by measuring the spin polarization of the secondary electrons depends on the energy E_S of the secondaries. This must come about because the probing depth depends on E_S. The changes of the probing depth are however small, of the order of 1 or 2 lattice spacings only. If hysteresis loops can change over such short distances, exchange and anisotropy must indeed be very different at the surface compared to the bulk. We have developed a simple model[5] with which one can

Springer Series in Surface Sciences Vol. 14
Solvay Conference on Surface Science
Editor: F.W. de Wette

interpret surface induced changes in hysteresis loops. One can in fact derive the direction and magnitude of the surface specific magnetic anisotropy K_s, and obtain estimates for the surface to bulk exchange stiffness $A_{1,2}$. It turns out that K_s and $A_{1,2}$ may be different from the bulk values by orders of magnitude. This has many implications, particularly on the interpretation of surface sensitive experiments like spin polarized photoemission which do not relate in a straightforward way to the temperature dependence of bulk magnetism as previously assumed[6].

References

1. M. Aeschlimann, F. Meier, M. Stampanoni and A. Valerlaus, private communication.
2. G. L. Bona, F. Meier, H. C. Siegmann and R. J. Gambino, *Appl. Phys. Lett.* **52,** 166 (1988).
3. M. Aeschlimann, G. L. Bona, F. Meier, M. Stampanoni, A. Valerlaus, H. C. Siegmann, E. E. Marinero and H. Notarys, to appear in *IEEE Trans. Mag.*, Feb. 1988.
4. R. Allenspach, M. Taborelli, M. Landolt and H. C. Siegmann, *Phys. Rev. Lett.* **56,** 953 (1986).
5. H. C. Siegmann, P. S. Bagus and E. Kay, *Z. Physik B* **62,** 485 (1988).
6. H. Hasegawa, *J. Phys. F: Met Phys.* **17,** 165 (1987).

Discussion

H. IBACH: Oepen and Kirschner have recently been able to measure the magnetic polarization in a domain wall at the surface and found it to be of Néel-type. Would that have a similar effect on the hysteresis as you reported?

H.C. SIEGMANN: Koike and collaborators have first confirmed an older theoretical prediction that there are no Bloch-walls at the surface, and that the Bloch-domain wall always transforms into a Néel wall. Oepen and Kirschner have confirmed this work with their very impressive new instrument. There will be effects on surface vs. bulk hysteresis loops noticeable mainly in thin films magnetized in the hard direction. Also, the structure of domain walls is obviously not as simple as anticipated from Lorentz-microscopy, averaging these effects.

M. BOUDART: Would it be a good way to study surface magnetism by getting rid of the bulk as done in cluster work? A few years ago, we studied by Mössbauer effect spectroscopy the surface magnetic anisotropy of Fe clusters on MgO support. Our chemical conclusions concerning these clusters were confirmed later by work in Somorjai's group on *large* single crystals of iron.

H.C. SIEGMANN: Cluster work will be the most important test of our understanding of magnetism. At this time, I do not see how one could do conclusive experiments on clusters with a defined number of atoms, and with well–defined surface conditions.

Near Edge X-Ray Absorption Fine Structure (NEXAFS) of Molecules Adsorbed on Surfaces

A.M. Bradshaw, J. Somers, and T. Lindner

Fritz-Haber-Institut der Max-Planck-Gesellschaft,
Faradayweg 4–6, D-1000 Berlin 33, Germany

In NEXAFS studies of adsorbed molecular species the number of allowed resonances, their degeneracy and polarisation dependence can be predicted by reference to the m.o. scheme and by consideration of the effective symmetry of the molecule and its immediate environment.

1. INTRODUCTION

The NEXAFS of molecules and molecular fragments adsorbed on surfaces shows characteristic resonances similar to those observed in the corresponding gas phase species [1] (Fig. 1). In a molecular orbital picture [2] they originate from the excitation of a core electron into unoccupied levels both above and below the ionisation threshold. In the former case we are concerned with transitions into anti-bonding m.o.'s, so-called continuum resonances, which are generally very broad. An alternative description of these features in terms of intramolecular scattering resonances has often been discussed in the literature [1,3]: the photoelectron is resonantly trapped by a potential barrier through which it eventually tunnels before being emitted into the continuum. The energy and width of the scattering resonance depends on the details, or shape, of the molecular potential - hence the frequently-used expression "shape" resonance. The m.o. and shape resonance pictures are equivalent, but different descriptions of the same physical phenomenon. The unoccupied levels *below* the ionisation threshold

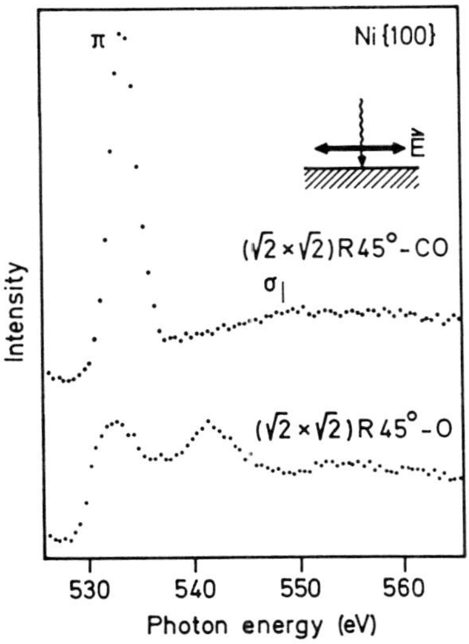

Figure 1: O 1s NEXAFS for CO adsorbed on a Ni{100} surface in the $(\sqrt{2}\times\sqrt{2})$ R45° structure. The spectrum is dominated by the sharp π resonance. Due to the experimental geometry (see Fig. 2) little or no intensity is present in the σ resonance. For comparison the NEXAFS from adsorbed oxygen in the $(\sqrt{2}\times\sqrt{2})$R45° structure under the same conditions is also shown. The two spectra have been normalised to the same edge jump height. Note the strong "substrate resonances" in the oxygen spectrum (see Sec. 4).

Springer Series in Surface Sciences Vol. 14
Solvay Conference on Surface Science
Editor: F.W. de Wette

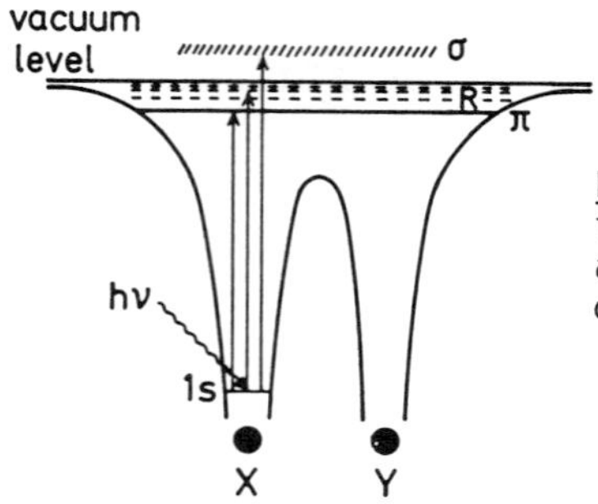

Figure 2: Schematic representation of the transitions observable at the absorption edge of a core level in a diatomic molecule.

Fi
th
ab
di

also consist of (usually anti-bonding) m.o.'s and/or molecular Rydberg states; in this case one speaks of bound resonances. These are sharp with linewidths below 1 eV and, even in the case of adsorption of the molecule on a surface, only slightly broader. Figure 2 shows a schematic representation of these two types of transition in a diatomic molecule. Excitation with photons (or with electrons under conditions of zero momentum transfer) requires that dipole selection rules are obeyed. If the molecule has one or more symmetry elements, i.e. belongs to a point group, the transitions will be polarised. Thus for an isolated diatomic XY (Fig. 2) the transition 1s → σ is polarised along the X-Y axis and, the transition 1s → π is polarised perpendicular to that axis. It is then immediately apparent that NEXAFS offers the same possibility of determining molecular orientation as in UV-VIS spectroscopy of molecular crystals. The intensity of the 1s → σ transition is given by [4]: $I(\sigma) \sim [\mathbf{E}\cdot\mathbf{M}_z]^2 = \cos^2\alpha$, where $\mathbf{M}_z$ is the matrix element and α the angle between the **E** vector and the X-Y axis. Similarly, $I(\pi) \sim [\mathbf{E}\cdot\mathbf{M}_{x,y}]^2 = \sin^2\alpha$. In Fig. 1, where the CO molecule has C_{4v} symmetry, the **E** vector is parallel to the surface giving maximum intensity in the π resonance.

NEXAFS of molecules on surfaces has enjoyed considerable attention in recent years: nearly a hundred adsorption systems have been investigated in varying degrees of detail since 1982 [5]. This has largely been due to the pioneering work of STÖHR and his colleagues (see [6,7] and references therein). The applications of the technique are threefold. It provides firstly a chemical "fingerprint" of the species on the surface (although, to be fair, one should add that it pales in comparison with surface vibrational spectroscopy!). Following from this an interesting, but controversial [3], empirical correlation has been derived linking bond length with the energy of σ resonances in various molecules. It has even been claimed that this represents a method of determining accurately bond lengths in adsorbed species on surfaces [8]. Secondly, by probing the unoccupied states of the adsorbate/substrate system NEXAFS necessarily provides information on the chemisorption bond. In fact, there are probably more details here than we have hitherto been able to extract. The third application, already briefly mentioned above, lies in the determination of molecular orientation. Although the experimental problem here reduces to one of varying the angle, θ, of the E vector of the incident radiation relative to the surface normal, it is first necessary to come to some understanding about the nature of the resonances involved and the symmetry properties of the surface molecule.

2. CURRENT NEXAFS PHILOSOPHY

The very rapid development of NEXAFS studies of adsorbed species has resulted in insufficient attention being paid to the molecular spectroscopy aspects of the technique. Current NEXAFS philosophy may be briefly summarised as follows [7]: The near edge structure can be described in terms of an m.o. picture in which there is a one-to-one correspondence between the observed σ and π resonances and the unoccupied molecular orbitals of the system. In the case of an adsorbed diatomic the polarisation dependence is given simply by $\sin^2\alpha$ and $\cos^2\alpha$ functions suitably modified to take into account the incomplete polarisation of the incident radiation and, if appropriate, the effect of azimuthal averaging. The surface merely serves to orient the molecule but does not lower its symmetry. The same approach can be applied to ring molecules except that the σ and π resonances are polarised parallel and perpendicular to the ring plane. Polyatomic molecules can then be "assembled" by linking diatomics or rings. Associated with each of these moieties are σ and π resonances which are symmetric and anti-symmetric, respectively, to a "local" symmetry plane. The polarisation dependences follow accordingly. This philosophy does admit of the possibility that interactions between adjacent bonds may lead to the m.o.'s becoming more extended but they can still be labelled σ or π with respect to the local symmetry plane. In special cases these interactions give rise to non-degenerate m.o.'s whose polarisation dependences are then no longer characteristic of the direction of the internuclear axes but rather of the "spatial orientation" of the delocalised orbitals. However, the sum of the corresponding resonance intensities are thought to follow the angular dependence characteristic of the internuclear axes. A similar kind of sum rule has also been used to extend the empirical correlation between σ resonance energy and bond length to polyatomic molecules.

It is our contention that this somewhat crude picture, whilst perhaps adequate in certain well-defined situations, cannot describe accurately essential features of the NEXAFS of polyatomic molecules such as the number of allowed transitions, their degeneracy and polarisation dependence.

3. THE "SPECTROSCOPY" APPROACH

In contrast to the valence bond-like approach above, molecular orbital theory starts from the premise that all orbitals extend spatially, or are delocalised, over the whole molecule. There is the possibility that for some orbitals the wave function may only have significant amplitude in certain parts of the molecule. Such "localised" bonding is, however, a special case fully accounted for in m.o. theory and does not invalidate the main thesis. The second point to emphasise is that these orbitals have symmetry properties dictated by the molecular symmetry. In the language of group theory they are said to belong to the irreducible representations of the molecular point group. For an adsorbed molecule the point group symmetry may be lower due to interaction with the surface [9]. If this interaction is weak, i. e. such as in physisorption or in cases where the electronic energy levels under investigation are essentially unaffected by bonding to the substrate, the effective symmetry may remain that of the free molecule. What ramifications does this have for NEXAFS on surfaces? Firstly we note that both the core levels and the anti-bonding orbitals are part of this description and possess symmetry governed by the effective point group. Excitation of core electrons into unfilled levels is then subject to dipole selection rules. Whilst we can still assume a correspondence between the NEXAFS resonances and the anti-bonding m.o.'s, it is not necessarily "one-to-one" because some transitions may be forbidden for symmetry reasons. A case in point is the carbonate species on Ag{110} which assumes a lying-down geometry [10]. At the C edge the σ-type resonance attributed to a transition into the $5a_1'$ orbital is absent. The effective symmetry remains D_{3h} with the result that the C 1s excitation $2a_1' \rightarrow 5e'$ is allowed (for E_x, E_y) but $2a_1' \rightarrow 5a_1'$ is forbidden. (The Cartesian co-ordinates are normally chosen such that the z axis lies in the surface normal.) At the oxygen edge both resonances are allowed and actually observed in the experimental spectrum. The analysis also reveals that the degeneracy of the 5e' resonance remains at the C edge but is lifted at the O edge, although the splitting cannot be experimentally resolved. The data in this example show relatively simple near edge structures which are easily interpretable. In other cases Rydberg transitions, many-body features and so-called substrate resonances complicate the picture. We return to this topic below.

The adsorbed carbonate species provides an obvious example of the molecular orbital treatment and of the use of selection rules. The result is nevertheless still somewhat surprising. Although the molecule is relatively strongly bound to the surface the effective symmetry, at least for the σ-type resonances, remains D_{3h}. There is, however, some indication that this is not the case for the π resonance: The E vector components parallel to the surface are to some extent effective in exciting this transition. There is thus evidence for *symmetry lowering* to C_{3v}. To illustrate this effect and, at the same time, to introduce the problem of orientation determination with a single example, we adsorb our diatomic XY in a lying-down configuration on a structureless (jellium-like) surface. As usual the Cartesian co-ordinates are chosen such that the z axis is perpendicular to the surface and the x axis is along the X-Y bond. If the symmetry remains $C_{\infty v}$, i.e. if the influence of the surface is not felt at all, the 1s → 2π transition (π resonance) can be excited with both E_z and E_y but the 1s → 6σ (σ resonance) only with E_x. If the effective symmetry is C_s, the 2π level splits into two components, 2π(a') and 2π(a"), the first of which can be excited with E_z *and* E_x and the second with E_y. Similarly, the σ resonance can be excited with both E_x and E_z. The extent to which the E_z component excites the transition into 6σ *depends on the strength of the interaction with the surface*. In earlier publications we have discussed such effects in photoemission and in vibrational spectroscopy of adsorbed species [9]. The bonding of the CN molecule to a metal surface provides an illustrative example. Vibrational spectroscopy and NEXAFS show that CN is adsorbed on Pd{111} such that the C-N axis is parallel, or very nearly parallel, to the surface. The 2π resonance is split and dominated by the a" component at all angles θ [11]. However, the a' component does not disappear, even at θ = 90°, thus showing that it can be excited by both E_z and E_x components. On the other hand, the σ resonance appears to be excited only by the E_x component indicating that here the effective point group remains $C_{\infty v}$. The equations for diatomics and rings given in Ref. [7], described as "the theoretical expressions needed to determine the accurate orientation of chemisorbed molecules" are formally only correct for isolated, oriented species. Their application to σ-type resonances in molecules on surfaces is probably admissible; for π-type resonances their use could lead to incorrect results.

To sum up the results of the last two paragraphs we note that the polarisation dependence is determined by the symmetry of the orbitals and thus by the effective point group of the molecule. There is no recipe for deciding *a priori* whether the symmetry of the free molecule or that of the surface species (molecule + substrate) pertains. There are, however, several examples where symmetry lowering is unimportant but where current NEXAFS philosophy still gives the wrong answer. Let us consider the example of the formate species on Cu(110) where the polarisation dependence of the π-type resonances at the O edge has been used to show that the molecular plane is oriented perpendicular to the surface and parallel to the <110> azimuth [12]. This fact, together with evidence from vibrational spectroscopy, shows that the formate has C_{2v} symmetry with two equivalent C-O bonds. The NEXAFS analysis then went on to determine the O-C-O bond angle using the polarisation dependence of the (single) σ resonance in the <110> azimuth.

This is based on the assumption that the function actually measured is a superposition of the polarisation dependences of two non-interacting C-O moieties. A glance at the m.o. orbital energy diagram for the formate shows that in fact one π and *two, possibly three* σ resonances are expected. Whereas at the oxygen edge only one resonance is observed in the "σ region", at the carbon edge two can be resolved (splitting ~2.5 eV) [13]. Here, the transition $2a_1 \rightarrow 7a_1$ can be excited by E_z and $2a_1 \rightarrow 5b_1$ by E_x, which is observed in practice. At the oxygen edge both resonances are excited by both E_z and E_x. There is, in short, no possibility of determining the angle of the C-O bonds relative to the surface. Although one might describe this particular example as having arisen during the early days of NEXAFS, the same type of analysis has recently been applied to determine the orientation of Langmuir-Blodgett films [14]. As might be expected, the resonances here are in fact polarised perpendicular and parallel to the chain direction: photoemission measurements, for example, demonstrate dispersion of the occupied m.o.'s in these quasi-one-dimensional systems [15].

4. SOME FURTHER REMARKS

In addition to the problems of assignment and effective symmetry, the analysis of NEXAFS data contains further pitfalls. Normalisation of the angle-dependent resonance intensites is, for example, very important. The *edge jump* height is usually taken because it is a measure of the strength of excitation into the continuum, which is independent of angle and proportional to the number of absorbing atoms. Normalisation to photoemission features has been used but there is doubt as to the correctness of this procedure. In principle, the safest approach would be to measure the polarisation dependence of resonance intensity *ratios* which obviates the necessity of normalisation. Both approaches assume, however, that the pure edge jump contribution and the resonance intensities can be separated i.e. that we know the form of the *background* under the resonances. The tendency has been to simply take a flat or smooth background above the edge jump. This can be a dangerous procedure because of the presence of so-called substrate resonances which derive from multiple scattering of the photoelectron at substrate atoms (essentially the first SEXAFS features). Figure 1 shows that these features can be very strong in the case of an atomic overlayer. Such structure can also be considered as due to anti-bonding-like levels associated with the bond to the substrate. The comparison with adsorbed CO in Fig. 1 is perhaps unfair since the oxygen chemisorption bond is much stronger (and the O-Ni distance much shorter), but it does show how important such effects could be for the background under molecular resonances. There are several examples where substrate resonances have clearly been shown to play a role, e.g. [11,16]. In one case, the methoxy species on Cu{100} [17], one can obtain quite different results for the inclination of the O-C bond relative to the surface depending whether the data is taken at the oxygen or at the carbon edge!

Two more problems deserve brief attention. The first concerns the correct assignment of Rydberg-like transitions. Current NEXAFS philosophy dismisses the possibility that these will be observed due to a "quenching" effect of the surface. There is, however, some evidence that they will survive for core levels which, due to the adsorption geometry, are not located immediately at the surface [13]. There is also the possibility that Rydberg transitions will mix with excitations into anti-bonding m.o.'s if they are of comparable energy. Assignment difficulties thus still exist, even for free molecules [18]. It is important to remember, however, that Rydberg states also belong to the irreducible representations of the point group and that the corresponding transitions obey dipole selection rules. Enormous intensity differences can arise between allowed Rydberg transitions; without calculation, no *a priori* prediction of relative strengths is possible. Another pitfall associated with the assignment of NEXAFS features is the presence of many-body effects [2]. By this we mean the occurrence of multi-electron excitations which can lead to satellites or even, in the case of strong correlation, to a complete disappearance of the single particle excitation. It has, for example, been recently proposed [19] that the $C1s^{-1}$, b_{2g}^{1} configuration in benzene is very nearly degenerate with the $C1s^{-1}$, e_{1g}^{-1}, e_{2u}^{2} configuration and that they mix strongly, leading to two features above and below the ionisation threshold. Since the second feature could be accidentally degenerate with other, σ-type resonances, problems could arise in the determination of orientation via the polarisation dependence.

In conclusion we should perhaps emphasise that, despite some of the misgivings expressed in this article, NEXAFS is still a useful method of examining molecular orientation on surfaces. However, the "accurate" orientation determination to within 10° that is claimed for the technique is probably somewhat optimistic except in high symmetry situations (e.g. a standing-up diatomic). Moreover, it must be applied with care, taking cognizance of the molecular orbital scheme and of the appropriate point group.

REFERENCES

1. J. L. Dehmer and D. Dill: J. Chem. Phys. **65**, 5327 (1976)
2. F. A. Gianturco, C. Gidotti and U. Lamanna: J. Chem. Phys. **57**, 840 (1972)
3. M. N. Pincastelli, D. W. Lindle, T. A. Ferrett and D. A. Shirley: J. Chem. Phys. **86**, 2765 (1987) and references therein
4. J. S. Somers, Th. Lindner, M. Surman, A. M. Bradshaw, G. P. Williams, C. F. McConville and D . P. Woodruff: Surface Sci. **183**, 576 (1987)
5. J. Stöhr and R. Jaeger: Phys. Rev. B**26**, 4111 (1982)
6. J. Stöhr and D. A. Outka: J. Vac. Sci. Technol. A**5**, 919 (1987)
7. J. Stöhr and D. A. Oukta: Phys. Rev. B**36**, 7891 (1987)
8. J. Stöhr, F. Sette and A. L. Johnson: Phys. Rev. Lett. **53**, 1684 (1984)
9. A. M. Bradshaw: Z. Phys. Chem (NF) **112**, 33 (1978); N. V. Richardson and A. M. Bradshaw: in *Electron Spectroscopy: Theory, Techniques and Applications*, ed. by A. Baker and C . R. Brundle, Vol. 4 (Academic Press, London 1981) p. 153
10. M. Bader, B. Hillert, A. Puschmann, J. Haase and A. M. Bradshaw: Europhys. Lett, in press
11. J. Somers, M. E. Kordesch, Th. Lindner, H. Conrad, A. M. Bradshaw and G. P. Williams: Surface Sci. **188**, L693 (1987); Spectrochim. Acta **43**A, 1561 (1987)
12. A. Puschmann, J. Haase, M. D. Crapper, C. E. Riley and D. P. Woodruff: Phys. Rev. Lett. **54**, 2250 (1985)
13. J. Somers, Th. Lindner and A. M. Bradshaw, to be published
14. D. A. Outka, J. Stöhr, J. P. Rabe, J. D. Swalen and H. H. Rotermund: Phys. Rev. Lett. **59**, 1321 (1987).
15. K. Seki, N. Ueno, U. O. Karlsson, R. Engelhardt and E. E. Koch: Chem. Phys. **105**, 247 (1986)
16. D. A. Outka, J. Stöhr, W. Jark, P. Stevens, J. Solomon and R. J. Madix: Phys. Rev. B**35**, 4119 (1987)
17. Th. Lindner, J. S. Somers, A. M. Bradshaw, D. Kilcoyne, C. McConville and D. P. Woodruff, to be published
18. I. Ishii, R. McLaren, A. P. Hitchcock and M. B. Robin: J. Chem. Phys. **87**, 4344 (1987)
19. W. H. E. Schwarz, T. C. Chang, U. Seeger and K. H. Hwang: Chem. Phys. **117**, 73 (1987)

Discussion

D.R. HAMANN: I believe that the NEXAFS resonances must be analyzed in terms of a symmetry group which is the symmetry of the molecule in the presence of the localized core hole. In general, the core hole Auger lifetime is too short compared to core hole hopping times to consider the core hole to be delocalized in a molecular orbital.

I agree with what you say, but I think it has to be elaborated by one further complication, and that is: You can't use a point group of molecule in its ground state; you have to consider the point group of the molecule in the presence of the localized core hole that you excited, because the entire resonance is shifted by that interaction.

A.M. BRADSHAW: Yes, this is correct; but in order to predict the allowed transitions the ground state molecular symmetry can still be used. Group theory will take care of the problem.

D.R. HAMANN: The resonance occurs according to the symmetry of the final state potential. And the final state potential does have that core hole interaction. That also gives you a big shift, if you have several holes in the resonance position.

A. M. BRADSHAW: Yes, but that does not affect the symmetry analysis using the full molecular point group.

D.R. HAMANN: Well, in a benzene molecule, you excite, for example, a core on one hydrogen atom and now, you know, out goes most of the symmetry; maybe you have a reflection point. What I'm saying is, use group theory, taking the core hole into account, because it breaks the symmetry that you have.

A.M. BRADSHAW: These are two ways of performing the symmetry analysis. Take the benzene case which you mention. By creating a core hole on one of the C atoms the symmetry is reduced from D_6h to C_2v. The NEXAFS resonances can then be analyzed on this basis. Alternatively, one can assume that the D_6h point group still pertains, and that the core levels from molecular orbitals...

D.R. HAMANN: No they don't.

A.M. BRADSHAW: ...and then perform the symmetry analysis. The results are identical, a fact which can be easily proved. Whether the core levels actually do form m.o.'s is not important, although quantum chemical calculations do predict a splitting of C1s in benzene, implying in turn that there is a finite probability of a core hole hopping from C atom to another.

D.R. HAMANN: We can disagree; because I think that on the timescale that's relevant to these core experiments and are actually within the Auger lifetime of the core hole—well let me put it this way: if the Auger lifetime of the core hole gives an energy that's broad compared to the band width in the core hole hopping, so that core hole's effectively localized, then it has to be analyzed.

High Resolution Core Level Shift Spectroscopy from Tungsten Single Crystal Planes

K.G. Purcell[1], *J. Jupille*[2], *and D.A. King*[3]

The Donnan Laboratories, University of Liverpool, Liverpool L69 3BX, UK
[1]Present address: Department of Chemistry, University of Washington, Seattle, WA 98195, USA
[2]Present address: Laboratoire Maurice Letort, BP 104, F-56400 Villers-les-Nancy, France
[3]Department of Physical Chemistry, Lensfield Road, Cambridge, UK

An extensive study of core level binding energy shifts from six different crystal planes of tungsten has recently been completed. Small shifts in the W4f core levels have been determined in the photon energy range 50 to 90 eV. The data have been carefully analysed to determine the lineshape of the components of each spectrum, to pinpoint the contribution from the bulk, and hence to produce a description of the total lineshape of each spectrum. All of the resolved peaks produce a linear relationship between the observed shift in the W4f binding energy from the bulk value and the effective coordination number of the emitting W atom, with a scatter of data points within ± 50 meV of this line, the largest shift being 400 meV.

1. Introduction

At clean metal surfaces, the electronic core-level binding energies of surface atoms are shifted with respect to those of bulk atoms, by as much as 300 to 400 meV in some instances. For both clean and adsorbate-covered surfaces the technique of Surface Core Level Spectroscopy (SCLS) has diverse applications.[1] Several models have been used to explain the origin of these surface core level shifts.[2-5] At a microscopic level, it is the change in potential at the surface, resulting from a charge transfer between surface and bulk due to the narrowing and displacement of the surface layer density of states compared to the bulk, which rigidly shifts the core level from bulk values. The narrowing of the surface density of states is a consequence of a reduction in coordination of surface atoms. In principle, therefore, SCLS can be used to distinguish between atoms in different coordination sites on a metal surface.

We have recently completed a series of studies on a range of crystal planes of tungsten, including a detailed study of the reversible phase transition on W{100}[6], both clean and with adsorbates.[7-9] Here, we briefly review our attempts to remove ambiguities in the analysis of line spectra containing more than two incompletely resolved components; and to establish a relationship between the coordination of surface atoms at different sites (e.g. step or terrace) or on different crystal planes and the core level shift. Such a relationship is a prerequisite for more detailed studies of surface phase changes and for the use of SCLS as a means of probing stepped surfaces.

2. Linewidth and shape: W4f spectra from W{111}

The surfaces of the bcc {111} metals contain three groups of surface and sub-surface atoms whose coordination number, up to next nearest

Springer Series in Surface Sciences Vol. 14
Solvay Conference on Surface Science
Editor: F.W. de Wette

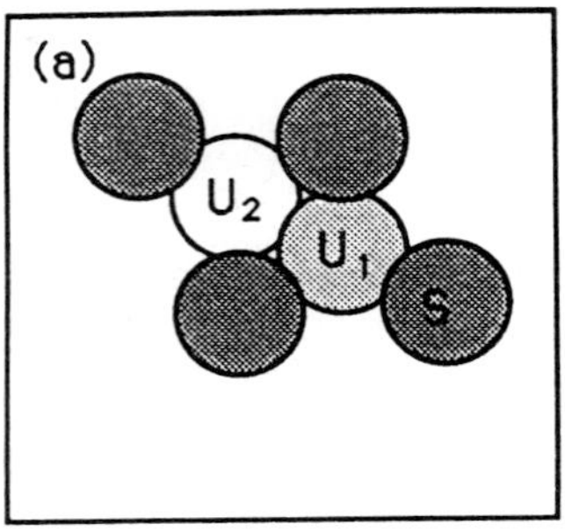

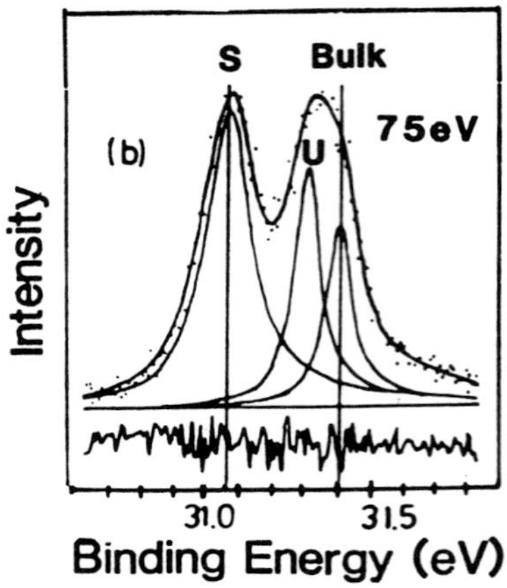

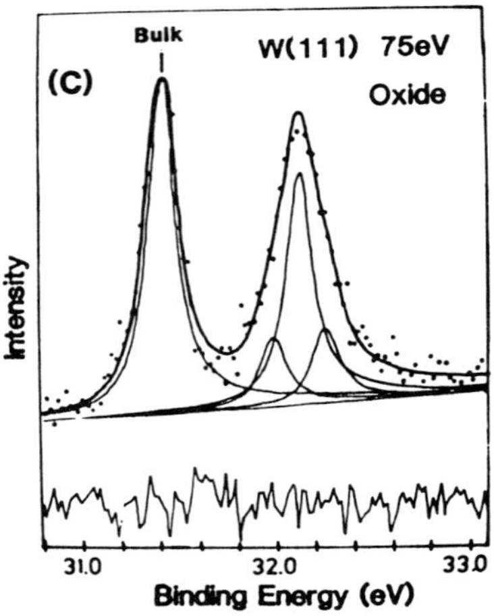

Figure 1

(a) The b.c.c. {111} surface.
(b) Photoemission spectrum at hν = 75 eV, $W4f_{7/2}$, clean W{111}.
(c) Spectrum after oxidation of W{111}.

neighbour (nnn), differs from that of the bulk, as illustrated in figure 1(a). However, in earlier work on W{111} the experimental data for the $W4f_{7/2}$ lineshape has been analysed in terms of only 2 surface component peaks, in addition to the bulk peak. Our data for the $W4f_{7/2}$ core level photoemission spectrum at a photon energy of 75 eV, are shown in figure 1 (b), indicating the fitting to three peaks.[8]

The lineshape parameters of the $4f_{7/2}$ core level were determined by fitting the isolated bulk peak in a spectrum taken from a heavily oxidised W{111} surface (figure 1 (c)). The spectrum shows the bulk peak completely separated from the surface peaks; all of the peaks from W atoms in the surface layers have been shifted by ≈ 1.3 eV relative to the clean surface peaks, due to charge transfer from the oxygen atoms. The bulk peak was well fitted by a single Doniach-Sunjic peak located at 31.38 ± 0.01 eV below E_f, with a lorentizian width (FWHM) 0.05 eV and asymmetry parameter 0.05.

3. Phonon Broadening: W4f on W{110}

On the densely packed {110} plane of tungsten, only two clearly resolved peaks contribute to the W4f photoemission spectrum (figure 2).

The coordination of second layer atoms is already sufficiently close to the bulk that the binding energy is effectively unshifted, leaving only

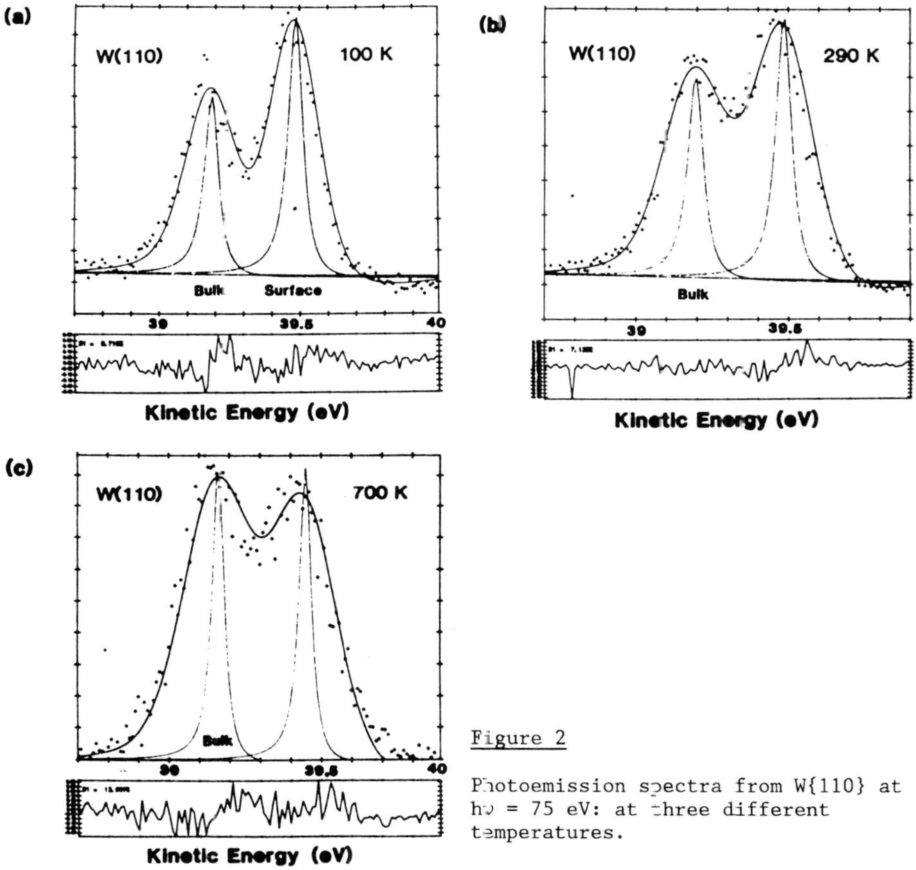

Figure 2

Photoemission spectra from W{110} at $h\nu$ = 75 eV: at three different temperatures.

surface layer and bulk contributions. The study of this surface is relatively unambiguous.

In the analysis of more complex surfaces, such as W{111}, it would be useful to establish whether or not the phonon broadening of the surface peak(s) is larger than that of bulk peaks, and is a significant parameter, as suggested, for example, by the fitting shown in figure 1 (b). We have obtained $W4f_{7/2}$ spectra from a W{110} surface at 100 K, 290 K, and 700 K to elucidate the extent of phonon broadening (figure 2). The broadening is visible as a loss in the depth of the valley between surface and bulk peaks at the higher temperatures. Fitting bulk and surface peaks to the lorentzian width and asymmetry parameter derived above, the two peaks are convoluted with a gaussian peak representing the sum in quadrature of the inhomogeneous broadening, phonon broadening, and the instrument response function. The latter was independently determined, as 0.16 ± 0.01 eV, while at 290 K. The phonon broadening contribution at this temperature was determined as 0.09 eV, compared with 0.07 eV at 100 K, in close agreement with a theoretical estimate for the bulk phonon broadening. There is no significant difference experimentally in the phonon broadening of bulk and surface peaks. We conclude that the analysis of core level lineshapes can, to a first approximation, be based on the assumption that all component peaks have the same shape and width as the bulk peak.

4. Analysis of W4f spectra from W{111}

The spectrum from W{111} was re-analysed[8] using four peaks; the bulk peak energy was obtained from the oxidised surface, and the line width and asymmetry of each of the component peaks were fixed to the values determined from the isolated bulk peak. A better overall fit to the spectrum is obtained (figure 3) and a four-peak spectrum is what is expected from a consideration of the coordination of atoms in the top three layers. The three surface peaks, labelled S, U_1 and U_2, are shifted by -0.45, -0.30 and -0.12 eV, respectively, relative to the bulk peak. These are assigned, respectively, to the surface layer, first underlayer, and second underlayer of the W{111} selvedge (as in figure 1 (a)).

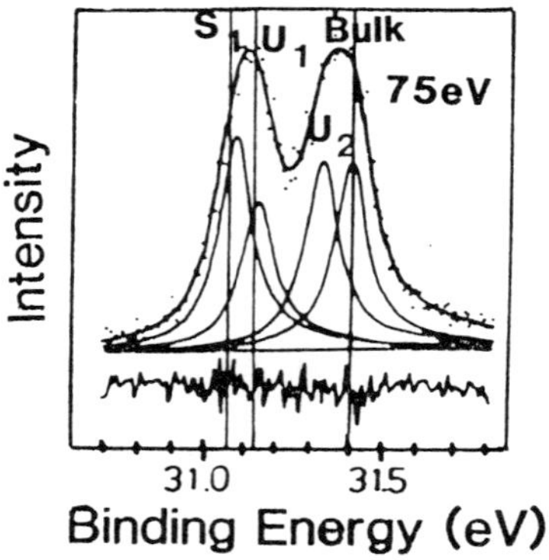

Figure 3

Fitting of W{111} spectrum to four peaks; lettering refers to figure 1 (a)

5. Dependence of Core Level Shifts on Coordination

Core-level photoemission data have been obtained from the singular W{111}, {100} and {110} crystal planes, and the step-terrace {310}, {320} and {610} planes (figure 4), and each set of data has been analysed in the manner described above for W{111}. To assist in the analysis, data were obtained at photon energies between 60 and 90 eV. From each surface, it was possible to assign peaks to selvedge atoms of a given coordination. The lowest coordination atoms are those in the surface layer of W{111}, with a core level shift of -0.43 eV, and a coordination of 4 nn and 3 nnn; the highest (non-bulk) coordination peak identified is attributed to underlayer atoms at steps, with a core level shift of -0.11 eV and a coordination of 8 nn and 5 nnn. All the data obtained from the six crystal planes studied is summarised in figure 5; here the coordination sum is obtained with a reciprocal distance dependence, i.e. $\sum_{j=1}^{3} \frac{N_j}{r_j}$ where N_j is the shell coordination number and r_j the shell spacing, normalised to the first shell spacing, where we have included 3 shells. We attach no particular importance to the linear dependence observed between core level shift and effective coordination. The implication, however, is clear: coordination is a major determining factor. In fact,

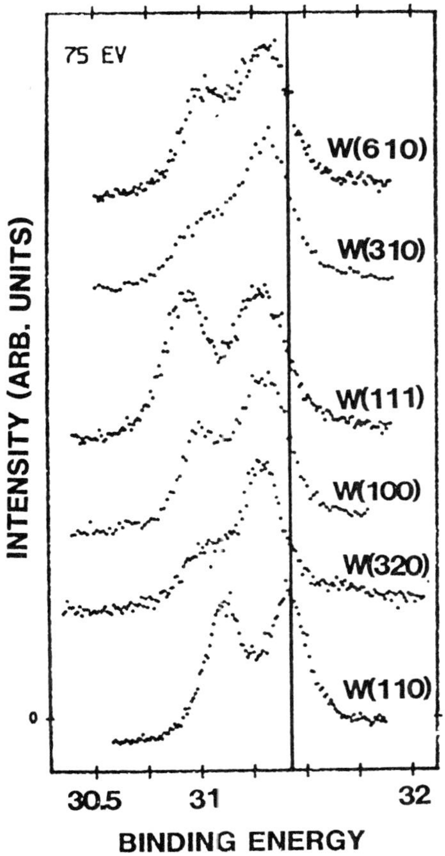

Figure 4

W4f$_{7/2}$ spectra at hν = 75 eV from 6 different crystal planes of tungsten; the vertical line indicates the position of the bulk peak.

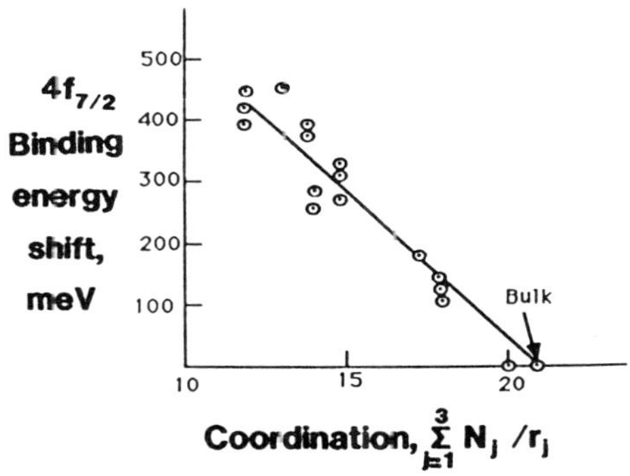

Figure 5

Dependence of binding energy shift on effective atomic coordination.

the scatter of data points is considered to be larger than the experimental error. Secondary effects, such as local reconstruction, are therefore also factors of importance, but apparently resulting in a maximum deviation of no more than ± 50 meV. This is pertinent, for example, in the investigation of core level shifts associated with the reversible reconstruction on W{100}.[6,9] On an empirical level, the relationship established here between core level shift and effective coordination is a useful basis for the analysis of core level spectra from clean, alloy and adsorbate covered surfaces.

Acknowledgement

We are grateful to the SERC for providing beamtime at the Daresbury Synchrotron Radiation Source, for a Ph.D. studentship to one of us (K.G.P.), and for a postdoctoral fellowship (J.J.).

References

1. D. Spanjaard, C. Guillot, M.S. Desjonqueres, G. Treglia and J. Lecante, Surface Science Reports, 5 (1985) 1; W. Egelhoff, Surface Science Reports 6 (1987) 253; D.E. Eastman, F.J. Himpsel and J.F. van der Veen, J. Vac. Sci. Techn. 20 (1982) 609.
2. M.C. Desjonqueres, D. Spanjaard, Y. Lassailly and C. Guillot, Solid State Comm. 34 (1980) 807.
3. P.H. Citrin and G.K. Wertheim, Phys. Rev. B27 (1983) 3176.
4. B. Johansson and N. Martensson, Phys. Rev. B21 (1980) 4427; B22 (1980) 3706.
5. W. Egelhoff, Phys. Rev. Lett. 50 (1983) 587.
6. K.G. Purcell, J. Jupille and D.A. King, to be published.
7. J. Jupille, K.G. Purcell and D.A. King, Solid State Commun. 58 (1986) 529.
8. K.G. Purcell, J. Jupille, G.P. Derby and D.A. King, Phys. Rev. B36 (1987) 1288.
9. J. Jupille, K.G. Purcell, G.P. Derby, J. Wendelken and D.A. King, Proceedings of International Conference on the Structure of Solid Surfaces (Amsterdam) 1987 (in press).

Discussion

M.G. LAGALLY: I missed whether the core level shifts are simply due to a changed coordination, or whether you would also be sensitive to small relaxations at step edges and other defects, which conceivably also could produce core level shifts.

D.A. KING: The correlation shown in Figure 5 of our paper clearly demonstrates a close link between effective coordination number and the observed binding energy shift. However, the scatter of data points is considered to be larger than the expermental error margin, and coordination number is therefore only a first order effect. For example, in studies of the W(100) phase transition, a very small W4f core level binding energy shift is observed through the transition, associated with lateral displacements of surface atoms. These shifts, and those associated with relaxation at step edges, for example, are likely to be in the region of the cohesive energy change associated with the relaxation, i.e., $\lesssim 100 meV$. I believe that SCLS provides a unique means of getting an experimental handle on these cohesive energy changes, as discussed in Reference 9 of our paper.

A.M. BRADSHAW: I would like to make a trivial point concerning SEXAFS analyses. The fact that LEED gives you the space group vastly decreases the number of parameters in your simulation.

D.A. KING: I think that LEED and SEXAFS combined are unbeatable in terms of surface crystallography. SEXAFS gives you local structure, while LEED gives you long range order. And, I mean the best example I've got of that is Cl on Ag(110) where LEED is telling you this is an incommensurate phase; SEXAFS tells you, 'but it's *locally* commensurate, it's locally got a well-defined structure.' You put the two together and you can get also the long-range structure of that system.

A.M. BRADSHAW: There's no multiple scattering in your calculations?

D.A. KING: There's no multiple scattering in our calculations, no.

K. TAKAYANAGI: When Cl/Ag structure has *incommensuration*, distances of Cl-Ag vary spatially. How can such long range variation of the interatomic distance be taken into account in SEXAFS analysis. How meaningful is the *higher shell calculation* made in your analysis?

D.A. KING: What you're saying would be absolutely true for a uniformly incommensurate structure, and in fact the SEXAFS, from a uniformly incommensurate structure, should be very washed out—for exactly the reason that you're indicating. Certainly anything beyond the first shell will be just washed out, because of the multiplicity of distances involved in an incommensurate structure. What we have is *non*-uniformly incommensurate structure, in fact, *proved* by the SEXAFS. It's a domain wall structure in which there are vacancies every two or three atoms on the cell, and it's these vacancies which generate the long-range incommensurate nature of the pattern. But *locally* it's commensurate.

K. TAKAYANAGI: How long is the period?

D.A. KING: It's incommensurate, so, ah. . .it's very long indeed, but it's almost a center 4×2 structure. Almost.

Surface Spectroscopy of High-Tc Y-Ba-Cu-O Superconductors

Z.Y. Hua [1], J.F. Deng [1], N.J. Chen [1], H.S. Cheng [1], L. Fei [1], X.L. Pan [1], A.R. Zhu [1], J. Zhuge [1], H.J. Ye [2], and X.C. Shen [2]

[1] Fudan University, Shanghai, People's Republic of China
[2] Institute of Technical Physics, Academia Sinica, Beijing, People's Republic of China

Various samples of Y-Ba-Cu-O system have been studied by Rutherford backscattering (RBS). At least within the selvage layer of 0.2 micron thickness, the real composition usually differs from its nominal composition, unless Y:Ba:Cu = 1:2:3. Surface infrared reflectance (SIR) under different temperatures from 4.2 K to 200K has been used for measuring energy gaps. Many electron spectroscopies are also measured under room temperature, but it is difficult to get reproducible and reliable results. Some preliminary results, however, are discussed.

INTRODUCTION

Materials of the Y-Ba-Cu-O system have recently generated great interest because of the observation of superconductivity with a zero-resistance temperature (Tc) > 90K [1]. Many physical and chemical methods have been used for the characterization of those superconductors. However, the electron and ion spectroscopies which are conventionally used for surface analysis have been scarcely reported. This is due to the following reasons:

1. Until recently all the superconductivity problems studied are bulk properties rather than surface effects.
2. Surface studies of oxide superconductors, unlike metallic or semiconductor surfaces, are very difficult because these materials are usually porous and uneven. Especially in the case of $YBa_2Cu_3O_{7-x}$, the oxygen content of samples can vary significantly if they are stored in a vacuum. But it is still worthwhile examining their capabilities and limitations in the research of high Tc superconductors.

In the present work preliminary results from various surface spectroscopies and surface infrared reflectance (SIR) are reported. Samples are prepared from oxide powders and sintered in oxygen flow. It is found that in this system any sample with a nominal composition (Y + Ba) : Cu = 1 ~ 2 could be superconductive with Tc > 90K if the fraction of Y = 15% ~ 40% [2].

AUGER ELECTRON SPECTROSCOPY (AES)

AES is measured by a scanning Auger microprobe, PHI-590, with cylindrical mirror analyzer and another homemade machine having a retarding grid analyzer. We find that it is difficult to get reproducible results because AES is extremely sensitive to C, S and O. The survey

Springer Series in Surface Sciences Vol. 14
Solvay Conference on Surface Science
Editor: F.W. de Wette

diagram of the surface constituents usually differs from one point to another. Yet all diagrams show a residual carbon. This is due to the starting material containing $BaCO_3$ for the source of BaO. Unlike the preparation of oxide-coated cathodes which are processed under 1250 °C and continuously evacuated in a vacuum system, now the thermal decomposition of $BaCO_3$ is complete under a lower temperature (950 °C) and atmospheric pressure. If, instead of $BaCO_3$, barium peroxide is used for a component of the starting material, carbon peaks are almost invisible.

AUGER ELECTRON APPEARANCE POTENTIAL SPECTROSCOPY (AEAPS)

AEAPS has no drawback of presenting unusual sensitibity for C and S. But unfortunately, it is also insensitive for Cu and a very weak signal is generated for Y. Only Ba has an acceptable signal-to-noise ratio.

Since AEAPS is very sensitive to lanthanide elements [3], the reason of low yield for Y is not clear and a further study seems necessary.

RUTHERFORD BACKSCATTERING (RBS)

RBS can measure the chemical elements contained in surface and near-surface regions of solids. Using the same facilities described elsewhere [4], 2 MeV He + ion beam is injected to the Y-Ba-Cu-O sample and the scattered ions are measured by a semiconductor detector. The ion beam is 1 mm in diameter and its scattering angle is 150°. Although oxygen can also be analyzed by RBS, yet it is difficult to produce good quantitative results for the light element component in a heavy element bulk. Therefore, for the oxygen content, a complementary method of nuclear reaction, $^{16}O(d,p)$, is used. It is obtained with a 900 KeV D+ beam and the proton yield was measured by a semiconductor detector, too, but with a Mylar film of 13 microns thickness for absorbing the scattered deuterium ions. SiO_2 was used as a standard sample for the calculation of O [4]. Figs. 1 and 2 are typical graphs. The total error of this experiment is < 5% which meets the results discussed by L'Ecuyer et al. [5].

Our results show that after sintering the real composition of a superconductive sample usually differs from the nominal one (as also mentioned by [6]), at least in the effective range of the incident helium ions, i.e., in the selvage with a thickness 0.2 micron from the surface, unless Y:Ba:Cu = 1:2:3 [7]. Also, we found by high resolution electron micrography that, in any superconductive sample, a phase of $YBa_2Cu_3O_7$ with oxygen-defect perovskite structure is always formed [2]. It implies that the Y:Ba:Cu = 1:2:3 is the most stable and easy-to-form superconductor in the Y-Ba-Cu-O family. This result also explains why single-phase Y-Ba-Cu-O superconductors can only be realized by a unique proportion Y:Ba:Cu = 1:2:3. It also implies that any superconductor in this family with higher Tc (> 100 K) should easily degenerate into a 90K-superconductor.

Thermogravimetry analysis (TGA) is generally used for measuring the oxygen content [8], but nuclear reaction is a nondestructive method. Most recently, a simple "wet chemistry" method [9] which can be used to check the non-destructive results is found to determine oxygen and trivalent copper (cu +++) content in Y-Ba-Cu-O superconductors.

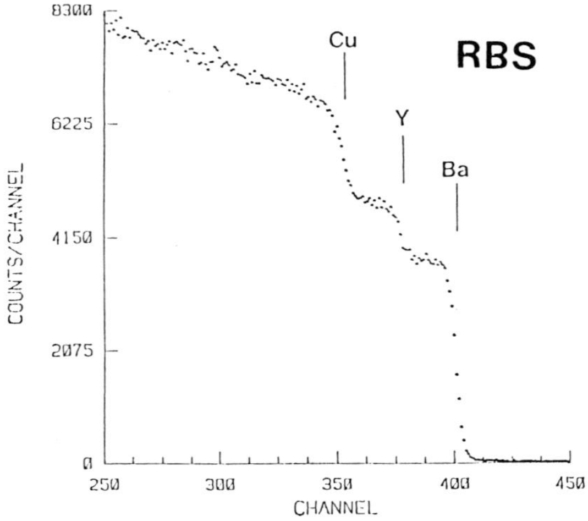

Fig. 1 RBS of a Y-Ba-Cu-O compound using He + beam for determination of Y, Ba and Cu contents.

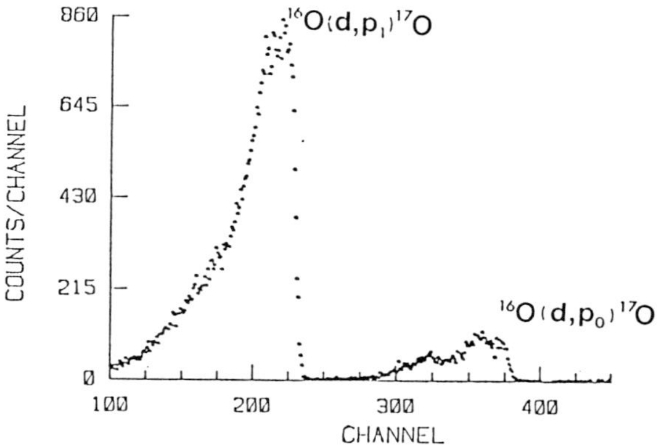

Fig. 2 The yield from a nuclear reaction for determining oxygen contents in a Y-Ba-Cu-O compound.

X-RAY PHOTOEMISSION SPECTROSCOPY (XPS)

For a $YBa_2Cu_3O_{7-x}$ sample, XPS shows the surface is inhomogeneous, quite chemically active and unstable. Carbon is also found even after Ar + sputtering. Spectra of Y, Ba and O usually vary from point to point, yet Cu (2p) is stable for different samples with a peak shape similar to cu-contained perovskites. The FWHM of this peak is 4 eV, close to the results obtained by Ihara et al. [10]. It also gives rise to a strong satellite structure, which is very similar to the satellites of Cu halides and oxides and can be explained as the many-body effect to the final state of the photoemission.

IONIZATION LOSS SPECTROSCOPY (ILS)

The results of ILS which leads to the density of empty state for the superconductors are not very reproducible. However, Fig. 3 presents a comparison between two samples: "sc" is a superconductive $YBa_2Cu_3O_7$ and "nc" is a nonconductive Y-Ba-Cu-O compound. The spectroscopies are obtained with PHI-590. Each curve is an accumulation of signals for 60 min with 1 scan/min. Evidently, they are quite different. Cu-L_3 of sc is a "peak" type with larger density of states and that of nc is a "step" type. Compared with the Cu spectra before and after oxidation (Fig. 4), sc is similar to the oxidized Cu surface and nc is similar to the non-oxidized Cu.

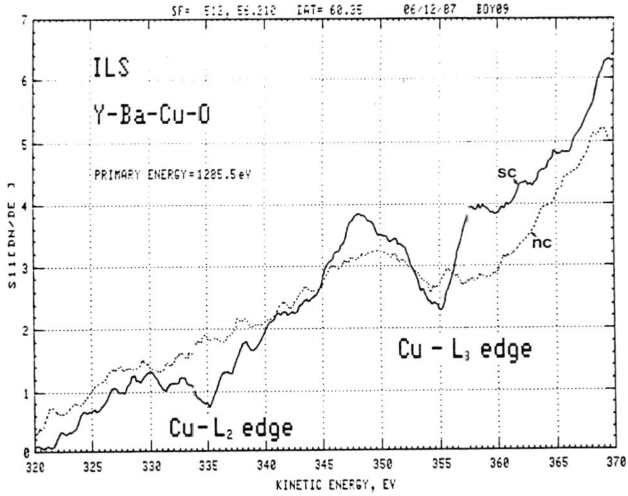

Fig. 3 ILS of two different samples.
sc - superconductive $YBa_2Cu_3O_7$
nc - nonconductive Y-Ba-Cu-O compound

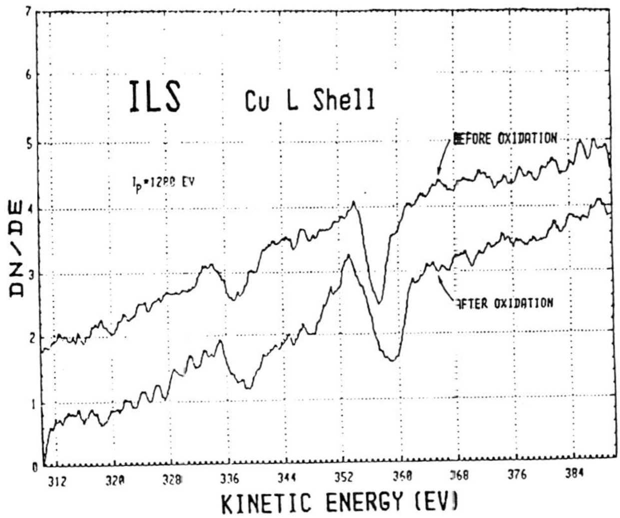

Fig. 4 ILS of Cu before and after surface oxidation

This indicates that in a Y-Ba-Cu-O system the coordination between Cu and O plays an important role in superconductivity.

SURFACE EXTENDED ENERGY-LOSS FINE STRUCTURE (SEELFS)

As an analogy to the principle of extended X-ray absorption fine structure (EXAFS), SEELFS also brings about the nearest neighbor distance [11]. With an incident beam energy < 2KeV, O-K edge is studied for measuring the nearest neighbor distance R of oxygen in a $YBa_2Cu_3O_7$ sample. After data processing [12], radial distribution function (RDF) is obtained (Fig. 5) and gives R = 1.32 Å. If a phase shift 0.18 Å is added, R = 1.50 Å. This value is quite smaller to the shortest Cu-O bond-length (= 1.83 Å [9]) measured by X-ray diffraction of a $YBa_2Cu_3O_7$ single crystal. The reason remains unclear whether this is due to the surface-to-bulk difference or not.

After electron bombardment for a sufficiently long period (e.g. 6 hrs), SEELFS shows that R increases to a value of 1.50 Å which gives a nearest neighbor distance 1.68 Å. The lattice deformation of superconductive phase $YBa_2Cu_3O_7$ under sustained 400 KeV electron bombardment has been reported previously [13]. Here is another example for structure deterioration under electron radiation of much lower energy.

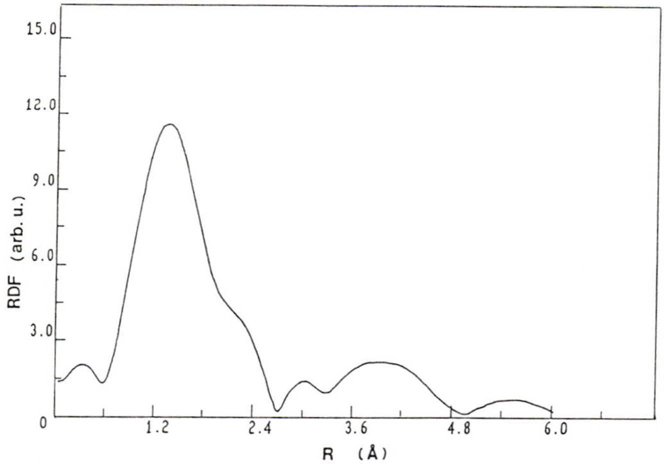

Fig. 5 Radial distribution function (RDF) vs. distance (R) curve

SURFACE INFRARED REFLECTANCE (SIR)

To determine the energy gap of a superconductor, SIR displays the advantages of good reproductivity and simplicity. The experiments are conducted in a 200 SXV Fourier IR Spectrometer with liquid helium cryostat bolometer. Four samples of single-phase $YBa_2Cu_3O_7$ are

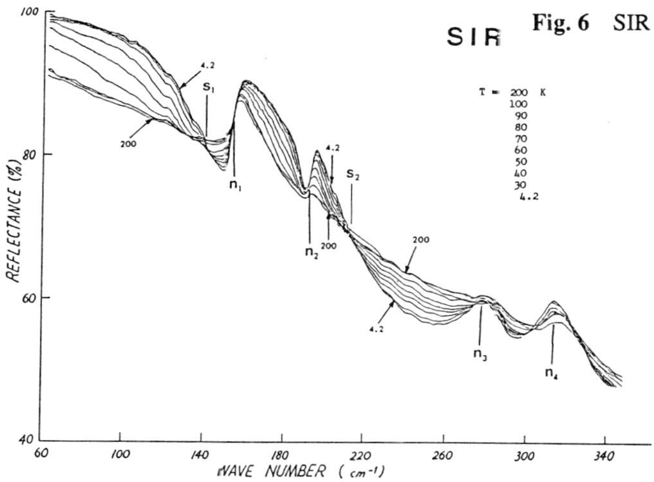

Fig. 6 SIR of $YBa_2Cu_3O_7$.

measured under different temperatures, 4.2 K to 200 K. Since Tc is about 95 K, it covers both superconductive and normal states. The reflection curves of different samples are nearly the same and one is shown in Fig. 6. Six features appeared in the range of wavenumber 60 ~ 360 cm^{-1} . s_1 and s_2 only appeared at T < Tc but the other four, n_1, n_2, n_4, appeared in all cases, disregarding whether the state is superconductive or not. The average values of 2 Δ/kT for s_1 and s_2 are 2.14 and 3.31, respectively, for four samples. The latter seems rather close to the value (3.5) predicted by BCS electron-phonon interactions and the former may be due to another mechanism.

CONCLUSION

Various ion, electron and photon spectroscopies are used to investigate high Tc Y-Ba-Cu-O superconductors, mostly under room temperature. Among them RBS and SIR seem to be the most preferable for compositional analysis and measurement of energy gaps, respectively. A further study in the capability of other methods is required.

Because of variable surface conditions, all measurements with single-phase and/or multiphase sample are only for reference and cannot be crucial. Therefore, for the most of physico-chemical studies, the preparation of single crystal of high Tc superconductors is of top importance.

REFERENCES

[1] Wu, M.K., Ashburn, J.R., Torng, C.T., Hor, P.H., Meng, R.L., Gao, L., Huang, Z.J., Wang, Y.Q. and Chu, C.W., Phys. Rev. Lett. 58, 908 (1987).

[2] Hua, Z.Y., Cheng, H.S., Jia, C.L., Jiang, A.R. and Miao, B.C., to be published in "Progress in High Temperature Superconductivity," vol. 1.

[3] Hua, Z.Y., Zhuge, J., Pan, X.L., Proc. Surface Science Symposium (Xerox PARC, 1983), p. 127.

[4] Cheng, H.S., Chen, J.X., Ren, C.G., Xu, Z.W., Yang, F.C., Zhao, G.Q. and Zhuo, Z.Y., Nucl. Instrum. Meth. 191, 391 (1981).

[5] L'Ecuyer, J., Davies, J.A. and Matsunami, N., Nucl. Instrum. Meth. 160, 337 (1979).

[6] Hatano, T., Matsushita, A., Nakamura, K., Honda, K., Matsumoto, T. and Ogawa, K., Jap. J. Appl. Phys. 26, L374 (1987).

[7] Hua, Z.Y., Jia, C.L., Cheng, H.S., Cai, Y.M. and Jiang, A.R., to be published in "Progress in High Tc Superconductivity," vol. 2.

[8] Greedan, J.E., O'Reilly, A.H. and Stager, C.V., Phys. Rev. 35 B, 8770 (1987).

[9] Tang, H.S. and Hua, Z.Y., to be published.

[10] Ihara, H., Hirabayashi, M., Terada, N., Kimura, Y., Senzaki, K., Akimoto, M., Bushida, K., Kawashima, F. and Uzuka, R., Jap. J. Appl. Phys. 26, L460 (1987).

[11] Hitchock, A.P. and Teng, C.H., Surf. Sci. 149, 558 (1985).

[12] Fei, L., Zheng, Y., Zhang, Q.J. and Hua, Z.Y., to be published in J. Vac. Sci. Technol. A.

[13] Hua, Z.Y., Cai, Y.M., Qiu, J.W., Jia, C.L., Miao, B.C., Chen, K.Y., Luo, J. and Ouyang, L., Chinese Phys. Lett. 4, 445 (1987).

Administrative Council and Scientific Committees of the Solvay Foundation

Administrative Council of the Instituts Internationaux de Physique et de Chimie, fondés par E. Solvay

J. Solvay	President of the Administrative Board of the Solvay Institutes
F. Bingen	Vice–President of the Administrative Board of the Solvay Institutes
I. Prigogine	Director of the Solvay Institutes
J. Philippot	Vice–Director of the Solvay Institutes
J. Reignier	Treasurer of the Administrative Board of the Solvay Institutes
G. Nicolis	Secretary of the Administrative Board of the Solvay Institutes
R. Debever	
P. Glansdorff	
A. Jaumotte	
A. Molitor	
G. Van Binst	

Scientific Committee for Physics

L. Van Hove *Chairman*	CERN, Division TH CH–1211 Genève 23 Switzerland
A. Abragam	CEN/Saclay Orme des Merisiers Boîte Postale No. 2 F–91190 Gif–sur–Yvette France

E. Amaldi	Instituto di Fisica "Guglielmo Marconi" Universita degli Studi Piazzale Aldo Moro 2 I-00185 Rome Italy
P.W. Anderson	AT&T Bell Laboratories 600 Mountain Avenue Murray Hill, New Jersey 07974 USA
I.M. Khalatnikoff	Institute for Theoretical Physics Vorobieskoje Chaussee 2 Moscow 117940, B–344 USSR
R. Kippenhahn	Max Planck Institut für Physik und Astrophyusik Karl–Schwarzschild–Strasse 1 D–8046 Garching bei München Federal Republic of Germany
R. Kubo	Nishina Memorial Foundation 28–45 Hon–Komagone 2 Chome Bunkyo–ku, Tokyo Japan
Y. Ne'eman	Tel Aviv University Mortimer and Raymond Sackler Institute of Advanced Study Ramat–Aviv, Tel Aviv Israel
A.B. Pippard	University of Cambridge Department of Physics Cavendish Laboratory Madingley Road, Cambridge CB3 0HE England
A. Sandage	Mount Wilson and Las Campanas Palomar Observatories 813 Santa Barbara Street Pasadena, CA 91101 USA
Tsin San Tsiang	Academia Sinica Beijing People's Republic of China

V.F. Weisskopf
MIT Department of Physics
Cambridge, Massachusetts 02139
USA

E.P. Wigner
Princeton University
P.O. Box 708
Princeton, New Jersey 08504
USA

J. Géhéniau
Université Libre de Bruxelles
Department de Mathématiques
Campus Plaine, CP-228
Boulevard du Triomphe
B-1050 Brussels
Belgium

Scientific Committee for Chemistry

A.R. Ubbelohde
President
Imperial College
Chemical Engineering
Prince Consort Road
London
England SW7 2BY

Stig Claesson
C/o Mrs. Claesson
45 Skeragatan
Helsingborg 25263
Sweden

M. Eigen
Max-Planck Institut für Biophysikalische Chemie
Nikolausberg, Postfach 968
D-3400 Göttingen
Federal Republic of Germany

B. Hess
Max-Planck Institut für Ernährungsphysiologie
Rheinlanddamm 201
D-4600 Dortmund 1
Federal Republic of Germany

E. Katchalski-Katzir
Department of Biophysics
Weizmann Institute of Science
Rehovot
Israel

William N. Lipscomb	Department of Chemistry Harvard University 12 Oxford Street Cambridge, Massachusetts 02138 USA
R. Martin	Drève des Gendarmes, 169 B–1180 Brussells Belgium
Yu. A. Ovchinikov	Director, Department of Protein Chemistry Shemyakin Institute of Biorganic Chemistry ul. Vavilova 32 Moscow, 117312 USSR
L. Pauling	Institute of Science and Medicine 2700 Sand Hill Roads Menlo Park, California 94025 USA
V. Prelog	ETH Lab. für Organische Chemie Universitätstrasse 16 CH–8092 Zürich Switzerland
Lord Todd	Professor, Christ's College Cambridge CB2 3BU England
A. Bellemans	Service de Chimie Physique Université Libre de Bruxelles CP–231, Campus Plaine Boulevard du Triomphe B–1050 Bruxelles Belgium

Organizing Committee of the Solvay Confernce

F. W. de Wette Chairman **L. Reichl** Secretary

J. L. Erskine L. Kleinman R. Smoluchowski J. M. White

List of Participants

Professor P. Antoniewicz
Physics Department
The University of Texas
Austin, TX 78712, USA

Professor A.J. Bard
Chemistry Department
The University of Texas
Austin, TX 78712, USA

Professor John Bardeen
Loomis Lab./Physics Department
University of Illinois
Urbana, IL 61801, USA

Professor Pierre Borckmans
Service de Chimie Physique
CP-231, Campus Plaine, ULB
Boulevard du Triomphe
B-1050 Brussels, Belgium

Professor Michel Boudart
Department of Chemical Engineering
Stanford University
Stanford, CA 94305, USA

Professor A.M. Bradshaw
Fritz-Haber-Institut der
Max-Planck-Gesellschaft
Faradayweg 4-6
D-1000 Berlin 33 (Dahlem)
Germany

Professor Al Campion
Department of Chemistry
The University of Texas
Austin, TX 78712, USA

Dr. Mark J. Cardillo
AT&T Bell Laboratories
Materials Research Laboratory
600 Mountain Avenue
Murray Hill, NJ 07974, USA

Prof. Dr. Manuel Cardona
Max-Planck-Institut
für Festkörperforschung
Heisenbergstrasse 1
7000 Stuttgart 80
Federal Republic of Germany

Professor S.T. Ceyer
Department of Chemistry, MIT
77 Massachusetts Ave.
Cambridge, MA 02139, USA

Professor J.G. Dash
Department of Physics
University of Washington
Seattle, Washington 98195, USA

Professor A. de Lozanne
Department of Physics
The University of Texas
Austin, TX 78712, USA

Dr. J.E. Demuth
IBM
Thomas J. Watson Research Center
P.O. Box 218
Yorktown Heights, NY 10598, USA

Professor Guy Dewel
Service de Chimie Physique
CP-231, Campus Plaine, ULB
Boulevard du Triomphe
B-1050 Brussels, Belgium

Professor F.W. de Wette
Department of Physics
The University of Texas
Austin, TX 78712, USA

Professor Michael Downer
Department of Physics
The University of Texas
Austin, TX 78712, USA

Dr. Charles B. Duke
Battelle
Pacific Northwest Laboratories
P.O. Box 999
Richland, WA 99352, USA

Professor Ivan Elskens
Dept. Scheikunde
Vrye Universiteit Brussel
Pleinlaan 2
B-1050 Brussels, Belgium

Professor J.L. Erskine
Department of Physics
The University of Texas
Austin, TX 78712, USA

Professor Gerhard Ertl
Fritz-Haber-Institut
der Max-Planck-Gesellschaft
Faradayweg 4-6
D-1000 Berlin 33
Germany

Dr. J.W. Gadzuk
Surface Science Division
National Bureau of Standards
Gaithersburg, MD 20899, USA

Professor J.A. Golovchenko
Department of Physics
Lyman Laboratory
Harvard University
Cambridge, MA 02138, USA

Professor Robert Gomer
Department of Chemistry
The James Franck Institute
The University of Chicago
5640 Ellis Avenue
Chicago, IL 60637, USA

Professor Arthur C. Gossard
University of California
Materials Department
Santa Barbara, CA 98106, USA

Professor Norman Hackerman
Department of Chemistry
The University of Texas
Austin, TX 78712, USA

Dr. D.R. Hamann
AT&T Bell Laboratories
Surface Physics Research
600 Mountain Avenue
Murray Hill, NJ 07974-2070, USA

Professor Robert Herman
Civil Engineering Department
The University of Texas
Austin, TX 78712, USA

Professor Hua Zhong-yi
Vacuum Physics Laboratory
Fudan University
Shanghai
People's Republic of China

Professor Harald Ibach
KFA Jülich, IGV
Postfach 1913
D-5170 Jülich
Federal Republic of Germany

Professor D.A. King
Department of Chemistry
The University of Cambridge
Lensfield Road
Cambridge CB2 1EW, England

Professor L. Kleinman
Physics Department
The University of Texas
Austin, TX 78712, USA

Professor Walter Kohn
Department of Physics
University of California
Santa Barbara, CA 93106, USA

Professor Max G. Lagally
Metallurgical and Mineral Engineering
University of Wisconsin-Madison
1509 University Avenue
Madison, WI 53706, USA

Dr. Norton D. Lang
IBM
Thomas J. Watson Research Center
P.O. Box 218
Yorktown Heights, NY 10598, USA

Professor Lin Zhangda
Institute of Physics
Chinese Academy of Sciences
Beijing
People's Republic of China

Professor Steven G. Louie
Department of Physics
University of California
Berkeley, CA 94720, USA

Dr. Paul Marcus
IBM
Thomas J. Watson Research Center
P.O. Box 218
Yorktown Heights, NY 10598, USA

Professor Jagdish Mehra
7830 Candle Lane
Houston, TX 77071, USA

Professor J.B. Pendry, FRS
Department of Physics
Blackett Laboratory
Prince Consort Road
London SW7 2BZ, England

Professor Y. Petroff
LURE, Bâtiment 209 D
Centre Universitaire Paris-Sud
91405 Orsay Cedex
France

Professor Earl W. Plummer
Department of Physics
University of Pennsylvania
Philadelphia, PA 19104, USA

Professor Linda E. Reichl
Department of Physics
The University of Texas
Austin, TX 78712, USA

Professor Stuart A. Rice
The James Franck Institute
The University of Chicago
5640 Ellis Avenue
Chicago, IL 60637, USA

Professor Hans C. Siegmann
Laboratorium für Festkörperphysik
ETH Zürich-Hongg
Zürich CH-8093, Switzerland

Professor R. Smoluchowski
Department of Physics
The University of Texas
Austin, TX 78712, USA

Professor Gabor A. Somorjai
Department of Chemistry
University of California
Berkeley, CA 94720, USA

Professor Victor Szebehely
Department of Aerospace Engineering
The University of Texas
Austin, TX 78712, USA

Professor Kunio Takayanagi
Department of Physics
Tokyo Institute of Technology
Oh-Okayama, Meguro
Tokyo, Japan

Professor Kenzi Tamaru
Department of Chemistry
Science University of Tokyo
Kagurazaka, Shinjuku-ku
Tokyo, Japan

Professor S.J. Teichner
Lab. de Thermodynamique
et Cinetique Chimiques
CNRS No. 231
Université Claude Bernard
43, Blvd. du 11 Novembre 1918
69622 Villeurbanne Cedex, France

Professor J. Peter Toennies
Max-Planck-Institut
für Strömungsforschung
Bunsenstrasse 10
3400 Göttingen
Federal Republic of Germany

Dr. E. Tosatti
SISSA/ICTP
Strada Costiera 11
I-34014 Trieste, Italy

Dr. J.C. Tully
AT&T Bell Laboratories
600 Mountain Avenue
Murray Hill, NJ 07974-2070, USA

Professor R.F. Wallis
Department of Physics
University of California
Irvine, CA 92717, USA

Professor J.M. White
Department of Chemistry
The University of Texas
Austin, TX 78712, USA

Professor Roy F. Willis
Department of Physics
Pennsylvania State University
State College, PA 16804, USA

Professor Xie Xide, President
Fudan University
Shanghai
People's Republic of China

Prof. John T. Yates, Jr.
Department of Chemistry
University of Pittsburgh
Pittsburgh, PA 15260, USA

Professor Zhang Kaiming
Fudan University
Shanghai
People's Republic of China

Index of Contributors